Advanced Fluoride-Based Materials for Energy Conversion

Advanced Fluoride-Based Materials for Energy Conversion

Edited by

Tsuyoshi Nakajima
Department of Applied Chemistry
Aichi Institute of Technology
Yakusa, Toyota, Japan

Henri Groult
Sorbonne Universités, UPMC Univ.
Laboratoire PHENIX
CNRS, Paris, France

ELSEVIER

AMSTERDAM • BOSTON • HEIDELBERG • LONDON • NEW YORK • OXFORD
PARIS • SAN DIEGO • SAN FRANCISCO • SINGAPORE • SYDNEY • TOKYO

Elsevier
Radarweg 29, PO Box 211, 1000 AE Amsterdam, The Netherlands
The Boulevard, Langford Lane, Kidlington, Oxford OX5 1GB, UK
225 Wyman Street, Waltham, MA 02451, USA

Copyright @ 2015 Elsevier Inc. All rights reserved.

No part of this publication may be reproduced or transmitted in any form or by any means, electronic or mechanical, including photocopying, recording, or any information storage and retrieval system, without permission in writing from the publisher. Details on how to seek permission, further information about the Publisher's permissions policies and our arrangements with organizations such as the Copyright Clearance Center and the Copyright Licensing Agency, can be found at our website: www.elsevier.com/permissions.

This book and the individual contributions contained in it are protected under copyright by the Publisher (other than as may be noted herein).

Notices
Knowledge and best practice in this field are constantly changing. As new research and experience broaden our understanding, changes in research methods, professional practices, or medical treatment may become necessary.

Practitioners and researchers must always rely on their own experience and knowledge in evaluating and using any information, methods, compounds, or experiments described herein. In using such information or methods they should be mindful of their own safety and the safety of others, including parties for whom they have a professional responsibility.

To the fullest extent of the law, neither the Publisher nor the authors, contributors, or editors, assume any liability for any injury and/or damage to persons or property as a matter of products liability, negligence or otherwise, or from any use or operation of any methods, products, instructions, or ideas contained in the material herein.

British Library Cataloguing in Publication Data
A catalogue record for this book is available from the British Library

Library of Congress Cataloging-in-Publication Data
A catalog record for this book is available from the Library of Congress

ISBN: 978-0-12-800679-5

For information on all Elsevier publications
visit our web site at http://store.elsevier.com/

Contents

Contributors   xiii
Preface   xvii

1. **High Performance Lithium-Ion Batteries Using Fluorinated Compounds**
   Zonghai Chen, Zhengcheng Zhang and Khalil Amine

1.1	Introduction	1
1.2	Stabilization of Lithiated Anodes	2
	1.2.1 Modification of SEI Layer	5
	1.2.2 Artificial SEI Layer	7
1.3	Fluorinated Redox Shuttles	13
1.4	High Voltage Electrolytes	19
1.5	Closing Remarks	29
	Acknowledgments	29
	References	30

2. **Electrochemical Behavior of Surface-Fluorinated Cathode Materials for Lithium Ion Battery**
   Susumu Yonezawa, Jae-Ho Kim and Masayuki Takashima

2.1	Surface Fluorination of $LiFePO_4$	33
	2.1.1 Introduction	33
	2.1.2 Experimental Details	34
	2.1.3 Characterization of Fluorinated $LiFePO_4$	35
	2.1.4 Electrochemical Properties of Fluorinated $LiFePO_4$	40
2.2	Surface Fluorination of $LiNi_{0.5}Mn_{1.5}O_4$	43
	2.2.1 Introduction	43
	2.2.2 Experimental Details	44
	2.2.3 Characterization of Surface-Fluorinated $LiNi_{0.5}Mn_{1.5}O_4$	44
	2.2.4 Electrochemical Properties of Surface-Fluorinated $LiNi_{0.5}Mn_{1.5}O_4$	46
2.3	Summary	49
	References	49

3. Fluoride Cathodes for Secondary Batteries
Munnangi Anji Reddy and Maximilian Fichtner

3.1	Introduction	51
3.2	Metal Fluorides for Electrochemical Energy Storage	52
3.3	Metal Fluorides for Lithium Batteries	53
	3.3.1 Introduction	53
	3.3.2 Binary Metal Fluorides	53
	3.3.3 Ternary Metal Fluorides (LiMFeF$_6$, Li$_3$MF$_6$)	63
3.4	Metal Fluorides for Fluoride Ion Batteries	65
	3.4.1 Introduction	65
	3.4.2 Solid Fluoride Ion Conductors	66
	3.4.3 Solid State Fluoride Ion Batteries (FIB)	69
	3.4.4 Fluoride Ion Batteries with Liquid Electrolyte	72
3.4	Perspective	73
	References	74

4. Fluorosulfates and Fluorophosphates As New Cathode Materials for Lithium Ion Battery
Christian M. Julien, Alain Mauger and Henri Groult

4.1	Introduction	77
4.2	General Considerations	79
	4.2.1 Average Voltage of Li Insertion Cathodes	79
	4.2.2 The Inductive Effect	80
4.3	Fluorophosphates	81
	4.3.1 Fluorine-Doped LiFePO$_4$	81
	4.3.2 LiVPO$_4$F	83
	4.3.3 LiMPO$_4$F (M = Fe, Ti)	87
	4.3.4 Li$_2$FePO$_4$F (M = Fe, Co, Ni)	88
	4.3.5 Li$_2$$MPO_4$F ($M$ = Co, Ni)	89
	4.3.6 Na$_3$V$_2$(PO$_4$)$_2$F$_3$ Hybrid Ion Cathode	90
	4.3.7 Other Fluorophosphates	91
4.4	Fluorosulfates	92
	4.4.1 LiFeSO$_4$F	92
	4.4.2 LiMSO$_4$F (M = Co, Ni, Mn)	95
4.5	Concluding Remarks	96
	References	97

5. Fluorohydrogenate Ionic Liquids, Liquid Crystals, and Plastic Crystals
Kazuhiko Matsumoto and Rika Hagiwara

5.1	Introduction	103
5.2	Structural Properties of Fluorohydrogenate Anions	104
5.3	Fluorohydrogenate Ionic Liquids	106

5.4	Fluorohydrogenate Ionic Liquid Crystals	116
5.5	Fluorohydrogenate Ionic Plastic Crystals	118
	References	121

6. Novel Fluorinated Solvents and Additives for Lithium-Ion Batteries

T. Böttcher, N. Kalinovich, O. Kazakova, M. Ponomarenko, K. Vlasov, M. Winter and G.-V. Röschenthaler

6.1	Introduction	125
6.2	Lithium Conductive Salts: Polyfluorinated Lithium Sulfonates	125
6.3	Solvents and Cosolvents for Electrolyte Systems	132
	6.3.1 Alkyl Tetrafluoro-2-(Alkoxy)Propionates as Cosolvents for Propylene-Carbonate-Based Electrolytes	132
	6.3.2 Derivatives of Hexafluorbutandiol as Cosolvents and SEI Additives	135
	6.3.3 Fluorinated 1,2,4-Thiadiazianan-3,5-dion-1,1-dioxides Useful SEI Additives in PC	138
6.4	Overcharge Protecting Agents: PF_5–Carbene Adducts	139
	References	143

7. Safety Improvement of Lithium Ion Battery by Organofluorine Compounds

Tsuyoshi Nakajima

7.1	Introduction	147
7.2	Organofluorine Compounds	148
7.3	Differential Scanning Calorimetry Study on the Thermal Stability of Fluorine Compound-Mixed Electrolyte Solutions	151
	7.3.1 Reactions of Metallic Lithium (Li) or Lithium-Intercalated Graphite (LiC_6) with Cyclic and Linear Carbonate Solvents	151
	7.3.2 Reactions of Metallic Lithium (Li) with Organofluorine Compounds and Those of Lithium-Intercalated Graphite (LiC_6) with Fluorine Compound-Mixed Electrolyte Solutions	154
	7.3.3 Influence of Molecular Structures of Fluorine Compounds on the Reactivity with Li	161
7.4	Electrochemical Oxidation Stability of Fluorine Compound-Mixed Electrolyte Solutions	161
7.5	Charge/Discharge Behavior of Natural Graphite Electrodes in Fluorine Compound-Mixed Electrolyte Solutions	165
7.6	Conclusions	171
	References	171

8. Artificial SEI for Lithium-Ion Battery Anodes: Impact of Fluorinated and Nonfluorinated Additives

D. Lemordant, W. Zhang, F. Ghamouss, D. Farhat, A. Darwiche, L. Monconduit, R. Dedryvère, H. Martinez, S. Cadra and B.Lestriez

8.1	Introduction	174
	8.1.1 Li-Ion Principle	174
	8.1.2 SEI Formation	175
	8.1.3 SEI Structure and Role	178
	8.1.4 SEI Components at the Electrode Surface	178
	8.1.5 SEI Components Dissolved in the Electrolyte	180
8.2	Application to TiSnSb Anodes	181
	8.2.1 Electrochemical Properties of TiSnSb Electrode	183
	8.2.2 Surface Analysis Results	186
	8.2.3 Discussion	194
	8.2.4 Experimental Section	198
8.3	Conclusion	200
	Acknowledgment	200
	References	201

9. Surface Modification of Carbon Anodes for Lithium Ion Batteries by Fluorine and Chlorine

Tsuyoshi Nakajima

9.1	Introduction	203
9.2	Effect of Surface Fluorination and Chlorination of Natural Graphite Samples	204
	9.2.1 Surface Fluorination of Natural Graphite Samples with Small Surface Areas	204
	9.2.2 Surface Fluorination of Natural Graphite Samples with Large Surface Areas	208
	9.2.3 Surface Chlorination of Natural Graphite Samples with Large Surface Areas	210
9.3	Effect of Surface Fluorination of Petroleum Cokes	214
9.4	Conclusions	219
	References	221

10. Application of Polyvinylidene Fluoride Binders in Lithium-Ion Battery

Ramin Amin-Sanayei and Wensheng He

10.1 Introduction	225
10.2 Fluorine-Containing Binder	226
10.3 Properties of Fluorinated Binder	228
10.4 Binder Swelling in Electrolyte Solvent	230
10.5 Electrochemical Stability	230
10.6 Electrode Preparation Method	231

10.7	Peel Strength Measurement	232
10.8	Electrode Performance Test	233
10.9	Fluorinated Waterborne Binders	234

11. Electrodeposition of Polypyrrole on CF$_x$ Powders Used as Cathode in Primary Lithium Battery

Henri Groult, Christian M. Julien, Ahmed Bahloul, Sandrine Leclerc, Emmanuel Briot, Ana-Gabriela Porras-Gutierrez and Alain Mauger

11.1	Introduction	237
11.2	Experimental Section	242
11.3	Results	243
	11.3.1 Preparation of PPy-CF$_x$ Samples in Acetonitrile	243
	11.3.2 Physical–Chemical Characterizations of PPy-Coated Samples	246
	11.3.3 Electrochemical Studies in Primary Lithium Battery	251
11.4	Conclusions	259
	Acknowledgments	259
	References	259

12. New Nano-C–F Compounds for Nonrechargeable Lithium Batteries

K. Guérin and M. Dubois

12.1	Introduction	261
12.2	Contribution of Nanomaterials to Enhance the Energy Density of a Primary Lithium Battery	263
12.3	New Fluorination Ways to Increase the Power Density of a Primary Lithium Battery	271
12.4	Increasing the Faradic Yield	277
12.5	Next-Generation Carbon Fluorides for Primary Lithium Batteries: Some Key Points	284
	References	285

13. Recent Advances on Quasianhydrous Fuel Cell Membranes

Benjamin Campagne, Ghislain David and Bruno Ameduri

13.1	Introduction	290
13.2	Fluorinated Copolymers Based on Nitrogen Heterocycles	295
	13.2.1 Introduction and Challenges	295
	13.2.2 Membranes Based on Fluorinated Copolymers Containing Triazole Groups and Sulfonated Poly(Ether Ether Ketone)	297
13.3	Proton Mobility in Membranes Based on Nitrogenous Heterocycles and s-PEEK	302

13.4 Crosslinking of Membranes Based on Nitrogenous
     Heterocycles   305
     13.4.1 Thermal Stabilities of Crosslinked Membranes
            Composed of s-PEEK-Na (B) and Poly(CTFE-*alt*-
            IEVE)$_{94\%}$-*g*-1H-1,2,4-triazole-3-thiol$_{90\%}$-*co*-poly(CTFE-
            alt-GCVE)$_{6\%}$ Terpolymer Crosslinked by DiA or TEPA   309
     13.4.2 Mechanical Properties of Crosslinked Membranes   309
     13.4.3 Protonic Conductivities of Membranes Composed
            of s-PEEK and Poly(CTFE-*alt*-GCVE)$_{6\%}$-poly(CTFE-*alt*-
            IEVE)$_{94\%}$-*g*-1H-1,2,4-triazole$_{90\%}$ Terpolymer Crosslinked
            by Diamines DiA or TEPA   315
     13.4.4 Comparison of Conductivities of A/B-*ret*-DiA and
            A/B-*ret*-TEPA Membranes with Uncured Membranes
            Composed of Poly(CTFE-*alt*-IEVE)-*g*-1H-1,2,4-triazole$_{95\%}$
            Copolymer/s-PEEK Blends   316
13.5 Conclusion   317
     Acknowledgments   320
     References   320

14. The Use of Per-Fluorinated Sulfonic Acid (PFSA) Membrane as Electrolyte in Fuel Cells

Madeleine Odgaard

14.1 Introduction   326
14.2 Polymer Electrolyte Fuel Cells   328
     14.2.1 Principle of Operation   328
     14.2.2 The Role of the Polymer Electrolyte   331
14.3 Properties of the PFSA Membrane   331
     14.3.1 Introduction   331
     14.3.2 Perfluorinated Membranes in General   332
     14.3.3 The Nafion® Membrane   333
     14.3.4 Morphology and Proton Conductivity   336
     14.3.5 Water and Methanol Transport a Technological Aspect   339
     14.3.6 Other Properties Facing the FC Requirements   340
     14.3.7 Degradation and Durability Aspects   341
14.4 Application and Performance of PFSA Membranes in FCs   345
     14.4.1 Introduction   345
     14.4.2 The PEMFC Stack   346
     14.4.3 Application of PEMFC   348
     14.4.4 DMFC and Their Applications   356
     References   364

15. Surface-Fluorinated Carbon Materials for Supercapacitor

Young-Seak Lee

15.1 Introduction   375
15.2 Fluorinated Activated Carbons for Supercapacitor   376

	15.3 F-AC Fibers for Supercapacitor	379
	15.4 Fluorinated Carbon Nanotubes for Supercapacitor	382
	References	385

16. Fluorine Chemistry for Negative Electrode in Sodium and Lithium Ion Batteries

Mouad Dahbi and Shinichi Komaba

16.1 Introduction	387
16.2 Why Na-Ion Battery?	388
16.3 Hard-Carbon as Potential Negative Electrode	390
16.4 Fluorinated Electrolyte and Additive	392
16.5 Poly Vinylidene Fluoride and CMC-Based Binder	397
16.6 Aluminum Corrosion Inhibitor	401
16.7 Na Alloys and Compounds	403
16.7.1 Tin and Antimony as Alloy Materials	404
16.7.2 Red and Black Phosphorus	405
16.8 Silicon for Lithium-Ion Battery	407
16.9 Conclusive Remarks	409
Acknowledgments	411
References	411

17. Application of Carbon Materials Derived from Fluorocarbons in an Electrochemical Capacitor

Soshi Shiraishi and Osamu Tanaike

17.1 Introduction	415
17.2 Synthesis Method and Basic Characterization of Porous Carbon from Fluorocarbon	417
17.3 Performance of Electric Double Layer Capacitance of Porous Carbon from Fluorocarbon	422
17.4 Conclusion	427
Acknowledgment	429
References	429

Index 431

Contributors

Bruno Ameduri Institut Charles Gerhardt, Ingénierie et Architectures Macromoléculaires, UMR CNRS 5253, Ecole Nationale Supérieure de Chimie de Montpellier, Montpellier, France

Khalil Amine Chemical Sciences and Engineering Division, Argonne National Laboratory, Lemont, IL, USA

Ramin Amin-Sanayei Arkema Inc, King of Prussia, PA, USA

Ahmed Bahloul Laboratoire des Matériaux et Systèmes Électroniques, Centre Universitaire de Bordj Bou Arréridj, Bordj Bou Arréridj, Algeria

T. Böttcher School of Engineering and Science, Jacobs University GmbH, Bremen, Germany

Emmanuel Briot Sorbonne Universités, UPMC Univ., Laboratoire PHENIX, CNRS, Paris, France

S. Cadra CEA/DAM, Le Ripault, Monts, France

Benjamin Campagne Institut Charles Gerhardt, Ingénierie et Architectures Macromoléculaires, UMR CNRS 5253, Ecole Nationale Supérieure de Chimie de Montpellier, Montpellier, France

Zonghai Chen Chemical Sciences and Engineering Division, Argonne National Laboratory, Lemont, IL, USA

Mouad Dahbi Department of Applied Chemistry, Tokyo University of Science, Tokyo, Japan; Elements Strategy Initiative for Catalysts and Batteries, Kyoto University, Kyoto, Japan

A. Darwiche IPREM-ECP CNRS UMR 5254, Pau, France

Ghislain David Institut Charles Gerhardt, Ingénierie et Architectures Macromoléculaires, UMR CNRS 5253, Ecole Nationale Supérieure de Chimie de Montpellier, Montpellier, France

R. Dedryvère ICG-AIME, Université Montpellier 2, Montpellier, France

M. Dubois Institute of Chemistry of Clermont-Ferrand, University of Blaise Pascal, Aubiere Cedex, France

D. Farhat PCM2E, Université F. Rabelais, Parc de Grandmont, Tours, France

Maximilian Fichtner Helmholtz Institute Ulm for Electrochemical Energy Storage (HIU), Helmholtzstr, Ulm, Germany

F. Ghamouss PCM2E, Université F. Rabelais, Parc de Grandmont, Tours, France

Henri Groult Sorbonne Universités, UPMC Univ., Laboratoire PHENIX, CNRS, Paris, France

K. Guérin Institute of Chemistry of Clermont-Ferrand, University of Blaise Pascal, Aubiere Cedex, France

Rika Hagiwara Department of Fundamental Energy Science, Graduate School of Energy Science, Kyoto University, Kyoto, Japan

Wensheng He Arkema Inc, King of Prussia, PA, USA

Christian M. Julien Sorbonne Universités, UPMC Univ., Laboratoire PHENIX, CNRS, Paris, France

N. Kalinovich School of Engineering and Science, Jacobs University GmbH, Bremen, Germany

O. Kazakova School of Engineering and Science, Jacobs University GmbH, Bremen, Germany

Jae-Ho Kim Headquarters for Innovative Society-Academia Cooperation, Fukui University, Fukui, Japan

Shinichi Komaba Department of Applied Chemistry, Tokyo University of Science, Tokyo, Japan; Elements Strategy Initiative for Catalysts and Batteries, Kyoto University, Kyoto, Japan

Sandrine Leclerc Sorbonne Universités, UPMC Univ., Laboratoire PHENIX, CNRS, Paris, France

Young-Seak Lee Department of Applied Chemistry and Biological Engineering, Chungnam National University, Daejeon, Republic of Korea

D. Lemordant PCM2E, Université F. Rabelais, Parc de Grandmont, Tours, France

B. Lestriez Institut des Matériaux Jean Rouxel (IMN), CNRS UMR 6502, Université de Nantes, Nantes, France

H. Martinez ICG-AIME, Université Montpellier 2, Montpellier, France

Kazuhiko Matsumoto Department of Fundamental Energy Science, Graduate School of Energy Science, Kyoto University, Kyoto, Japan

Alain Mauger Sorbonne Universités, UPMC Univ., Laboratoire PHENIX, CNRS, Paris, France; UPMC Univ. Paris 06, Institut de Minéralogie et Physique de la Matière Condensée, Paris, France

L. Monconduit IPREM-ECP CNRS UMR 5254, Pau, France

Tsuyoshi Nakajima Department of Applied Chemistry, Aichi Institute of Technology, Yakusa, Toyota, Japan

Madeleine Odgaard IRD Fuel Cells A/S, Kullinggade, Svendborg, Denmark

M. Ponomarenko School of Engineering and Science, Jacobs University GmbH, Bremen, Germany

Ana-Gabriela Porras-Gutierrez Sorbonne Universités, UPMC Univ., Laboratoire PHENIX, CNRS, Paris, France

Munnangi Anji Reddy Helmholtz Institute Ulm for Electrochemical Energy Storage (HIU), Helmholtzstr, Ulm, Germany

G.-V. Röschenthaler School of Engineering and Science, Jacobs University GmbH, Bremen, Germany

Soshi Shiraishi Graduate School of Science and Technology, Gunma University, Kiryu, Japan

Masayuki Takashima Department of Materials Science & Engineering, Faculty of Engineering, University of Fukui, Fukui, Japan

Osamu Tanaike Research Center for Compact Chemical System, National Institute of Advanced Industrial Science and Technology, Sendai, Japan

K. Vlasov School of Engineering and Science, Jacobs University GmbH, Bremen, Germany

M. Winter MEET—Münster Electrochemical Energy Technology, Westfälische Wilhelms-Universität, Münster, Germany

Susumu Yonezawa Headquarters for Innovative Society-Academia Cooperation, Fukui University, Fukui, Japan; Department of Materials Science & Engineering, Faculty of Engineering, University of Fukui, Fukui, Japan

Zhengcheng Zhang Chemical Sciences and Engineering Division, Argonne National Laboratory, Lemont, IL, USA

W. Zhang ICG-AIME, Université Montpellier 2, Montpellier, France

Preface

The present book summarizes recent progress of fluoride-based materials for lithium and sodium batteries, fuel cells, and capacitors. These electric energy-generating devices have been widely used for many objectives. In particular, they are important energy sources not only for hybrid and electric vehicles, but also for our daily lives. To save petroleum resources and suppress CO_2 generation, the social demand for electrochemical devices generating electric energy has been increasing. Hybrid cars are now popular in our society and fuel cell cars will be soon sold. Fuel and solar cells are also found in many buildings and houses. Lithium batteries are of course important energy sources for many kinds of electronic devices. Among various materials used as electrodes, electrolytes, membranes, and so on for lithium and sodium batteries, fuel cells and capacitors, fluorine-containing compounds have exhibited high functions not found in other materials. Fluorine atom with a small size has the highest electronegativity and small polarizability, making strong and stable chemical bonds with other elements. Since fluorine gas (F_2) has a small dissociation energy ($155\,kJ\,mol^{-1}$), the reactivity of F_2 with other compounds and single substances is very high. Not only F_2, but also NF_3 and ClF_3 are important fluorinating agents used for the preparation of many kinds of fluorine compounds. Because of these properties of fluorine, inorganic and organic fluorine compounds and fluoropolymers are now employed for lithium batteries, fuel cells, and capacitors. The use of graphite fluoride as a cathode material of primary lithium battery was realized about 40 years ago. Since then, graphite fluoride has been used for primary lithium battery as an excellent cathode. Commercialization of $Li/(CF)_n$ battery revealed the usefulness of fluorine compounds as energy materials. Polytetrafluoroethylene (PTFE) has been also used for fuel cells to control the surface property of electrodes. Recently, many kinds of fluorine-containing materials are being examined as new electrodes, electrolytes, additives, membranes, and binders. Surface modification is one of the convenient methods to improve the functions of electrode materials because F_2 and other fluorinating agents have high reactivity and can easily modify the surface structures and composition of solid materials. Nowadays many inorganic, organic, and polymer fluorine chemists are working to develop new materials for batteries and capacitors. "Energy" is now one of the important discussion themes at international fluorine symposiums and conferences. The present book provides advanced information on fluorinated materials used as energy conversion materials, being quite useful for researchers, graduate students, and engineers in universities, research institutes, and industries.

Tsuyoshi Nakajima (Aichi Institute of Technology, Japan)
Henri Groult (University of Pierre and Marie Curie, CNRS, France)

Chapter 1

High Performance Lithium-Ion Batteries Using Fluorinated Compounds

Zonghai Chen, Zhengcheng Zhang and Khalil Amine
Chemical Sciences and Engineering Division, Argonne National Laboratory, Lemont, IL, USA

Chapter Outline

1.1 Introduction	1	1.3 Fluorinated Redox Shuttles	13
1.2 Stabilization of Lithiated Anodes	2	1.4 High Voltage Electrolytes	19
		1.5 Closing Remarks	29
1.2.1 Modification of SEI Layer	5	Acknowledgments	29
		References	30
1.2.2 Artificial SEI Layer	7		

1.1 INTRODUCTION

Lithium-ion batteries have been the dominant energy storage technology for powering modern portable electronics. There is also a global effort on R&D of advanced lithium-ion batteries for propulsion and stationary applications in smart grids. The major technological barriers that hinder the realization of these emerging applications include (1) high cost, (2) insufficient life, (3) insufficient energy density, and (4) intrinsically poor safety characteristics. These barriers can be partially tackled by developing advanced electrode materials with a better structural stability, developing materials with higher energy density, and/or using low-cost starting materials and manufacturing processes. High energy-density materials are of great interest since they can lead to an overall reduction in the battery size, resulting in a large saving on other materials like the electrolyte and separator. These barriers can also be addressed by developing functionalized electrolytes that suppress the side reactions between the electrode materials and the electrolytes; these side reactions are the major contributors to the degradation of battery performance.

The current lithium-ion batteries generally use graphitic carbons as the anode material, a lithium transition metal oxide as the cathode material, and a solution of $LiPF_6$ in a blend solvent of alkyl carbonates as the electrolyte. The electrolyte in the battery is used as the lithium-ion conducting medium to transport lithium ions between the anode and cathode. The chemical–electrochemical reactions leading to performance degradation, as well as a potential safety hazard, mostly occur at the interface between the electrode material and the nonaqueous electrolyte, where the electrolyte components can act as either the reactant or the dilution medium that promotes detrimental reactions [1,2]. In an ideal system, the solvent should have a combination of several physical–chemical properties. First, it should be able to dissolve a fairly high concentration of lithium salts for high lithium-ion conductivity. Second, the energy level of the highest occupied molecular orbital (HOMO) of the solvents should be low enough for good resistance to oxidation by the delithiated cathode. Third, the energy level of the lowest unoccupied molecular orbital (LUMO) of the solvents should be high enough to prevent the reduction by lithiated anode materials. So far, an electrolyte that meets the above three requirements has not been identified. For instance, the LUMOs of currently used carbonates are substantially lower than the Fermi energy level of lithium, and hence, they are thermodynamically incompatible with lithium metal and lithiated graphite. The long-term stability of lithiated graphite with the presence of nonaqueous electrolytes can only be kinetically achieved with the presence of the solid-electrolyte interphase (SEI) [3–5], which is a thin layer of an organic–inorganic composite deposited on the surface of a graphitic electrode and acts as a kinetic barrier to protect the lithiated graphite from rapid reaction with the nonaqueous electrolyte. Recently, a massive effort has been devoted to developing cathode materials with high specific capacity and high voltage to meet the energy requirements for plug-in hybrid electric vehicles and full electric vehicles [6–10]. These R&D efforts have pushed the working potential of the cathode materials beyond the thermodynamic limit of the carbonate solvents, and an advanced electrolyte is highly desired to enable high-voltage cathodes [11,12]. Even within a well-characterized lithium-ion chemistry, the working potential of electrode materials can be driven beyond the electrochemically stable window of solvents during overcharge abuse, which can occur during the normal operation of an off-balance lithium-ion battery pack [13–16].

In this chapter, emphasis will be placed on advanced electrolytes with fluorinated components, including (1) advanced electrolyte additives that stabilize lithiated anodes; (2) fluorinated redox shuttles for overcharge protection and automatic capacity balance of the lithium-ion battery pack; and (3) advanced high-voltage electrolytes comprising fluorinated solvents.

1.2 STABILIZATION OF LITHIATED ANODES

Graphitic materials have been the dominant anode material for state-of-the-art lithium-ion technology. However, it is also well known that the lower cutoff potential of lithiated graphite can be as low as 0.1 V versus Li^+/Li, which is far below the standard redox potential of carbonates used for lithium-ion batteries.

The long-term compatibility between the lithiated graphite and the electrolyte solvents is kinetically achieved with the presence of an SEI layer that acts as a physical barrier to prevent the direct exposure of the lithiated graphite to non-aqueous electrolytes. Figure 1.1 schematically shows the SEI formation mechanism during the initial lithiation of a graphitic anode. In general, the open circuit potential of the anode in a freshly prepared lithium-ion cell is about 3.0 V versus Li$^+$/Li. During the initial charge (also called the formation stage) of a fresh lithium-ion cell, the potential of the graphitic anode decreases with the lithiation process. The lithium salt, typically LiPF$_6$, starts to decompose at a potential below 1.5 V versus Li$^+$/Li [17,18]. Part of the decomposition product, mostly inorganic components, deposits on the graphite surface while other components like PF$_5$ promote the electrically triggered polymerization reaction of ethylene carbonate (EC) that forms a flexible layer, mostly organic components, on top of the inorganic layer. This layer of organic–inorganic thin film, or SEI layer, gives a long life to lithium-ion batteries using graphitic anodes. The thermal and electrochemical stability of this SEI layer will determine the electrochemical performance of the batteries using graphitic anodes. Zheng et al. [5] investigated the capacity loss of half cells comprising mesocarbon microbeads (MCMB) electrodes during storage experiments at various temperatures, and found that the capacity loss increased exponentially with the storage temperature. Kinetics study revealed that the process leading to the capacity loss had a fairly low activation energy of 39.7 kJ mol^{-1}. It was concluded that the thermal instability of the SEI layer was a contributor to the capacity loss in lithium-ion cells stored or operated at elevated temperatures. Alternatively, Levi et al. [19] investigated the same issue by tracing the self-discharge process of the graphitic anodes; this process was attributed to the continuous reaction between the lithiated graphite

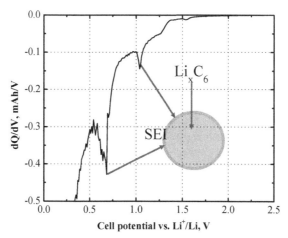

FIGURE 1.1 Schematic showing the formation of solid electrolyte interphase during the initial lithiation of a graphitic anode. SEI, solid-electrolyte interphase.

and the nonaqueous electrolyte, leading to the continuous loss of lithium in the material. Even with the presence of the SEI layer, the self-discharge current measured at 45 °C was about 0.57 mA g^{-1} of the active material. The activation energy measured for the self-discharge was about 31.9 kJ mol^{-1}, which is very close to that reported by Zheng et al. [5]. Studying the harvested anode materials from charged lithium cells, Holzapfel et al. suggested that the thermal decomposition of the SEI layer could start at a temperature as low as 60 °C at a steady state [17].

Figure 1.2(a) shows a typical differential scanning calorimetry (DSC) profile of a sample containing about 3 mg of lithiated MCMB and ~3 μL of electrolyte, i.e., 1.2 M LiPF$_6$ in a mixture solvent of EC and ethyl methyl carbonate (EMC) with a mass ratio of 3:7. An extended exothermal peak was observed at above 110 °C, which was attributed to the thermal decomposition of the metastable SEI layer [1,20]. The activation energy for this process was determined to be about 53 kJ mol^{-1} [1]. The new fresh surface area generated from the decomposition of the original SEI provides active sites for electrolyte components to continuously react with the lithium hosted in the MCMB anode. The in situ high-energy X-ray diffraction (HEXRD) patterns in Figure 1.2(b) clearly show a continuous decrease of intensity for LiC$_6$ and LiC$_{12}$ peaks when the temperature was higher than 107 °C. The (002) peak indicates that graphite appeared when the temperature was above 246 °C. This suggests that the loss of peak intensity for LiC$_6$ and LiC$_{12}$ is primarily due to the chemical reaction between the lithiated MCMB and the electrolyte, which was enabled after the thermal decomposition of the SEI layer at about 107 °C. It is also believed that the slow

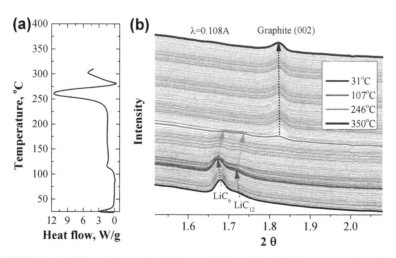

FIGURE 1.2 (a) Differential scanning calorimetry (DSC) profile of a lithiated mesocarbon microbeads with the presence of electrolyte and (b) evolution of in situ high-energy X-ray diffraction patterns during the thermal ramping of a DSC sample. The electrolyte was 1.2 LiPF$_6$ in ethylene carbonate/ethyl methyl carbonate (3:7 ratio, by weight).

reaction between the lithiated MCMB and the electrolyte led to the extended and low-intensity exothermic peak between 110 °C and 230 °C. This finding indicates that the thermal stability of the SEI layer is the key to improve the electrochemical performance of lithium-ion batteries using graphitic anodes.

1.2.1 Modification of SEI Layer

Figure 1.3 shows a conceptual model of the SEI layer on the surface of graphite. The SEI layer is an organic/inorganic composite thin film composed of LiF, Li_2O, Li_2CO_3, polyolephines, semicarbonates, and maybe other unidentified species [18,21–23]. Those inorganic species are mostly generated from the decomposition of lithium salts, and the organic components are mostly from the reduction of solvents. The composition of the nonaqueous electrolyte can significantly affect the composition, as well as physical–chemical properties, of the SEI layer. Note that the relative concentration of different species in Figure 1.3 has no physical meaning since the SEI layer is not actually uniformly distributed on the surface of the graphite. For instance, Bar-Tow et al. reported that the SEI layer on the basal plane is substantially different from that on the edge of the graphite; the SEI layer on the basal plane is dominated by the organic components while that on the edge is rich in inorganic compounds [21]. The discrepancy between the basal plane and the edge can originate from the difference of the chemical environment, which was confirmed by surface analysis of the SEI layer on different carbonaceous materials. For instance, Eshkenazi et al. compared the chemical composition of the SEI layer on soft carbon against that on hard carbon, and found that the substrate has a more pronounced influence on SEI formation on carbonaceous materials than does the electrolyte [18]. Similarly, Chen et al. deployed DSC and in situ HEXRD to investigate the kinetics of the thermal decomposition reaction of SEI layers on different carbonaceous substrates [1]. They found that the SEI on natural graphite was thermally more stable than that on MCMB, and the increased stability was attributed to the presence of 3R graphite in the natural

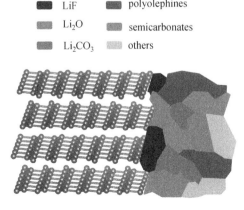

FIGURE 1.3 Conceptual model of solid-electrolyte interphase layer formed on the surface of graphitic anodes.

graphite. In addition, Chen et al. tried to modify the surface chemistry of soft carbon using 3,3,3-trifluoropropyltrimethoxysilane (TFPTMS) (see Figure 1.4) and found that the electrochemical performance of lithium-ion cells using soft carbon was greatly improved by reacting some acidic surface groups like –OH with TFPTMS [24].

An alternative, or more effective, way to modify the SEI layer is to use an anion receptor as an electrolyte additive. The anion receptor mentioned here is a chemical that can complex with or bind to the anion of the electrolyte salt and, hence, can improve its solubility, as well as the transference number. The anion receptors of particular interest are boron-based compounds that have an empty π orbital on the boron center to accept donated electrons from electron-rich atoms like oxygen or fluorine (see Figure 1.5). Yang et al. successfully demonstrated that boron-based anion receptors are capable of coordinating with F^- and help to dissolve LiF in nonaqueous solvents [25–27]. They demonstrated that such a solution could be directly used as an electrolyte for lithium-ion cells without the addition of common lithium salts like $LiPF_6$.

As mentioned above, the SEI layer on the surface of graphitic anodes is composed of a large amount of insulating inorganic components like LiF and

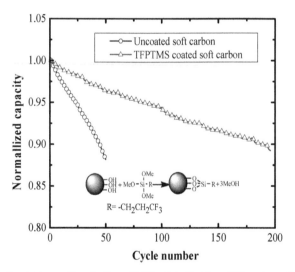

FIGURE 1.4 Nominal capacity retention of $LiMn_2O_4$/soft carbon lithium-ion cells showing the positive effect of 3,3,3-trifluoropropyltrimethoxysilane (TFPTMS) on the soft carbon. The cells were cycled at 25 °C.

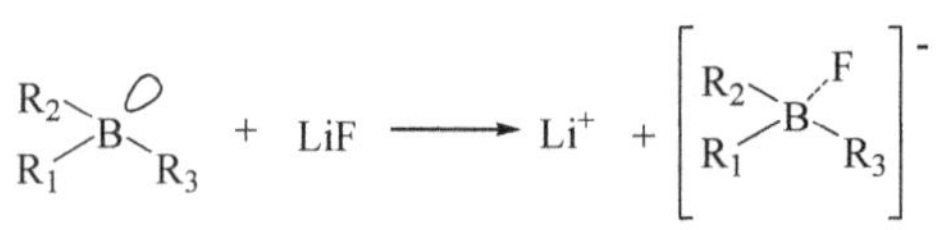

FIGURE 1.5 Generalized coordination reaction between boron-based anion receptors and LiF.

Li$_2$O, which can impede the transportation of lithium ions through the SEI layer. Therefore, the addition of a small amount of anion receptor to the nonaqueous electrolyte can improve both the rate capabilities [28] and the capacity retention [28–31] of lithium-ion cells.

In the general formula shown in Figure 1.5, the group R can be alkyl, alkoxyl, or aryl groups. It was reported that the type of substitution group R has a significant effect on the electrochemical performance of lithium-ion cells using anion receptors as the electrolyte additive. Chen et al. found that the addition of a small amount of tripropyl borate (TPB, R=–OC$_3$H$_7$) (≤0.1 wt%) could lead to improved capacity retention, as well as a slight reduction of the cell impedance [31]. However, a high concentration of TPB degraded cell performance, which was speculated to originate from the oxidation of TPB at elevated temperature. In addition, the alkoxyl group has a strong tendency to donate part of the lone pair electrons on oxygen to the empty π orbital of the boron center, leading to a decreased affinity between the boron center and other anions like F$^-$ or O^{2-}. Density function theory (DFT) calculations confirmed that boranes had a higher bonding strength toward F$^-$ than borates with a similar structure [32]. Based on the results of the DFT calculations, a group of highly fluorinated boron-based anion receptors was selected for detailed electrochemical characterization. It turned out that the fluorine affinity of the anion receptors has a strong effect, but not monotonic, on the electrochemical performance of lithium-ion cells [30]. Some strong anion receptors even lead to a decrease in reversible capacity of the cells, which was attributed to the difficulty in controlling the amount of anion receptor in the electrolyte; excess strong anion receptors can also promote the decomposition of LiPF$_6$ into LiF and PF$_5$. The extra PF$_5$ can then promote the electrochemical polymerization of the carbonates and a thicker SEI layer that impedes the transportation of lithium ions through the SEI layer. Among the fluorinated anion receptors tested, tris(pentafluorophenyl)borane (TPFPB) was found the best additive to enhance both the life and rate capability of lithium-ion cells. A substantial reduction of the cell impedance was reported when up to 1 wt% of TPFPB was added to the electrolyte. When the concentration of TPFPB was higher than 1 wt%, the cell impedance increased quickly due to the reaction of excess TPFPB with LiPF$_6$. Regardless of the impact on the cell impedance, the capacity retention of the cells improved monotonically with the content of TPFPB in the electrolyte [28,29].

1.2.2 Artificial SEI Layer

A more effective way to improve the compatibility between the lithiated graphite and the nonaqueous electrolyte is to build a more stable artificial SEI layer below the naturally occurring SEI layer. The artificial SEI layer can serve as the physical barrier to prevent the direct exposure of lithiated graphite to nonaqueous electrolytes, even when the naturally occurring SEI layer decomposes at a temperature above 107 °C (see Figure 1.2). This strategy can be implemented by developing electrolyte additives that can be reduced to trigger the polymerization reaction on the surface of anodes. Of particular interest are those with a reduction potential higher than 1.3 V versus Li^+/Li so that the formed artificial SEI layer can be placed under the natural SEI layer. It has been reported that vinylene carbonate (VC) [33–35], vinyl ethylene carbonate (VEC) [33], 3,9-divinyl-2,4,8,10-tetraoxaspiro[5,5] undecane (TOS) [36], and others are promising electrolyte additives to form a thermally more stable SEI layer and, hence, improve the capacity retention of lithium-ion cells. Most importantly, the improved capacity retention was achieved at the expense of a reduced rate capability because these additives have a strong tendency to form a thick and resistive film that also impedes the transportation of lithium ions through the artificial SEI layer.

Of particular interest here are two fluorinated additives, lithium difluoro[oxalato]borate (LiDFOB) [20,37–39] and lithium tetrafluoro[oxalato] phosphate (LiTFOP) [40] (see Figure 1.6(b) and (c) for detailed molecular structures). Both additives can form a thermally more stable SEI layer on the surface of graphite to extend the life of graphite-based lithium-ion cells, but without adding extra interfacial impedance to the cells. Shown in Figure 1.6(a) is the structure of lithium bis[oxalato]borate (LiBOB), which is also a good electrolyte additive to protect lithiated graphite, but with substantial increase in the cell impedance [37].

Figure 1.7(a) shows the differential capacity profiles (dQ/dV) of Li/MCMB half cells showing the characteristic reduction peak introduced by the addition of LiBOB and LiDFOB to the $LiPF_6$-based electrolyte. The black line with solid circles illustrates the profile of the control cell without any additive. A clear reduction peak appears at about 0.7 V versus L^+/Li, which was attributed to the reduction of ethylene carbonate to form the organic component of the SEI layer. The small peak related to $LiPF_6$ decomposition at ~1.2 V versus Li^+/Li was barely visible due to the scale used. With the addition of LiBOB to the electrolyte, a new reduction peak appears at about 1.7 V versus Li^+/Li. The peak was assigned to the formation of the artificial SEI layer with the participation of LiBOB. With the formation of the new artificial SEI layer, the reduction peak of ethylene carbonate at ~0.7 V versus Li^+/Li remained unchanged, meaning that the SEI layer for $LiPF_6$-based electrolytes was formed on top of the artificial one. The LiBOB-based artificial SEI layer is robust and has better thermal

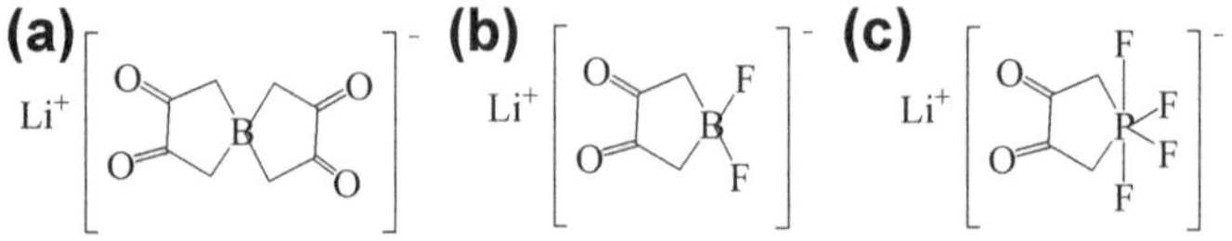

FIGURE 1.6 Molecular structures of oxalato-based electrolyte additives to form an artificial solid-electrolyte interphase layer, (a) lithium bis(oxalato)borate (LiBOB), (b) lithium difluoro(oxalato) borate (LiDFOB), and (c) lithium tetrafluoro(oxalato)phosphate (LiTFOP).

stability than the natural layer and, hence, results in a substantial improvement in the capacity retention. However, the major impedance rise due to the thick artificial SEI layer leads to significant decrease of rate capabilities and the power density. Hence, the understanding on the reduction/passivation mechanism became a key to enable high-rate and high-power applications.

Both empirical speculation [41] and DFT calculation [37] indicate that the BOB$^-$ anion can accept one electron, leading to a ring-open reaction by breaking

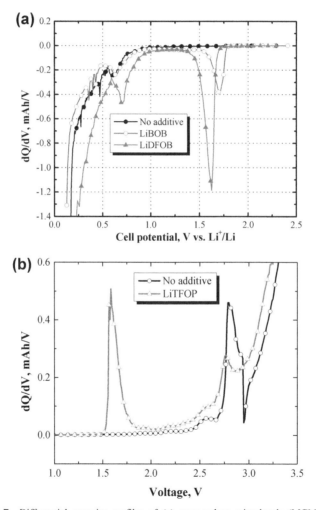

FIGURE 1.7 Differential capacity profiles of (a) mesocarbon microbeads (MCMB)/Li cells with no additive and lithium bis[oxalato]borate (LiBOB) and lithium difluoro[oxalato]borate (LiDFOB) additives and (b) Li$_{1.1}$[Mn$_{1/3}$Ni$_{1/3}$Co$_{1/3}$]$_{0.9}$O$_2$/MCMB cell with no additive and lithium tetrafluoro[oxalato]phosphate (LiTFOP) additive.

a B–O bond; the resulting radical can then attack another BOB⁻ anion to initiate the polymerization reaction. The resulting polymer will deposit on the surface of the anode as an artificial SEI layer. Meanwhile, the second oxalato-based five-member ring can also accept one electron to initiate a cross-linking reaction between the polymer chains. Therefore, we expect that the artificial SEI layer is primarily composed of dense cross-linked 3D polymers that impede the transportation of lithium ions. LiDFOB is a variation of LiBOB by replacing one of the oxalato-based groups with two F's so that the cross-linking reaction can be inhibited. To confirm the effectiveness of the substitution strategy, the electrochemical impedance plots of lithium-ion cells using MCMB as the anode and $Li_{1.1}[Ni_{1/3}Mn_{1/3}Co_{1/3}]_{0.9}O_2$ as the cathode are compared in Figure 1.8; the electrolyte salts for those cells were $LiPF_6$, LiBOB, and LiDFOB, respectively. Clearly, the impedance of the cells using LiDFOB was very close to that of the cell using $LiPF_6$, but was far lower than that of the cell using LiBOB. When LiDFOB is used as an electrolyte additive for $LiPF_6$-based electrolytes, we can expect only a small change on the cell impedance regardless of the content of LiDFOB.

Figure 1.9 shows the normalized capacity loss of lithium-ion cells using $Li_{1.1}[Ni_{1/3}Mn_{1/3}Co_{1/3}]_{0.9}O_2$ as the cathode and MCMB as the anode with and without 2 wt% LiDFOB as the electrolyte additive. The electrolytes were 1.2 M $LiPF_6$ in a mixture solvent of ethylene (EC), propylene carbonate (PC), and dimethyl carbonate (DMC) with a mass ratio of 1:1:3. The cell without LiDFOB lost about 22% of its initial capacity after 100 cycles, while the cell containing 2% LiDFOB had an improved capacity retention, only 14% capacity loss [39]. The positive role of LiDFOB as an electrolyte additive is further confirmed with the kinetics study of the thermal decomposition of the artificial

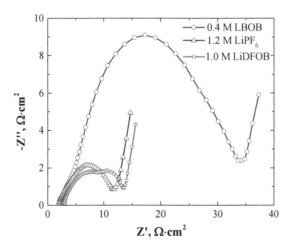

FIGURE 1.8 AC impedance of $Li_{1.1}[Mn_{1/3}Ni_{1/3}Co_{1/3}]_{0.9}O_2$/mesocarbon microbeads cells showing the impact of lithium bis[oxalato]borate (LiBOB) and lithium difluoro[oxalato]borate (LiDFOB) additives.

SEI layer formed with the participation of LiDFOB. Figure 1.10 shows DSC profiles of the lithiated MCMB with and without the addition of LiDFOB to the electrolyte. The scanning rate was 10 °C/min. When no LiDFOB was added to the electrolyte, a broad exothermic peak appeared at a temperature as low as 110 °C; this peak was attributed to the thermal decomposition of the metastable SEI film [1]. When LiDFOB was added to the electrolyte, the decomposition temperature of the SEI film was increased to 150 °C. These data suggest that

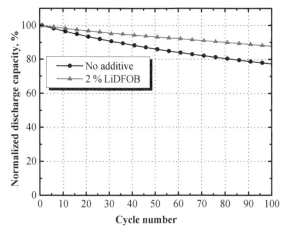

FIGURE 1.9 Normalized discharge capacity of $Li_{1.1}[Mn_{1/3}Ni_{1/3}Co_{1/3}]_{0.9}O_2$/mesocarbon microbeads cells showing the positive effect of lithium difluoro[oxalato]borate (LiDFOB) additive.

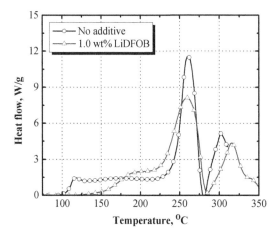

FIGURE 1.10 Differential scanning calorimetry profiles of lithiated mesocarbon microbeads with the presence of a non-aqueous electrolyte showing that the addition of lithium difluoro[oxalato]borate (LiDFOB) as the electrolyte additive increases the onset temperature of solid-electrolyte interphase decomposition.

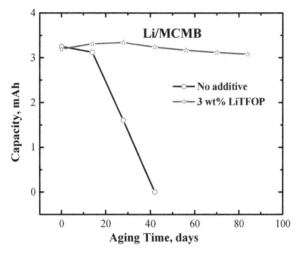

FIGURE 1.11 Capacity of mesocarbon microbeads (MCMB)/Li cell after storage at 55 °C and 1 mV showing that lithium tetrafluoro[oxalato]phosphate (LiTFOP) additive significantly improves electrochemical performance of the cell during storage.

the addition of LiDFOB can significantly suppress the kinetics of the thermal decomposition of lithiated MCMB in nonaqueous electrolyte. Since no new exothermic peak occurred with the addition of LiDFOB, one can conclude that the lithiated MCMB was stabilized by the deposition of new SEI film with the participation of LiDFOB, which acted as a chemical barrier for the reaction between lithiated MCMB and the electrolyte [20].

The compound LiTFOP (see Figure 1.6(c) for its molecular structure) is another successful fluorinated electrolyte additive that fulfills the above design philosophy: it leaves only one oxalato-based five-member ring for the electrochemically initiated polymerization reaction and attaches four F atoms to the P atom to avoid any potential cross-linking reaction. Figure 1.7(b) shows the differential capacity profiles of lithium-ion cells using $Li_{1.1}[Ni_{1/3}Mn_{1/3}Co_{1/3}]_{0.9}O_2$ as the cathode and MCMB as the anode. The baseline electrolyte was 1.2 M $LiPF_6$ dissolved in the mixture solvent of EC/EMC with a ratio of 3:7 by volume, with or without additive. During the initial charging of the cell without additive, two small peaks appeared in the voltage range between 2.5 and 3.0 V, both of which were assigned to the formation of SEI on the graphite surface in the $LiPF_6$-based electrolyte. When LiTFOP was added as an electrolyte additive, an extra peak appeared at about 1.65 V, which is attributed to the formation of an artificial SEI layer [40].

Figure 1.11 shows the charge capacity of Li/MCMB cells at 25 °C after the cell had been aged at 1 mV and 55 °C for different periods of time. After being aged for 30 days at 55 °C, the cell with pristine electrolyte had only half of its reversible capacity left, and it completely died after 40 days of aging. However, the cell with 3 wt% LiTFOP as the electrolyte additive showed excellent

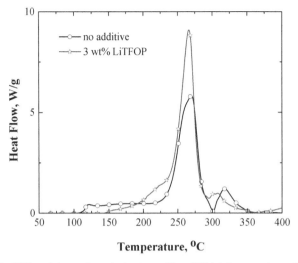

FIGURE 1.12 Differential scanning calorimetry profiles of lithiated mesocarbon microbeads with the presence of a non-aqueous electrolyte showing that the addition of lithium tetrafluoro[oxalato] phosphate (LiTFOP) additive increases the onset temperature of solid-electrolyte interphase decomposition.

capacity retention of almost 100%, even after more than 80 days of aging. It is clear that the artificial SEI layer introduced by LiTFOP successfully protects the lithiated MCMB from attack by the electrolyte components. The DSC study also showed that the onset temperature for the SEI decomposition was increased from about 110 °C to >150 °C with the addition of LiTFOP as an electrolyte additive (see Figure 1.12) [40].

1.3 FLUORINATED REDOX SHUTTLES

Lithium-ion batteries are sensitive to thermal and overcharge abuse and pose significant fire hazards. Overcharge of lithium-ion cells can lead to chemical and electrochemical reactions between battery components [42,43], gas release [42–44], and rapid increase of the cell temperature [42–44]. It can also trigger self-accelerating reactions in the batteries, which can lead to thermal runaway and possible explosion. Overcharge generally occurs during the charge of a battery pack with multiple lithium-ion cells connected in series, as shown in Figure 1.13. When the battery pack is charged, the charger generally continuously monitors the voltage of the battery pack to roughly estimate the state of charge (SOC) of the battery pack; it does not monitor each cell and so assumes that each cell in the pack is identical to the others in terms of capacity and SOC. However, this assumption is difficult to validate and maintain in real operation. There is always chance that one or more cells in the battery pack has less capacity than the others, as shown in Figure 1.13(a), where the cell with less capacity is called the "weak cell" (the second cell in Figure 1.13(a)). When this battery pack is charged, the weak cell will reach its top SOC first while the others are still not fully charged (see Figure 1.13(b)). At this point, the voltage of the whole pack is still lower than the expected value, and the charger will continue to charge the pack and overcharge the weak cell. Thus, the overcharge protection must operate at the cell level to assure safe operation of the battery pack. The cell-level overcharge protection can also reduce the need for costly cell capacity balancing during battery manufacturing, maintenance, and repair. With the cell-level overcharge protection mechanism, the battery pack can be charged as a whole. When a cell reaches its top SOC during charge, the cell voltage can electrically trigger the overcharge protection mechanism, and the excess current can be handled by the incorporated overcharge protection mechanism without causing overcharge to the cell. Under this mechanism, the charging process of the battery pack can continue until all the cells are fully charged.

Redox shuttles [45] are a class of electrolyte additives that can be reversibly oxidized/reduced at a characteristic potential and provide intrinsic overcharge protection for lithium-ion batteries that neither increases the complexity and weight of the control circuit nor permanently disables the cell when activated. The redox shuttle molecule (S) has a defined redox potential at which it can be oxidized on the cathode and form a radical cation ($S^{\bullet+}$):

$$S \rightarrow S^{\bullet+} + e^- \qquad (1.1)$$

The radical cation then travels to the anode through the electrolyte and is reduced in accordance with.

$$S^{\bullet+} + e^- \rightarrow S \qquad (1.2)$$

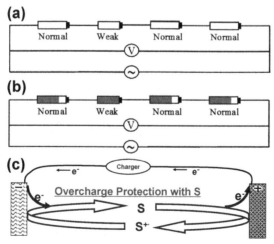

FIGURE 1.13 Schematics showing (a,b) the effect of potential overcharge abuse on the weak cell in a series-connected battery pack, and (c) overcharge protection mechanism using a redox shuttle.

The redox shuttle molecule then diffuses back to the cathode for the next redox cycle. During normal operation, the redox potential of the redox shuttle is not reached, and the molecules stay inactive. When the cell is overcharged, the cathode potential increases, and the redox cycle of the redox shuttle molecules is then activated (see Figure 1.13(c)). The net reaction of the redox cycle is to shuttle the charge forced by the external circuit through the lithium-ion cell without fostering intercalation/deintercalation of lithium in the electrodes.

Chen et al. reported the first stable redox shuttle, 2,5-di-terbutyl-1,4-dimethoxybenzene (DDB), for overcharge protection of 4-V class lithium-ion batteries [16]. However, the redox potential of DDB is about 3.92 V versus Li$^+$/Li; therefore, it can only work with a LiFePO$_4$ cathode. A much higher redox potential is needed to enable the applications of redox shuttles for dominant cathode materials like LiCoO$_2$ and LiNi$_{0.8}$Co$_{0.15}$Al$_{0.05}$O$_2$. In general, the oxidation of the redox shuttle involves the removal of an electron from the highly conjugated π orbitals. Replacing some side groups around the π orbitals with strong electron-withdrawing groups like –F or –CF$_3$ can substantially increase the redox potential. Here we will discuss two types of highly fluorinated redox shuttles, both of which have redox potentials high enough to provide overcharge protection of the dominant 4-V cathode materials.

The first fluorinated redox shuttle introduced here is 2-(pentafluorophenyl)-tetrafluoro-1,3,2-benzodioxaborole (PFPTFBB, see Figure 1.14(a) for its molecular structure). Figure 1.15 shows that PFPTFBB has a reversible redox reaction at about 4.43 V versus Li$^+$/Li. The onset potential of PFPTFBB is about 4.3 V versus Li$^+$/Li, which is high enough to provide overcharge protection for most 4-V class cathode materials. Chen et al. reported that lithium-ion cells using PFPTFBB as a redox shuttle additive could be overcharge-abused for up to 150 cycles before the redox shuttle mechanism stopped [29]. In addition, the boron center incorporated in PFPTFPBB can also act as an anion receptor to

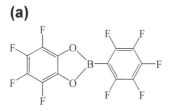

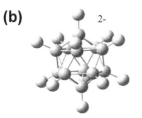

FIGURE 1.14 Molecular structures of (a) 2-(pentafluorophenyl)-tetrafluoro-1,3,2-benzodioxaborole and (b) B$_{12}$F$_{12}^{2-}$.

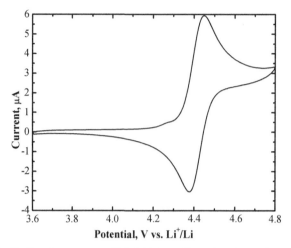

FIGURE 1.15 Cyclic voltammogram of 0.05 M 2-(pentafluorophenyl)-tetrafluoro-1,3,2-benzodioxaborole and 1.2 M LiPF$_6$ in ethylene carbonate/ethyl methyl carbonate (3:7 ratio, by weight) using a Pt/Li/Li three-electrode cell.

improve the capacity retention of lithium-ion cells. It was also demonstrated that the binding between the boron center with F$^-$ does not affect the electrochemical reactivity of the aromatic ring; hence, PFPTFPBB can be considered bifunctional, i.e., both a redox shuttle and an anion receptor.

The second class of fluorinated redox shuttle is fluorinated lithium borate cluster salt, Li$_2$B$_{12}$F$_{12-x}$H$_x$. The molecular structure of a fully fluorinated anion is shown in Figure 1.14(b). This class of compounds is also bifunctional. It can be easily dissolved into the carbonate-based solvent up to 0.4 M; the concentration of Li$^+$ is about 0.8 M. Hence, it can be used as an independent salt to prepare non-aqueous electrolytes [46]. Moreover, the anions of this salt have unique reversible redox capability: the redox potential of the anion increases monotonically with the degree of fluorination. Figure 1.16 shows the cyclic voltammogram (CV) of two members, Li$_2$B$_{12}$F$_9$H$_3$ and Li$_2$B$_{12}$F$_{12}$, whose reversible redox potential are about 4.5 and 4.6 V versus Li$^+$/Li, respectively. Clearly, this class of salts can also act as redox shuttle for overcharge protection.

Compared to the conventional salt LiPF$_6$, lithium borate cluster salts are insensitive toward water present in the electrolyte [38]. Hence, the SEI layer formed in Li$_2$B$_{12}$F$_{12-x}$H$_x$-based electrolyte is generally weak, and an additional electrolyte additive is needed for a robust SEI layer to prevent the continuous reaction between the lithiated graphite and the solvent.

Figure 1.17 shows the discharge capacity of MCMB/Li$_{1.156}$Mn$_{1.844}$O$_4$ cells that were cycled at 25 °C between 3.0 and 4.2 V with a constant current of C/3 (~0.5 mA). The lithium-ion cell using Li$_2$B$_{12}$F$_9$H$_3$ could only be charged and discharged for a few cycles with rapid capacity fade. Hence, a better SEI can be crucial to improve the capacity retention of such cells. Another cell using

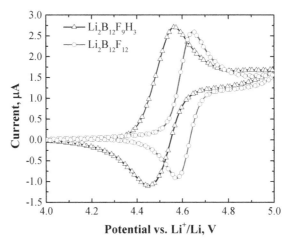

FIGURE 1.16 Cyclic voltammogram of 0.05 M lithium boron cluster salts and 1.2 M $LiPF_6$ in ethylene carbonate/ethyl methyl carbonate (3:7 ratio, by weight) using a Pt/Li/Li three-electrode cell.

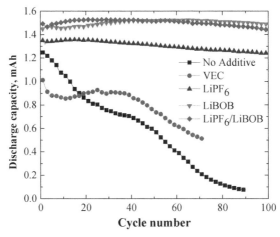

FIGURE 1.17 Discharge capacity of $Li_{1.156}Mn_{1.944}O_4$/mesocarbon microbeads cells using $Li_2B_{12}F_9H_3$-based electrolytes containing different additives. LiBOB, lithium bis[oxalato]borate; VEC, vinyl ethylene carbonate.

1.0 wt% VEC as the electrolyte additive to form an artificial SEI layer on the graphite surface showed an improvement in the capacity retention. However, the initial reversible capacity was about 1/3 lower than that of the cells without any additive, probably due to the highly resistive SEI layer that formed with the added VEC. A high initial reversible capacity and good capacity retention were achieved when either $LiPF_6$ or LiBOB was added as the electrolyte additive [47]. When both LiBOB and TPFPB were added to the $Li_2B_{12}F_9H_3$-based electrolyte, the lithium-ion cell could be cycled for 1200 cycles with 85% capacity

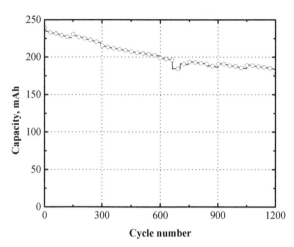

FIGURE 1.18 Discharge capacity of $Li_{1.1}[Ni_{1/3}Mn_{1/3}Co_{1/3}]$/mesocarbon microbeads cell using $Li_2B_{12}F_9H_3$-based electrolyte with lithium difluoro[oxalato]borate additive. The cell maintained 74% of its initial capacity after being tested at 55 °C for 1200 cycles.

retention. The superior performance of $Li_2B_{12}F_9H_3$-based electrolyte was also demonstrated with $Li_{1.1}[Mn_{1/3}Ni_{1/3}Co_{1/3}]_{0.9}O_2$/MCMB pouch cells. Figure 1.18 shows the discharge capacity of a pouch cell that was cycled with a constant current of C/3 at 55 °C. The electrolyte used was 0.4 M $Li_2B_{12}F_9H_3$ in mixture solvent of EC/EMC (3:7 by weight) and 2 wt% LiDFOB as an electrolyte additive to form a robust artificial SEI layer. The cell was able to maintain 74% of its initial capacity after being continuously tested for 1200 cycle at 55 °C [38].

The most striking feature of lithium borate cluster salts is that the characteristic redox chemistry of their anions can offer overcharge protection. Figure 1.19 shows the voltage profile of a graphite/$Li_{1.1}[Ni_{1/3}Mn_{1/3}Co_{1/3}]_{0.9}O_2$ cell during an overcharge abuse test. The electrolyte was 0.4 M $Li_2B_{12}F_9H_3$ and 0.1 M LiBOB in a mixture solvent of EC/EMC with a ratio of 3:7 by weight; 1.0 wt% of PFPT-FBB was added as a supplemental electrolyte additive. During the overcharge abuse test, the cell was charged for 4 h with a constant current of C/3 before being constant-current discharged to 3.0 V at the same rate. Figure 1.19 shows that the voltage of the cell was capped at about 4.5 V when $B_{12}F_9H_3^{2-}$ was activated, and this overcharge protection mechanism was maintained up to about 450 cycles (2000 h), even though the discharge capacity of the cell declined with the cycle number [38].

While $LiPF_6$ is sensitive to moisture, lithium borate cluster salts are not, which can significantly improve the thermal stability of delithiated cathodes [2,38]. In situ high-energy X-ray diffraction was used to trace the structural change of delithiated $Li_{1.1}[Mn_{1/3}Ni_{1/3}Co_{1/3}]_{0.9}O_2$ and showed that the decomposition of $LiPF_6$ in the presence of moisture can generate a trace amount of proton, which reduce the onset temperature of the decomposition of

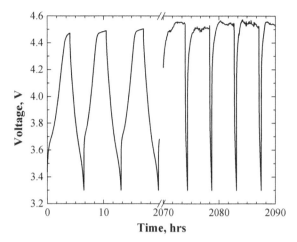

FIGURE 1.19 Voltage profile of a graphite/Li$_{1.1}$[Ni$_{1/3}$Mn$_{1/3}$Co$_{1/3}$]$_{0.9}$O$_2$ cell during continuous overcharge test at low rate (~C/3, 0.6 mA). The electrolytes consisted of 0.4 M Li$_2$B$_{12}$F$_9$H$_3$ and 0.1 M lithium bis[oxalato]borate in ethylene carbonate/ethyl methyl carbonate, with additive of 1 wt% 2-(pentafluorophenyl)-tetrafluoro-1,3,2-benzodioxaborole.

Li$_{1.1-x}$[Ni$_{1/3}$Mn$_{1/3}$Co$_{1/3}$]$_{0.9}$O$_2$ to about 198 °C. The onset temperature was increased by simply replacing LiPF$_6$ with a lithium borate cluster salt [2,38].

1.4 HIGH VOLTAGE ELECTROLYTES

Of all the possible high oxidation organic solvents available, fluorinated organic solvents attract our attention for high-voltage Li-ion batteries. They generally have higher oxidation potentials than their nonfluorinated counterparts due to the strong electron-withdrawing effect of the fluorine atom or fluorinated alkyl. Fluorinated organics may be good candidates for electrolyte solvents. Fluorinated cyclic and linear carbonate compounds possess desirable physical properties as lithium-ion battery electrolyte, which are imparted by the presence of the fluorine substituents, such as low melting point, increased oxidation stability, and less flammability [48,49]. In fact, fluorinated cyclic carbonate has been reported as a cosolvent by McMillan [50] and Nanbu [51], and as an SEI formation additive for graphite [52] and silicon [53] anodes. A series of fluorinated linear carbonates was designed and synthesized by Smart et al. [54] at NASA's Jet Propulsion Laboratory as new electrolyte components to improve the low temperature performance of the lithium-ion battery for deep space applications. Achiha et al. [49,55] and Arai [56] separately reported the fluorinated carbonates and fluorinated ethers as nonflammable electrolytes or as new solvents for nonaqueous lithium-ion battery electrolytes.

Here, Amine and coworkers conducted an electrochemical investigation of these fluorinated carbonates and fluorinated ether in high-voltage LiNi$_{0.5}$Mn$_{1.5}$O$_4$ (LNMO) cells [11,12]. The chemical structures of the studied fluorinated carbonates and ethers are listed in Figure 1.20. The initial prediction of the oxidation stability of the fluorinated compounds shown in Figure 1.20 was carried out by DFT calculations, and the results are reported in Table 1.1. Oxidation potentials were calculated by subtracting the absolute free energies of the neutral species from those of the corresponding cations, and 1.46 V was subtracted to obtain values relative to a Li^+/Li reference electrode [57]. Calculations of the HOMO and LUMO energies with the B3LYP/6-311+G(3df,2p) basis set are also included in Table 1.1, since it has been shown that B3LYP HOMO and LUMO energies can be successfully correlated with molecular properties such as ionization potential and electron affinity in a semiquantitative manner [58].

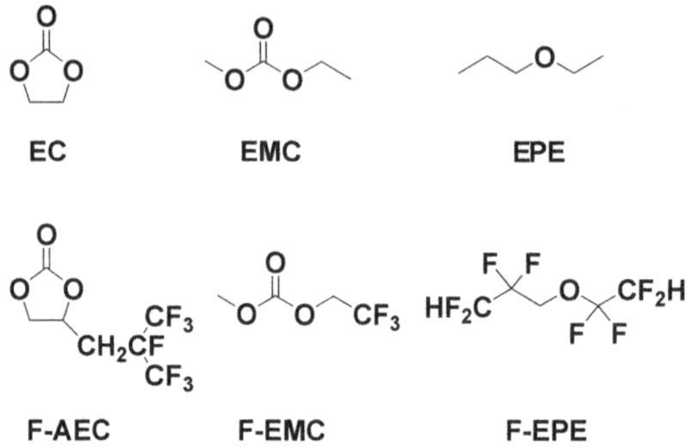

FIGURE 1.20 Chemical structures of the baseline carbonate (ethylene carbonate (EC) and ethyl methyl carbonate (EMC)), ethyl propyl ether (EPE), fluorinated cyclic carbonate (F-AEC), fluorinated linear carbonate (F-EMC), and fluorinated ether (F-EPE). *Reprinted with permission from Ref. [11]. © 2013 Royal Society of Chemistry.*

TABLE 1.1 Structure, Oxidation Potential (P_{ox}), and HOMO/LUMO Energies of Carbonates, Ethers, Fluorinated Carbonates, and Fluorinated Ethers

Molecule	Structure	P_{ox} (V, Theory)	HOMO (au)	LUMO (au)
EC	fx1	6.91 (6.83 open)	−0.31005	−0.01067
EMC	fx2	6.63	−0.29905	0.00251
EPE	fx3	5.511	−0.26153	0.00596
F-AEC	fx4	6.98	−0.31780	−0.01795
F-EMC	fx5	7.01	−0.31946	−0.00363
F-EPE	fx6	7.24	−0.35426	−0.00356

Reprinted with permission from Ref. [12]. © 2013 Royal Society of Chemistry.

Based on the DFT results shown in Table 1.1, fluorine substitution in organic carbonate and ether lowers both HOMO and LUMO levels, resulting in simultaneously higher oxidation stability and higher reduction potential. The results also indicate that the fluorinated electrolytes are thermodynamically more stable than their nonfluorinated counterparts under certain high voltage conditions. It is worth noting that the oxidation potential of the linear carbonate EMC can be greatly increased, from 6.63 to 7.01 V, through fluorine substitution (Table 1.1). The calculated oxidation potentials of the fluorinated carbonates and ether are much higher compared with other high voltage electrolytes. Shao and coworkers [59] reported the oxidation potential of ethyl methyl sulfone, calculated with the DFT method, to be around 6.0 V.

To investigate the stability of various electrolytes against oxidation under high voltage conditions, we formulated six new fluorinated electrolytes to compare with the Gen 2 electrolyte, 1.2 M $LiPF_6$ in EC/EMC (3:7), as follows:

E1: 1.2 M $LiPF_6$ in EC/EMC/F-EPE (2/6/2);
E2: 1.2 M $LiPF_6$ in EC/EMC/F-EPE (2/5/3);
E3: 1.2 M $LiPF_6$ in F-AEC/EMC/F-EPE (2/6/2);
E4: 1.2 M $LiPF_6$ in F-AEC/EC/EMC/F-EPE (1/1/6/2);
E5: 1.2 M $LiPF_6$ in F-AEC/F-EMC/F-EPE (2/6/2);
E6: 1.2 M $LiPF_6$ in EC/F-EMC/F-EPE (2/6/2).

Results in Table 1.1 suggest that E5 be the most stable electrolyte since its components (F-AEC, F-EMC, and F-EPE) have the highest oxidation potentials. For electrolyte E6, the EC component forms an open cation radical at 6.83 V, which is lower than F-AEC (6.98 V) in E5, suggesting oxidative decomposition in E6 at lower potential than in E5.

The ambient conductivities of these fluorinated electrolytes are comparable with those of the Gen 2 electrolyte, and the dependence of conductivity on temperature is illustrated in Figure 1.21.

The electrochemical floating test was performed with the fluorinated electrolytes and the Gen 2 electrolyte from 5.3 to 6.1 V versus Li^+/Li. As shown in Figure 1.22, while the Gen 2 electrolyte and the fluorinated electrolytes E1 to E6 remain stable at 5.3 V with minimal leakage current, Gen 2 electrolyte shows much larger leakage current and maintains this current when the potential is raised to 5.7 V. In contrast, the fluorinated electrolytes show minimal increase in leakage current at this potential. At this high potential, the fluorinated ether F-EPE enhances the electrochemical stability, as evidenced by the fact that the only major difference between E1 and Gen 2 is the addition of F-EPE in the former. The EMC-based electrolyte compositions E1 to E4 show a similar amount of increase in the leakage current, while the F-EMC-based E5 and E6 retain a minimal leakage current. This finding clearly indicates that the predominant potential-limiting factor arises from the linear carbonate rather than the cyclic carbonate. However, this does not mean that the cyclic carbonate plays a trivial role in the electrolyte oxidative decomposition under high potentials.

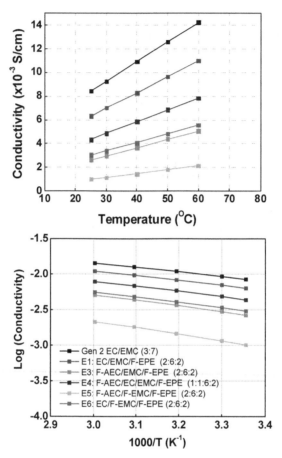

FIGURE 1.21 Conductivity measurement of Gen 2 and E1 to E6 electrolytes.

A close comparison between E5 and E6 suggests that substituting the EC component with F-AEC further increases the potential limit of the electrolyte in the high potential range (Figure 1.23). Given that E5 can tolerate the potential up to at least 6.8 V, it should be a good candidate for high voltage cells if SEI formation is not considered.

To validate the DFT calculations and electrochemical floating test results, we conducted additional electrolyte tests using the $LiNi_{0.5}Mn_{1.5}O_4$ (LNMO)/ $Li_4Ti_5O_{12}$ (LTO) couple cycled galvanostatically between 2.0 and 3.45 V at a C/2 rate and 55 °C. In terms of capacity retention, the E5 and E6 electrolyte cells perform much better than Gen 2, with negligible capacity loss after 80 cycles (Figure 1.24). This experimental result agrees with the oxidation potentials and HOMO energies calculated by the DFT method. By comparison, the LNMO/ LTO cell with Gen 2 electrolyte shows more than 70% capacity loss in the same cycling span.

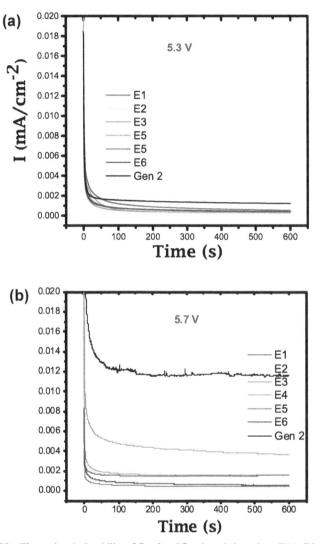

FIGURE 1.22 Electrochemical stability of Gen 2 and fluorinated electrolytes E1 to E6 at (a) 5.3 V and (b) 5.7 V using a 3-electrode electrochemical cell. *Reprinted with permission from Ref. [11]. © 2013 Royal Society of Chemistry.*

In addition, the new fluorinated electrolytes E3, E5, and E6 are able to sustain the high rate (C/2) under this testing condition, although their conductivities are lower than that of the Gen 2 electrolyte (Figure 1.21). The rate capability of another high voltage system was reported in the literature for a sulfone-based electrolyte [55,60]; however, this electrolyte can only stand up to a C/5 rate due to its high viscosity and poor compatibility with both the electrode and the

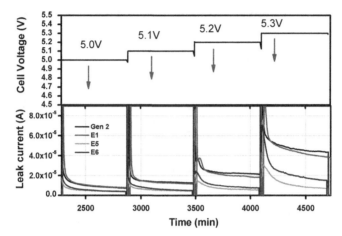

FIGURE 1.23 Potentiostatic profiles (constant voltage charge curve) of $LiNi_{0.5}Mn_{1.5}O_4$/Li half cells maintained at 5.0, 5.1, 5.2, and 5.3 V for 10 h at room temperature. *Reprinted with permission from Ref. [11]. © 2013 Royal Society of Chemistry.*

separator. In contrast, the fluorinated carbonate-based electrolytes can tolerate up to the 2C rate due to their low viscosity and high lithium-ion conduction.

Since LUMO energy levels are also lowered by fluorine substitution, the fluorinated electrolytes have higher reduction potential, resulting in instability on the anode side of the high-voltage graphite/LNMO cells [55]. However, whether the predicted reductive decomposition can be prevented or not depends on whether a protective yet conductive SEI can form on the anode surface. For the EC-based electrolyte, the SEI formation starts at 0.6 V versus Li/Li^+ [61]. If the fluorinated molecules in the electrolyte formulation decompose prior to this potential and are incapable of forming an effective SEI, this effect will cause continuous decomposition, leading to the failure of the battery. As shown in Figure 1.25, the addition of the fluorinated ether F-EPE in E1 and E2 does not pose such a threat. However, when EC is partially (E4) or fully replaced (E3 and E5) by the fluorinated cyclic carbonate F-AEC, a low initial discharge capacity (E5) or low capacity retention (E3 and E4) was observed due to the extensive decomposition of F-AEC with tremendous lithium loss before the SEI formation that prevents or slows down the decomposition. An efficient SEI additive or another cyclic carbonate that works like ethylene carbonate is necessary for the new electrolytes to stabilize the graphite anode and electrolyte interphase.

This study demonstrated that the fluorinated cyclic carbonate (F-AEC), fluorinated linear carbonate (F-EMC), and fluorinated ether (F-EPE) have superior anodic stability compared with the EC/EMC-based Gen 2 electrolyte. The substitution of EMC with F-EMC and EC with F-AEC greatly improves the voltage limits of the electrolyte, and this result is supported by the enhanced cycling performance of the all-fluorinated electrolyte in LNMO/LTO cells at elevated temperature. However, to realize the real 5 V energy output with the

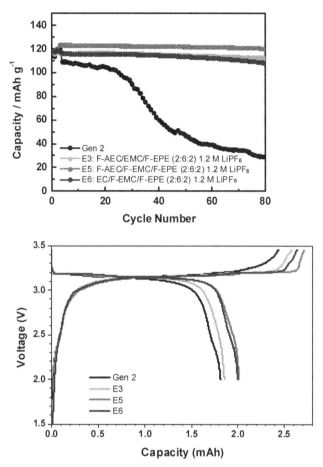

FIGURE 1.24 Cycling capacity retention of $LiNi_{0.5}Mn_{1.5}O_4/Li_4Ti_5O_{12}$ cells operated at 55 °C with baseline electrolyte Gen 2 and fluorinated electrolyte E3, E5, and E6. *Reprinted with permission from Ref. [11]. © 2013 Royal Society of Chemistry.*

LNMO/graphite system, further development is needed. We have developed a fluorinated high voltage electrolyte (HVE), one component of which is fluoroethylene carbonate (FEC) due to its capability of SEI formation on graphitic anodes by reductive decomposition at a potential higher than that of EC. The HVE formulation consists of 1.0 M $LiPF_6$ in FEC/F-EMC/F-EPE at 3/5/2 ratio by volume.

The cycling performance of the HVE cell compared to the Gen 2 cell at room temperature (RT) and 55 °C is shown in Figure 1.26. At 55 °C, the HVE cell exhibits considerable improvement in capacity retention and coulombic efficiency over the Gen 2 cell, indicating that electrolyte decomposition is mitigated with HVE. At RT the improvement is insignificant because the conventional electrolyte did not decompose severely at the tested condition, evidenced by the

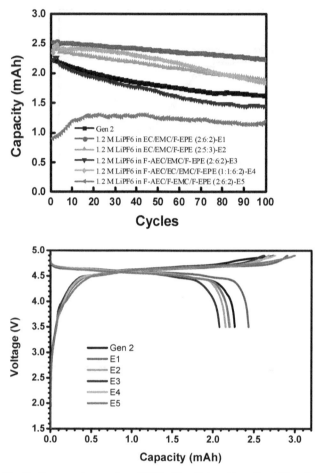

FIGURE 1.25 Cycling capacity retention of $LiNi_{0.5}Mn_{1.5}O_4$/graphite cells with baseline electrolyte Gen 2 and fluorinated electrolyte E1, E2, E3, E4, and E5. *Reprinted with permission from Ref. [11]. © 2013 Royal Society of Chemistry.*

coulombic efficiency of Gen 2 reaching above 99% after around 30 cycles and remaining there for the test duration (100 cycles). At 55 °C, however, the decomposition of Gen 2 electrolyte remains severe as the cycling advances, and the coulombic efficiency of the Gen 2 cell remains below 98%. For the HVE cell, at both RT and 55 °C, the coulombic efficiency remains above 99.5% for the majority of cycles, signaling much less decomposition even at the elevated temperature. Longer cycling data show that the HVE cell is able to retain 50% of its initial capacity over 600 cycles at RT and 250 cycles at 55 °C and a rate of C/3. These results indicate that although other cell components affect the high-voltage cell performance to a certain degree, the electrolyte plays the most significant role.

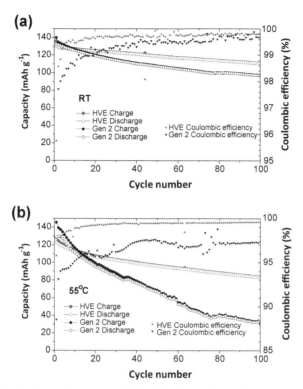

FIGURE 1.26 Cycling performance of $LiNi_{0.5}Mn_{1.5}O_4$/graphite cells at (a) room temperature (RT) and (b) 55 °C with high voltage electrolyte (HVE) electrolyte and Gen 2 electrolyte. (Cutoff voltages of 3.5–4.9 V, rate of C/3.) *Reprinted with permission from Ref. [12]. © 2013 Elsevier.*

To understand the mechanism behind this improvement, we undertook electrochemical diagnosis and characterizations of the harvested electrodes and electrolytes. Differential capacity analysis (dQ/dV) of graphite/Li cells with the HVE and Gen 2 electrolyte indicated that formation of an SEI layer occurs at a voltage of about 0.6 V versus Li^+/Li with the Gen 2 cells and at the much higher voltage of 1.5 V versus Li^+/Li with the HVE cell. This difference is due to the higher reduction potential of the FEC in the electrolyte, which may contribute to a more stable SEI on the graphite anode at elevated temperature and at high charging voltage.

The SEM images of the harvested LNMO cathodes from the LNMO/graphite cells cycled at 55 °C are shown in Figure 1.27. While the pristine cathode shows mostly clean LNMO crystals and carbon black networks in between, the cycled cathode using Gen 2 electrolyte shows a heavy coating on the LNMO particles as well as carbon black. In contrast, the HVE cathode looks almost as clean as the pristine cathode. This is a clear demonstration of the stability of the HVE against oxidation on the cathode surface.

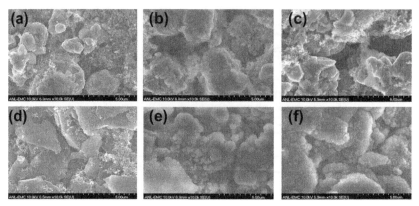

FIGURE 1.27 SEM images of (a) pristine $LiNi_{0.5}Mn_{1.5}O_4$ (LNMO) cathode, (b) harvested LNMO cathode from Gen 2 cell, (c) harvested LNMO cathode from high voltage electrolyte (HVE) cell, (d) pristine graphite anode, (e) harvested graphite anode from Gen 2 cell, and (f) harvested graphite anode from HVE cell. (Cutoff voltages of 3.5–4.9 V, rate of C/3, temperature of 55 °C.) *Reprinted with permission from Ref. [12]. © 2013 Elsevier.*

Figure 1.28 shows the Fourier transform infrared (FT-IR) spectra of the harvested LNMO cathodes and graphite anodes. The FT-IR characterization agrees with the SEM characterization. On the harvested LNMO cathode from Gen 2 electrolyte, while no obvious organic buildup can be identified at RT, at 55 °C a prominent peak appears at around the 1720 cm^{-1} region (Figure 1.28(a)), which corresponds to the carbonyl function group. This peak signifies considerably more decomposition compared with room temperature cycling, and the decomposition products may be carbonates or carboxylate derivatives. In the case of HVE, although the trend is similar to that with Gen 2 electrolyte, the change is much less dramatic, and the carbonyl peak is trivial as shown in Figure 1.28(b). The more interesting case actually lies in the anode. From RT to 55 °C, the graphite anode with Gen 2 electrolyte undergoes a significant increase in decomposition (Figure 1.28(c)). This finding indicates a significant morphology change of the SEI layer under high temperature operation, which agrees with the SEM data shown in Figure 1.27(e). In comparison, the HVE anode shows less organic residues in the FT-IR spectrum (Figure 1.28(d)), indicating the presence of an SEI comprising more inorganic species on the graphite surface, as observed in the SEM (Figure 1.27(f)). No visual difference was observed between the RT and 55 °C data. These results demonstrate the superior high temperature stability of the SEI layer formed by the HVE on the graphite anode during the high-voltage charging and discharging.

In summary, a new electrolyte based on fluorinated solvents was studied in a high voltage Li-ion cell using graphite as the anode and 5 V spinel $LiNi_{0.5}Mn_{1.5}O_4$ as the cathode. The electrolyte shows significantly enhanced voltage stability compared with the conventional electrolytes at elevated temperature (55 °C). Post-test study of the harvested cathode using FT-IR and SEM indicated that the buildup of organic decomposition product on the cathode surface is negligible for the fluorinated electrolyte.

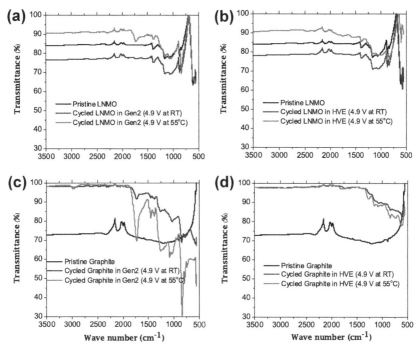

FIGURE 1.28 Fourier transform infrared spectra of pristine and harvested cathodes from cycled LiNi$_{0.5}$Mn$_{1.5}$O$_4$ (LNMO)/graphite cells with (a) Gen 2 and (b) high voltage electrolyte (HVE) electrolytes and harvested anodes from cycled LNMO/graphite cells with (c) Gen 2 and (d) HVE electrolytes. All cells were cycled at room temperature (RT) and 55 °C. *Reprinted with permission from Ref. [12].* © 2013 Elsevier.

1.5 CLOSING REMARKS

Fluorinated compounds have been widely considered as promising components for advanced functionalized electrolytes for next-generation lithium-ion batteries due to the unique chemical stability of X-F (X is nonmetal element) bonds and strong electronegativity of F. As demonstrated in this chapter, the fluorinated electrolyte has a wide spectrum of applications, including (1) tuning the chemical composition of the SEI layer, (2) forming a robust artificial SEI layer to maximize the protection of graphitic anodes, (3) providing overcharge protection and automatic capacity balancing of the battery pack, and (4) acting as stable solvents for high voltage electrolytes. It is anticipated that more applications of fluorinated compounds will emerge with the advance of battery research.

ACKNOWLEDGMENTS

Research was funded by US Department of Energy, Vehicle Technologies Office. Argonne National Laboratory is operated for the US Department of Energy by UChicago Argonne, LLC, under contract DE-AC02-06CH11357.

REFERENCES

[1] Z.H. Chen, Y. Qin, Y. Ren, W.Q. Lu, C. Orendorff, E.P. Roth, K. Amine Wner, Environ. Sci. 4 (2011) 4023–4030.
[2] Z.H. Chen, Y. Ren, E. Lee, C. Johnson, Y. Qin, K. Amine, Adv. Ener. Mater. 3 (2013) 729–736.
[3] S. Yamaguchi, H. Asahina, K.A. Hirasawa, T. Sato, S. Mori, Mol. Cryst. Liquid Cryst Sci. Tech. 322 (1998) 239–244.
[4] M.N. Richard, J.R. Dahn, J. Electrochem. Soc. 146 (1999) 2068–2077.
[5] T. Zheng, A.S. Gozdz, G.G. Amatucci, J. Electrochem. Soc. 146 (1999) 4014–4018.
[6] J. Cabana, S.H. Kang, C.S. Johnson, M.M. Thackeray, C.P. Grey, J. Electrochem. Soc. 156 (2009) A730–A736.
[7] S.H. Kang, M.M. Thackeray, Electrochem. Commun. 11 (2009) 748–751.
[8] H. Kawaura, D. Takamatsu, S. Mori, Y. Orikasa, H. Sugaya, H. Murayama, K. Nakanishi, H. Tanida, Y. Koyama, H. Arai, Y. Uchimoto, Z. Ogumi, J. Power Sources 245 (2014) 816–821.
[9] D. Liu, W. Zhu, J. Trottier, C. Gagnon, F. Barray, A. Guerfi, A. Mauger, H. Groult, C.M. Julien, J.B. Goodenough, K. Zaghi, RSC Adv. 4 (2014) 154–167.
[10] X.B. Wu, S.H. Wang, X.C. Lin, G.M. Zhong, Z.L. Gong, Y. Yang, J. Mater. Chem. A 2 (2014) 1006–1013.
[11] Z.C. Zhang, L.B. Hu, H.M. Wu, W. Weng, M. Koh, P.C. Redfern, L.A. Curtiss, K. Amine, Energy Environ. Sci. 6 (2013) 1806–1810.
[12] L.B. Hu, Z.C. Zhang, K. Amine, Electrochem. Commun. 35 (2013) 76–79.
[13] Z.H. Chen, Y. Qin, K. Amine, Electrochim. Acta 54 (2009) 5605–5613.
[14] Z.H. Chen, Q.Z. Wang, K. Amine, J. Electrochem. Soc. 153 (2006) A2215–A2219.
[15] C. Buhrmester, J. Chen, L. Moshurchak, J.W. Jiang, R.L. Wang, J.R. Dahn, J. Electrochem. Soc. 152 (2005) A2390–A2399.
[16] J. Chen, C. Buhrmester, J.R. Dahn, Electrochem. Solid State Lett. 8 (2005) A59–A62.
[17] M. Holzapfel, F. Alloin, R. Yazami, Electrochim. Acta 49 (2004) 581–589.
[18] V. Eshkenazi, E. Peled, L. Burstein, D. Golodnitsky, Solid State Ionics 170 (2004) 83–91.
[19] M.D. Levi, C. Wang, D. Aurbach, J. Electrochem. Soc. 151 (2004) A781–A790.
[20] Z.H. Chen, Y. Qin, J. Liu, K. Amine, Electrochem. Solid State Lett. 12 (2009) A69–A72.
[21] D. Bar-Tow, E. Peled, L. Burstein, J. Electrochem. Soc. 146 (1999) 824–832.
[22] D. Aurbach, J.S. Gnanaraj, M.D. Levi, E.A. Levi, J.E. Fischer, A. Claye, J. Power Sources 97–98 (2001) 92–96.
[23] D. Aurbach, B. Markovsky, I. Weissman, E. Levi, Y. Ein-Eli, Electrochim. Acta 45 (1999) 67–86.
[24] Z.H. Chen, Q.Z. Wang, K. Amine, Electrochim. Acta 51 (2006) 3890–3894.
[25] L.F. Li, H.S. Lee, H. Li, X.Q. Yang, K.W. Yang, W.S. Yoon, J. McBreen, X.J. Huang, J. Power Sources 184 (2008) 517–521.
[26] L.F. Li, B. Xie, H.S. Lee, H. Li, X.Q. Yang, J. McBreen, X.J. Huang, J. Power Sources 189 (2009) 539–542.
[27] B. Xie, H.S. Lee, H. Li, X.Q. Yang, J. McBreen, L.Q. Chen, Electrochem. Commun. 10 (2008) 1195–1197.
[28] Z.H. Chen, K. Amine, J. Electrochem. Soc. 153 (2006) A1221–A1225.
[29] Z.H. Chen, K. Amine, Electrochem. Commun. 9 (2007) 703–707.
[30] Y. Qin, Z.H. Chen, H.S. Lee, X.Q. Yang, K. Amine, J. Phys. Chem. C 114 (2010) 15202–15206.
[31] Z. Chen, J. Liu, K. Amine, Electrochim. Acta 53 (2008) 3267–3270.
[32] Z.H. Chen, K. Amine, J. Electrochem. Soc. 156 (2009) A672–A676.

[33] R. Petibon, E.C. Henry, J.C. Burns, N.N. Sinha, J.R. Dahn, J. Electrochem. Soc. 161 (2014) A66–A74.
[34] I.A. Profatilova, C. Stock, A. Schmitz, S. Passerini, M. Winter, J. Power Sources 222 (2013) 140–149.
[35] J.C. Burns, R. Petibon, K.J. Nelson, N.N. Sinha, A. Kassam, B.M. Way, J.R. Dahn, J. Electrochem. Soc. 160 (2013) A1668–A1674.
[36] Y. Qin, Z.H. Chen, W.Q. Lu, K. Amine, J. Power Sources 195 (2010) 6888–6892.
[37] K. Amine, Z.H. Chen, Z. Zhang, J. Liu, W.Q. Lu, Y. Qin, J. Lu, L. Curtis, Y.K. Sun, J. Mater. Chem. 21 (2011) 17754–17759.
[38] Z.H. Chen, Y. Ren, A.N. Jansen, C.K. Lin, W. Weng, K. Amine, Nat. Commun. (2013) 4.
[39] J. Liu, Z.H. Chen, S. Busking, K. Amine, Electrochem. Commun. 9 (2007) 475–479.
[40] Y. Qin, Z.H. Chen, J. Liu, K. Amine, Electrochem. Solid State Lett. 13 (2010) A11–A14.
[41] K. Xu, U. Lee, S.S. Zhang, J.L. Allen, T.R. Jow, Electrochem. Solid State Lett. 7 (2004) A273–A277.
[42] T. Ohsaki, T. Kishi, T. Kuboki, N. Takami, N. Shimura, Y. Sato, M. Sekino, A. Satoh, J. Power Sources 146 (2005) 97–100.
[43] R.A. Leising, M.J. Palazzo, E.S. Takeuchi, K.J. Takeuchi, J. Electrochem. Soc. 148 (2001) A838–A844.
[44] R.A. Leising, M.J. Palazzo, E.S. Takeuchi, K.J. Takeuchi, J. Power Sources 97–98 (2001) 681–683.
[45] M. Adachi, K. Tanaka, K. Sekai, J. Electrochem. Soc. 146 (1999) 1256–1261.
[46] Z. Chen, J. Liu, A.N. Jansen, G. GirishKumar, B. Casteel, K. Amine, Electrochem. Solid State Lett. 13 (2010) A39–A42.
[47] Z.H. Chen, A.N. Jansen, K. Amine, Energy Environ. Sci. 4 (2011) 4567–4571.
[48] N. Nanbu, M. Takehara, S. Watanabe, M. Ue, Y. Sasaki, Bull. Chem. Soc. Jap. 80 (2007) 1302–1306.
[49] T. Achiha, T. Nakajima, Y. Ohzawa, M. Koh, A. Yamauchi, M. Kagawa, H. Aoyama, J. Electrochem. Soc. 156 (2009) A483–A488.
[50] R. McMillan, H. Slegr, Z.X. Shu, W.D. Wang, J. Power Sources 81 (1999) 20–26.
[51] N. Nanbu, K. Takimoto, M. Takehara, M. Ue, Y. Sasaki, Electrochem. Commun. 10 (2008) 783–786.
[52] Z.C. Wang, J. Xu, W.H. Yao, Y.W. Yao, Y. Yang, Fluoroethylene carbonate as an electrolyte additive for improving the performance of mesocarbon microbead electrode, in: M.K. Sunkara et al. (Ed.), Rechargeable Lithium and Lithium Ion Batteries, , 2012, pp. 29–40.
[53] N.S. Choi, K.H. Yew, K.Y. Lee, M. Sung, H. Kim, S.S. Kim, J. Power Sources 161 (2006) 1254–1259.
[54] M.C. Smart, B.V. Ratnakumar, V.S. Ryan-Mowrey, S. Surampudi, G.K.S. Prakash, J. Hub, I. Cheung, J. Power Sources 119 (2003) 359–367.
[55] T. Achiha, T. Nakajima, Y. Ohzawa, M. Koh, A. Yamauchi, M. Kagawa, H. Aoyama, J. Electrochem. Soc. 157 (2010) A707–A712.
[56] J. Arai, J. Electrochem. Soc. 150 (2003) A219–A228.
[57] J.M. Vollmer, L.A. Curtiss, D.R. Vissers, K. Amine, J. Electrochem. Soc. 151 (2004) A178–A183.
[58] C.G. Zhan, J.A. Nichols, D.A. Dixon, J. Phys. Chem. A 107 (2003) 4184–4195.
[59] N. Shao, X.G. Sun, S. Dai, D.E. Jiang, J. Phys. Chem. B 116 (2012) 3235–3238.
[60] A. Abouimrane, I. Belharouak, K. Amine, Electrochem. Commun. 11 (2009) 1073–1076.
[61] K. Xu, Chem. Rev. 104 (2004) 4303–4418.

Chapter 2

Electrochemical Behavior of Surface-Fluorinated Cathode Materials for Lithium Ion Battery

Susumu Yonezawa,[1,2] Jae-Ho Kim[1] and Masayuki Takashima[2]
[1]*Headquarters for Innovative Society-Academia Cooperation, Fukui University, Fukui, Japan;*
[2]*Department of Materials Science & Engineering, Faculty of Engineering, University of Fukui, Fukui, Japan*

Chapter Outline

- 2.1 Surface Fluorination of LiFePO$_4$ — 33
  - 2.1.1 Introduction — 33
  - 2.1.2 Experimental Details — 34
  - 2.1.3 Characterization of Fluorinated LiFePO$_4$ — 35
  - 2.1.4 Electrochemical Properties of Fluorinated LiFePO$_4$ — 40
- 2.2 Surface Fluorination of LiNi$_{0.5}$Mn$_{1.5}$O$_4$ — 43
  - 2.2.1 Introduction — 43
  - 2.2.2 Experimental Details — 44
  - 2.2.3 Characterization of Surface-Fluorinated LiNi$_{0.5}$Mn$_{1.5}$O$_4$ — 44
  - 2.2.4 Electrochemical Properties of Surface-Fluorinated LiNi$_{0.5}$Mn$_{1.5}$O$_4$ — 46
- 2.3 Summary — 49
- References — 49

2.1 SURFACE FLUORINATION OF LiFePO$_4$

2.1.1 Introduction

Lithium secondary battery has been widely studied because of a large terminal voltage and a large energy density. Lithium-containing transition metal oxides, LiCoO$_2$, LiNiO$_2$, LiMn$_2$O$_4$, LiFePO$_4$, and their derivatives have been investigated to obtain the high performance cathode active materials of lithium secondary battery. Among these cathode materials, LiFePO$_4$ has flat voltage profile, good electrochemical and thermal stability, relatively low cost for synthesis, and environmental compatibility with less toxicity than other cathode materials. Due to these advantages, LiFePO$_4$ has been attracting much attention as a promising new cathode electrode material for lithium-ion batteries. However, the low electric conductivity and low diffusion coefficient of Li$^+$ are the main shortcomings that limit its application in industry. Extensive efforts have been performed to improve its electrochemical performance over decades, which can be classified into the following categories: particle size/shape optimizing [1–6], metal doping [7], and mixing with the electronically conductive materials like carbon, metal, and metal oxide [8–10]. Among these methods used for improving electrochemical properties of LiFePO$_4$, carbon coating is one of the most frequently used techniques to improve the specific capacity, rate performance, and cycling life [11,12]. However, it should be further discussed on the optimum conditions with the thickness and formation of carbon layer.

The surface modification of a cathode active material gives strong effects to the battery performance because electrochemical reaction takes place at the interface among active material, carbon as an electro-conductive material and electrolyte. The surface modification of LiMn$_2$O$_4$ with fluorine/carbon nanocomposite was already reported [13,14]. Charge/discharge capacity and cycle ability of LiMn$_2$O$_4$ as a cathode material are enhanced by optimizing the arrangement of nanothickness carbon film and surface fluorination, which is superior to the carbon-coated LiMn$_2$O$_4$. In the present section, the effects of surface fluorination on the electrochemical properties and thermal stability of carbon-coated LiFePO$_4$ are summarized.

2.1.2 Experimental Details

Carbon-coated LiFePO$_4$ particles (SLFP-PD60, anatase; 98% purity) were obtained from Hohsen Corp. Nitrogen trifluoride gas (NF$_3$, 99.5% purity) was supplied by Central glass Co., Ltd. Details of the fluorination apparatus are given in our previous papers [15,16]. Fluorinated LiFePO$_4$ (F-LiFePO$_4$) particles were prepared by direct fluorination using NF$_3$ gas under the following conditions. Reaction temperature, NF$_3$ pressures, and reaction time were set at 298 K, 6.67, and 5066 kPa, and 1 h, respectively.

The structure and chemical bonds of the samples were investigated using powder X-ray diffraction (XRD, XD-6100) and X-ray photoelectron spectroscopy (XPS, XPS-9010). The surface morphology of the samples was observed using a scanning electron microscope (SEM, s-2400; Hitachi Ltd.). Fluorine contents in F-LiFePO$_4$ were determined using an ion chromatography (IC; SD-8022, Tosoh Co.) after dissolving in distilled water.

As shown in Figure 2.1, two-electrode test cell (TOM Cell) was used for the electrochemical measurements. The cathode mixture consists of LiFePO$_4$ sample, acetylene black (AB), and polyvinylidene difluoride (PVDF) in weight ratios of 8:1:1. The mixture was rolled spread to a film with 0.1 mm thickness and the film was cut into a disk with 13 mm$^\phi$. It was then pressed onto a titanium mesh welded on the bottom of SUS304 container (20 mm$^\phi$ × 3 mm^t). The cathode was fully dried under vacuum (~10^{-1} Pa) for 12 h at room temperature prior to use. The mixture of propylene carbonate (PC) and dimethoxyethane (DME) (1:1 vol.) containing 1.0 mol dm^{-3} LiPF$_6$ was used as an electrolyte solution. Li metal foil (0.2 mm^t Kyokuto Kinzouku Co. Ltd.) was used as the reference and counter electrodes. After assembling the test cell, AC impedance measurements of test cell

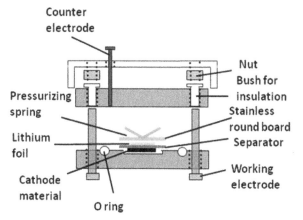

FIGURE 2.1 Schematic illustration of two-electrode test cell.

containing each sample were carried out using a potentio/galvanostat (1287; Solartron Analytical) and frequency response analyzer (1287; Solartron Analytical). The frequency range was 100 mHz to 10 kHz and the current amplitude was 1 mA cm^{-2}.

Charge–discharge test was carried out at the currents of 0.2 °C (discharge rate) and 0.2 °C (charge rate) (Hokuto Denko Co., HJ101SM6). The temperature of cathode was controlled at 25 °C. Cut-off potentials were 3.0 V (discharge) and 4.0 V (charge).

The thermal behavior of the delithiated samples prepared after charging at 4.0 V was investigated using differential scanning calorimetry (DSC, DSC6300; Seiko Instruments Inc.) at a heating rate of 4 K min^{-1} up to 723 K. The electrode mixture was taken from the cell, washed with tetrahydrofuran, and dried under vacuum, and part of it (approximately 5 mg) was used for the DSC measurement.

2.1.3 Characterization of Fluorinated LiFePO$_4$

Sample names, reaction conditions, total fluorine contents (wt%), and surface fluorine contents are summarized in Table 2.1, in which fluorinated LiFePO$_4$ (F-LiFePO$_4$) samples prepared under different reaction conditions with NF$_3$ gas are named as F1-LiFePO$_4$ and F2-LiFePO$_4$. The total fluorine contents in F-LiFePO$_4$ were evaluated from the IC measurements (SD-8022, Tosoh Corporation). The surface fluorine contents (x) were calculated from XPS spectra as shown in Figure 2.2.

Figure 2.3 shows XRD profiles of untreated (a), fluorinated at 6.67 kPa (b) and 50.66 kPa (c). All XRD profiles agree with that of phosphor-olivine LiFePO$_4$ [17]. There is no extra peak in the profiles even after LiFePO$_4$ was allowed to react with 50.66 kPa of NF$_3$ gas. There is also no change in the intensity and full width at half maximum (FWHM) of the peak for untreated and fluorinated LiFePO$_4$.

SEM images of untreated and fluorinated LiFePO$_4$ samples are presented in Figure 2.4. No change is observed in the shape and morphology of the surface of

TABLE 2.1 Reaction Conditions of LiFePO$_4$ Particles Treated with NF$_3$ Gas and the Fluorine Contents

Sample Name	Temperature (K)	Time (h)	NF$_3$ Pressure (kPa)	Total Fluorine Contents (wt%)	Surface Fluorine Contents (Li:F = 1:x)[a]
Untreated LiFePO$_4$	—	—	—	0.00	0.00
F1-LiFePO$_4$	298	1	6.67	0.02	0.16
F2-LiFePO$_4$	298	1	50.66	0.05	0.22

[a] Surface fluorine contents (Li:F = 1:x) were evaluated from XPS results shown in Figure 2.2.

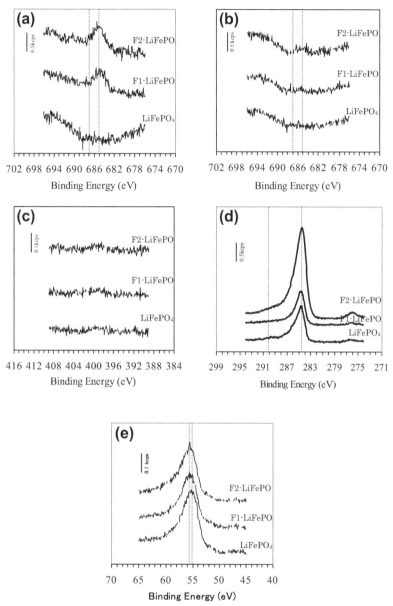

FIGURE 2.2 XPS spectra of (a) F 1s, (b) F 1s after Ar^+ ion etching, (c) N 1s, (d) C 1s, and (e) Li 1s for untreated $LiFePO_4$ and fluorinated $LiFePO_4$ (F1–$LiFePO_4$ and F2–$LiFePO_4$) samples.

$LiFePO_4$ particles. The powders have an average size between 0.1 and 0.5 μm. The fluorination effect on the surface morphology of $LiFePO_4$ particles is not detected in the SEM images. However, the existence of fluorinated surface layer of $LiFePO_4$ particles was confirmed by XPS spectra.

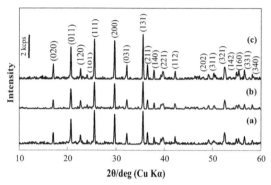

FIGURE 2.3 XRD profiles of (a) untreated LiFePO$_4$, (b) F1–LiFePO$_4$, and (c) F2–LiFePO$_4$.

FIGURE 2.4 SEM images of (a) untreated LiFePO$_4$, (b) F1–LiFePO$_4$, and (c) F2–LiFePO$_4$.

Figure 2.2 indicates XPS spectra of F 1s electron for untreated and fluorinated LiFePO$_4$ samples. All binding energies are calibrated to the C 1s peak at 284.8 eV of carbon. An F 1s peak located at the binding energy (BE) of 685 eV was detected in the fluorinated samples as shown in Figure 2.2(a). Even at room temperature (298 K), the surface of LiFePO$_4$ reacts with NF$_3$ gas because fluorine is introduced at its surface. NF$_3$ is reactive at around 503 K due to the dissociation into ·NF$_2$ radical and F atom, but fluorination reaction is likely to take place at LiFePO$_4$ surface even at room temperature (298 K). The BE of 685 eV indicates the existence of ionic fluorine, F$^-$ such as alkali metal fluorides in the solid samples [18]. Therefore the peak may correspond to F$^-$ having interaction with Li$^+$ in the sample. However, any peak corresponding to FeF$_2$ was not confirmed at 687 eV in Figure 2.2(a). The LiFePO$_4$ belongs to the olivine family of lithium *ortho*-phosphates with an orthorhombic structure in the space group *Pnma* [19]. The structure consists of corner-shared FeO$_6$ octahedral

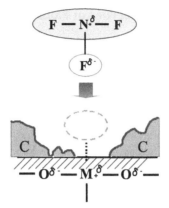

FIGURE 2.5 Schematic layout of the surface fluorination between C/LiFePO$_4$ and NF$_3$.

and edge-shared LiO$_6$ octahedral running parallel to the b-axis, which is linked together by PO$_4$ tetrahedral. Especially oxygen atoms are strongly bonded to both Fe and P atoms, which leads to low ionic diffusivity and poor electronic conductivity of LiFePO$_4$ [20]. Thus fluorine preferentially attacks LiO$_6$ edge sites rather than FeO$_6$ or PO$_4$ corner sites. However, the fluorine peak at 685 eV almost disappeared after Ar$^+$ ion etching as shown in Figure 2.2(b). Because the Li–F bond at 685 eV is not a strong one, they can be easily eliminated by low-energy argon ion etching (300 V, 5 s) [21]. In addition, due to the weak bond between Li$^+$ and F$^-$, the fluorinated parts existing at the sample surface can be dissolved in distilled water. As a result, the total fluorine contents of F-LiFePO$_4$ were analyzed as shown in Table 2.1. In the case of N 1s spectra, as same as the untreated sample, any nitrogen peak was not found in the fluorinated samples as indicated in Figure 2.2(c). Regarding the reaction between fluorine and carbon coated on LiFePO$_4$ (C-LiFePO$_4$), any C–F bond was not found at around 290 eV in Figure 2.2(d) (F1-LiFePO$_4$) except C–C peak at 284.5 eV. However, a small C–F peak at 290 eV is found in the case of F2-LiFePO$_4$. In the Li 1s spectra, the BE is located at 55.1 eV in the original sample. Furthermore the Li–F bond seems to appear at 55.7 eV in the fluorinated samples (F-LiFePO4) as shown in Figure 2.2(e). It suggests that fluorine makes a chemical bond not only with Li$^+$, but also with carbon on LiFePO$_4$ with increasing NF$_3$ contents.

Figure 2.5 shows the schematic layout of the surface fluorination between C/LiFePO$_4$ and NF$_3$. It is known that fluorination of organic compounds such as hydrocarbons by elemental fluorine proceeds via electrophilic reaction, that is F$^{\delta+}$ attacks C$^{\delta-}$ with higher electron density [22]. In the case of NF$_3$, however, F$^{\delta-}$ released from F$_2$N$^{\delta+}$–F$^{\delta-}$ may prefer to attack M$^{\delta+}$(M=Li, Fe) sites in LiFePO$_4$ rather than the C$^{\delta-}$ sites with higher electron density, as shown in Figure 2.5 because an exchange reaction between fluorine and oxygen may be slightly easier than the formation of covalent C–F bond. When NF$_3$ content is increased, fluorine seems to react with both LiFePO$_4$ and carbon as shown in the XPS data.

2.1.4 Electrochemical Properties of Fluorinated LiFePO$_4$

Figure 2.6 shows the discharge curves of (a) untreated and (b and c) fluorinated LiFePO$_4$ samples with cycle number. The shape of discharge curves is similar to each other. However, the discharge capacity was changed by the surface fluorination. The discharge capacity of F1-LiFePO$_4$ increased by about 10% during five cycles, compared with untreated sample. However, the discharge capacity of F2-LiFePO$_4$ was similar to that of untreated one. This seems that an excess fluorination causes the formation of a resistive film which consists of some fluorides at the surface [19].

To better characterize the capacity fading mechanism of LiFePO$_4$ cathodes, electrochemical impedance spectra (EIS) of fluorinated samples were measured, which helps to clarify the ohmic and charge transfer resistances coincident with capacity loss. According to EIS studies of lithium-ion cells by Chen et al. [23], the cell impedance is primarily attributed to cathode impedance and charge transfer resistance. The influence of the surface fluorination on the conductivity of LiFePO$_4$ was investigated. Figure 2.7 shows the Nyquist plots of LiFePO$_4$/Li cells containing fluorinated cathode samples. All the EIS measurements were

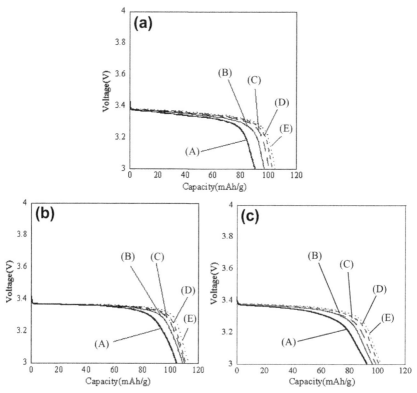

FIGURE 2.6 Discharge curves of (a) untreated LiFePO$_4$, (b) F1–LiFePO$_4$, and (c) F2–LiFePO$_4$ with cycle number. ((A) 1st cycle, (B) 2nd cycle, (C) 3rd cycle, (D) 4th cycle, (E) 5th cycle).

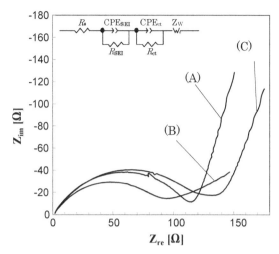

FIGURE 2.7 Nyquist plots for LiFePO$_4$/Li half cells containing (a) untreated LiFePO$_4$, (b) F1–LiFePO$_4$, and (c) F2–LiFePO$_4$.

carried out at the lower terminal voltage, 2.5V, i.e., at fully discharged state at room temperature. The impedance spectra of all samples consist of depressed semicircles in high frequency region and straight lines in low frequency region as shown in Figure 2.7. Impedance on the Z_{re} at high-frequency region corresponds to ohmic resistance (R_s), which represents the resistance of the electrolyte and other physical resistances. The depressed semicircle in high frequency relates to the Li-ion migration resistance (R_{SEI}) through SEI film formed on cathode surface. The semicircle in the middle frequency range indicates the charge transfer resistance (R_{ct}). The inclined line in the low frequency represents the Warburg impedance (Z_w), which is associated with lithium-ion diffusion in the LiFePO$_4$ particles. A simplified equivalent circuit model (inset in Figures 2.2–2.7) was constructed to analyze the impedance spectra. A constant phase element CPE was placed to represent the double layer capacitance.

Table 2.2 shows the parameters of the equivalent circuit for untreated LiFePO$_4$, F1-LiFePO$_4$, and F2-LiFePO$_4$ obtained from computer simulations using the Zview 2.0 software. The resistances of electrolyte/electrode (R_s) and SEI film (R_{SEI}) are similar for all samples. However, the charge transfer resistance (R_{ct}) is much lower for F1-LiFePO$_4$ sample. Exchange current density of the electrode (i_0) was also calculated according to the following equation [24] as shown in Table 2.2.

$$i_0 = \frac{RT}{nFR_{ct}} \quad (2.1)$$

It can be seen that the exchange current density (i_0) of the F1-LiFePO$_4$ sample is higher than those of other samples. These i_0 values are in good

TABLE 2.2 Electrode Kinetic Parameters of Samples Obtained Form Equivalent Circuit in Figure 2.7

Samples	R_s (Ω)	R_{SEI} (Ω)	R_{ct} (Ω)	i_0 (mA)
Untreated LiFePO$_4$	1.21	11.77	51.54	0.49
F1-LiFePO$_4$	1.47	12.47	35.55	0.72
F2-LiFePO$_4$	2.30	12.91	59.11	0.43

agreement with the results of discharge capacity as indicated in Figure 2.6. Both electron conduction and lithium diffusion are significant factors in the polarization of the electrode. Thus the electrode polarization is strongly related with the ohmic loss, charge transfer and mass transfer [25]. Among these samples, F1-LiFePO$_4$ showed the lowest resistance and largest exchange current density. The reason may be the improvement of wettability of electrolyte with cathode material by introduction of fluorine to the surface of oxides because fluorine has the larger electronegativity than oxygen. Since the metal–fluorine (M–F) bond is more polarized than the metal–oxygen (M–O) bond, the affinity for polar electrolyte solution may be higher at the fluorinated surface than oxide.

Although LiFePO$_4$ has a good thermal stability, it should be further improved to meet the requirements for specified application of lithium batteries at elevated temperatures. For instance, large-scale lithium batteries for electric vehicles or loading systems may be operated at elevated temperatures. Figure 2.8 presents DSC curves of (a) untreated LiFePO$_4$, (b) F1-LiFePO$_4$, and (c) F2-LiFePO$_4$ electrodes in fully delithiated state. As reported by Yamada [1], three clear peaks are observed in the DSC curves of LiFePO$_4$ electrodes. The first peak was found at 481.1 K, the second peak at 526.1 K, the third peak at 538.6 K. Comparing with untreated LiFePO$_4$ (a), the decomposition peaks of fluorinated LiFePO$_4$ (b and c) shifted to the higher temperatures. Especially in the case of first peak, the decomposition temperature (495 K) of F1-LiFePO$_4$ moved to the higher than those of other samples. It may be attributed to that the cathode materials having fluorinated surface may be stable in the electrolyte solution containing LiPF$_6$. Because hydrogen fluoride (HF) is produced by hydrolysis of supporting salts such as LiPF$_6$ with a small amount of residual water contained in the electrolyte solution, the HF attacks the M–O bond in cathode materials and water is then reproduced, so that these reactions may be repeated alternatively. In contract, the M–F bond in fluorinated cathode materials is unable to be attacked by HF. Furthermore, it may be considered that the fluorine existing at the surface of cathode materials makes stable the higher oxidation state of Fe in LiFePO$_4$ in fully charged state. These DSC data indicate that surface fluorination can improve the thermal stability of LiFePO$_4$.

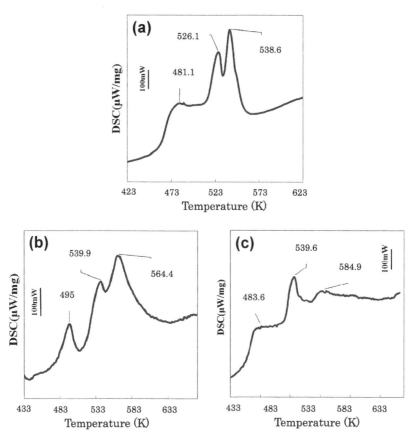

FIGURE 2.8 DSC curves of (a) untreated LiFePO$_4$, (b) F1–LiFePO$_4$, and (c) F2–LiFePO$_4$ electrodes in fully delithiated state.

2.2 SURFACE FLUORINATION OF LiNi$_{0.5}$Mn$_{1.5}$O$_4$

2.2.1 Introduction

Lithium-containing transition metal oxides, LiCoO$_2$, LiNiO$_2$, LiMn$_2$O$_4$, and their derivatives have been investigated as the high performance cathode active materials of lithium secondary battery [26–31]. To fulfill these usages, LiMn$_2$O$_4$ spinel has been attractive due to its good thermal stability and fast Li$^+$-ion diffusivity among various cathode materials [32–36]. However, the dissolution of manganese from the LiMn$_2$O$_4$ spinel structure leads to the destruction of the structure, which is initiated at the interface between electrolyte and electrode, resulting in rapid capacity fade during charge/discharge cycles [37,38]. To improve this structural instability, partial substitution of transition metals for manganese sites in LiMn$_2$O$_4$ [LiM$_x$Mn$_{2-x}$O$_4$ (M = transition metals)] has been intensively investigated [39–41]. LiNi$_{0.5}$Mn$_{1.5}$O$_4$ spinel exhibits the potential plateau at ~4.7V in the charge/discharge curves due to the presence of a Ni^{2+}/Ni^{3+} and Ni^{3+}/Ni^{4+} redox pairs [42,43]. $LiM_xMn_{2-x}O_4$ suffers from poor cycling performance, which is believed to be caused by the large lattice strain during cycling [44,45], and the decomposition of electrolyte by the high operating voltage [46]. In addition, the synthesis of $LiNi_xMn_{2-x}O_4$ often encounters the formation of NiO impurity. Surface fluorination is effective for improving the electrochemical properties and thermal stability of $LiNi_{0.5}Mn_{1.5}O_4$.

2.2.2 Experimental Details

Fluorinated $LiNi_{0.5}Mn_{1.5}O_4$ (Tanaka Chemical Co. Ltd.) particles were prepared by direct fluorination using F_2 gas under the following conditions. Reaction temperature, fluorine pressure, and reaction time were set at 298 and 373 K, 6.67 kPa, and 1 h, respectively. The structural and electronic properties of the samples were investigated using powder X-ray diffraction (XRD, XRD-6100) and X-ray photoelectron spectroscopy (XPS, XPS-9010). Surface morphology of the samples was observed using a scanning electron microscope (SEM, s-2400; Hitachi Ltd.). Fluorine contents in fluorinated $LiNi_{0.5}Mn_{1.5}O_4$ were determined using an ion chromatography (SD-8022, Tosoh Co.) after dissolving in distilled water. Two-electrode test cell (TOM cell) was used for the electrochemical measurements. The cathode mixture consisted of $LiNi_{0.5}Mn_{1.5}O_4$ samples, AB and PVDF in weight ratios of 8:1:1. The mixture was rolled spread to film with 0.1 mm thickness and the film was cut into a disk with 13 mm$^\phi$. It was then pressed onto a titanium mesh welded on the bottom of SUS304 container (20 mm$^\phi$ × 3 mm^t). The cathode was completely dried under vacuum (~10^{-1} Pa) for 12 h at room temperature prior to use. Mixture of ethylene carbonate (EC) and dimethyl carbonate(DMC) (3:7 vol.) containing 1.0 mol dm^{-3} $LiPF_6$ was used as an electrolyte solution. Li metal foil (0.2 mm^t Kyokuto Kinzouku Co., Ltd.) was used as the counter electrode. Charge/discharge test was carried out at 0.1 °C at 298 K in 1–10 cycles and 0.5 °C at 333 K in 11–50 cycles (Hokuto Denko Co., HJ101SM6). The amount of Mn deposited on Li metal after charge/discharge test was measured by atomic absorption analysis (Hitachi Z-5300, Hitachi Co, Ltd).

2.2.3 Characterization of Surface-Fluorinated $LiNi_{0.5}Mn_{1.5}O_4$

Figure 2.9 shows SEM images of untreated and fluorinated samples. It seems that there is no change in the shape and morphology of $LiNi_{0.5}Mn_{1.5}O_4$ surface even when it was treated with F_2 gas at 373 K. The effect of fluorination on the surface morphology of $LiMn_2O_4$ particles was not detected in SEM images. These results are similar to the SEM images of fluorinated $LiMn_2O_4$ [18]. Figure 2.10 shows XRD profiles of untreated and fluorinated samples. There is no extra peak in the

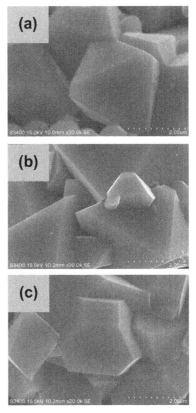

FIGURE 2.9 SEM images of untreated and fluorinated samples. (a) Untreated, (b) 6.67 kPa at 298 K, (c) 6.67 kPa at 373 K.

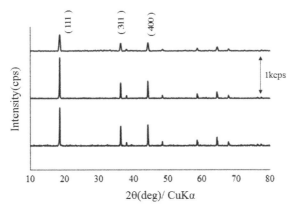

FIGURE 2.10 X-ray diffraction patterns of various samples. (a) Untreated, (b) 6.67 kPa at 298 K, (c) 6.67 kPa at 373 K.

TABLE 2.3 Ratio of Peak Intensity and FWHM

Sample	Ratio I(400)/(111)	Ratio I(311)/(111)	FWHM (111)
Untreated	70	59	0.158
6.67 kpa at 298 K	54	45	0.153
6.67 kpa at 373 K	53	49	0.410

profiles even after fluorination of $LiNi_{0.5}Mn_{1.5}O_4$ with 6.67 kPa-F_2 gas at 373 K. But the intensity and peak shape slightly changed after fluorination at 373 K as shown in Figure 2.10(c). The peak intensity ratios for (400)/(111) and (311)/(111), and FWHM values obtained from XRD patterns (Figure 2.10) are given in Table 2.3. As shown in Table 2.3, the peak intensity ratios and FWHM values of fluorinated samples are smaller and broader than those of untreated sample. It may be considered that surface fluorination changes the crystallinity of surface region of $LiNi_{0.5}Mn_{1.5}O_4$. XPS spectra of F 1s electron for $LiNi_{0.5}Mn_{1.5}O_4$ treated by 6.67 kPa-F_2 at 298 and 373 K are shown in Figure 2.11. Even at 298 K, fluorine is introduced to the surface of $LiNi_{0.5}Mn_{1.5}O_4$. The peak positions of $LiNi_{0.5}Mn_{1.5}O_4$ fluorinated at 373 K (Figure 2.11(b)) are different from that of $LiNi_{0.5}Mn_{1.5}O_4$ fluorinated at 298 K. The peak in Figure 2.11(b) is composed of two peaks at 685.4 and 687.6 eV. The 685.4 eV is the BE for ionic F^- in the solid samples such as alkali metal fluorides. Therefore, the peak at 685.4 eV may correspond to F^- having interaction with Li^+ in the sample. On the other hand, the peak at around 687.6 eV may be fluorine atoms bonded to transition metal ions such as Mn^{n+} and Ni^{n+}. After argon ion etching, the peak intensity of fluorinated samples decreased as shown in Figure 2.11. Especially the peak at 687.6 eV almost disappeared. However, the peak at 685.4 eV still remained for all fluorinated samples. Considering the results by XRD together, it seems that fluorine diffuses into the inner part of $LiNi_{0.5}Mn_{1.5}O_4$ particles. These results are similar to those of $LiMn_2O_4$ fluorinated under the same conditions [18].

2.2.4 Electrochemical Properties of Surface-Fluorinated $LiNi_{0.5}Mn_{1.5}O_4$

Figure 2.12 shows first discharge curves of $LiNi_{0.5}Mn_{1.5}O_4$ fluorinated under different conditions. The shape of the discharge curves is similar to each other. But the discharge capacity changed by fluorination. As shown in Figure 2.12, the fluorination of $LiNi_{0.5}Mn_{1.5}O_4$ surface with F_2 causes the increase of the discharge capacity. However, the discharge capacity of the sample fluorinated at 373 K was less than that of untreated one. This seems because excess fluorination causes the formation of a resistive film consisting of some fluorides at the surface.

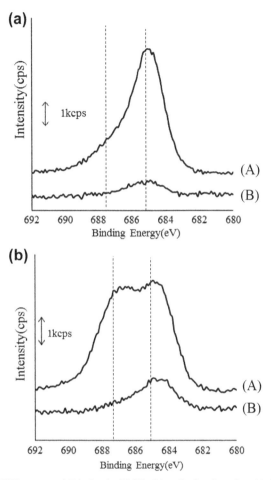

FIGURE 2.11 XPS spectra of F 1s level of $LiNi_{0.5}Mn_{1.5}O_4$ fluorinated at (a) 6.67 kPa at 298 K and (b) 6.67 kPa at 373 K. [(A) before etching, (B) after Ar^+ etching].

The discharge capacity of the sample fluorinated at 298 K was about 4% larger than that of untreated one. The change in the discharge capacities during 50 cycles is shown in Figure 2.13. Comparing with untreated sample, the loss of discharge capacity of the sample fluorinated at 298 K is improved. However, the cycle stability of the sample fluorinated at 373 K is inferior to untreated one because of the formation of a resistive fluoride film. These results show that the electrochemical properties of $LiNi_{0.5}Mn_{1.5}O_4$ are improved by introducing fluorine to the sample surface without formation of a resistive fluoride film. This may be because the wettability of electrolyte with cathode is improved by the introduction of fluorine to the oxide surface since fluorine has the larger electronegativity than oxygen.

Manganese (Mn) dissolution of untreated and fluorinated samples is shown in Figure 2.14. The Mn dissolution in $LiNi_{0.5}Mn_{1.5}O_4$ is restrained by surface

48 Advanced Fluoride-Based Materials for Energy Conversion

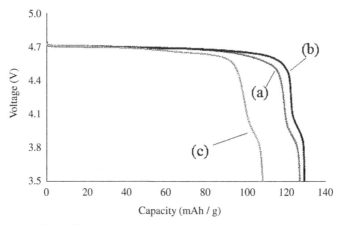

FIGURE 2.12 Charge/discharge curves of various $LiNi_{0.5}Mn_{1.5}O_4$ samples. (a) Untreated, (b) 6.67 kPa at 298 K, (c) 6.67 kPa at 373 K.

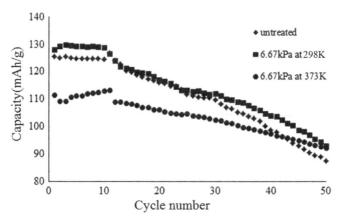

FIGURE 2.13 Discharge capacity versus cycle number of untreated and fluorinated samples.

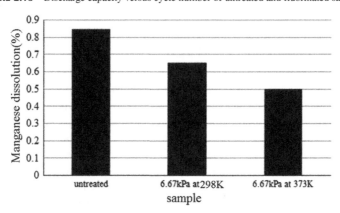

FIGURE 2.14 Manganese dissolution of untreated and fluorinated $LiNi_{0.5}Mn_{1.5}O_4$ samples.

fluorination. The main reason for the suppression of Mn dissolution may be the creation of Mn^{3+}. The fluorine existing at the sample surface may play a role to stabilize the higher oxidation state of manganese, which would improve the cycle stability and stability of $LiNi_{0.5}Mn_{1.5}O_4$.

2.3 SUMMARY

The effects of surface fluorination on the electrochemical properties and stability of $LiFePO_4$ and $LiNi_{0.5}Mn_{1.5}O_4$ are shown in this chapter. Fluorinated $LiFePO_4$ (F-$LiFePO_4$) was successfully prepared by direct fluorination using NF_3 gas. XRD and SEM indicated that the surface fluorination with NF_3 gas did not affect the crystal structure and particle morphology of carbon-coated $LiFePO_4$. However, XPS proved the existence of fluorinated surface layer at the surface of $LiFePO_4$. $F^{\delta-}$ released from $F_2N^{\delta+}-F^{\delta-}$ reacts with $M^{\delta+}$(M=Li, Fe) site in $LiFePO_4$. Li–F bond is mainly detected at the surface. In the case of F1-$LiFePO_4$ without any resistive film, the discharge capacity was 10% higher than that of untreated $LiFePO_4$. EIS also showed the lowest resistance ($R_{ct}=35.55\,\Omega$) and largest exchange current density ($i_0=0.72\,mA$). DSC proved that the surface fluorination improves the thermal stability of $LiFePO_4$. Thus surface fluorination improves the electrochemical properties and thermal stability of $LiFePO_4$ as a promising cathode in high-power lithium-ion cells for hybrid electric vehicle (HEVs).

Fluorinated $LiNi_{0.5}Mn_{1.5}O_4$ was successfully prepared by direct fluorination using F_2 gas. XRD and SEM indicate that the surface fluorination with F_2 gas does not affect the crystal structure and particle morphology of $LiNi_{0.5}Mn_{1.5}O_4$. However, XPS has proved that the fluorinated surface layer exists at the surface of $LiNi_{0.5}Mn_{1.5}O_4$. Especially Li–F bond is mainly detected at the surface. In the case of $LiNi_{0.5}Mn_{1.5}O_4$ fluorinated at 298 K, the discharge capacity was higher than that of untreated $LiNi_{0.5}Mn_{1.5}O_4$. Cycle stability and Mn dissolution of $LiNi_{0.5}Mn_{1.5}O_4$ are also improved by the surface fluorination. The surface fluorination improves the electrochemical properties of $LiNi_{0.5}Mn_{1.5}O_4$ as a promising cathode in high-power lithium-ion cells for HEVs.

REFERENCES

[1] A. Yamada, S.C. Chung, K. Hinokuma, J. Electrochem. Soc. 148 (2001) A224–A229.
[2] D. Kim, J. Lim, V. Mathew, B. Koo, Y. Paik, D. Ahn, S.M. Paek, J. Kim, J. Mater. Chem. 22 (2012) 2624–2631.
[3] Y. Wu, Z. Wen, J. Li, Adv. Mater. 23 (2011) 1126–1129.
[4] O. Waser, R. Buchel, A. Hintennach, P. Novak, S.E. Pratsinis, J. Aerosol Sci. 42 (2011) 657–667.
[5] M. Cuisinier, J.F. Martin, N. Dupre, R. Kanno, D. Guyomard, J. Mater. Chem. 21 (2011) 18575–18583.
[6] S.W. Kim, J. Ryu, C.B. Park, K. Kang, Chem. Commun. 46 (2010) 7409–7411.
[7] S.Y. Chung, J.T. Bloking, Y.M. Chiang, Nat. Mater. 1 (2002) 123–128.

[8] H. Huang, S.C. Yin, L.F. Nazar, Electrochem. Solid-State Lett. 4 (2001) A170–A172.
[9] F. Croce, A.D. Epifanio, J. Hassoun, A. Deptula, T. Olczac, B. Scrosati, Electrochem. Solid-State Lett. 5 (2002) A47–A50.
[10] A. D'Epifanio, F. Croce, P. Reale, L. Settimi, B. Scrosati (Eds.), Proceedings of the 203rd ECS Meeting, Paris, vol. 27. 2003(Abstract No. 1114).
[11] H. Li, H. Zhou, Chem. Commun. 48 (2012) 1201–1217.
[12] M.L. Trudeau, D. Laul, R. Veillette, A.M. Serventi, A. Mauger, C.M. Julien, K. Zaghib, J. Power Sources 196 (2011) 7383–7394.
[13] M. Takashima, S. Yonezawa, M. Ozawa, Mol. Cryst. Liq. Cryst. 388 (2002) 153–159.
[14] S. Yonezawa, M. Ozawa, M. Takashima, Tanso 205 (2002) 260–262.
[15] M. Takashima, Y. Nosaka, T. Unishi, Eur. J. Solid State Inorg. Chem. 29 (1992) 691–703.
[16] J.H. Kim, H. Umeda, M. Ohe, S. Yonezawa, M. Takashima, Chem. Lett. 40 (2011).
[17] JCPDS card No. 40–1499.
[18] S. Yonezawa, M. Yamasaki, M. Takashima, J. Fluorine Chem. 125 (2004) 1657–1661.
[19] W.J. Zhang, J. Power Sources 196 (2011) 2962–2970.
[20] X.C. Tang, L.X. Li, Q.L. Lai, X.W. Song, L.H. Jiang, Electrochem. Acta 54 (2009) 2329–2334.
[21] J.H. Kim, F. Nishimura, S. Yonezawa, M. Takashima, J. Fluorine Chem. 144 (2012) 165–170.
[22] T. Nakajima, Solid State Sci. 9 (2007) 777–784.
[23] C.H. Chen, J. Liu, K. Amine, J. Power Sources 96 (2001) 321–328.
[24] A.J. Bard, L.R. Faulkner, Electrochemical Methods, Wiley, New York, 1980, p. 213.
[25] J.H. Ryu, J. Electrochem. Sci. Tech. 2 (2011) 136–142.
[26] W. Ebner, D. Fouchard, L. Xie, Solid State Ionics 69 (1994) 238–256.
[27] D. Guyomard, J.M. Tarascon, Solid State Ionics 69 (1994) 222–237.
[28] M.M. Thackeray, A. deKock, M.H. Rossouw, D. Liles, R. Bittihn, D. Huge, J. Electrochem. Soc. 139 (1992) 363–366.
[29] Y. Todorov, C. Wang, B.I. Banov, M. Yoshio, Electrochem. Soc. Proc., Paris 97-18 (1997) 176–184.
[30] N. Hayashi, H. Ikuta, M. Wakihara, J. Electrochem. Soc. 146 (1999) 1351–1354.
[31] T. Ohzuku, A. Ueda, J. Electrochem. Soc. 141 (1994) 2972–2977.
[32] M.M. Thackeray, P.J. Johnson, L.A. Depicciotto, P.G. Bruce, J.B. Good-enough, Mater. Res. Bull. 19 (1984) 179.
[33] M.M. Thackeray, A. deKock, J. Solid State Chem. 74 (1988) 414.
[34] J. Cho, M.M. Thackeray, J. Electrochem. Soc. 146 (1999) 3577.
[35] J.Y. Luo, Y.G. Wang, H.M. Xiong, Y. Xia, Chem. Mater. 19 (2007) 4791.
[36] S. Lim, J. Cho, Electrochem. Commun. 10 (2008) 1478.
[37] Y. Xia, M. Yoshio, J. Electrochem. Soc. 143 (1996) 825.
[38] R.J. Gummow, A. deKock, M.M. Thackeray, Solid State Ionics 69 (1994) 59.
[39] A.D. Robertson, S.H. Lu, W.F. Averill, W.F. Howard, J. Electrochem. Soc. 144 (1997) 3500.
[40] K. Amine, H. Tukamoto, H. Yasuda, Y. Fujia, J. Electrochem. Soc. 143 (1996) 1607.
[41] Q. Zhong, A. Bonakdarpour, M. Zhang, Y. Gao, J.R. Dahn, J. Electrochem. Soc. 144 (1997) 205.
[42] H. Kawai, M. Nagata, H. Tukamoto, A.R. West, J. Power Sources 81 (1999) 67.
[43] T. Ohzuku, S. Takeda, M. Iwanaga, J. Power Sources 81 (1999) 90.
[44] S. Mukerjee, X.Q. Yang, X. Sun, S.J. Lee, J. McBreen, Y. Ein-Eli, Electrochem. Acta 49 (2004) 3373.
[45] T.A. Arunkumar, A. Manthiram, Electrochem. Solid-State Lett. 8 (2005) A403.
[46] Y.K. Sun, K.J. Hong, J. Prakash, K. Amine, Electrochem. Commun. 4 (2002) 344.

Chapter 3

Fluoride Cathodes for Secondary Batteries

Munnangi Anji Reddy and Maximilian Fichtner
Helmholtz Institute Ulm for Electrochemical Energy Storage (HIU), Helmholtzstr, Ulm, Germany

Chapter Outline
- 3.1 Introduction 51
- 3.2 Metal Fluorides for Electrochemical Energy Storage 52
- 3.3 Metal Fluorides for Lithium Batteries 53
  - 3.3.1 Introduction 53
  - 3.3.2 Binary Metal Fluorides 53
    - 3.3.2.1 Metal Trifluorides 53
    - 3.3.2.2 Metal Difluorides (MF_2) 61
  - 3.3.3 Ternary Metal Fluorides ($LiMFeF_6$, Li_3MF_6) 63
- 3.4 Metal Fluorides for Fluoride Ion Batteries 65
  - 3.4.1 Introduction 65
- 3.4.2 Solid Fluoride Ion Conductors 66
  - 3.4.2.1 Fluorite-Type Fluoride Ion Conductors 67
  - 3.4.2.2 Tysonite-Type Fluoride Ion Conductors 68
- 3.4.3 Solid State Fluoride Ion Batteries (FIB) 69
- 3.4.4 Fluoride Ion Batteries with Liquid Electrolyte 72
- 3.4 Perspective 73
- References 74

3.1 INTRODUCTION

The international goal to gradually replace the current fossil and nuclear-based energy supply by renewable energy necessitates further advancements in research and developmental efforts. While in principle there exists substantial energy resources, the fluctuating nature of wind and solar systems require scientific and technological advances in electrochemical energy storage systems for large-scale applications (stationary and mobile), though already there are commercially viable utilities such as consumer electronics, power tools, hybrid electric vehicles, and a smaller number of full-electric vehicles. The complexity of modern electrochemical storage systems requires strategies in research to gain in-depth understandings of the fundamental processes occurring in the electrochemical cell in order to apply this knowledge to develop new conceptual electrochemical energy storage systems [1].

On a mid- and long-term perspective, the development of batteries with new chemistries is needed which could offer higher energy density, Li-free operation, better safety, use of abundant materials, lower cost, and operational flexibility at different temperatures. This pursuit has led to increasing research activities in new types of batteries which are based on alternative anionic (F^-, Cl^-) and cationic (Na^+, Mg^{2+}) shuttles.

Recently, fluoride-based materials have been introduced in a variety of energy applications, not only as electrodes for solar cells, electrolytes for fuel cells, but also as electrodes for aqueous batteries. Moreover, it has been demonstrated that the application of fluorine containing materials may also have beneficial effects in high-energy lithium nonaqueous batteries, as functional coatings, additives, electrolytes, or as active materials [2,3].

3.2 METAL FLUORIDES FOR ELECTROCHEMICAL ENERGY STORAGE

In this chapter, we review and discuss on the application of fluorides as active electrode materials in batteries. Since the earliest years of twenty-first century, metal fluorides have gained considerable interest in the development of energy materials. In particular, fluorides have been introduced as electrode materials in Li-ion batteries due to the high electronegativity of fluorine and the large free energy associated with the formation of fluorides [2,4]. Moreover, using metal fluorides as active materials in conversion electrodes offer the opportunity to make use of various oxidation states of a metal leading to the development of electrodes with both high potentials and high capacity. As an example, the theoretical gravimetric capacity of FeF_3 with a theoretical average discharge potential of 2.9 V is 712 mA h g^{-1} which is five times above that of the theoretical capacity of the currently used lithium cobalt oxide. Such properties make fluorides particularly attractive as high capacity cathode materials. Nevertheless, the fact that they are electrical insulators makes it difficult to integrate them in an environment which converts them into electrochemically active materials.

A variety of synthesis methods have been developed to produce fluoride based conversion materials, using either high energy ball milling of the active materials with conductive additives or by new routes based on solid-state chemistry. Despite the simple one-step procedures used in the latter case, the products with complex hierarchical structures consisting of both nanoscale active particles and a conductive carbon matrix can be prepared. Such nanocomposites are electrochemically active and exhibit better cyclic stability than materials produced by mixing active material and carbon, in a high-energy ball milling method [5,6].

Lately, secondary batteries based on fluoride ion shuttle have been realized. After early conceptual studies, where the anion shuttle was proposed for solid state batteries [7] and PbF_2 and fluorite type materials were investigated as thin film electrolytes [8,9] first data was presented in 2011 which showed reversibly working fluoride ion batteries based on pressed powder pellets [10]. Such type of battery has attracted considerable interest in the meantime due to its very high theoretical volumetric energy densities around 5800 Wh L^{-1}, which is 50% beyond that of the Li–O_2 couple. Table 3.1 summarizes electrochemical properties of the couples between metal fluoride cathode and a light metal anode. These values were calculated using thermodynamic data and materials properties of the different compounds.

3.3 METAL FLUORIDES FOR LITHIUM BATTERIES

3.3.1 Introduction

Metal fluorides tower in their high energy density values compared to insertion based materials for energy storage applications. But, their full potential remains unexplored, because they are electrical insulators exhibiting sluggish kinetics and large polarization between discharge and charge processes in lithium batteries. As already mentioned, the size of the metal fluoride particles can be reduced down to few nanometers to shorten the lithium ion and electron diffusion path lengths, thereby facilitating each of the particles to advocate electronic conductivity. Further, conversion reactions are generally accompanied by large volume changes during discharge and charge processes. Incorporation of electronically conductive backbone should be able to buffer these volume changes and provide undisrupted electronic conductivity. Due to the above mentioned reasons, electrochemical properties of metal fluorides can be strongly influenced by the synthetic routes and electrode structuring. A variety of metal fluorides have already been examined for their reversible lithium storage properties. In the following section, we review the application of binary metal fluorides and ternary metal fluorides and we discuss the influence of electrochemical performance of metal fluorides with respect to various synthetic approaches.

3.3.2 Binary Metal Fluorides

3.3.2.1 Metal Trifluorides

Most of the transition metal trifluorides (MF_3) crystallize in ReO_3 type open frame work structure. The MF_6 octahedra are corner shared three dimensionally leaving big cavity, which is suitable for the accommodation of cations.

3.3.2.1.1 FeF_3

Among the various metal fluorides, iron fluorides have gained importance due to their low cost and low toxicity. Therefore, most of the recent studies were centered on iron trifluoride.

TABLE 3.1 Theoretical Properties of Electrochemical Couples Based on Metal Fluorides

MF_X	Capacity (mAhg^{-1})	Li EMF (V)	Li Energy Density (Wh kg^{-1})	Li Energy Density (Wh L^{-1})	Na EMF (V)	Na Energy Density (Wh kg^{-1})	Na Energy Density (Wh L^{-1})	Mg EMF (V)	Mg Energy Density (Wh kg^{-1})	Mg Energy Density (Wh L^{-1})
TiF_3	766.8	1.386	886.5	1501.9	0.957	442.5	723.6	0.845	480.9	1210.0
VF_3	744.9	1.853	1157.3	2096.1	1.424	647.3	1109.6	1.313	731.0	1988.8
CrF_3	737.7	2.279	1411.5	2707.6	1.850	835.8	1490.6	1.738	961.0	2814.8
MnF_2	576.8	1.909	957.8	2073.2	1.479	570.9	1120.9	1.368	625.5	1964.2
MnF_3	718.3	2.637	1597.1	3002.6	2.208	981.3	1728.2	2.097	1136.0	3204.9
FeF_2	571.2	2.626	1306.8	2876.5	2.197	842.3	1674.3	2.086	946.3	3027.6
FeF_3	2.733	712.5	1644.1	3224.4	2.304	1018.9	1847.7	2.193	1180.8	3516.4
CoF_2	553.0	2.737	1324.1	3074.0	2.308	865.7	1789.6	2.197	971.3	3296.7
CoF_3	693.6	3.607	2120.8	4211.1	3.178	1382.0	2530.5	3.067	1618.1	4848.5
NiF_2	554.4	2.961	1435.2	3407.8	2.531	951.1	1996.3	2.420	1072.1	3753.9
CuF_2	527.9	3.541	1644.7	3796.5	3.112	1130.9	2336.4	3.001	1278.2	4234.3

Continued

ZnF$_2$	518.5	2.395	1094.6	2725.5	1.966	705.4	1538.1	1.854	778.3	2832.9
ZrF$_4$	641.1	1.402	770.6	1674.3	0.972	402.2	786.5	0.861	427.7	1404.6
NbF$_5$	713.2	2.569	1546.7	2558.3	2.140	947.0	1524.8	2.029	1093.4	2600.4
AgF	211.3	4.153	831.8	3209.8	3.724	666.0	2199.3	3.612	696.4	3376.9
SnF$_2$	342.1	2.974	934.6	2644.5	2.545	673.1	1670.0	2.434	720.7	2702.3
MgF$_2$	860.4	0.540	380.3	632.8	0.111	55.2	88.9	—	—	—
AlF$_3$	957.5	1.147	880.0	1395.5	0.718	377.4	587.9	0.606	404.9	1014.5
CaF$_2$	686.6	−0.001	−0.6	−1.1	−0.430	−185.9	−320.4	−0.542	−283.5	−753.3

Arai et al. studied lithium insertion into FeF_3, considering its high theoretical voltage versus Li [11]. In this work, FeF_3 showed a reversible capacity of 80 mA h g^{-1} at an average discharge voltage of 3.0 V in the voltage range of 4.5–2.0 V. This is only 30% of the theoretical specific capacity 237 mA h g^{-1}, considering one electron reaction. This work generated considerable interest on metal fluorides to be possible host materials for lithium storage applications.

Later, Amatucci's group pioneered the work on metal fluorides. They synthesized carbon metal fluoride nanocomposites (CMFNCs) through high energy mechanical milling process. In the case of CMFNCs containing FeF_3 they observed a reversible capacity of 200 mA h g^{-1} in the insertion range between 4.5 and 2.0 V with low capacity fade over cycling [12]. They have also examined the FeF_3 containing CMFNCs in the 1.3–4.5 V, where they observed further reduction of Li_xFeF_3 to LiF and Fe metal. A high capacity of 600 mA h g^{-1} was obtained at 70 °C (theoretical 712 mA h g^{-1}) due to a conversion reaction of FeF_3 to LiF and Fe which occurs after the insertion at lower voltage. However, the capacity was fading rapidly with cycling [13]. To simplify the discussion we defined the insertion range as 2.0–4.5 V and conversion range as 4.5–1.3 V. The synthesis of CMFNCs was tuned by coupling the mechanical milling with the redox reaction between carbon fluoride (CF) and FeF_2, to form carbon–FeF_3 nanocomposite. These nanocomposites showed a stable capacity of 500 mA h g^{-1} [14]. The cycling stability was attributed to the improved electronic conductivity in the nanocomposites. Heat treating the ball milled carbon-FeF_3 nanocomposites at 350 °C significantly improved the cycling stability in the 4.5–2.0 V region [15].

Apart from Amatucci's group work on CMFNCs, there are many reports available on FeF_3 application, both in the insertion and in the conversion range. For instance, Wu et al. studied the effect of mixed conducting matrix on the electrochemical performance of FeF_3 in the insertion range [16]. They synthesized FeF_3/V_2O_5 nanocomposites through mechanical milling FeF_3 with V_2O_5 at different durations. It has been found that the addition of V_2O_5 significantly improved the electrochemical performance of FeF_3. The FeF_3/V_2O_5 composite synthesized by milling for 3 h exhibited a high reversible capacity of 192 mA h g^{-1} even after 30 cycles. Li et al. fabricated mesoporous $FeF_3 \cdot 0.33H_2O$ by novel one-step ionic liquid assisted synthesis. The compound was reported to show a reversible capacity of 130 mA h g^{-1} after 30 cycles at a current density of 14 mA g^{-1} [17]. In situ wiring of these nanoparticles with single-walled carbon nanotubes substantially improved the capacity and rate capability of the composite. A reversible capacity of 220 mA h g^{-1} at 0.1 °C and 80 mA h g^{-1} at 10 °C was observed [18].

Kim et al. synthesized FeF_3 nanoflowers decorated on carbon nanotubes (CNT). The nanocomposites showed excellent cyclability and high rate capability. Figure 3.1 shows the electrochemical performance of the FeF_3-CNT nanocomposites after one precycle in the 1.5–4.5 V. A specific capacity

Fluoride Cathodes for Secondary Batteries Chapter | 3 57

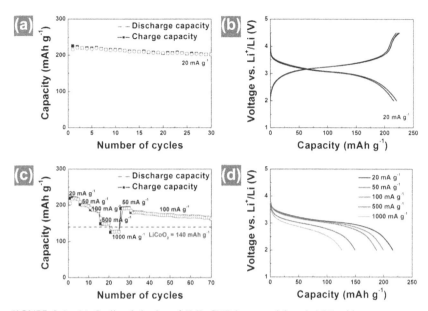

FIGURE 3.1 (a) Cycling behavior of FeF$_3$-CNT between 2.0 and 4.5V with a current rate of 20 mA g^{-1} after one precycle at a voltage range of 1.5–4.5V, and (b) corresponding charge–discharge profiles of the initial three cycles. (c) Specific capacity of FeF$_3$–CNT at a voltage range of 2.0–4.5V with different current rates from 20 to 1000 mA/g after the precycle, and (d) corresponding discharge profiles at each current rate. The horizontal dashed line in (c) indicates the practical capacity of the conventional LiCoO$_2$ cathode. Reproduced with permission [19]. *Copyright © 2010 WILEY-VCH Verlag GmbH & Co.*

of 210 mA h g^{-1} is observed at current density of 20 mA g^{-1}. A reversible capacity of 150 mA h g^{-1} is observed at 500 mA g^{-1}. The excellent rate capability was attributed to the intimate contact between the active material (FeF$_3$) and the conductive agent (CNT) as well as fast electron transport along the CNT [19].

Ma et al. synthesized three-dimensionally ordered macroporous FeF$_3$ by using polystyrene colloidal crystals as hard template and subsequently coated with poly(3,4-ethylenedioxythiophene). The resulting nanostructure showed high reversible capacity of 210 mA h g^{-1} and 120 mA h g^{-1} at a current density of 1000 mA g^{-1} [20]. Doping of cobalt in FeF$_3$ appeared to significantly improve the discharge capacity and cycling stability. While FeF$_3$/C composites exhibited the discharge capacity of 102 mA h g^{-1} at 1 °C, the 5% cobalt doped Fe$_{0.95}$Co$_{0.05}$F$_3$/C showed a reversible capacity of about 140 mA h g^{-1} at 1 °C after 100 cycles [21].

Recently, few groups employed graphene as supporting matrix for iron fluoride. Graphene wrapped FeF$_3$ nanocomposites were prepared by the fluorination of Fe$_3$O$_4$/G composite with HF vapor at 120 °C [22]. The electrochemical properties of FeF$_3$/G nanocomposites were studied in the conversion range. A reversible capacity of 185 mAh g^{-1} was observed at a current density of 20.8 mA g^{-1}. In a different approach, iron fluoride nanocrystals were uniformly precipitated on reduced graphene sheets. Figure 3.2 shows the schematic approach for the synthesis of FeF$_3$ anchored graphene nanosheets [23]. The nanocomposites showed a high reversible capacity of 210 mAh g^{-1} at 0.2 °C rate in the insertion range and 490 mAh g^{-1} in the conversion range (1.5–4.5 V).

Ionic liquid assisted synthesis of iron fluoride/graphene nanosheet (GNS) hybrid nanostructures was developed. The nanostructure exhibit high capacity (230 mAh g^{-1} at 0.1 C) and remarkable rate performance (74 mAh g^{-1} at 40 °C). Further, FeF$_3$·0.33H$_2$O/GNS prepared by this method showed extended cycling stability and a reversible capacity of 115 mAh g^{-1} was achieved even after 250 cycles at a high current density of 2000 mA g^{-1} [24]. Microwave assisted fluorolytic sol–gel route was adopted for the synthesis of FeF$_3$·0.33H$_2$O and this material was subsequently supported on reduced graphene oxide. The composite showed a reversible capacity of 150 mAh g^{-1} after 50 cycles in the voltage range of 1.5–4.5 V [25]. Li et al. fabricated a binder free flower like hexagonal tungsten bronze type FeF$_3$·0.33H$_2$O on titanium foil using hydrothermal method. The hierarchical nanostructure reported to exhibit high rate capability and a reversible capacity of 126 mAh g^{-1} at 3 °C rate [26].

One of the constraints with the employment of FeF$_3$ is that it is necessary to use lithium as anode or prelithiated material as anode. Hence, attempts

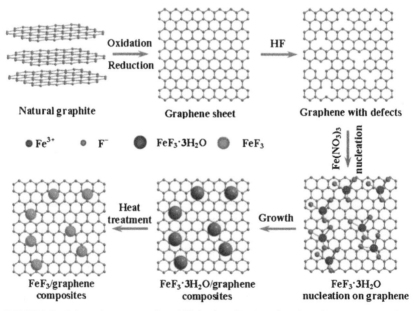

FIGURE 3.2 Schematic representation of fabrication of FeF$_3$ anchored graphene nanocomposites. Reproduced with permission [23]. *Copyright © 2013, Royal Society of Chemistry.*

were made to prelithiate the cathode. Amatucci et al. synthesized xLiF+Fe or nanocomposites by solid state reduction of FeF_2 and FeF_3 using Li_3N as a reducing agent. The reaction was facilitated by high energy mechanical milling. The nanocomposites prepared by this method were electrochemically active and the 2LiF+Fe showed a reversible capacity of 284 mA h g^{-1} was achieved at 8.8 mA g^{-1} for the first few cycles. However, no cycling data was reported [27]. Prakash et al. employed ferrocene as both carbon and iron source and synthesized carbon–iron lithium fluoride nanocomposite. In a typical synthesis, a mixture of ferrocene and lithium fluoride was heated at 700 °C for 2 h in closed Swagelok cell reactor. The resulting nanocomposites composed of carbon encapsulated iron nanoparticles which were agglomerated in larger particles, while the agglomerates were electronically wired by carbon nanotubes. These nanocomposites showed a very stable capacity of 170 mA h g^{-1} even after 200 cycles in the voltage range of 1.3–4.3 V [28]. When the synthesis was performed in a rotating oven, more uniform nanocomposite was produced and it showed significant improvement in the reversible capacity. Figure 3.3 shows the charge discharge curves and cycling behavior of nanocomposites obtained with modified synthesis. A stable capacity of 230 mA h g^{-1} was achieved. A remarkable feature of these nanocomposites is their extended cycling behavior. For the first time the authors showed that such a stable cycling was possible using metal fluoride as active material [29]. Das et al. studied the effect of mixed conducting matrix on the cyclability of LiF+Fe nanocomposites. They synthesized $LiF/Fe/V_2O_5$ nanocomposites by high energy mechanical milling. The resulting nanocomposites were electrochemically active and a stable capacity of 270 mA h g^{-1} was achieved after 50 cycles. The heat treated nanocomposites showed an even improved capacity of 310 mA h g^{-1} after 50 cycles [30]. By far, this is the best capacity achieved for prelithiated cathode materials based on metal fluorides.

3.3.2.1.2 Other Metal Trifluorides (BiF_3, TiF_3, VF_3, MnF_3, and CrF_3)

Among other trifluorides, BiF_3 was studied as cathode material for lithium ion batteries. Although the theoretical specific capacity is lower (300 mA h g^{-1}), it has a huge volumetric energy density of 5500 Wh L^{-1}, which may be relevant for certain applications. Bervas et al. synthesized the carbon–BiF_3 nanocomposites by high energy ball milling and studied as cathode material for lithium batteries [31]. The nanocomposites delivered a specific capacity of 230 mA h g^{-1}. Amorphous $AlPO_4$ coating was found to improve the electrochemical performance of the BiF_3. $AlPO_4$ coating significantly reduces the formation of solid electrolyte interface (SEI) and improves initial discharge capacity and the reversible capacity. Initial discharge capacity of about 272 mA h g^{-1} and a reversible capacity of 209 mA h g^{-1} was achieved at a current density of 30 mA g^{-1} [32]. Amatucci et al. also synthesized Bi+LiF composites by the solid state reduction of BiF_3 with Li_3N similar to LiF+Fe composites, and studied as cathode material for lithium ion batteries [27]. The Bi+LiF nanocomposites delivered a reversible capacity of 199 mA h g^{-1}.

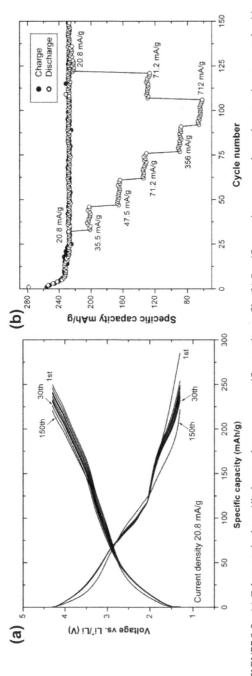

FIGURE 3.3 (a) Galvanostatic charge/discharge voltage versus specific capacity profile. (b) Specific capacity versus cycle number at various current densities. The nanocomposite was cycled between 4.3 and 1.3 V. Reproduced with permission [29]. *Copyright © 2011 Elsevier B.V.*

In addition to FeF_3, Arai et al. studied lithium insertion into TiF_3, VF_3 and MnF_3 in the insertion range considering their high theoretical voltage. Lithium can be inserted into TiF_3 and VF_3 in the voltage range between 2.5 and 2.2 V respectively. A discharge capacity of 80 mA h g^{-1} was observed similar to FeF_3 [11]. Li et al. studied lithium insertion into TiF_3 and VF_3 in the conversion range. High reversible capacity of about 550 mA h g^{-1} was observed in the case of TiF_3 when cycled between 0.02 and 3.5 V. VF_3 showed a large overvoltage in the first discharge. However, from the second discharge a plateau was observed at 1.7 V, which is consistent with its theoretical voltage. A reversible capacity of 500 mA h g^{-1} was obtained in the full voltage range of 0.02–4.3 V [33]. Despite the structural similarity of MnF_3 to other trifluorides, lithium insertion did not precede due to high overvoltage. CrF_3-containing CMFNCs were also prepared by high energy mechanical milling of CF with CrF_2 [14]. The CrF_3 containing CMFNCs showed initial discharge capacity of 682 mA h g^{-1} and a reversible capacity of 440 mA h g^{-1} in the first few cycles. No extended cycling data was reported, however.

3.3.2.2 Metal Difluorides (MF_2)

Apart from metal trifluorides, difluorides have extensively been studied as cathode material for lithium batteries [5,13,33–47]. Transition metal difluorides adopt rutile type structure. The structure is built of MF_6 octahedra that share edges along the *c*-direction and corners in the *ab*-plane leaving narrow channels in the *c*-direction. These channels contain possible sites for lithium insertion. However, in the case of metal difluorides lithium insertion is not expected as transition metal ions could not be reduced to 1$^+$ oxidation state, rather they directly reduced to metal.

Among transition metal difluorides, FeF_2 was studied extensively as cathode materials for lithium batteries. The theoretical specific capacity of FeF_2 is 571 mA h g^{-1} with an *emf* of 2.66 V, which leads to the theoretical gravimetric energy density or specific energy of 1519 Wh kg^{-1}. This makes FeF_2 an attractive cathode material for lithium batteries. Pereira et al. synthesized nanocrystalline FeF_2 by thermal decomposition of $FeSiF_6 \cdot 6H_2O$ in argon atmosphere at 200 °C. $FeSiF_6 \cdot 6H_2O$ was prepared by the reaction of H_2SiF_6 with Fe metal. As prepared FeF_2 was subsequently ball milled with carbon and tested as cathode material for lithium batteries [5]. The nanocomposites showed high initial capacity of 430 mA h g^{-1} but faded rapidly upon cycling. In the following, few studies on C-FeF_2 were dedicated to the lithiation and delithiation mechanism in nanocomposites [34,35]. Recently, Reddy et al. synthesized CF_x derived C–FeF_2 nanocomposites in one-step reaction between CF_x and $Fe(CO)_5$ at 250 °C, in a closed Swagelok reactor [36], in a gas–solid state reaction. CF_x is a solid source for carbon and fluoride. $Fe(CO)_5$ is a liquid source of iron metal. During the course of reaction $Fe(CO)_5$ evaporates at 103 °C and diffuses into the layers of the CF_x structure. At 150 °C it starts to decompose and forms highly reactive iron metal clusters inside the carbon fluoride. These clusters react with the fluorine in CF_x and form FeF_2 leaving behind a reduced and well conductive carbon matrix.

The as-formed FeF_2 is clamped in carbon layers. During this process, carbon is reduced from sp^3 to sp^2 state. Figure 3.4 shows the elemental mapping of the as-synthesized nanocomposites. FeF_2 crystallites are well distributed in carbon matrix. These $C-FeF_2$ nanocomposites synthesized by this method showed very low electrical resistivity, leading to superior electrochemical performance. The optimized composite showed a reversible capacity of $325\,mA\,h\,g^{-1}$ at $25\,°C$ and $418\,mA\,h\,g^{-1}$ at $40\,°C$. Further, optimization is possible by changing the amount of fluoride to carbon ratio, the $C-FeF_2$ nanocomposites synthesized from $CF_{1.0}$ shows superior electrochemical performance compared to the samples prepared from $CF_{1.1}$ [37]. Core–shell structured $C-FeF_2$ nanocomposites with rod-shaped FeF_2 core and graphitized carbon as shell was prepared by a one-step thermal reaction using a mixture of ferrocene and polyvinylidene fluoride as precursor. The composite delivered stable reversible capacity of $217\,mA\,h\,g^{-1}$ over 50 cycles [38]. These results show that fabrication of appropriate and stable microstructure in $C-FeF_2$ nanocomposites is important to achieve a high reversible capacity.

Copper (II) fluoride being one of the highest energy density cathode among transition metal fluorides was widely investigated for primary lithium batteries [39]. Two major drawbacks of CuF_2 limit its applicability as cathode for rechargeable Li batteries. The poor reversibility of the cell that are resulted due to the cathode dissolution in liquid phase followed by oxidation of Cu during the reconversion [40] and secondly due to its low voltage profile (~2V) compared to its theoretical potential (~3.5V). CuF_2 embedded in either metal oxides or sulfides or carbon exhibited capacities close to theoretical capacities mostly at theoretical voltages [40]. Among the nanocomposites, the best performance was shown for CuF_2 embedded in MoO_3. However, the reversibility of such

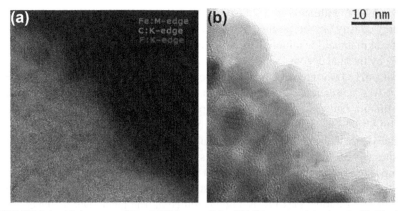

FIGURE 3.4 (a) An energy filtered TEM map of the $C(FeF_2)_{0.55}$ material using the Fe M-edge (red), C K-edge (green), and F K-edge (blue). (b) Corresponding High-resolution transmission electron microscopy image. Reproduced with permission [36]. *Copyright © 2012 WILEY-VCH Verlag GmbH & Co. (For interpretation of the references to color in this figure legend, the reader is referred to the online version of this book.)*

nanocomposites was not mentioned. The enhancement in capacity of CuF_2 in CuF_2-metal oxide nanocomposite was reported partially due to a limited fluorine abstraction or ion exchange with oxygen.

Cobalt (II) fluoride, CoF_2 is another interesting conversion candidate with a theoretical capacity of 553 mA h g^{-1} at a theoretical electrode potential of 2.85 V versus Li/Li$^+$. Thin films of amorphous CoF_2 fabricated by a pulsed laser deposition (PLD) process exhibited high initial capacities of 600 mA h g^{-1} with limited cycling stability [41]. Similarly, CoF_2 obtained by ball milling exhibited initial capacities of 600 mA h g^{-1} at 1.0 V versus Li/Li$^+$ with significant capacity decline over 25 cycles [42]. Later, much higher initial capacities of 650 mA h g^{-1} at a potential of 1.2 V versus Li/Li$^+$ were obtained for supercritical-fluid synthesized crystalline CoF_2 [43]. However, the capacity faded with subsequent charge–discharge cycles yielding ~100 mA h g^{-1} after 10 cycles. The loss in capacity in subsequent cycles may be due to the unfavorable re-conversion mechanism in addition to electrolyte incompatibility with the CoF_x.

Nickel (II) fluoride, NiF_2 is yet another positive electrode candidate for rechargeable lithium batteries with a theoretical specific capacity of 554 mA h g^{-1} at an electrode potential of 2.96 V versus Li/Li$^+$. Reversible conversion reactions of NiF_2 with Li were reported involving Li_2NiF_4 as an intermediate. Nanostructured NiF_2 thin films fabricated by PLD exhibited a first discharge capacity of 650 mA h g^{-1} [44]. The conversion potential (0.7 V) obtained in NiF_2 thin films was much lower than theoretical potential but with capacities close to 500 mA h g^{-1} after 40 cycles. To improve the electrochemical properties of such insulating, high voltage metal fluoride, compositing was performed with conductive carbon matrices [13,45] or the incorporation of covalent M–O bonds [46]. NiF_2/C nanocomposites fabricated by high energy milling exhibited an initial discharge capacity of 1100 mA h g^{-1}, which drastically faded in subsequent cycles eventually to a capacity of 200 mA h g^{-1} in the tenth cycle. The initial discharge capacity of 1100 mA h g^{-1} was higher than the theoretical value of 554 mA h g^{-1} and could be attributed to an interfacial interaction of lithium within the Ni/LiF matrix, possibly leading to a distinct local charging [33], in addition to the formation of a SEI film on electrode surface [47]. The conversion reactions of NiO-doped NiF_2–C composite was investigated and compared with pristine NiF_2 with the aim to improving electronic conductivity by introducing covalent M–O bonds [46]. The NiO-doped NiF_2–C composite exhibited a slight improvement in conversion potential and reversibility compared to pristine NiF_2.

3.3.3 Ternary Metal Fluorides ($LiMFeF_6$, Li_3MF_6)

Lithium containing transition metal fluorides are of considerable interest due to their prelithiated state. As the materials are present in the discharged state, lithium free anodes can be used in principle. Further, lithium insertion–deinsertion kinetics would be different due to different crystal structure. Rutile structure related $LiMFeF_6$ and cryolite-type Li_3MF_6 compounds have been studied extensively.

Liao et al. synthesized trirutile-type $LiFe_2F_6$ and $LiMgFeF_6$ by high energy mechanical milling and their electrochemical activity was studied toward lithium. The theoretical specific capacity of $LiFe_2F_6$ is $126\,mAh\,g^{-1}$ assuming extraction of one lithium. It was found that $LiFe_2F_6$ is electrochemically active between 2.0 and 4.5 V, whereas $LiMgFeF_6$ is almost unreactive. In situ XRD studies showed that in the case of stoichiometric $LiFe_2F_6$, 0.5 Li can be extracted from the metal frame work and 0.6 Li can be reversibly inserted into the channels of the structure. In the case of nonstoichiometric $Li_{1.2}Fe_2F_{6.2}$, 0.5 Li can be extracted from the framework and 0.9 Li can be reversibly inserted. With optimized nonstoichiometric composition ($Li_{1.5}Fe_2F_{6.5}$) a reversible capacity of $139\,mAh\,g^{-1}$ was achieved in the voltage range 2.0–4.5 V, after 28 cycles [48]. Lieser et al. developed a sol–gel process for the synthesis of $LiNiFeF_6$ by using trifluoroacetic acid as fluorine source. However, ball-milling of $LiNiFeF_6$ with carbon is necessary to obtain an electrochemical active nanocomposite. In the case of $LiNiFeF_6$, unlike $LiFe_2F_6$, lithium in the metal frame work cannot be deintercalated due to the high oxidation potential of Ni^{2+} to Ni^{3+} or Fe^{3+} to Fe^{4+}. The nanocomposites were first discharged to 2.0 V and then charged to 4.5 V. The first discharge capacity of $95\,mAh\,g^{-1}$ is observed. A reversible capacity of $88\,mAh\,g^{-1}$ was achieved after 20 cycles [49].

Lithium insertion was studied in cryolite-type low temperature monoclinic α-Li_3FeF_6 and high temperature orthorhombic β-Li_3FeF_6. In both cases, the as synthesized phases are electrochemically inactive toward lithium in the insertion range. When ball milled with carbon at 500 rpm for 5 h, both phases become electrochemically active. Figure 3.5 shows the first discharge-charge curves of α-Li_3FeF_6 and β-Li_3FeF_6 ball milled with carbon for 5 and 12 h. In the case of α-Li_3FeF_6 about 0.43 Li was inserted and 0.11 Li was inserted in β-Li_3FeF_6. When milling time increased to 12 h the inserted lithium increased to 0.7 Li and 0.55 Li in the case of α- and β-phase respectively. Also here, the crystallite size was found important in determining the electrochemical activity of

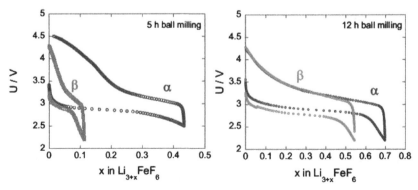

FIGURE 3.5 First discharge–charge profiles of α-Li_3FeF_6 and β-Li_3FeF_6 ball milled with carbon for 5 and 12 h. Reproduced with permission [6]. *Copyright © 2010 Electrochemical Society.*

Li_3FeF_6 irrespective of nature of the phase [6]. Basa et al. synthesized nano-sized β-Li_3VF_6 by low temperature solution method. The as-synthesized sample delivered only 30 mA h g^{-1}, but upon ball milling with carbon it could deliver 144 mA h g^{-1} which corresponds to the theoretical specific capacity by assuming the reduction of V^{3+} to V^{2+}. A reversible capacity of 90 mA h g^{-1} was achieved upon cycling. The extraction of lithium ions from Li_3VF_6 by oxidizing V^{3+} to V^{4+}, was proved to be irreversible [50].

A fluorolytic method was developed as a general synthesis procedure for a series of metal fluorides Li_3MF_6 (M = Cr, V, and Fe) and Li_2MnF_5, starting from metal acetylacetonates as metal salts and HF in ethanol as fluorinating agent [51]. However, no data was presented regarding the electrochemical activity of these compounds.

3.4 METAL FLUORIDES FOR FLUORIDE ION BATTERIES

3.4.1 Introduction

As already mentioned, fluorides are fascinating electrode materials due to their high electronegativity and relatively low weight. The chemical reaction of fluorine with metals leads to the formation of metal fluorides which is typically accompanied by a large change in free energy yielding high theoretical voltages. Fluoride anion (F$^-$) may be utilized as an electrochemically stable transport ion between two electrodes. A bivalent or trivalent metal can react with several fluoride ions and thereby more than one electron can be reversibly stored per metal atom resulting in high theoretical specific capacity. Thus, by choosing appropriate metal/metal fluoride systems combined with suitable fluoride ion conducting media, high energy density electrochemical cells can be built. Figure 3.6 represents the architecture of fluoride ion battery.

At the anode, electrons are generated in a thermodynamically feasible, spontaneous redox reaction during the discharge process. These electrons travel through an external circuit to combine with the cathode material whereby the metal fluoride is reduced to metal, M. The released fluoride anions shuttle from the cathode, migrate through the electrolyte and reacts with metal M′ of the anode to form metal fluoride M′F$_x$, ensuring the electroneutrality. The charge process is opposite to the discharge process. The following reactions are anticipated at cathode and anode during the discharge process and opposite reactions are expected during the charge process.

$$\text{At cathode: } xe^- + MF_x \rightarrow M + xF^- \quad (3.1)$$

$$\text{At anode: } xF^- + M' \rightarrow M'F_x + xe^- \quad (3.2)$$

The successful realization of fluoride ion battery depends on suitable fluoride transporting electrolyte. There are several fluoride ion conducting compounds. However, fluoride conductors based on low electropositive elements may not be suitable for battery applications. For example, the fluoride ion

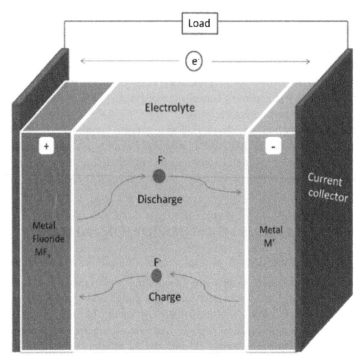

FIGURE 3.6 Representation of fluoride ion battery. Reproduced with permission [10]. *Copyright © 2011, Royal Society of Chemistry.*

conductivity of PbSnF$_4$ is ~1×10^{-3} S cm^{-1} at room temperature. However, its electrochemical stability is limited and high electropositive elements like Li, Ca, Na, Mg, etc. cannot be used as anodes in combination with such electrolyte. Instead, metal fluorides based on rare earth, alkaline earth, and alkaline metal based fluorides may be suitable for electrolyte applications. In the following we will only discuss those fluoride conductors which are electrochemically stable.

3.4.2 Solid Fluoride Ion Conductors

Fluoride conducting ceramics are known for a long time [9,52]. In principle, the fluoride conductivity depends strictly on the crystal structure of the compounds as the F anions or vacancies should have a high diffusion rate in the crystals [53,54]. Two types of structures exhibit particularly high fluoride conductivity: fluorite (*Fm-3m*) or tysonite (*P-3c1*) type structures. Alkaline-earth (Ca, Ba, Sr) and lead fluorides (MF$_2$) have a fluorite structure and some rare-earth fluorides (RF$_3$) have a tysonite structure (La, Ce, Pr, Nd). The fluoride conductivity of pure compounds is low and needs to be improved for battery applications.

The most widely adopted method to increasing the conductivities of fluorides is homogeneous doping the host lattice by monovalent (AF, A for alkaline metals, mostly Na or K), divalent (MF_2) or trivalent cations (RF_3). A series of reports have been published in particular for the fluorides having the fluorite or the tysonite structure. It was found possible to obtain various solid solutions by mixing AF, MF_2 and RF_3 in different amount. The most common solid solutions studied can be written as with $M_{1-x}R_xF_{2+x}$ or $M_{1-x}A_xF_{2-x}$ for compounds with the fluorite structure and $R_{1-y}M_yF_{3-y}$ for compounds with the tysonite structure.

3.4.2.1 Fluorite-Type Fluoride Ion Conductors

The fluorite structure is rather open and can accommodate relatively large amount of dopant [55]. For example, it is possible to obtain solid solution of 55 mol% LaF_3 in BaF_2 keeping the fluorite structure. Ivanov-Shits and coworkers [56] prepared $Ba_{1-x}R_xF_{2+x}$ with different rare earth elements R with various compositions. For all rare-earth elements, the conductivity increases with x but sometimes it saturated at high concentration only (probably above the percolation threshold). This was found to be related to the decrease (and saturation) of the activation enthalpy. The highest conductivities (and lowest activation enthalpies) have been obtained for lanthanum doped BaF_2 with for example $\sigma_{473K} = 1.9 \times 10^{-4}\,S\,cm^{-1}$ for $Ba_{0.6}La_{0.4}F_{2.4}$. Similar results have been reported by Sorokin and Breiter [57]. Doping BaF_2 with alkaline fluorides (NaF, KF) found to increase the ionic conductivity. However, they have very limited solubility in BaF_2 [58].

Sorokin and Breiter [59] studied the conductivity of doped-SrF_2 single crystals. The temperature behavior was similar to what was obtained with doped-BaF_2. The highest conductivities were obtained for SrF_2 doped with La, Ce or Pr. The highest value reported for $Sr_{0.6}La_{0.4}F_{2.4}$ was around $2.2 \cdot 10^{-4}\,S\,cm^{-1}$ at 200°C. The minimum activation energy was found to be 0.64 eV (Pr). The same authors performed a similar study on doped CaF_2 single crystals [60]. The conductivities were smaller than for doped SrF_2 crystals ($7.3 \times 10^{-6}\,S\,cm^{-1}$ at 200°C for $Ca_{0.8}Gd_{0.2}F_{2.2}$) which is mostly related to the much higher activation energies measured (minimum 0.71 eV for Dy doping). Similar values were given by Ivanov-Shits et al. [61] for doped-CaF_2 or doped SrF_2 single crystals. Doping with a mixture of rare-earth fluorides does not provide any improvement in the conductivity mostly because the lattice mismatch is increased by mixing different cations without providing more defects or new conduction paths [62]. Wang and Grey [63] studied YF_3-doped CaF_2 by ^{19}F MAS NMR spectroscopy and could identify the F anions in different interstitial positions. They found four resonances lines in addition to the main resonance line of F anion in the fluorite structure (normal position) corresponding to four different interstitials positions.

Rongeat et al. studied the fluoride conductivity of LaF_3 doped BaF_2 nanocrystalline fluorides synthesized by mechanical milling [64]. They found that the conductivity of nanocrystalline samples is high compared to that of single crystals with same compositions. The highest conductivity was measured for $Ba_{0.6}La_{0.4}F_{2.4}$ and reached $1.9 \times 10^{-4}\,S\,cm^{-1}$ at 160 °C. The increase in ionic conductivity was attributed to the decrease in the activation energy along the grain boundaries.

An improvement of the ionic conductivity has been reported when mixing two fluorite-types compounds: CaF_2–SrF_2 [65] or BaF_2–CaF_2 [66]. Especially for the mixture CaF_2 with BaF_2 by ball milling, Düvel et al. [67] obtained a metastable solid solution with a fluorite structure demonstrating a much higher conductivity than ball milled pure BaF_2 or CaF_2. This was explained by a significant decrease of the activation enthalpy. Such an improvement has not been observed when mixing BaF_2–SrF_2 or CaF_2–SrF_2.

Puin et al. [68] showed that the conductivity can be improved by using nanocrystalline CaF_2 in which grain boundaries provide fast conduction pathways. It has also been demonstrated that the improvement of the conductivity of CaF_2 is possible by activating the grain boundaries using Lewis acids [69]. The adsorbed SbF_5 or BF_3 at the boundaries attract F atoms and create vacancies that can facilitate the F anion mobility. The conductivity of CaF_2 and BaF_2 can also be improved by mixing with an insulator, for example Al_2O_3 [70].

3.4.2.2 Tysonite-Type Fluoride Ion Conductors

Most of the rare-earth fluorides show a tysonite-type structure (P-$3c1$) with F anion in three nonequivalent positions F1, F2, and F3. The highest conductivities have been reported for LaF_3- and CeF_3-based fluorides, and hence we focus our discussion on the results pertaining only to these two types of compounds. For LaF_3, the conductivity is slightly anisotropic with a higher conductivity parallel to the c-axis than perpendicular to this c-axis [71]. Usually, the activation energy decreases above 400–450 K (130–180 °C) indicating a change in the conductivity mechanism from a vacancy mechanism within the F1 fluorine subsystem to a mechanism with exchange of F vacancies between the F1 subsystem and the F2–F3 subsystem [72]. The anisotropy disappears for higher temperature or after doping the LaF_3.

Similar to fluorite-type compounds, the conductivities are drastically improved when doping with aliovalent fluorides. The most common doping mechanism in tysonite-type fluorides is $R_{1-y}M_yF_{3-y}$ (R = La or Ce and M = Ba, Sr, or Ca). The solubility limit of dopant in tysonite is 10–15 mol% [73]. For low concentration of dopant, the conductivity increases with y and the activation energy decreases. This is mainly related to the formation of an increasing number of defects that are free for migration. The conductivity reaches a maximum at the percolation threshold (c. $y = 0.07$ for BaF_2 in LaF_3). From this value, the activation energy remains somehow constant but the vacancy migration is hindered by the presence of too many vacancies. Takahashi et al. [74] have obtained higher conductivity values for doped CeF_3 compared to LaF_3. The conductivity is improved greatly by the addition of 5 mol% of BaF_2, SrF_2 or CaF_2. The highest conductivity was obtained for $Ce_{0.95}Ca_{0.5}F_{2.95}$ with $1.2 \times 10^{-2}\,S\,cm^{-1}$. By contrast with fluorite-type fluorides as CaF_2, the conductivity has not been improved with nanoparticles of LaF_3, the conductivity decreases compared to single crystal. The introduction of bivalent cation should lead to the formation of vacancies. Moreover, the formation of F interstitials in tysonite-type crystal is unlikely because the interstitials sites in the tysonite structure are very small (0.84 Å and for F anion, radius is 1.19 Å) [75]. The intrinsic defects in the tysonite structure should be the Schottky defects (cation vacancy associated with anion vacancy) and the anion conduction is a vacancy mechanism. The conduction mechanism is then different to what has been described for doped BaF_2.

Reddy et al. [10,76] synthesized and studied the conductivity of nanocrystalline $La_{1-y}Ba_yF_{2-y}$ ($0 \leq y \leq 0.15$). They found that the conductivity of BaF_2 doped LaF_3 sample is lower compared to that of single crystals of the same composition. This observation was in contrast to the behavior found in fluorite type $Ba_{1-x}La_xF_{2+x}$. The observed low conductivity in tysonite-type fluorides was attributed to the blocking effect of the grain boundaries. Consequently, the conductivity could be increased by sintering at 800 °C for 2 h. However, the conductivity of nanocrystalline $La_{0.9}Ba_{0.1}F_{2.9}$ is still sufficient ($2.8 \times 10^{-4}\,S\,cm^{-1}$ at 160 °C) to build solid state fluoride ion batteries.

3.4.3 Solid State Fluoride Ion Batteries (FIB)

A few examples of FIB have already been reported in literature mostly for primary cells (only discharge). For example, using thin film geometry, Kennedy and Hunter [8] prepared a primary cell using Pb as anode, PbF_2 as electrolyte, and CuF_2 as cathode (mixed with PbF_2 to ensure good ionic conductivity). They were able to discharge this cell at room temperature to a maximum of 40% of the theoretical capacity. However, the voltage was rather low (0.4–0.5 V). Similar results [77] were reported for a cell $BiO_{0.09}F_{2.82}$ as cathode instead of CuF_2. It was also reported that better discharge can be obtained using doped-Ca anode [78,79] instead of pure Ca and that a voltage of c. 3 V can be obtained with $BiO_{0.1}F_{2.8}$ as cathode. Higher capacities were reported by Potanin [80] using La or Ce as anode, doped LaF_3 or CeF_3 as electrolyte and doped PbF_2 or BiF_3 as cathodes. The discharge delivered c. $130\,mA\,h\,g^{-1}$ (PbF_2) or ca. $200\,mA\,h\,g^{-1}$ (BiF_3) at around 2–2.5 V but using rather high temperatures (500 °C). Baukal [81] was able to build a rechargeable FIB using Mg as anode, doped-CaF_2 as electrolyte and NiF_2 or CuF_2 as cathode. This battery was operated at high temperature (400–500 °C) and it was necessary to carefully prepare the electrodes to ensure both electronic and ionic conductivity. However, no data was presented regarding the reversibility of the system. Table 3.2 summarizes a few examples of FIB presented in the literature.

TABLE 3.2 Type of Fluoride Ion Batteries Studied

Type[a]	Structure	Anode	Electrolyte	Cathode	Operating Temp. (°C)	Maximum Capacity (mAh g^{-1})	Voltage (V)	References
P	Thin films	Pb	PbF$_2$	CuF$_2$	25	108	0.4–0.5	[8]
P	Films	Ca Ca$_{0.85}$La$_{0.15}$ Ca$_{1-x}$Y$_x$	Ca$_{0.68}$U$_{0.02}$Ce$_{0.30}$F$_{2.34}$ Ba$_{0.55}$La$_{0.45}$F$_{2.45}$	BiO$_{0.1}$F$_{2.8}$	100	—	3	[79]
P	Films	Pb	PbF$_2$ + AgF	BiO$_{0.09}$F$_{2.82}$	25	—	0.33	[77]
P	Deposited layers	La Ce	La$_{0.94}$Ba$_{0.06}$F$_{2.94}$ Ce$_{0.94}$Sr$_{0.06}$F$_{2.94}$	PbF$_2$ (+0.06 KF) BiF$_3$ (+0.06 KF)	500	130 (PbF$_2$) 200 (BiF$_3$)	2.3–2.7	[80]
S	Layers	Mg	Doped CaF$_2$	NiF$_2$ CuF$_2$	400–500	—	—	[81]
S	Layers	Ce	La$_{0.9}$Ba$_{0.1}$F$_{2.9}$	CuF$_2$ BiF$_3$ SnF$_2$ KBiF$_4$	150	320	2–2.7	[10]

[a] P: primary and S: secondary battery.

Recently, we have demonstrated also secondary fluoride ion battery with various metal fluorides as cathode material (see Table 3.2), cerium metal as anode and tysonite-type $La_{0.9}Ba_{0.1}F_{2.9}$ as electrolyte. Figure 3.7 shows the first discharge curves of various composite cathode materials obtained at 150 °C. Only 50–70% theoretical specific capacities were observed. Ex situ XRD studies performed on the cathode side of the pellet revealed the reduction of metal fluorides to corresponding metals, confirming the feasibility of fluoride ion batteries. Figure 3.8 shows the discharge–charge curves of BiF_3 and its cycling behavior. The capacity fades gradually with cycling which

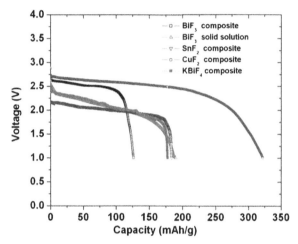

FIGURE 3.7 Voltage-composition profiles of various metal fluorides versus Ce metal. The discharge curves were obtained at 150 °C with a current density of 10 mA cm^{-2}. Reproduced with permission [10]. *Copyright © 2011, Royal Society of Chemistry.*

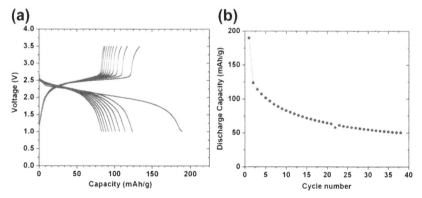

FIGURE 3.8 (a) Voltage–composition profiles of BiF_3 solid solution electrode for the first 10 cycles, and (b) cycling behavior. Reproduced with permission [10]. *Copyright © 2011, Royal Society of Chemistry.*

is typical for nonoptimized conversion electrodes. It should be possible to further improve the initial discharge capacity as well as the cycling behavior of the cells by improving the fabrication of cathode-electrolyte-anode pellets and/or employing thin electrolyte layers.

The performance of the solid state fluoride ion battery is limited by several factors. One reason is that the fluoride conductivity of solid electrolyte is not high enough. Moreover, the ceramic electrolytes cannot follow the volume changes due to low elasticity. As a consequence, large volume changes during discharge–charge reactions may lead to contact loss between solid electrolyte and electrode material, as well as the electronic conductor and electrode material. Further, the volume changes may lead to loss of contact at the cathode–electrolyte interface and at the anode–electrolyte interface. The above mentioned problems can be mitigated by the use of fluoride transporting liquid electrolyte.

3.4.4 Fluoride Ion Batteries with Liquid Electrolyte

In spite of its high scientific importance it is interesting to note that work on fluoride conducting liquid electrolyte is still in its infancy. One of the challenges is that fluoride is a small anion with high charge density and is a strong hydrogen bond acceptor. Further, polar aprotic solvents do not solvate anions significantly due to the presence of lone pair of electrons, in which case the fluoride is left unsolvated (naked). Unsolvated fluoride ion is a strong base, can deprotonate even weak acidic protons and thus potentially generate HF_2^- species. This unique property of fluoride makes it difficult to design fluoride conducting liquid electrolyte based on naked F^- ions.

In a recent patent, Weiss et al. showed that 1-methyl, 1-propyl piperidinium fluoride (MPPF) dissolved in EC: DMC mixture can act as a fluoride transporting media [82]. They showed a Bi/1M MPPF in EC: DMC/ CuF_2 cell which operates at 0.2V and capacity fades rapidly with cycling. However, there are no further reports available on this particular electrolyte and especially the compatibility of this electrolyte with highly electropositive elements is unknown. They also reported ionic liquids as electrolytes and with various organic fluoride salts [83]. Though these electrolytes show high ionic conductivity, their electrochemical behavior was not satisfactory, according to the authors.

Very recently, Gschwind et al. reported ammonium bifluoride-doped polyethylene glycol (PEG) dissolved in acetonitrile as fluoride transporting electrolyte. Figure 3.9 shows a drawing of PEG stabilized NH_4HF_2. Due to the strong bonding between the PEG moiety and ammonium bifluoride the thermal stability of NH_4HF_2 increased from 130 °C to 340 °C. This complex dissolved in acetonitrile can act as a fluoride transporting electrolyte. The electrochemical test cells were built with BiF_3 composite as cathode and Li as anode shows ammonium bifluoride doped PEG dissolved in acetonitrile as electrolyte. It was shown that the

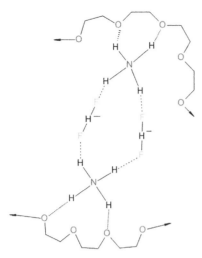

FIGURE 3.9 Shows the drawing of the ammonium bifluoride doped-polyethylene glycol (PEG) matrix. Reproduced with permission [19]. Copyright © 2013, Royal Society of Chemistry.

electrochemical performance of the cells depend on the concentration of bifluoride–PEG complex. Further studies are under way toward the improvement of electrolyte and electrode materials [84].

3.4 PERSPECTIVE

Metal fluorides are potential high energy density systems. However, further technical improvements will be necessary before they can be used in commercial lithium batteries. In particular, reaction rates should be improved as in the current state the obvious presence of kinetic barriers lead to polarization with considerable voltage hysteresis. Moreover, although great improvements have been made, the long-term cyclic stability needs to be addressed in case stationary applications are envisaged. To obtain higher capacities, significant progress was made regarding synthesis of carbon-metal fluoride nanocomposites which can deliver high energy at reasonable power density. Recent advances in lithium anode technology generated renewed interest on metal fluorides as cathode materials.

In principle, carbon–metal fluoride nanocomposites developed for lithium batteries could be used for fluoride ion batteries as well. There are some additional advantages with the use of metal fluorides as electrodes in fluoride ion batteries compared to lithium batteries. It is possible to achieve near equilibrium voltages, less polarization between charge–discharge, and fast kinetics with metal fluorides and fluoride transporting electrolyte, as there is a direct interfacial transfer of fluoride between electrode and electrolyte. The benefits are realized and research and development of fluoride ion batteries is initiated and growing worldwide. Still, the development of liquid fluoride transporting electrolyte is key for the development of fluoride ion batteries.

REFERENCES

[1] C.-X. Zu, H. Li, Energy Environ. Sci. 4 (2011) 2614–2624.
[2] G.G. Amatucci, N. Pereira, J. Fluorine Chem. 128 (2007) 243–262.
[3] J. Cabana, L. Monconduit, D. Larcher, M.R. Palacín, Adv. Mater. 22 (2010) E170–E192.
[4] Y. Koyama, I. Tanaka, H. Adachi, J. Electrochem. Soc. 147 (2000) 3633–3636.
[5] N. Pereira, F. Badway, M. Wartelsky, S. Gunn, G.G. Amatucci, J. Electrochem. Soc. 156 (2009) A407–A416.
[6] E. Gonzalo, A. Kuhn, F. García-Alvarado, J. Electrochem. Soc. 157 (2010) A1002–A1006.
[7] W. Baukal, Ger. Offen. (1971) GWXXBX DE 2017128.
[8] J.H. Kennedy, J.C. Hunter, J. Electrochem. Soc. 123 (1) (1976) 10–14.
[9] P. Hagenmuller, J.M. Reau, C. Lucat, S. Matar, G. Villeneuve, Solid State Ionics 3 (4) (1981) 341–345.
[10] M. Anji Reddy, M. Fichtner, J. Mater. Chem. 21 (2011) 17059–17062.
[11] H. Arai, S. Okada, Y. Sakurai, J. Yamaki, J. Power Sources 68 (1997) 716.
[12] F. Badway, N. Pereira, F. Cosandey, G.G. Amatucci, J. Electrochem. Soc. 150 (2003) A1209.
[13] F. Badway, F. Cosandey, N. Pereira, G.G. Amatucci, J.Electrochem. Soc. 150 (9) (2003) A1318–A1327.
[14] I. Plitz, F. Badway, J. Al-Sharab, A. DuPasquier, F. Cosandey, G.G. Amatucci, J. Electrochem. Soc. 152 (2005) A307–A315.
[15] N. Yabuuchi, M. Sugano, Y. Yamakawa, I. Nakai, K. Sakamoto, H. Muramatsu, S. Komaba, J. Mater. Chem. 21 (2011) 10035–10041.
[16] W. Wu, Y. Wang, X. Wang, Q. Chen, X. Wang, S. Yang, X. Liu, J. Guo, Z. Yang, J. Alloys Compd. 486 (2009) 93–96.
[17] C. Li, L. Gu, S. Tsukimoto, P.A. van Aken, J. Maier, Adv. Mater. 22 (2010) 3650–3654.
[18] C. Li, L. Gu, J. Tong, J. Maier, ACS Nano 5 (2011) 2930–2938.
[19] S.-W. Kim, D.-H. Seo, H. Gwon, J. Kim, K. Kang Adv, Mater. 22 (2010) 5260–5264.
[20] D.-L. Ma, Z.-Y. Cao, H.-G. Wang, X.-L. Huang, L.-M. Wang, X.-B. Zhang, Energy Environ. Sci. 5 (2012) 8538.
[21] L. Liu, M. Zhou, L. Yi, H. Guo, J. Tan, H. Shu, X. Yang, Z. Yang, X. Wang, J. Mater. Chem. 22 (2012) 17539.
[22] R. Ma, Z. Lu, C. Wang, H.-E. Wang, S. Yang, L. Xia, J.C.Y. Chung, Nanoscale 5 (2013) 6338–6343.
[23] J. Liu, Y. Wan, W. Liu, Z. Ma, S. Ji, J. Wang, Y. Zhou, P. Hodgson, Y. Li, J. Mater.Chem. A 1 (2013) 1969–1975.
[24] B. Li, D.W. Rooney, N. Zhang, K. Sun, ACS Appl. Mater. Interfaces 5 (2013) 5057–5063.
[25] L.D. Carlo, D.E. Conte, E. Kemnitz, N. Pinna, Chem. Commun. 50 (2014) 460.
[26] B. Li, Z. Cheng, N. Zhang, K. Sun, Nano Energy 4 (2014) 7–13.
[27] G.G. Amatucci, N. Pereira, F. Badway, M. Sina, F. Cosandey, M. Ruotolo, C. Cao, J. Fluorine Chem. 132 (2011) 1086–1094.
[28] R. Prakash, A.K. Mishra, A. Roth, Ch Kübel, T. Scherer, M. Ghafari, H. Hahn, M. Fichtner, J. Mater. Chem. 20 (2010) 1871–1876.
[29] R. Prakash, C. Wall, A.K. Mishra, Ch Kübel, M. Ghafari, H. Hahn, M. Fichtner, J. Power Sources 196 (2011) 5936–5944.
[30] B. Das, A. Pohl, V.S.K. Chakravadhanula, C. Kübel, M. Fichtner, J. Power Sources 267 (2014) 203–211.
[31] M. Bervas, F. Badway, L.C. Klein, G.G. Amatucci, Electrochem. Solid-State Lett. 8 (2005) A179.

[32] B. Hu, X. Wang, Y. Wang, Q. Wei, Y. Song, H. Shu, X. Yang, J. Power Sources 218 (2012) 204–211.
[33] H. Li, G. Richter, J. Maier, Adv. Mater. 15 (2003) 736–739.
[34] F. Wang, R. Robert, N.A. Chernova, N. Pereira, F. Omenya, F. Badway, X. Hua, M. Ruotolo, R. Zhang, L. Wu, V. Volkov, D. Su, B. Key, M.S. Whittingham, C.P. Grey, G.G. Amatucci, Y. Zhu, J. Graetz, J. Am. Chem. Soc. 133 (2011) 18828–18836.
[35] F. Wang, H.-C. Yu, M.-H. Chen, L. Wu, N. Pereira, K. Thornton, A. Van der Ven, Y. Zhu, G.G. Amatucci, J. Graetz, Nat. Commun. 3 (2012) 1201.
[36] M. Anji Reddy, B. Breitung, V.S.K. Chakravadhanula, C. Wall, M. Engel, C. Kübel, A.K. Powell, H. Hahn, M. Fichtner, Adv. Energy Mater. 3 (2013) 308–313.
[37] B. Breitung, M. Anji Reddy, V.S.K. Chakravadhanula, M. Engel, C. Kübel, A.K. Powell, H. Hahn, M. Fichtner, Beilstein J. Nanotechnol. 4 (2013) 705–713.
[38] Y. Zhang, L. Wang, J. Lia, L. Wen, X. He, J. Alloys Compd. 606 (2014) 226–230.
[39] [a] G. Pistoia, J. Electrochem. Soc. 118 (1) (1971) 153–158.
[b] M. Hughes, N.A. Hampson, S.A.G.R. Karunathilaka, J. Power Sources 12 (1984) 83–144.
[40] F. Badway, A.N. Mansour, N. Pereira, J.F. Al-Sharab, F. Cosandey, I. Plitz, G.G. Amatucci, Chem. Mater. 19 (2007) 4129.
[41] Z.-W. Fu, C.-L. Li, W.-Y. Liu, J. Ma, Y. Wang, Q.-Z. Qin, J. Electrochem. Soc. 152 (2) (2005) E50–E55.
[42] C. Wall, R. Prakash, C. Kübel, H. Hahn, M. Fichtner, J. Alloys Compd. 530 (2012) 121–126.
[43] M.J. Armstrong, A. Panneerselvam, C. O'Regan, M.A. Morris, J.D. Holmes, J. Mater. Chem. A 1 (2013) 10667–10676.
[44] Z-W. Fu, Y-N.Zhou, Q. Sun, H. Zhang, Solid State Sciences. 10 (9) (2008) 1166–1172.
[45] Y.L. Shi, M.F. Shen, S.D. Xu, X.Y. Qiu, L. Jiang, Y.H. Qiang, Q.C. Zhuang, S.G. Sun, Int. J. Electrochem. Sci. 6 (2011) 3399–3415.
[46] D.H. Lee, K.J. Carroll, S. Calvin, S. Jin, Y.S. Meng, Electrochim. Acta 59 (1) (2012) 213–221.
[47] S. Laruelle, S. Grugeon, P. Poizot, M. Dolle, L. Dupont, J.M. Tarascon, J. Electrochem. Soc. 149 (2002) A627–A634.
[48] P. Liao, J. Li, J.R. Dahn, J. Electrochem. Soc. 157 (2010) A355–A361.
[49] G. Lieser, C. Dräger, M. Schroeder, S. Indris, L. de Biasi, H. Geßwein, S. Glatthaar, H. Ehrenberg, J.R. Binder, J. Electrochem. Soc. 161 (2014) A1071–A1077.
[50] A. Basa, E. Gonzalo, A. Kuhn, F. García-Alvarado, J. Power Sources 207 (2012) 160–165.
[51] J. Kohl, D. Wiedemann, S. Nakhal, P. Bottke, N. Ferro, T. Bredow, E. Kemnitz, M. Wilkening, P. Heitjanse, M. Lerch, J. Mater. Chem. 22 (2012) 15819.
[52] [a] W. Bollmann, Phys. Status Solidi A 18 (1973) 313–321.
[b] W. Bollmann, and H.Henniger, Phys. Status Solidi A, 11 (1972) 367–371.
[53] V. Trnovcová, L.S. Garashina, A. Skubla, P.P. Fedorov, R. Cicka, E.A. Krivandina, B.P. Sobolev, Solid State Ionics 157 (2003) 195–201.
[54] N.I. Sorokin, B.P. Sobolev, Crystallogr. Rep. 52 (2007) 842–863.
[55] E.G. Ippolitov, L.S. Garashina, A.G. Maklachkov, Inorg. Mater. 3 (1967) 59–62.
[56] A.K. Ivanov-Shits, N.I. Sorokin, P.P. Fedorov, B.P. Sobolev, Solid State Ionics 31 (1989) 269–280.
[57] N.I. Sorokin, M.W. Breiter, Solid State Ionics 99 (1997) 241–250.
[58] [a] W. Bollmann, Phys. Status Solidi A 18 (1973) 313–321.
[b] P.W.M. Jacobs, S.H. Ong, Crystal Lattice Defects 8 (1980) 177–184.
[59] N.I. Sorokin, M.W. Breiter, Solid State Ionics 104 (1997) 325–333.
[60] N.I. Sorokin, M.W. Breiter, Solid State Ionics 116 (1999) 157–165.

[61] [a] A.K. Ivanov-Shits, N.I. Sorokin, Solid State Ionics 36 (1989) 7–13
[b] A.K. Ivanov-Shits, N.I. Sorokin, P.P. Fedorov, and B.P. Sobolev, Solid State Ionics, 31 (1989) 253–268.
[62] V. Trnovcová, P.P. Fedorov, I.I. Buchinskaya, M. Kubliha, Russ. J. Electrochem. 47 (2011) 639–642.
[63] F. Wang, C.P. Grey, Chem. Mater. 10 (1998) 3081–3091.
[64] C. Rongeat, M. Anji Reddy, R. Witter, M. Fichtner, J. Phys. Chem. C 117 (2013) 4943–4950.
[65] M.V. Subrahmanya Sarma, S.V. Suryanarayana, Solid State Ionics 42 (1990) 227–232.
[66] B. Ruprecht, M.Wilkening, A. Feldhoff, S. Steuernagel, P. Heitjans, Phys. Chem. Chem. Phys. 11 (2009) 3071–3081.
[67] A. Düvel, B. Ruprecht, P. Heitjans, M. Wilkening, J. Phys. Chem. C 115 (2011) 23784–23789.
[68] W. Puin, S. Rodewald, R. Ramlau, P. Heitjans, J. Maier, Solid State Ionics 131 (2000) 159–164.
[69] Y. Saito, J. Maier, J. Electrochem. Soc. 142 (1995) 3078–3083.
[70] S. Fujitsu, M. Miyayama, K. Koumoto, H. Yanagida, T. Kanazawa, J. Mater. Sci. 20 (1985) 2103–2109.
[71] J. Schoonman, G. Oversluizen, K.E.D. Wapenaar, Solid State Ionics 1 (1980) 211–221.
[72] N.I. Sorokin, E.A. Krivandina, Z.I. Zhmurova, B.P. Sobolev, M.V. Fominykh, V.V. Fistul, Phys. Solid State 41 (1999) 573–575.
[73] B.P. Sobolev, Crystallogr. Rep. 57 (2012) 434–454.
[74] T. Takahashi, H. Iwahara, T. Ishikawa, J. Electrochem. Soc. 124 (1977) 280–284.
[75] A. Roos, F.C.M. Van de Pol, R. Keim, J. Schoonman, Solid State Ionics 13 (1984) 191–203.
[76] C. Rongeat, M. Anji Reddy, R. Witter, M. Fichtner, ACS Appl. Mater. Interfaces 6 (2014) 2103–2110.
[77] J. Schoonman, J. Electrochem. Soc. 123 (1976) 1772.
[78] J. Schoonman, A. Wolfert, J. Electrochem. Soc. 128 (1981) 1522–1523.
[79] J. Schoonman, A. Wolfert, Solid State Ionics 3-4 (1981) 373–379.
[80] A.A. Potanin, Russ. Chem. J. 45 (2001) 61–66.
[81] W. Baukal, R. Knodler, W. Kuhn, Chem. Ingenieur Technik 50 (1978) 245–249.
[82] C.M. Weiss, S.C. Jones, A. Tiruvannamalai, I. Darolles, M.M. Alam, S. Hossain, US20110143219A1.
[83] I. Darolles, C.A. Weiss, M.M. Alam, A. Tiruvannamalai, S.C. Jones, US20120164541A1.
[84] F. Gschwind, Z. Zao-Kargera, M. Fichtner, J. Mater. Chem. A 2 (2014) 1214–1218.

Chapter 4

Fluorosulfates and Fluorophosphates As New Cathode Materials for Lithium Ion Battery

Christian M. Julien,[1] Alain Mauger[1,2] and Henri Groult[1]
[1]*Sorbonne Universités, UPMC Univ., Laboratoire PHENIX, CNRS, Paris, France;* [2]*UPMC Univ. Paris 06, Institut de Minéralogie et Physique de la Matière Condensée, Paris, France*

Chapter Outline

4.1 Introduction	77
4.2 General Considerations	79
4.2.1 Average Voltage of Li Insertion Cathodes	79
4.2.2 The Inductive Effect	80
4.3 Fluorophosphates	81
4.3.1 Fluorine-Doped $LiFePO_4$	81
4.3.2 $LiVPO_4F$	83
4.3.3 $LiMPO_4F$ (M=Fe, Ti)	87
4.3.4 Li_2FePO_4F (M=Fe, Co, Ni)	88
4.3.5 Li_2MPO_4F (M=Co, Ni)	89
4.3.6 $Na_3V_2(PO_4)_2F_3$ Hybrid Ion Cathode	90
4.3.7 Other Fluorophosphates	91
4.4 Fluorosulfates	92
4.4.1 $LiFeSO_4F$	92
4.4.2 $LiMSO_4F$ (M=Co, Ni, Mn)	95
4.5 Concluding Remarks	96
References	97

4.1 INTRODUCTION

Following the discovery of the high mobility of lithium ions in transition metal oxides and the first commercialization in the early 1990s of the lithium–cobalt oxide/graphite cell, Li-ion batteries (LIBs) have been widely developed for various scaling applications. LIBs are used in portable electronic devices, ~58% of sale on mobile phones, ~32% on notebook computers, and ~7% on camera. Presently, their application has also been extended for green transportation (hybrid electric vehicle) and grid storage systems at a sustainable level. Considering the principle of LIBs based on insertion reaction, a source of lithium is requested. In classical Li-ion cell, the Li^+ ions are supplied by the positive electrode (usually named cathode), typically a layered transition metal oxide or a spinel or an olivine framework, while the negative electrode (usually named anode) is a carbonaceous material that accepts lithium up to the reduced LiC_6 form in the full charge state of the battery. Chronologically, one can mention oxides such as $LiCoO_2$ (LCO), $LiMn_2O_4$ (LMO), $LiNi_{0.8}Co_{0.15}Al_{0.05}O_2$ (NCA), $LiNi_{1/3}Mn_{1/3}Co_{1/3}O_2$ (NMC), and $LiNi_{0.5}Mn_{1.5}O_4$ (LNM) as positive electrodes for Li-ion cells [1].

During the past decade, numerous studies have been devoted to replace oxides by materials with a polyanion-based framework that are considered as safe alternatives for the traditional oxide electrodes. For instance, $M_x(SO_4)_y$ sulfate-based and $M_x(PO_4)_y$ phosphate-based compounds (M is a transition metal ion) that house interstitial Li^+ ions such as $LiFe_2(SO_4)_3$ [2], $LiFePO_4$ [3], $Li_3V_2(PO_4)_3$ [4], $Li_{2.5}V_2(PO_4)_3$ [5], $LiVOPO_4$ [6,7], and $LiVP_2O_7$ [8,9] have been all considered as thermally stable. All these materials usually exhibit excellent stability on long-term cycling compared to lithium metal oxides and essentially no release of oxygen from the lattice or reactivity with the electrolyte. However, the materials were found to be poor electronic conductors [3].

The search of new cathode materials aiming to maintain a good mix of properties, with focus on electrochemical and safety parameters, has resulted in the improvement of the electrochemical performance using two strategies: (1) substitution of fluorine for oxygen or (2) fluorine coating of the active particles. As a result, fluorinated compounds display several advantages such as high-voltage redox reactions, stabilization of the host lattice, protection of the electrode particle surface from HF attack and electrolyte decomposition that impedes a side reaction, and easy transport of mobile Li^+ ions [10–15]. Accordingly, anion substitution is expected as an effective way to enhance the electrochemical performance of spinel and layered compounds, especially for NMC materials [16] due to the strong electronegativity of the F^- anion, which will make the structure more stable. Among the metal fluorides as surface fluorination (coating) agents of oxide-based cathode particles, the most popular are ZrF_x [17], AlF_3 [18], CaF_2 [19], and LiF [20]. While these materials are treated in the Chapters 1 and 3, attention hereunder is focused on technological developments of fluorophosphates and fluorosulfates [21–23]. The present chapter gives the state of the art in the understanding of the properties of these F-containing materials. Owing to the progress in this field, these compounds are promising active cathode elements for the next generation of LIBs to improve the technology of energy storage and electric transportation. This chapter is organized as follows. The preliminary considerations are dedicated to a brief recall of the energetic properties of these compounds. The second section is devoted to the structural and electrochemical properties of fluorophosphate materials. Then the fluorosulfates are treated in the Section 3. Finally, some concluding remarks are given with emphasis on the quality that ensures the reliability and the optimum electrochemical performance of these materials.

4.2 GENERAL CONSIDERATIONS

4.2.1 Average Voltage of Li Insertion Cathodes

The open-circuit voltage for a Li insertion reaction of a polyanionic compound $LiM(XO_4)$, where M is a transition metal cation and X is a counter anion, is given by [24],

$$V = -\frac{\mu_{Li}^{Cathode} - \mu_{Li}^{Anode}}{zF}, \quad (4.1)$$

where z is the number of electrons transferred and F is Faraday's constant. For the case of Li^+ ions being the charge carrier, z is unity. The first term in the numerator (the chemical potential of Li in the cathode) is a function of the Li content. However, the average voltage for the complete delithiation, which is the constant voltage that results in the same total capacity, is obtained from the difference between the Gibbs free energy of the lithiated $LiM(XO_4)$ and that of delithiated $M(XO_4)$ states. In that case, the average voltage is given by,

$$\overline{V} = -\frac{1}{F}\left(\frac{\partial \Delta G(x)}{\partial x}\right)_{T,P} = -\frac{1}{F}\left(\frac{\partial \lfloor G_{LiM(XO_4)} - G_{M(XO_4)} \rfloor}{\partial x}\right)_{T,P}. \quad (4.2)$$

As the average voltage of an intercalation compound is directly related to the energies of the end states (charged and discharged), the voltage depends on the structure of the cathode element. Several computer simulation techniques have been employed to address the voltage, diffusion, and nanostructural properties of cathode materials [25,26]. Hubbard-corrected Density Functional Theory (DFT+U) methods have also been used recently to investigate the structural and electronic properties of both tavorite and triplite $LiFeSO_4F$ polymorphs. Islam and Fisher [25] have pointed out that the difference in voltage is mainly due to the difference in the stabilities of the delithiated states $FeSO_4F$, which can be rationalized in terms of the Fe^{3+}–Fe^{3+} repulsion in the edge-sharing geometry of the triplite structure.

Considering the crystal field concept developed by Goodenough [27], the redox potential of an insertion electrode material is governed by the ionocovalency of the M–X bond: more ionic the bonds, higher the potential. Thus, the redox potential strongly depends on the electronegative ion: substitution of fluorine F^- for oxygen O^{2-} results in higher potential, while the opposite is observed for sulfur. As an example, the redox potential of the $LiFePO_4F$ tavorite phase is 750 mV higher than that of the $LiFePO_4$ olivine phase, because the ionicity in Fe–F bonds is larger than that of Fe–O bonds. A computational investigation on fluorinated polyanionic compounds has demonstrated that the more ionic M–F bond and the resulting stabilization of the energy of the antibonding $3d$ orbitals of the transition metal ion by fluorine substitution for oxygen is a general law acting on the electrochemical properties [28].

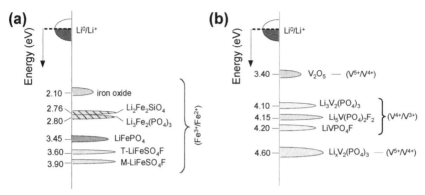

FIGURE 4.1 Energy of the redox couples of iron (a) and vanadium (b) phosphate frameworks relative to the Fermi level of metallic lithium.

4.2.2 The Inductive Effect

Another aspect of tuning the redox potential of an electrode material has been demonstrated by Goodenough et al. [3,29]. They have shown that the use of polyanions $(XO_4)^{n-}$ such as $(SO_4)^{2-}$, $(PO_4)^{3-}$, $(AsO_4)^{3-}$, or even $(WO_4)^{2-}$ lowers $3d$ metals' redox energy to useful levels compared to the Fermi level of the Li anode. Thus, the most attractive key point of the polyanion frameworks can be seen in the strong X–O covalency, which results in a decrease of the Fe–O covalency. This inductive effect is responsible for a decrease of the redox potential in comparison to the oxides [29,30]. The polyanion PO_4^{3-} unit stabilizes the olivine structure of $LiFePO_4$ and lowers the Fermi level of the $Fe^{2+/3+}$ redox couple through the Fe–O–P inductive effect which results in a higher potential for the olivine material. The discharge voltage 3.45 V is almost 650 mV higher than that of $Li_3Fe_2(PO_4)_3$ [3]. It is also 350 mV higher than that of $Fe_2(SO_4)_3$ [2], which is consistent with the stronger Bronsted acidity of sulfuric vs phosphoric acid. In the case of Li_2FeSiO_4, the lower electronegativity of Si vs P results in a lowering of the $Fe^{2+/3+}$ redox couple [31]. On the other hand, the higher thermal stability of the phosphoolivines and their lower tendency to release oxygen is explained by the strong X–O covalency and the rigid $(XO_4)^{n-}$ units decreasing the safety risks. However, $AMXO_4$ compounds and $AM(XO_4)_3$ as well (A is an alkali ion) exhibit a very low electronic conductivity because of the separation between MO_6 octahedra and XO_4 tetrahedra that induces a large polarization effect during charge–discharge reaction [32]. Figure 4.1 illustrates the changes in redox energies relative to the Fermi level of Li for the $Fe^{2+/3+}$ and $V^{n+/(n+1)+}$ couples. For instance, the electrochemical insertion properties of lithium vanadium fluorophosphate, $LiVPO_4F$, indicate that the V^{3+}/V^{4+} redox couple in $LiVPO_4F$ is located at a potential around 0.3 V higher than in the lithium vanadium phosphate, $Li_3V_2(PO_4)_3$ [32]. This property characterizes the impact of structural fluorine on the inductive effect of the PO_4^{3-} polyanion. The electrochemical characteristics of various lithiated compounds with polyanionic framework are listed in Table 4.1 [32–36].

TABLE 4.1 Electrochemical Characteristics of Lithiated Compounds With Polyanionic Framework

Compounds Lithiated State	Compounds Delithiated State	Redox Potential (V)	Capacity (mAh g^{-1})	Ref.
LiVPO$_4$F	VPO$_4$F	4.2	115	[32]
Li$_2$VPO$_4$F	LiVPO$_4$F	1.8	130	[33]
Li$_2$FePO$_4$F	LiFePO$_4$F	2.9	288	[34]
Li$_{1+x}$TiPO$_4$F	LiTiPO$_4$F	1.7	145	[35]
LiFeSO$_4$F	FeSO$_4$F	3.6	140	[35]
LiNiSO$_4$F	NiSO$_4$F	5.4 (?)	142	[36]

Most of these compounds crystallize in a structure similar to tavorite LiFe(PO$_4$)(OH). Using high-throughput density functional theory calculations, Mueller et al. [37] have evaluated tavorite-structured oxyphosphates, fluorophosphates, oxysulfates, and fluorosulfates. The activation energies for lithium diffusion through the tavorite frameworks of LiVO(PO$_4$), LiV(PO$_4$)F, and Li$_2$V(SO$_4$)F showed that these materials are capable of reversibly inserting two lithium ions per redox active metal at very high-rates.

4.3 FLUOROPHOSPHATES

Numerous inorganic material factors including the nature of the transition metal ion and its number of valence states affect the structural and transport properties. For instance, lithium fluorophosphates glasses formed by P$_2$O$_5$ glass former and LiF glass modifier with addition of dopant are well known to exhibit enhancement of their conductivity and decrease of their glass-transition temperature without altering the phosphate network [38,39]. The crystallized materials with the framework-structured Li-containing fluorophosphates of 3d metals, described by general formulas LiMPO$_4$F and Li$_2M$PO$_4$F, have been explored as perspective high-voltage cathode materials for rechargeable lithium batteries [22]. The crystallographic parameters of LiMPO$_4$F (M=Fe, Co, Ni) compounds are listed in Table 4.2.

4.3.1 Fluorine-Doped LiFePO$_4$

With the goal of increasing the voltage of the redox reaction, Liao et al. [44] were among the first groups who investigated the effects of fluorine substitution on electrochemical behavior of LiFePO$_4$/C composite electrode material. LiFe(PO$_4$)$_{1-x}$F$_{3x}$/C (x=0.01, 0.05, 0.1, 0.2) was synthesized by a solid-state carbothermal reduction (CTR) route at 650 °C using NH$_4$F as dopant. F-doped LiFePO$_4$/C nanoparticles were prepared either via a low temperature

TABLE 4.2 Crystallographic Parameters of LiMPO$_4$F (M = Fe, Co, Ni) Compounds

Compound	Space Group	a (Å)	b (Å)	c (Å)	α (°)	β (°)	γ (°)	V (Å^3)	Ref.
VPO$_4$F	C2/c	7.1553(2)	7.1014(1)	7.1160(2)	90	118.089(1)	90	319.00(8)	[35]
LiVPO$_4$F	P$\bar{1}$	5.1730(8)	5.3090(6)	7.2500(3)	72.479(4)	107.767(7)	84.375(7)	174.35(0)	[32]
LiFePO$_4$F	P$\bar{1}$	5.15510(3)	5.3044(3)	7.2612(4)	107.357(5)	107.855(6)	98.618(5)	1732.91(2)	[40]
LiTiPO$_4$F	P$\bar{1}$	5.1991(2)	5.3139(2)	7.2428(3)	106.975(3)	108.262(4)	97.655(4)	176.10(2)	[40]
Li$_2$VPO$_4$F	C2/c	7.2255(1)	7.9450(1)	7.3075(1)	90	116.771(1)	90	374.53(7)	[37]
Li$_2$FePO$_4$F	P$\bar{1}$	5.3746(3)	7.4437(3)	5.3256(4)	109.038(2)	94.423(6)	108.259(9)	189.03(4)	[34]
Li$_2$FePO$_4$F	Pbcn	5.0550(2)	13.5610(2)	11.0520(3)	90	90	90	757.62(1)	[41]
Li$_2$CoPO$_4$F	Pnma	10.4520(2)	6.3911(8)	10.8740(2)	90	90	90	726.40(3)	[42]
Li$_2$NiPO$_4$F	Pnma	10.4730(3)	6.2887(8)	10.8460(1)	90	90	90	714.33(2)	[43]

hydrothermal reaction followed by high temperature treatment at 750°C for 5 h under Ar atmosphere [45,46] or via sol–gel process using LiF [47]. Nanostructured C–LiFePO$_{3.98}$F$_{0.02}$ composite was synthesized by an aqueous precipitation of precursor material in molten stearic acid [47]. The excessively F-substituted LiFe(PO$_4$)$_{0.9}$F$_{0.3}$/C composite showed the most attractive high-rate performance and cycling life at high temperatures ($T > 50$°C) due to the fact that the more ionic Fe–F bond stabilizes the energy of the antibonding 3d orbitals of the transition metal ion. However, two proposals about the fluorine ion occupy were given: the first one suggested that 3F$^-$ ions replace PO$_4^{3-}$ group as a whole [44,45] and the second one declared that F$^-$ could only be replaced at the oxygen sites [46,47]. Considering the second proposition, there are three nonequivalent randomly occupied O sites, i.e., the O(1), O(2), and O(3) site, in the crystalline elementary cell of LiFePO$_4$. As compared with the pure olivine material, its discharge capacity at 10C rate was 110 mAh g^{-1} with a flat discharge voltage plateau of 3.3–3.0 V vs Li$^+$/Li. Lu et al. [46] reported that the length of Li–O bonds increased and those of P–O bonds decreased due to F doping. This implies that Li$^+$ diffusion between the lithiated phase and the delithiated phase could be improved, due to the F doping that weakens the Li–O bonds. Fluorine ions preferably occupy specific O(2) oxygen sites. It seems that fluorine doping leads to additional electron density on the lithium sites indicating the formation of Fe$_{Li}$+ − Li$_{Fe}$− antisite pairs. Such fluorine doping closes the gap in the electronic structure, which results in a finite density of states at the Fermi level. Enhanced high-rate performance and improved cycle stability of F-doped olivine phosphate was also reported in more recent studies [46,47]. C-LiFePO$_{3.98}$F$_{0.02}$ composite delivered the capacity of 164 mAh g^{-1} at C/10 rate. The first principles computational results indicate that electronic properties, the lithium insertion voltage, and general electrochemical behavior are very sensitive to the placement of fluorine ions in the structure of this compound [48]. It was suggested that LiVSiO$_4$F and Li$_{0.5}$FePO$_{3.5}$F$_{0.5}$ lithium deinsertion causes a too large M–F distance (indicative of M–F bond breaking), being the predicted lithium insertion voltage about 0.3 V lower than that of the parent compound.

4.3.2 LiVPO$_4$F

The vanadium-containing phosphate polyanion LiVPO$_4$F has been initially proposed and described in detail by Barker et al. [32,49–59] as a novel 4-V positive electrode material for LIBs. Initial testing of LiVPO$_4$F has shown that this material is substantially safer than the traditional oxide materials, and therefore should also be considered as a cathode replacement for the current generation of Li-ion cells [32]. LiVPO$_4$F is isostructural with the native mineral tavorite, which belongs to the lithium-bearing pegmatite family with favorite LiFe^{3+}(PO$_4$)•(OH), amblygonite (Li,Na)AlPO$_4$•(F,OH), and montebrasite LiAlPO$_4$•(F,OH) polymorphs. The tavorite phase crystallizes in a triclinic structure (P$\bar{1}$ space group) and its framework consists of V^{3+}O$_4$F$_2$ octahedra linked by fluorine vertices forming $(V^{3+}O_4F_2)_\infty$ chains along the c-axis (Figure 4.2). The connection of these chains with corner-sharing PO_4 tetrahedra form an open three-dimensional lattice with wide tunnels along the a, b, and c directions that accommodate the Li-ions in two sites, i.e., Li(1) is five coordinated with low occupancy (~18%) and Li(2) adopts a six coordination with high occupancy (~82%) on the $2i$ Wyckoff positions [60–63].

In the early work of Barker et al. [32] the synthesis of $LiVPO_4F$ was made by a two-step approach involving the initial carbothermal preparation of a VPO_4 intermediate using V_2O_5, $NH_4H_2PO_4$, and particulate carbon (super P grade) precursors, followed by a simple LiF incorporation process. The single

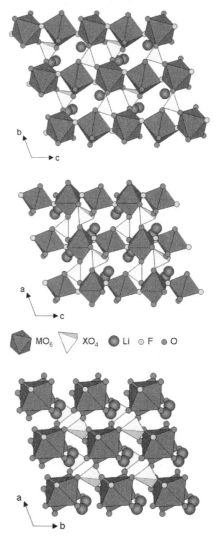

FIGURE 4.2 Schematic view of the tavorite structure ($P\bar{1}$ space group) along the a-, b-, and c-crystallographic directions. *Adapted from [37]*.

tavorite phase displays lattice parameters such as $a = 5.1730\,(8)$ Å, $b = 5.3090(6)$ Å, $c = 7.2500(3)$ Å, $\alpha = 72.479(4)°$, $\beta = 107.767(7)°$, $\gamma = 81.375(7)°$, and $V = 174.35(0)$ Å^3. Preliminary electrochemical studies of the LiVPO$_4$F cathode material demonstrate a reversible lithium extraction/insertion reaction based on the V$^{3+/4+}$ redox couple that indicates: (1) a discharge potential centered at around 4.19 V vs Li^0/Li$^+$, (2) a two-phase reaction mechanism coupled to phase nucleation behavior, and (3) a reversible specific capacity around 115 mAh g^{-1}, a performance roughly equivalent to cycling of x = 0.74 in Li$_{1-x}$VPO$_4$F. In Figure 4.3(a), we present the charge–discharge profile of LiVPO$_4$F//Li cell cycled at different C-rates with 1 mol L^{-1} LiPF$_6$ solution in ethylene carbonate (EC)-diethyl carbonate (DEC) with EC:DEC (1:1) as electrolyte. The first charge curve (red) is shown for comparison. The variation of the derivative capacity ($-dQ/dV$) as a function of the voltage during extraction (charge) and insertion (discharge) of Li$^+$ ions into the LiVPO$_4$F host is shown in Figure 4.3(b). In the charge profile, the two distinct peaks with a potential separation ~60 mV correspond to the V$^{3+/4+}$ redox couple with the two Li(1) and Li(2) site occupancies. Typical thermal response of the fully delithiated Li$_{1-x}$VPO$_4$F phase obtained by differential scanning calorimetry shows that, with a heat flow of ~205 J g^{-1}, the safety characteristics of LiVPO$_4$F are vastly superior to those of known oxide cathode materials (for instance, 345 J g^{-1} for LiMn$_2$O$_4$) [58].

Taking advantage of the multivalency of vanadium ions, an additional lithium insertion reaction at around 1.8 V vs Li^0/Li$^+$ was reported to be associated with the V$^{3+/2+}$ redox couple [49,56,57]. LiVPO$_4$F has two redox electric potentials based on the V$^{4+/3+}$ (LiVPO$_4$F→VPO$_4$F reaction) and V$^{3+/2+}$ (LiVPO$_4$F→Li$_2$VPO$_4$F reaction) couples, which offers the possibility of using this material for the cathode as well as the anode, as shown in the diagram presented in Figure 4.4. Ellis et al. [40] have investigated the end phases combining

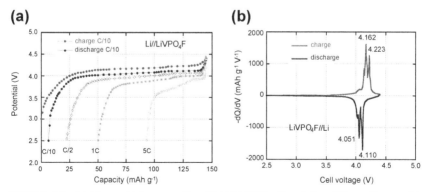

FIGURE 4.3 (a) Charge–discharge profile of LiVPO$_4$F//Li cell cycled at different C-rates with 1 mol L^{-1} LiPF$_6$ solution in EC-DEC (1:1) as electrolyte. The first charge curve (red) is shown for comparison. (b) Variation of the derivative capacity ($-dQ/dV$) as a function of the voltage during extraction (charge) and insertion (discharge) of Li$^+$ ions into the LiVPO$_4$F host. (For interpretation of the references to color in this figure legend, the reader is referred to the online version of this)

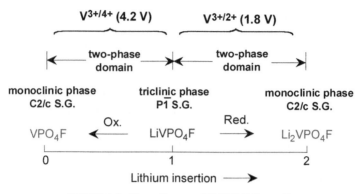

FIGURE 4.4 Phase diagram of Li_xVPO_4F ($0 \leq x \leq 2$).

X-ray and neutron diffraction for VPO_4F and Li_2VPO_4F prepared by chemical oxidation and reduction of the parent compound. The crystal chemistry of these end phases is listed in Table 4.2. The delithiated phase VPO_4F adopts a monoclinic structure (*C2/c* space group) in which corner-shared ($V^{4+}O_4F_2$) octahedral chains are interconnected by PO_4 tetrahedra. The lithiated Li_2VPO_4F phase displays the same structural change to monoclinic symmetry (*C2/c* space group) in which lithium ions occupy two sites, Li(1) and Li(2), filled with equal probability [64]. The redox activity between the two $LiVPO_4F$–Li_2VPO_4F compositions is very facile and occurs with an 8% change in volume and displays a stable specific capacity of 145 mAh g^{-1} [34]. More recently, Ellis et al. [40] and Plashnitsa et al. [65] and have utilized $LiVPO_4F$ as both cathode and anode for fabrication of a symmetric Li-ion $LiVPO_4F$//$LiVPO_4F$ cell with a nonflammable ionic liquid $LiBF_4$–$EMIBF_4$ (1-ethyl-3-methyl-imdazolium tetrafluoroborate) electrolyte. This symmetrical cell displays a potential window of 2.4 V with a reversible specific capacity of 130 mAh g^{-1} and has shown to be stable and safe at high temperature up to 80 °C.

The incorporation of aluminum into the $LiV_{1-y}Al_yPO_4F$ framework prepared by the two-step CTR method has generated some interesting properties: (1) an almost linear decrease of the discharge capacity on the $V^{3+/4+}$ redox couple with Al substitution, (2) a lower polarizability, (3) a gradual upshift in the $V^{3+/4+}$ redox peak of 90 mV, and (4) the relatively constant ratio of the two charge plateaus for $0 \leq y \leq 0.25$ [59].

The reactivity of $LiVPO_4F$ charged at C/5 rate with 1 mol L^{-1} $LiPF_6$ solution in EC:DEC electrolyte was tested by accelerating rate calorimetry at high temperature. The self-heating rate (dT/dt) of the CTR-synthesized $LiVPO_4F$ reacting with electrolyte is lower or about the same as that of $LiFePO_4$ olivine over the entire test temperature range 50–350 °C for powders of identical surface area (15 m^2 g^{-1}) [66]. Ma et al. [67] have investigated the effects of oxidation on the structure and performance of $LiVPO_4F$ as a cathode material. Two two-phase structural evolutions were detected upon Li$^+$ electrochemical extraction at average potentials at 4.26 and 4.30 V corresponding to the continuous transformation of LiVPO$_4$F→Li$_{0.72}$VPO$_4$F→VPO$_4$F in the first charge process, while the discharge process occurs without the appearance of the intermediate phase. Changes in the Li environment and ion mobility in the tavorite LiVPO$_4$F lattice have been observed by multinuclear solid-state $^{6/7}$Li and ^{31}P nuclear magnetic resonance spectroscopies on chemically and electrochemically delithiated samples [68]. The ionic conductivity determined from electrochemical impedance spectroscopy of tavorite-structured LiVPO$_4$F is $\sigma_t = 0.6 \times 10^{-7}$ S cm^{-1} at room temperature with the activation energy of $E_a = 0.75$ eV determined from the Arrhenius plot [69]. These values differ from those of LiFePO$_4$F $\sigma_t = 7 \times 10^{-11}$ S cm^{-1} and $E_a = 0.99$ eV reported by Recham et al. [35].

Various vanadium-based fluorophosphates were evaluated as cathode material for LIBs [35,70–73]. Wang et al. [71] reported the electrochemical performance of LiVPO$_4$F/C composite prepared by heating a precursor, obtained through ball milling with slurry of H$_2$C$_2$O$_4$•2H$_2$O as reacting agent and carbon source, NH$_4$H$_2$PO$_4$, NH$_4$VO$_3$, and LiF. This material shows specific discharge capacities of 151 and 102 mAh g^{-1} at 0.1C and 10C, respectively, in the voltage range of 3.0–4.4 V and displays a discharge coulombic efficiency 90.4% at 10C rate after 50 cycles. Reddy et al. [73] have examined the long-term behavior in the potential range 3.0–4.5 V at 0.92C-rate of LiVPO$_4$F prepared by CTR method. The capacity degrades slowly over 800–1260 cycles and the total loss is ~14%.

4.3.3 LiMPO$_4$F (M = Fe, Ti)

Following the discovery of Barker et al. [46], who described the preparation of LiM$_{1-y}$M$'_y$PO$_4$F materials (M and M' are transition metals having a +3 oxidation state) several attempts were made to substitute transition metal ions of the first row for vanadium such as Fe and Ti. DFT calculations using plane-wave methods were performed for Li$_2$MPO$_4$F, LiMPO$_4$F, and MPO$_4$F (M = V, Mn, Fe, Co, Ni) to address their feasibility as high-voltage cathode materials (>3.5 V relative to Li metal) for LIBs [74]. Average open-circuit voltages of 4.9, 5.2, and 5.3 V were calculated for Mn, C, and Ni, respectively.

LiFePO$_4$F synthesized by CRT method or by a hydrogen reduction reaction crystallizes in the triclinic structure (P$\bar{1}$ space group) with lattice parameters $a = 5.1528$ Å, $b = 5.3031$ Å, $c = 7.4966$ Å, $\alpha = 67.001°$, $\beta = 67.164°$, $\gamma = 81.512°$, and cell volume $V = 173.79$ Å^3. LiCrPO$_4$F prepared by the same methods crystallizes with the P$\bar{1}$ space group with lattice parameters $a = 4.996$ Å, $b = 5.307$ Å, $c = 6.923$ Å, $\alpha = 77.600°$, $\beta = 100.71°$, $\gamma = 78.546°$ and cell volume $V = 164.54$ Å^3. Recham et al. [35] experienced the growth of crystalline LiFePO$_4$F by either solid-state or ionothermal techniques. For the solid-state synthesis, the mixture of Li$_3$PO$_4$ and FeF$_3$ was heated at 700 °C for 24 h inside a platinum tube. For the ionothermal method, similar mixture stabilized with an ionic liquid and triflate produced nanostructured particles (~20 nm). This material shows a reversible specific capacity of ~145 mAh g^{-1} with the average potential at 2.8 V corresponding to the $Fe^{3+} \rightarrow Fe^{2+}$ redox couples. However, LiFePO$_4$F cannot be oxidized, which limits its use in LIBs [21]. The phase transition and electrochemistry of the LiFePO$_4$F–Li$_2$FePO$_4$F have been studied by Ramesh et al. [34]. The fully lithium-inserted phase, Li$_2$FePO$_4$F, adopts a triclinic tavorite-type structure that is closely related to the parent phase (see Table 4.2). Despite the presence of two crystallographic distinct Fe sites (1a and 1c Wyckoff positions) in the P$\bar{1}$ lattice, the electrochemical features of LiFePO$_4$F→Li$_2$FePO$_4$F show an overall potential at 3 V with a reversible capacity of 0.96 Li corresponding to a capacity of 145 mAh g^{-1}.

Ti-based LiMPO$_4$F is another tavorite structure built by distorted TiO$_4$F$_2$ octahedra linked by fluorine ions, which was synthesized by either solid-state or ionothermal techniques [35]. Single-phase LiTiPO$_4$F powders prepared at low temperature (260 °C) have lattice parameters $a = 5.1991$ Å, $b = 5.3139$ Å, $c = 7.2428$ Å, $\alpha = 106.975°$, $\beta = 108.262°$, $\gamma = 97.655°$, and cell volume $V = 176.10$ Å^3. Li extraction/insertion reaction from/into the Li$_{1+x}$TiPO$_4$F framework occurs in the range $-0.5 \leq x \leq 0.5$ with the appearance of two pseudoplateaus centered at 2.9 and 1.7 V corresponding to the Ti^{3+}→Ti^{4+} and Ti^{3+}→Ti^{2+} redox couples, respectively. However, due to the sensitivity of electrochemical features upon mild modifications of synthesis, Barpanda and Tarascon [21] conclude that LiTiPO$_4$F is a poor cathode material for battery applications.

4.3.4 Li$_2$FePO$_4$F (M = Fe, Co, Ni)

Fluorophosphates of general formula A_2MPO$_4$F (A = Li, Na and M = Fe, Mn, Co, Ni) crystallize in three structure types, which differ in the connectivity of (MO$_4$F$_2$) octahedra: face shared (Na$_2$FePO$_4$F), edge shared (Li$_2M$PO$_4$F, M = Co, Ni), and corner-shared (Na$_2$MnPO$_4$F) [42,75–83]. The Li$_2M$PO$_4$F (M = Fe, Co, Mn, Ni) crystallizes in three different structures types; triclinic (tavorite) and two-dimensional orthorhombic (*Pbcn* space group) and tunnel-like monoclinic (*P*2$_1$/*n* space group) [75]. In their prior work, Ellis et al. [41] have shown that A_2FePO$_4$F (A = Na, Li) could serve as a cathode in Li-ion cells. This compound possesses facile two-dimensional pathways for Li$^+$ motion, and the structural modifications on redox reactions are minimal with a volume change of only 3.7% that contributes to the absence of distinct two-phase behavior. Single-phase Li$_2$FePO$_4$F was prepared by ion exchange of Na$_2$FePO$_4$F in 1 mol L^{-1} LiBr acetonitrile solution. The material LiNaFePO$_4$F (Li/Na ratio 1:1) was obtained by reducing NaFePO$_4$F with LiI in acetonitrile for 6 h. Li$_2$FePO$_4$F crystallizes in the orthorhombic structure (*Pbcn* space group) with lattice parameters $a = 5.055$ Å, $b = 13.561$ Å, $c = 11.0526$ Å, $\beta = 90°$, and cell volume $V = 757.62$ Å^3. The open-circuit voltage is lower than the LiFePO$_4$ olivine (3.0 V vs 3.45 V); 80% of the theoretical capacity (135 mAh g^{-1}) is attainable on the first oxidation cycle. Recently, the same group [33] reported the crystal structure and electrochemical properties of Li$_2M$PO$_4$F fluorophosphates (M = Fe, Mn, Co, Ni) synthesized by solid-state and hydrothermal synthetic routes. New forms of lithium fluorophosphate cathode materials have been reported such as Li_2FePO_4F obtained by cycling the orthorhombic $NaLiFePO_4F$ phase (*Pnma* space group) in an electrochemical lithium cell [75], single-phase sub-stoichiometric nanocrystalline $Li_{1-x}Fe_{1-y}M_yPO_4$ (M=Fe, Co) [76], and $Li_{2-x}Na_xFe[PO_4]F$ tavorite structure synthesized by ion exchange using LiBr in ethanol [84].

4.3.5 Li_2MPO_4F (M = Co, Ni)

Several groups reported on high-voltage electrochemical performance of Li_2CoPO_4F and Li_2NiPO_4F [77–86]. Both $LiCoPO_4$ and Li_2CoPO_4F (isostructural with Li_2NiPO_4F) crystallize in the orthorhombic system (space group *Pnma*, Z=8). Nevertheless, there are remarkable differences between the structures from a crystallographic point of view. $LiCoPO_4$ has CoO_6 octahedra, LiO_6 octahedra, and PO_4 tetrahedra (Figure 4.2). In contrast, Li_2CoPO_4F has Co_4F_2 octahedra instead of CoO_6 octahedra. In addition, Li_2CoPO_4F has two kinds of Li sites, $4c$ and $8d$ [43]. It was confirmed that Li_2CoPO_4F is a new class of 5-V cathode material similar to $LiCoPO_4$ [13,42,79]. Dumont Botto et al. [86] have pointed out that, contrary to the Na phases which are quite simple to obtain, the synthesis of Li_2MPO_4F remains difficult and requires either the ion exchange of the Na counterparts or a lengthy solid-state reaction (at least 10-h heat treatment) [83]. In the search for an unconventional way to prepare Li_2CoPO_4F, a shorter reaction down to nine min was done by spark plasma sintering, which favors the formation of submicrometric particles (0.8 μm).

A considerable theoretical upper limit of approximately 310 mAh g^{-1} is expected for Li_2CoPO_4F and Li_2NiPO_4F. The theoretical estimation of the intercalation voltage of ~4.9 V for Li_2CoPO_4F cathode [78] is in good agreement with the voltage plateau observed at *ca.* 5 V [42]. A fault of both Li_2CoPO_4F and lithiated cobalt phosphate is a high irreversible capacity (especially in the first cycles), which is related to decomposition of electrolyte at high anodic potentials. Experimentally, $LiNiPO_4F$ discharge voltage is demonstrated to be close to 5.3 V [78]. Khasanova et al. [79] have investigated the electrochemical performance and structural properties of the high-voltage cathode material Li_2CoPO_4F. The cyclic voltammetry and coulometry under potential step mode in the voltage range 3.0–5.1 V vs Li revealed a structural transformation at potentials above 4.8 V. This transformation occurring upon Li extraction appears to be irreversible: the subsequent Li insertion does not result in restoration of the initial structure, but takes place within a new "modified" framework. According to the structure refinement, this modification involves the mutual rotations of (CoO_4F_2) octahedra and (PO_4) tetrahedra accompanied by the considerable unit cell expansion, which is expected to enhance the Li transport upon subsequent cycling. The new framework demonstrates a reversible Li insertion/extraction in a solid solution regime with stabilized discharge capacity at around 60 mAh g^{-1}. The Li_2CoPO_4F is prepared by a two-step solid-state method, followed by the application of wet coating containing various amounts of ZrO_2. Among the samples, the 5 wt.%-ZrO_2-coated

Li$_2$CoPO$_4$F material shows the best performance with an initial discharge capacity of up to 144 mAh g^{-1} within the voltage range of 2–5.2 V vs Li at current density 10 mA g^{-1} [83]. X-ray diffraction (XRD) patterns of Li$_2$CoPO$_4$F electrodes at different stages of cycling during the first charge–discharge cycle were investigated by Wang et al. [83]. Electrodes discharged from 5.0 V maintain the same trend as the fresh electrodes; however, a slight difference indicates that the structural relaxation of the framework of Li$_2$CoPO$_4$F occurs at a voltage greater than 5.0 V, which is consistent with the cyclic voltammetry (CV) measurements.

4.3.6 Na$_3$V$_2$(PO$_4$)$_2$F$_3$ Hybrid Ion Cathode

The sodium vanadium fluorophosphate materials have demonstrated reversible Li-ion insertion behavior [87,88]. The Na$_3$V$_2$(PO$_4$)$_2$F$_3$ phase prepared using a solid-state CTR approach involving the precursors VPO$_4$ and NaF crystallizes in the tetragonal space group $P4_2/mnm$ with lattice parameters $a = 0.0378(3)$ Å, $c = 10.7482(4)$ Å, and $V = 877.94(6)$ Å^3. The framework structure is best described in terms of (V$_2$O$_8$F$_3$) bi-octahedra and (PO$_4$) tetrahedral. The bi-octahedra are linked by one of the fluorine atoms, whereas the oxygen atoms are all interconnected through the PO$_4$ units. Electrochemical properties of Li$_x$Na$_{3-x}$V$_2$(PO$_4$)$_2$F$_3$ carried out in a metallic Li half-cell in the potential range 3.0–4.6 V vs Li reveal that mobile Na is rapidly exchanged for Li (Figure 4.5). The voltage response corresponded to the reversible cycling of two alkali ions per formula unit. The associated specific capacity was around 120 mAh g^{-1}, at an average discharge voltage of around 4.1 V vs Li metal anode [87]. The differential capacity data for the cell cycled at C/20 rate for charge and discharge is shown in Figure 4.5(a). Thus, when fully charged to 4.6 V vs Li, a cathode composition approximating to NaV$_2$(PO$_4$)$_2$F$_3$ is produced, a condition in which all the vanadium has been oxidized to V^{4+}. Such a hybrid ion

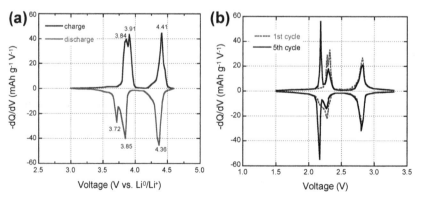

FIGURE 4.5 Differential capacity (dQ/dV) curves of (a) Li//Na$_3$V$_2$(PO$_4$)$_2$F$_3$ cell and (b) the first and fifth cycle for a Li$_{4/3}$Ti$_{5/3}$O$_4$//Na$_3$V$_2$(PO$_4$)$_2$F$_3$ hybrid ion cell. The electrolyte was composed of 1 mol L^{-1} LiPF$_6$ solution in ethylene carbonate/dimethyl carbonate (2:1 w.%).

cathode was used in LIB with $Li_{4/3}Ti_{5/3}O_4$ as Li source. The differential capacity ($-dQ/dV$) curves of the first and fifth cycle for a $Li_{4/3}Ti_{5/3}O_4//Na_3V_2(PO_4)_2F_3$ hybrid ion cell cycle are displayed in Figure 4.5(b). During the initial charge of the cell sodium ions are extracted from the $Na_3V_2(PO_4)_2F_3$ cathode. These results show that the $-dQ/dV$ response for the first cycle is broad and symmetrical, confirming the energetic reversibility of the cell chemistry with a capacity of ~120 mAh g^{-1}. However, sodium insertion into $Li_{4/3}Ti_{5/3}O_4$ is not favored. On subsequent cycles, the 2.27/2.23 V redox peaks shift by 100 mV to lower potential due to the Li$^+$/Na$^+$ ionic exchange that is now characterized by sharp redox peaks. Thus, change to a predominate lithium insertion mechanism occurred at the cathode [88]. Variation of the Li/V ratio in $Li_xNa_{3-x}V_2(PO_4)_2F_3$ as a function of cycle number showing the Li–Na exchange process is presented in Figure 4.6 As a result, the almost entire Li/Na exchange was completed after 10 cycles. Similar results for Na_2FePO_4F have been reported by Ellis et al. [31]. By ion exchange reaction, the entire Na content in Na_2FePO_4F may be replaced with Li to yield Li_2FePO_4F, which exhibits a slightly higher redox potential than the parent compound because of the more electronegative nature of Li compared to Na [43]. Later on, the same authors have shown that $LiVPO_4F$//graphite Li-ion cell delivered a capacity 130 mAh g^{-1} and an average discharge voltage of 4.06 V. Long-term cycling at C/5 rate displays a coulombic efficiency around 90% after 500 cycles [48].

4.3.7 Other Fluorophosphates

To follow the initial investigation of Barker et al., several compounds of the same structure have been proposed. Park et al. [89] have prepared the

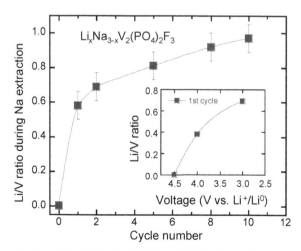

FIGURE 4.6 Variation of the Li/V ratio in $Li_xNa_{3-x}V_2(PO_4)_2F_3$ as a function of cycle number showing the Li–Na exchange process. Insert presents the change during the first cycle.

$Li_{1.1}Na_{0.4}VPO_{4.8}F_{0.7}$ phase from the pseudolayered structure $Na_{1.5}VPO_5F_{0.5}$ using an ion exchange process between Na^+ and Li^+ in 1-hexanol at its boiling point (160 °C) under reflux with LiBr as the lithium source. The lattice is made up of VO_5F octahedral and PO_4 tetrahedra units, where two VO_5F are linked by a bridging fluorine ion to form $V_2O_{10}F$ bi-octahedron. These units are repeatedly connected in the ab plane via PO_4 units sharing oxygen to construct an open framework having a layerlike spacing into which Na^+/Li^+ ions are inserted [90]. This compound shows reversible extraction and insertion of ~1.1 Li^+ ions at an ideal 4 V vs Li^0/Li^+ to provide a capacity of 156 mAh g^{-1} with coulombic efficiency 98% after 100 cycles at 60 °C. Among fluorophosphates used as electrode materials, the layered structure $Li_5M(PO_4)_2F_2$ with M = V, Cr operates as 4-V cathodes for LIBs [91,92], but the capacity is lower than 100 mAh g^{-1}.

4.4 FLUOROSULFATES

Recently, the concept of inductive effect have been applied to replace $(PO_4)^{3-}$ for $(SO_4)^{2-}$ in polyanionic cathode materials [35,36,93–119]. Table 4.3 summarizes the structural properties of some fluorosulfate compounds. The fluorosulfates LiMSO$_4$F constitute a wide family showing a good mix of properties, especially, both electrochemical and safety issues. However, we notice that the electroactive compounds appeared only in 2010 after the synthesis of the newest member of the tavorite family LiFe^{2+}(SO$_4$)F [36]. For instance, a simple substitution in Nasicon $Li_xM_3(XO_4)_3$ networks increases the redox potential by 800 mV independently of the 3d transition metal ion [2]. Recent review by Rousse and Tarascon [119] deals with the crystal chemistry and structural–electrochemical relationship of new fluorosulfate polyanionic LiMSO$_4$F electrode materials. The lithiated fluorosulfates present an interesting family from the view point of crystal chemistry with the three main types of structure that depend on the nature of the transition metal ion: tavorite (M=Fe), triplite (M=Mn), or sillimanite (M=Zn). Figure 4.7 shows the difference between the tavorite (P$\bar{1}$ space group) and triplite (P$\bar{1}$ space group) structure.

4.4.1 LiFeSO$_4$F

Single-phase tavorite LiFeSO$_4$F cannot be prepared by typical solid-sate methods because its low thermodynamic stability imposes that the crystallization must be effected at low (<400 °C) temperatures in hydrophobic ionic liquids [36,100]. Tripathi et al. [99] reported the easy synthesis of LiFeSO$_4$F by reaction in hydrophobic tetraethylene glycol at 220 °C to give a highly electrochemically active material. Reversible Li insertion is easy because of the close structural similarity of the lattice with that of the tavorite-type monoclinic FeSO$_4$F ($C2_1/c$ space group). LiFeSO$_4$F can be readily oxidized to produce the empty host FeSO$_4$F but the volume contraction (from 182.4 to 164.0 Å^3) is greater than that of the olivine LiFePO$_4$ framework [99]. The temperature dependence of

TABLE 4.3 Crystallographic Parameters of LiMSO$_4$F (M = Fe, Co, Ni) Compounds

Compound	Space Group	a (Å)	b (Å)	c (Å)	α (°)	β (°)	γ (°)	V (Å^3)	Ref.
LiFeSO$_4$F	P$\bar{1}$	5.1747(3)	5.4943(3)	7.2224(3)	106.522(3)	107.210(3)	97.791(3)	182.559(16)	[36]
LiFeSO$_4$F	C2/c	13.0238(6)	6.3957(3)	9.8341(5)	90	119.68(5)	90	711.64(1)	[107]
LiNiSO$_4$F	P$\bar{1}$	5.1430(6)	5.3232(7)	7.1404(7)	106.802(9)	107.512(8)	98.395(6)	172.56(4)	[100]
LiCoSO$_4$F	P$\bar{1}$	5.1721(7)	5.4219(7)	7.1842(8)	106.859(6)	107.788(6)	97.986(5)	177.80(4)	[100]
LiMnSO$_4$F	C2/c	13.2701(5)	6.4162(2)	10.0393(4)	90	120.586(2)	90	735.85(5)	[102]
LiMgSO$_4$F	P$\bar{1}$	5.1623(7)	5.388(1)	7.073(1)	106.68(1)	107.40(1)	97.50(1)	174.72(5)	[93]
LiZnSO$_4$F	Pnma	7.4035(7)	6.3299(5)	7.4201(6)	90	90	90	347.74(0)	[96]

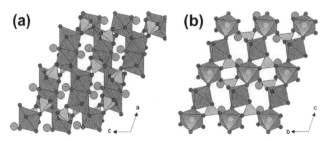

FIGURE 4.7 Schematic representation of the crystal chemistry of the triplite (a) compared with the tavorite phase (b).

the magnetic susceptibility gives evidence of a transition to long-range antiferromagnetic order at $T_N = 100$ K in $FeSO_4F$, while such ordering appears at 25 K in $LiFeSO_4F$ [108]. This large increase of T_N upon delithiation is not only linked to the change of valence of the iron ions, since it is not observed in the case of $LiFePO_4$, for instance. Instead, it gives evidence of the shortening of the bonding path responsible for the superexchange interactions and is thus another evidence of the large contraction of the lattice.

An ionothermal route (soft chemistry) has been developed for the synthesis of fluorosulfates [35,100], in which the nucleation is facilitated by ionic liquids decomposing at temperatures ~300 °C. In such a synthesis route the $FeSO_4 \cdot H_2O$ monohydrate is employed because of the structural similarity with the $LiMgSO_4F$ tavorite [93]. The galvanostatic cycling of ionothermally synthesized $LiFeSO_4F$ at C/10 rate shows a reversible capacity 130 mAh g^{-1} involving the $Fe^{2+/3+}$ redox reaction at 3.6 V vs Li^0/Li^+ [35]. Note that the cell voltage is enhanced by 150 mV over $LiFePO_4$. Ultrarapid microwave synthesis of triplite $LiFeSO_4F$ has been performed by Tripathi et al. [120]. Using the solid-state reaction method, the triplite was obtained upon long annealing at 320 °C in water-containing autoclaves; after 6 days the complete tavorite → triplite conversion was done [110]. The Fe triplite intercalates Li at 3.9 V, which is 0.3 V higher than its ordered tavorite analog, attributable to longer Fe–O bond length [113]. The origin of the voltage difference between the tavorite and the triplite phase has been discussed by Ben-Yahia et al. [113] from results of DFT+U calculations. The voltage increase originates from the difference in the anionic networks of the two polymorphs, due to the change in the electrostatic repulsions induced by the configuration of the fluorine atoms around the transition metal cations. The potential difference between the two $LiFeSO_4F$ polymorphs is illustrated in Figure 4.8.

Recently, Ati et al. [110] have discussed the nucleation of the triplite phase of $LiFeSO_4F$ using X-ray diffraction and transmission electron microscopy (TEM) studies. Besides preparing triplite $LiFeSO_4F$ from dry precursors, it was found that this phase could be obtained from the tavorite via a heating process at 320 °C for few days. Sobkowiak et al. [117] demonstrated the dependence of the electrochemical features of the tavorite $LiFeSO_4F$ on the synthesis conditions.

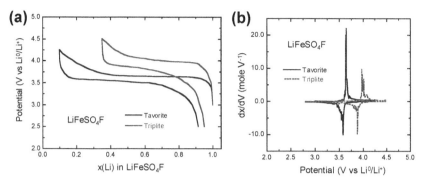

FIGURE 4.8 (a) Charge–discharge profiles and (b) derivative $-dx/dV$ for the tavorite and triplite LiFeSO$_4$F polymorphs.

The importance of the surface chemistry has been pointed out and the optimized cycling performance can be achieved by removing the unwanted residues and applying a conducting polymer coating such as poly(3,4-éthylènedioxythiophène) (PEDOT) film.

4.4.2 LiMSO$_4$F (M = Co, Ni, Mn)

The redox potentials of the fluorosulfates LiMSO$_4$F (M=Co, Ni, Mn) are expected to exhibit redox potentials of 4.25, 4.95, and 5.25 V, respectively. Barpanda et al. [100] succeeded in preparing the Li(Fe$_{1-x}M_x$)SO$_4$F solid solutions only when we used the corresponding monohydrate solid solutions as precursors, hence providing further evidence for the topotacticity of the synthesis reaction. None of these fluorosulfates was shown to present any electrochemical activity up to 5 V. Evidently, Co^{2+}/Co^{3+}, Ni^{2+}/Ni^{3+}, and Mn^{2+}/Mn^{3+} redox reactions do not occur within the explored galvanostatic cycling potential window, contrary to the case of the Fe^{2+}/Fe^{3+} redox reaction. Subsequently, the electrochemical activity of pure LiMSO$_4$F phases were tested at different rates (C/2–C/10) by cycling up to 5 V using an aluminum plunger for the cathode. Therefore, even at such high-voltages prone to electrolyte oxidation/decomposition, the activity of the corresponding M^{2+}/M^{3+} redox couples was not triggered. Among the Co-based polyanionic insertion compounds, Li$_2$CoP$_2$O$_7$ [121] is considered as a 4.9-V cathode. This pyrophosphate crystallizes in the monoclinic structure ($P2_1/c$ space group), in which Li occupies five sites; two are tetrahedrally coordinated, one forms bipyramidal sites, and two Li share their occupancy with Co bipyramids. The material synthesized using a two-step solid-state method delivered a discharge capacity of c. 80 mAh g^{-1} at C/20 rate. Sillimanite-structured LiZnSO$_4$F was produced by low-temperature synthesis (<300 °C). The Li$^+$ ion conductivity has been enhanced by a monolayer of ionic liquid grafting [115]. On another hand, it has been pointed out that the LiMnSO$_4$F polymorph is electrochemically inactive [103]. However, the effect of Zn substitution in the $LiFe_{1-y}Zn_ySO_4F$ forms an electrochemically active solid solution. The Fe-rich phase is obtained in the triplite structure, while the Zn-rich phase crystallizes in the sillimanite structure. The redox potentials are observed at 3.6 and 3.9 V vs Li^0/Li^+ for $LiFe_{0.8}Zn_{0.2}SO_4F$ (sillimanite) and $LiFe_{0.9}Zn_{0.1}SO_4F$ (triplite), respectively. In the same way, while $LiMnSO_4F$ that crystallizes in the triplite phase is inactive, $LiFe_{1-x}Mn_xSO_4F$ can be electrochemically active. Fe-rich solid solution ($0 < x < 0.2$) tavorite polymorphs have a redox activity at 3.6 V close to that of $LiFeSO_4F$ and characteristic of the Fe^{2+}/Fe^{3+} redox potential in the tavorite structure, with a standard two-phase voltage–composition curve [102]. On another hand, Fe-rich $LiFe_{1-x}Mn_xSO_4F$ triplite solid solution polymorphs show a redox activity at 3.9 V. This value characteristic of the Fe^{2+}/Fe^{3+} redox potential in the triplite structure is large. Unfortunately, only the iron shows an activity, so that the capacity decreases linearly with x (120 mAh g^{-1} for x = 0.2 [21]).

4.5 CONCLUDING REMARKS

The combination of high electronegativity of F with the inductive effect of polyanion allows for the tuning of the redox potential of many fluorine-based polyanionic compounds in the electrolytic window, which makes them promising electrodes for LIBs. Among them $LiVPO_4F$ can deliver a capacity of 145–150 mAh g^{-1}, which is comparable to the capacity delivered by $LiFePO_4$, but the operating voltage (4.1 V) is larger than that of the olivine (3.45 V). The $LiVPO_4F//LiVPO_4F$ cell works at 2.4 V with a reversible capacity 130 mAh g^{-1} and is safe at 80 °C, which is comparable to $LiFePO_4//Li_4Ti_5O_{12}$ that also works at this temperature [122]. However, the performance of $LiVPO_4F$ at high C-rates is smaller than that of $LiFePO_4$, since the capacity reduces to 100 mAh g^{-1} at 10C in the best case (preparation by the CRT method). Even at 1C a capacity loss of 14% is observed between 800 and 1260 cycles, while that of $LiFePO_4$ is stable under such conditions. Among the other fluorophosphates that we have reviewed, the best results are obtained with $LiFePO_4F$, which, however, cannot be oxidized. Again, the $LiFePO_4F \rightarrow Li_2FePO_4F$ reaction gives a reversible capacity of 145 mAh g^{-1}, but it takes place at an overall potential at 3 V only. An ideal 4-V battery is obtained with $Li_{1.1}Na_{0.4}VPO_{4.8}F_{0.7}$, providing a capacity of 156 mAh g^{-1} with coulombic efficiency 98% after 100 cycles at 60 °C. The corresponding energy density is thus better than that of $LiFePO_4$ owing to the larger operating voltage. Again, however, the advantage of $LiFePO_4$ will be its high-rate performance that allows for higher power densities, and the cycling life that exceeds 30,000 cycles at room temperature [122].

Among the fluorosulfates, $LiFeSO_4F$ in the triplite phase is promising, owing to the high Fe^{2+}/Fe^{3+} redox potential at 3.9 V, but the kinetics is very sluggish, and only a small fraction of the lithium has been extracted so far. The voltage is decreased to 3.6 V in the tavorite phase, but then the capacity is raised to 130 mAh g^{-1}, and this smaller capacity with respect to $LiFePO_4$ is compensated by the larger voltage. The capacity of LiFeSO$_4$F tavorite, however, decreases at C-rate lager than 1C, and this cathode has not seen to operate at 10C rate so far.

At the present time, fluorophosphates and fluorosulfates cannot compete in energy density with lamellar compounds, and cannot compete in power density and cycling life with C-LiFePO$_4$. It should be noted, however, that these materials have been studied for LIB applications only recently, while it took about 15 years to optimize C-LiFePO$_4$ and bring it to the position of winning a part of the market. In addition, the structural relationship between the precursor and the final product underlying the topotactic reaction makes possible the synthesis of many fluorophosphates and sulfates, and many of them have still to be discovered. Therefore, the research in this field will be very active in the years to come, justified by the fact that this family of materials is promising for applications in electrochemical energy storage.

REFERENCES

[1] C.M. Julien, A. Mauger, K. Zaghib, H. Groult, Inorganics 2 (2014) 132–154.
[2] A. Manthiram, J.B. Goodenough, J. Power Sources 26 (1989) 403–408.
[3] A.K. Padhi, K.S. Nanjundaswamy, J.B. Goodenough, J. Electrochem. Soc. 144 (1997) 1188–1194.
[4] M.Y. Saidi, J. Barker, H. Huang, J.L. Swoyer, G. Adamson, Electrochem. Solid-State Lett. 5 (2002) A149–A151.
[5] C. Yin, H. Grondey, P. Strobel, L.F. Nazar, Chem. Mater. 16 (2004) 1456–1465.
[6] B.M. Azmi, T. Ishihara, H. Nishiguchi, Y. Takita, J. Power Sources 146 (2005) 525–528.
[7] J. Gaubicher, T. Le Mercier, Y. Chabre, J. Angenault, M. Quarton, J. Electrochem. Soc. 146 (1999) 4375–4379.
[8] G. Rousse, C. Wurm, M. Morcrette, J. Rodriguez-Carvajal, J. Gaubicher, C. Masquelier, Int. J. Inorg. Mater. 3 (2001) 881–887.
[9] J. Barker, R.K.B. Gover, P. Burns, A. Bryan, Electrochem. Solid-State Lett. 8 (2005) A446–A448.
[10] G.-H. Kim, S.-T. Myung, H.J. Bang, J. Prakash, Y.-K. Sun, Electrochem. Solid State Lett. 7 (2004) A477–A480.
[11] J.T. Son, H.G. Kim, J. Power Sources 147 (2005) 220–226.
[12] Q. Luo, T. Muraliganth, A. Manthiram, Solid State Ionics 180 (2009) 703–707.
[13] K.R. Stroukoff, A. Manthiram, J. Mater. Chem. 21 (2011) 10165–10170.
[14] P. Yue, Z. Wang, H. Guo, X. Xiong, X. Li, Electrochim. Acta 92 (2013) 1–8.
[15] P. Yue, Z. Wang, X. Li, X. Xiong, J. Wang, X. Wu, H. Guo, Electrochim. Acta 95 (2013) 112–118.
[16] J.W. Fergus, J. Power Sources 195 (2010) 939–954.
[17] S.H. Yun, K.S. Park, Y.J. Park, J. Power Sources 195 (2010) 6108–6115.
[18] B.C. Park, H.B. Kim, S.T. Myung, K. Amine, I. Belharouak, S.M. Lee, Y.K. Sun, J. Power Sources 178 (2008) 826–831.
[19] K. Xu, Z. Jie, R. Li, Z. Chen, S. Wu, J. Gu, J. Chen, Electrochim. Acta 60 (2012) 130–133.
[20] S.J. Shi, J.P. Tu, Y.Y. Tang, Y.Q. Zhang, X.Y. Liu, X.L. Wang, C.D. Gu, J. Power Sources 225 (2013) 338–346.
[21] P. Barpanda, J.M. Tarascon, Fluorine-based polyanionic compounds for high-voltage electrode materials, in: B. Scrosati, W. Van Schalkwijk, K.M. Abraham, J. Hassoun (Eds.), Lithium Batteries: Advanced Technologies and Applications, John Wiley & Sons, 2013 (Chapter. 7).

[22] C.M. Julien, A. Mauger, Ionics 19 (2013) 951–988.
[23] M. Hu, X. Pang, Z. Zhou, J. Power Sources 237 (2013) 229–242.
[24] J.B. Goodenough, Solid State Ionics 69 (1994) 184–198.
[25] M.S. Islam, C.A.J. Fisher, Chem. Soc. Rev. 43 (2013) 185–204.
[26] M. Saubanère, M. Ben-Yahia, F. Lemoigno, M.-L. Doublet, ECS Meet. Abstr. MA2013-02 (2013) 840.
[27] J.B. Goodenough, Oxide cathodes, in: W. van Schalkwijk, B. Scrosati (Eds.), Advances in Lithium-ion Batteries, Kluwer Academic/Plenum, New York, 2002. Chap. 4.
[28] M.E. Arroyo y de Dompablo, U. Amador, J.M. Tarascon, J. Power Sources 174 (2007) 1251–1257.
[29] K.S. Nanjundaswamy, A.K. Padhi, J.B. Goodenough, S. Okada, H. Ohtsuka, H. Arai, J. Yamaki, Solid State Ionics 92 (1996) 1–10.
[30] A.K. Pahdi, M. Manivannan, J.B. Goodenough, J. Electrochem. Soc. 145 (1998) 1518–1520.
[31] A. Nyten, A. Abouimrane, M. Armand, T. Gustafsson, J.O. Thomas, Electrochem. Commun. 7 (2005) 156–160.
[32] J. Barker, M.Y. Saidi, J.L. Swoyer, J. Electrochem. Soc. 150 (2003) A1394–A1398.
[33] B.L. Ellis, W.R.M. Makahnouk, W.N. Rowan-Weetaluktuk, D.H. Ryan, L.F. Nazar, Chem. Mater. 22 (2010) 1059–1070.
[34] T.N. Ramesh, K.T. Lee, B.L. Ellis, L.F. Nazar, Electrochem. Solid State Lett. 13 (2010) A43–A47.
[35] N. Recham, L. Dupont, M. Courty, K. Djellab, D. Larcher, M. Armand, J.-M. Tarascon, Chem. Mater. 22 (2009) 1142–1148.
[36] N. Recham, J.-N. Chotard, L. Dupont, C. Delacourt, W. Walker, M. Armand, J.-M. Tarascon, Nat. Mater. 9 (2010) 68–74.
[37] T. Mueller, G. Hautier, A. Jain, G. Ceder, Chem. Mater. 23 (2011) 3854–3862.
[38] B.V.R. Chowdari, K.F. Mok, J.M. Xie, R. Gopalakrishnan, Solid State Ionics 76 (1995) 189–198.
[39] B. Sreedhar, M. Sairam, D.K. Chattopadhyay, K. Kojima, Mater. Chem. Phys. 92 (2005) 492–498.
[40] B.L. Ellis, T.N. Ramesh, L.J.M. Davis, G.R. Govard, L.F. Nazar, Chem. Mater. 23 (2011) 5138–5148.
[41] B.L. Ellis, W.R.M. Makahnouk, Y. Makimura, K. Toghill, L.F. Nazar, Nat. Mater. 6 (2007) 749–753.
[42] S. Okada, M. Ueno, Y. Uebou, J.I. Yamaki, J. Power Source 146 (2005) 565–569.
[43] M. Dutreilh, C. Chevalier, M. El-Ghozzi, D. Avignant, J.M. Montel, J. Solid State Chem. 142 (1999) 1–5.
[44] X.Z. Liao, Y.S. He, Z.F. Ma, X.M. Zhang, L. Wang, J. Power Sources 174 (2007) 720–725.
[45] M. Pan, X. Lin, Z. Zhou, J. Solid State Electrochem 16 (2011) 1615–1621.
[46] F. Lu, Y. Zhou, J. Liu, Y. Pan, Electrochim. Acta 56 (2011) 8833–8838.
[47] F. Pan, W. Wang, J. Solid State Electrochem 16 (2011) 1423–1427.
[48] M. Milovic, D. Jugovic, N. Cvjeticanin, D. Uskokovic, A.S. Milosevic, Z.S. Popovic, F.R. Vukajlovic, J. Power Sources 241 (2013) 70–79.
[49] Barker, J. Saidi, M.Y. and Swoyer, J.L. International Patent WO01/084,655 (2001).
[50] Barker, J. Saidi, M.Y. and Swoyer, J.L. US Patent 6387568 B1 (2002).
[51] J. Barker, M.Y. Saidi, J.L. Swoyer, Electrochem. Solid-State Lett. 6 (2003) A1–A4.
[52] J. Barker, M.Y. Saidi, J.L. Swoyer, J. Electrochem. Soc. 151 (2004) A1670–A1677.
[53] Barker, J. US Patent 6890686 B1 (2005).

[54] J. Barker, R.K.B. Gover, P. Burns, A.J. Bryan, Electrochem. Solid-State Lett. 8 (2005) A285–A287.
[55] J. Barker, R.K.B. Gover, P. Burns, A. Bryan, M.Y. Saidi, J.L. Swoyer, J. Electrochem. Soc. 152 (2005) A1776–A1779.
[56] Barker, J. Saidi, M.Y. and Swoyer, J.L. US Patent 6855462 B2 (2005).
[57] J. Barker, R.K.B. Gover, P. Burns, A. Bryan, M.Y. Saidi, J.L. Swoyer, J. Power Sources 146 (2005) 516–520.
[58] R.K.B. Gover, P. Burns, A. Bryan, M.Y. Saidi, J.L. Swoyer, J. Barker, Solid State Ionics 177 (2006) 2635–2638.
[59] J. Barker, M.Y. Saidi, R.K.B. Gover, P. Burns, A. Bryan, LiVPO$_4$F, J. Power Sources 174 (2007) 927–931.
[60] M.L. Lindberg, W.T. Pecora, Tavorite and barbosalite, Am. Mineral. 40 (1955) 952–966.
[61] A.C. Roberts, P.J. Dunn, J.D. Grice, D.E. Newbury, E. Dale, W.L. Roberts, Powder Diffr. 3 (1988) 93–95.
[62] L.A. Groat, M. Raudseep, F.C. Hawthorne, T.S. Ercit, B.L. Sherriff, J.S. Hartman, Am. Mineral. 75 (1990) 992–1008.
[63] J.L. Pizarro-Sanz, J.M. Dance, G. Villeneuve, M.L. Arriortuz-Marcaida, Mater. Lett. 18 (1994) 327–330.
[64] L.J.M. Davis, B.L. Ellis, T.N. Ramesh, L.F. Nazar, A.D. Bain, G.R. Govard, J. Phys. Chem. C 115 (2011) 22603–22608.
[65] L.S. Plashnitsa, E. Kobayashi, S. Okada, J.I. Yamaki, Electrochim. Acta 56 (2011) 1344–1351.
[66] F. Zhou, X. Zhao, J.R. Dahn, Electrochem. Commun. 11 (2011) 589–591.
[67] R. Ma, L. Shao, K. Wu, M. Shui, D. Wang, N. Long, Y. Ren, J. Shu, J. Power Sources 248 (2014) 874–885.
[68] L.J. Davis, L.S. Cahill, L.F. Nazar, G.R. Goward, ECS Meet. Abstr. MA-2010-01 (2010) 626.
[69] M. Prabu, M.V. Reddy, S. Selvasekarapandian, G.V. Subba Rao, B.V.R. Chowdari, Electrochim. Acta 85 (2012) 572–578.
[70] Mukainakano, Y. Ishii, R. and Shiozaki, R. Eur. Patent EP 2573046 A1 (2013).
[71] J.X. Wang, Z.X. Wang, L. Shen, X.H. Li, H.J. Guo, W.J. Tang, Z.G. Zhu, Trans. Nonferrous Met. Soc. China 23 (2013) 1718–1722.
[72] Q.-M. Zhang, Z.-C. Shi, Y.-X. Li, D. Gao, G.-H. Chen, Y. Yang, Acta Phys. Chim. Sin. 27 (2011) 267–274.
[73] M.V. Reddy, G.V. Subba-Rao, B.V.R. Chowdari, J. Power Sources 195 (2010) 5768–5774.
[74] J. Yu, K.M. Rosso, J.-G. Zhang, J. Liu, J. Mater. Chem. 21 (2011) 12054–12058.
[75] N.R. Khasanova, O.A. Drozhzhin, D.A. Storozhilova, C. Delmas, E.V. Antipov, Chem. Mater. 24 (2012) 4271–4273.
[76] S.P. Badi, T.N. Ramesh, B. Ellis, K.T. Lee, L.F. Nazar, ECS Meet. Abstr. MA2009-02 (2009) 397.
[77] S. Okada, M. Ueno, Y. Uebou, J.I. Yamaki, in: IMLB-12, 2004, p. 301. Abstracts.
[78] M. Nagahama, N. Hasegawa, S. Okada, J. Electrochem. Soc. 157 (2010) A748–A752.
[79] N.R. Khasanova, A.N. Gavrilov, E.V. Antipov, K.G. Bramnik, H. Hibst, J. Power Sources 196 (2011) 355–360.
[80] X. Wu, Z. Gong, S. Tan, Y. Yang, J. Power Sources 220 (2012) 122–129.
[81] N.V. Kosova, E.T. Devyatkina, A.B. Slobodyuk, Solid State Ionics 225 (2012) 570–574.
[82] K. Karthikeyan, S. Amaresh, K.J. Kim, S.H. Kim, K.Y. Chung, B.W. Cho, Y.S. Lee, Nanoscale 5 (2013) 5958–5964.

[83] S. Amaresh, K. Karthikeyan, K.J. Kim, M.C. Kim, K.Y. Chung, B.W. Cho, Y.S. Lee, J. Power Sources 244 (2013) 395–402.
[84] H. Ben-Yahia, M. Shikano, S. Koike, H. Sakaebe, M. Tabuchi, H. Kobayashi, J. Power Sources 244 (2013) 87–93.
[85] D. Wang, J. Xiao, W. Xu, Z. Nie, C. Wang, G. Graff, J.-G. Zhang, J. Power Sources 196 (2011) 2241–2245.
[86] E. Dumont-Botto, C. Bourbon, S. Patoux, P. Rozier, M. Dolle, J. Power Sources 196 (2011) 2274–2278.
[87] R.K.B. Gover, A. Bryan, P. Burns, J. Barker, Solid State Ionics 177 (2006) 1495–1500.
[88] J. Barker, R.K.B. Gover, P. Burns, A.J. Bryan, J. Electrochem. Soc. 154 (2007) A882–A887.
[89] Y.-U. Park, D.-H. Seo, B. Kim, K.-P. Hong, H. Kim, S. Lee, R.A. Shakoor, K. Miyasaka, J.-M. Tarascon, K. Kang, Sci. Rep. 2 (2012) 704–711.
[90] F. Sauvage, E. Quarez, J.M. Tarascon, E. Baudrin, Solid State Sci. 8 (2006) 1215–1221.
[91] S.C. Yin, R. Edwards, N. Taylor, P.S. Herle, L.F. Nazar, Chem. Mater. 18 (2006) 1745–1752.
[92] Y. Makimura, L.S. Cahill, Y. Iriyama, G.R. Goward, L.F. Nazar, Chem. Mater. 20 (2008) 4240–4248.
[93] L. Sebastian, J. Gopalakrishnan, Y. Piffard, J. Mater. Chem. 12 (2002) 374–377.
[94] M. Ati, M.T. Sougrati, N. Recham, P. Barpanda, J.-B. Leriche, M. Courty, M. Armand, J.-C. Jumas, J.M. Tarascon, J. Electrochem. Soc. 157 (2010) A1007–A1015.
[95] M. Ati, W.T. Walker, K. Djellab, M. Armand, N. Recham, J.-M. Tarascon, Electrochem, Solid-State. Lett. 13 (2010) A150–A153.
[96] P. Barpanda, J.-N. Chotard, C. Delacourt, M. Reynaud, Y. Filinchuk, M. Armand, M. Deschamps, J.-M. Tarascon, Angew. Chem. Int. Ed. 50 (2010) 2526–2531.
[97] P. Barpanda, J.-N. Chotard, N. Recham, C. Delacourt, M. Ati, L. Dupont, M. Armand, J.-M. Tarascon, Inorg. Chem. 49 (2010) 7401–7413.
[98] R. Tripathi, T.N. Ramesh, B.L. Ellis, L.F. Nazar, Angew. Chem. Int. Ed. 49 (2010) 8738–8742.
[99] R. Tripathi, T.N. Ramesh, B.L. Ellis, L.F. Nazar, Angew. Chem. Int Ed. 122 (2010) 8920–8924.
[100] P. Barpanda, N. Recham, J.N. Chotard, K. Djellab, W. Walker, M. Armand, J.M. Tarascon, J. Mater. Chem. 20 (2010) 1659–1668.
[101] C. Frayret, A. Villesuzanne, N. Spaldin, E. Bousquet, J.-N. Chotard, N. Recham, J.-M. Tarascon, Phys. Chem. Chem. Phys. 12 (2010) 15512–15522.
[102] P. Barpanda, M. Ati, B.C. Melot, G. Rousse, J.-N. Chotard, M.-L. Doublet, M.T. Sougrati, S.A. Corr, J.-C. Jumas, J-M. Tarascon, Nat. Mater. 10 (2011) 772–779.
[103] M. Ati, B.C. Melot, G. Rousse, J.-N. Chotard, P. Barpanda, J.-M. Tarascon, Angew. Chem. Int. Ed. 50 (2011) 10574–10577.
[104] M. Ramzan, S. Lebegue, T.W. Kang, R. Ahuja, J. Phys. Chem. C 115 (2011) 2600–2603.
[105] L. Liu, B. Zhang, X.-J. Huang, Prog. Nat. Sci.-Mater. Int. 21 (2011) 211–215.
[106] R. Tripathi, G.R. Gardiner, M.S. Islam, L.F. Nazar, Chem. Mater. 23 (2011) 2278–2284.
[107] M. Ati, B.C. Melot, J.-N. Chotard, G. Rousse, M. Reynaud, J.-M. Tarascon, Electrochem. Commun. 13 (2011) 1280–1283.
[108] B.C. Melot, G. Rousse, J.-N. Chotard, M. Ati, J. Rodríguez-Carvajal, M.C. Kemei, J.-M. Tarascon, Chem. Mater. 23 (2011) 2922–2930.
[109] R. Tripathi, G. Popov, B.L. Ellis, A. Huq, L.F. Nazar, Energy Environ. Sci. 5 (2012) 6238–6246.
[110] M. Ati, M. Sathiya, S. Boulineau, M. Reynaud, A. Abakumov, G. Rousse, B. Melot, G. Van Tendeloo, J.-M. Tarascon, J. Am. Chem. Soc. 134 (2012) 18380–18387.

[111] M. Ati, M.-T. Sougrati, G. Rousse, N. Recham, M.-L. Doublet, J.-C. Jumas, J.-M. Tarascon, Chem. Mater. 24 (2012) 1472–1485.
[112] N. Recham, G. Rousse, M.T. Sougrati, J.-N. Chotard, C. Frayret, S. Mariyappan, B.C. Melot, J.-C. Jumas, J.-M. Tarascon, Chem. Mater. 24 (2012) 4363–4370.
[113] M. Ben-Yahia, F. Lemoigno, G. Rousse, F. Boucher, J.-M. Tarascon, M.-L. Doublet, Energy Environ. Sci. 5 (2012) 9584–9594.
[114] A.V. Radha, J.D. Furman, M. Ati, B.C. Melot, J.M. Tarascon, A. Navrotsky, J. Mater. Chem. 22 (2012) 2446–2452.
[115] P. Barpanda, R. Dedryvère, M.P. Deschamps, C. Delacourt, M. Reynaud, A. Yamada, J.-M. Tarascon, J. Solid State Electrochem. 16 (2012) 1743–1751.
[116] R. Tripathi, PhD Thesis, Univ. of Waterloo, Ontario, Canada (2013).
[117] A. Sobkowiak, M.R. Roberts, R. Younesi, T. Ericsson, L. Häggström, C.-W. Tai, A.M. Andersson, K. Edström, T. Gustafsson, F. Björefors, Chem. Mater. 25 (2013) 3020–3029.
[118] J. Dong, X. Yu, Y. Sun, L. Liu, X. Yang, X. Huang, J. Power Sources 244 (2013) 716–720.
[119] G. Rousse, J.M. Tarascon, Chem. Mater. 26 (2014) 394–406.
[120] R. Tripathi, G. Popov, X. Sun, D.H. Ryan, L.F. Nazar, J. Mater. Chem. A 1 (2013) 2990–2994.
[121] H. Kim, S. Lee, Y.U. Park, H. Kim, J. Kim, S. Jeon, K. Kang, Chem. Mater. 23 (2011) 3930–3937.
[122] K. Zaghib, M. Dontogny, A. Guerfy, J. Trottier, J. Hamel-Paquet, V. Gariepy, K. Galoutov, P. Hovington, A. Mauger, H. Groult, C.M. Julien, J. Power Sources 216 (2012) 192–200.

Chapter 5

Fluorohydrogenate Ionic Liquids, Liquid Crystals, and Plastic Crystals

Kazuhiko Matsumoto and Rika Hagiwara
Department of Fundamental Energy Science, Graduate School of Energy Science, Kyoto University, Kyoto, Japan

Chapter Outline

5.1 Introduction	103	5.4 Fluorohydrogenate Ionic Liquid Crystals	116
5.2 Structural Properties of Fluorohydrogenate Anions	104	5.5 Fluorohydrogenate Ionic Plastic Crystals	118
5.3 Fluorohydrogenate Ionic Liquids	106	References	121

5.1 INTRODUCTION

Fluorohydrogenate anions formulated as $(FH)_nF^-$ are formed by Lewis acid–base reactions of F^- and HF, generalized as Eqn (5.1):

$$F^- + n\text{HF} \to (FH)_nF^- \tag{5.1}$$

$(FH)_nF^-$ acts as a base to react with another HF to form a less basic fluorohydrogenate anion as formulated in Eqn (5.2):

$$(FH)_nF^- + \text{HF} \to (FH)_{n+1}F^- \tag{5.2}$$

Various cations are known as countercations to form $(FH)_nF^-$ salts, including alkali metal, alkali earth metal, and alkylammonium. Some of them are industrially important [1]: Na[FHF] is used as a fluorine source, bleaching agent, and additive for metal plating; $[NH_4][FHF]$ is used as an etchant of glass and metals; and $K[(FH)_nF]$ is especially important as an electrolyte for electric production of fluorine gas [2,3]. The KF-HF phase diagram was reported in 1934 and the liquidus line therein descends with an increase in the HF content through several eutectic points, as shown in Figure 5.1 [4]. At an integer

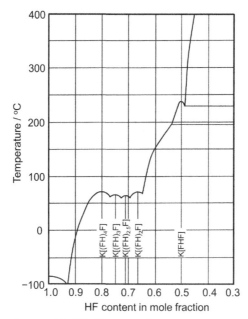

FIGURE 5.1 Phase diagram of the KF–HF system [4].

of n and at $n = 2.5$ in $(FH)_nF^-$, complex salts such as $K[(FH)_2F]$ are formed [5,6]. In a modern industrial plant, fluorine gas is electrochemically produced by electrolysis of the $K[(FH)_2F]$ molten salt at 363 K by considering the melting temperature and HF dissociation pressure. These fluorohydrogenate salts are generally prepared by reaction between binary fluoride and HF.

Organic fluorohydrogenate salts such as triethylammonium and pyridinium fluorohydrogenate have been known as fluorinating reagents [7,8]. Our recent studies revealed that many quaternary organic cations form low-melting salts with $(FH)_nF^-$. Some of them have extremely low melting temperatures and high ionic conductivities which surpass the values of popular ionic liquids. This chapter first describes the structural characteristics of $(FH)_nF^-$ which are important in understanding the unique properties of $(FH)_nF^-$ salts. Then, structures, properties, and applications of organic fluorohydrogenate ionic liquids, liquid crystals, and plastic crystals are described.

5.2 STRUCTURAL PROPERTIES OF FLUOROHYDROGENATE ANIONS

Structures of $(FH)_nF^-$ are shown in Figure 5.2. The $(FH)_nF^-$ anions are regarded as an ion where F^- is bound to one or more HF molecules. The H–F bond in $(FH)_nF^-$ is known as "strong hydrogen bond" [9] and has been a target of theoretical and computational studies [10-13] as well as X-ray diffractional and vibrational spectroscopic studies. The smallest fluorohydrogenate

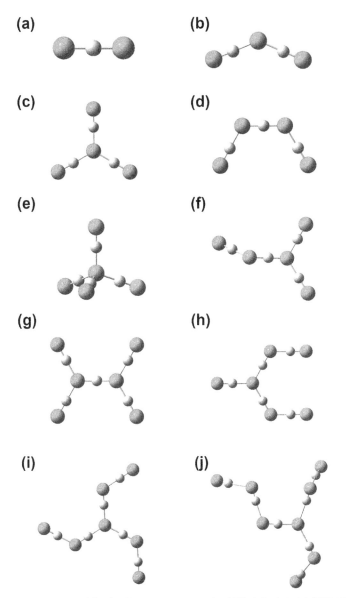

FIGURE 5.2 Structures of $(FH)_nF^-$. (a) FHF^- ($D_{\infty h}$), (b) $(FH)_2F^-$ (C_{2v}), (c) $(FH)_3F^-$ (D_{3h}), (d) $(FH)_3F^-$ (C_{2v}), (e) $(FH)_4F^-$ (T_d), (f) $(FH)_4F^-$ (C_{2v}), (g) $(FH)_5F^-$ (D_{2h}), (h) $(FH)_5F^-$ (C_2), (i) $(FH)_6F^-$ (C_{3h}), and (j) $(FH)_7F^-$ (C_1).

ion is FHF^- with $D_{\infty h}$ symmetry which is determined in a number of crystal structures (Figure 5.2 (a)) [5,14-22]. The F···F distance is slightly shorter than 2.3 Å and the H atom is located exactly at the center. As the number of HF molecules increases, the $(FH)_nF^-$ anion gives a variety of structures. The

$(FH)_2F^-$ anion has C_{2v} symmetry with two HF units bound to the central F atom (Figure 5.2 (b)) [12,23–28]. The F⋯F distance (~2.35 Å) is larger than that in FHF$^-$ and the position of the H atom is deviated from the middle of two F atoms to the terminal F atom. The $(FH)_nF^-$ anions with $n \geq 3$ have isomers; $(FH)_3F^-$ adopts C_{3v} (Figure 5.2 (c)) [12,24,26,27,29–31] and C_{2v} (Figure 5.2 (d)) [26,32] symmetries and $(FH)_4F^-$ adopts T_d (Figure 5.2 (e)) [6,28,30,31] and C_1 (Figure 5.2 (f)) [28] symmetries. Although $(FH)_4F^-$ with C_1 symmetry is known in the crystal structure, the stable structure in gas phase is considered to belong to C_s symmetry with the F⋯F⋯F⋯F dihedral angle of 90°. Two isomers of $(FH)_5F^-$ were determined in crystal structures. The D_{2h} isomer has a structure where four HF units are symmetrically bound to the central FHF$^-$ (Figure 5.2 (g)) [27,28]. The other isomer has a $(FH)F[(HF)(HF)]_2^-$ structure, where the terminal HF units in the two [(HF)(HF)] parts slightly deviate from the plane made by the other atoms to avoid steric repulsion, resulting in C_2 symmetry (Figure 5.2 (h)) [28]. Only the isomer known for $(FH)_6F^-$ has a shape like a three-bladed pinwheel with nearly C_{3h} symmetry (Figure 5.2 (i)) [28]. The largest $(FH)_nF^-$ determined in a crystal structure is $(FH)_7F^-$ which is best described as $[(FH)(FH)]_2F[(HF)(HF)(HF)]$ with C_1 symmetry [30].

In these ions, the F⋯F distance involving the terminal F atom tends to become longer from ~2.25 Å to ~2.50 Å as the number of HF molecules increases. According to quantum mechanical calculations, the $\Delta G°$ value of the reaction expressed by Eqn (5.2) at MP2/6-311 + G(d,p) decreases with an increase in n (-157.3 kJ mol^{-1} for $n = 0$, -84.2 kJ mol^{-1} for $n = 1$, -53.5 kJ mol^{-1} for $n = 2$, and -27.3 kJ mol^{-1} for $n = 3$) due to the decrease in fluorobasicity of fluorohydrogenate anion, suggesting that the dissociation of HF from $(FH)_nF^-$ becomes easier as n becomes larger.

The $(FH)_nF^-$ anions are best identified by infrared spectroscopy [12]. The FHF$^-$ anion gives infrared-active bands around 1250 cm^{-1} (Π_u mode) and 1500 cm^{-1} (Σ_u mode). The anions larger than FHF$^-$ give somewhat similar infrared spectra; 470, 1020, 1800, and 2500 cm^{-1} for $(FH)_2F^-$ and 1000 and 1800 cm^{-1} for $(FH)_3F^-$.

5.3 FLUOROHYDROGENATE IONIC LIQUIDS

Since the first report of a fluorohydrogenate ionic liquid (1-ethyl-3-methylimidazolium fluorohydrogenate, $([C_2C_1im][(FH)_{2.3}F])$) [33], various combinations of onium cations and $(FH)_nF^-$ have been attempted to obtain fluorohydrogenate ionic liquids [34-42]. The cations which form fluorohydrogenate ionic liquids are summarized in Figure 5.3. Fluorohydrogenate salts are synthesized by reaction of halide salts with a large excess of anhydrous HF. The following Eqn (5.3) is an example of this reaction for $[C_2C_1im][(FH)_nF]$:

$$[C_2C_1im][Cl] + (n+1)HF \rightarrow [C_2C_1im][(FH)_nF] + HCl \qquad (5.3)$$

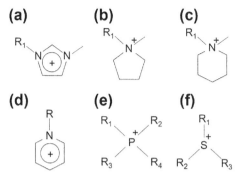

FIGURE 5.3 Structures of cations which form fluorohydrogenate ionic liquids. (a) 1-alkyl-3-methylimidazolium ($C_nC_1im^+$), (b) N-alkyl-N-methylpyrrolidinium ($C_nC_1pyrr^+$), (c) N-alkyl-N-methylpiperidinium ($C_nC_1pip^+$), (d) N-alkylpyridinium (C_npyr^+), (e) tetraalkylphosphonium (P_{nmlk}^+), and (f) trialkylsulfonium (S_{nml}^+), where C_n in $C_nC_1im^+$, $C_nC_1pyrr^+$, $C_nC_1pip^+$, and C_npyr^+ denotes alkyl chain with the carbon number of n and n, m, l, and k in P_{nmlk}^+ and S_{nml}^+ denote the carbon number of the alkyl chains on the P and S atoms, respectively.

Excess HF is thoroughly removed with the byproduct HCl under vacuum. This method gives products with low water contents (~100 ppm) as long as the starting materials are thoroughly dried. The vacuum-stable HF content at room temperature (n in $(FH)_nF^-$) differs depending on the cationic structure. In the cases of 1-alkyl-3-methylimidazolium, N-alkylpyridinium, and N-alkyl-N-methylpyrrolidinium, n is around 2.3 after removal of volatiles at room temperature, whereas noncyclic structure and long alkyl chain leads to $n < 2.3$. The n value is reversibly changed by removal of HF at elevated temperatures and addition of HF, corresponding to the formation of different $(FH)_nF^-$. In the phase diagram of $[C_2C_1im][(FH)_nF]$ ($1.0 \leq n \leq 2.3$), the liquidus line descends steeply from $n = 1.0$ to 1.8 and moderately in the range of $n \geq 2$, and there is a drop of the liquidus line around $n = 1.9$ [43]. Thoroughly vacuum-dried fluorohydrogenate ionic liquids do not etch Pyrex glass at ambient conditions.

Selected physical properties of the fluorohydrogenate ionic liquids are summarized in Table 5.1. The most prominent property of fluorohydrogenate ionic liquids is their high ionic conductivities. It is known that ionic liquids obey the Walden rule as in the cases of electrolytic solutions of bulky ions [44–46]. The Walden plot (molar conductivities against the viscosities) of selected ionic liquids including fluorohydrogenate ionic liquids is shown in Figure 5.4. The plot roughly exhibits linearity, suggesting that fluorohydrogenate ionic liquids do not have a special conduction mechanism such as proton hopping (see references [34,36,39,41,42,47–52] for the detailed physical property data). The highest ionic conductivity in this series is observed for $[S_{111}][(FH)_{1.9}F]$ (131 mS cm^{-1}, S_{111}^+ = trimethylsulfonium) [41]. Because the Walden rule connects viscosity not with ionic conductivity but with molar conductivity, both the low viscosity and large number of ions per volume for $[S_{111}][(FH)_{1.9}F]$ contribute to its high ionic conductivity. Although fluorohydrogenate ionic liquids

TABLE 5.1 Selected Physical Properties of Fluorohydrogenate Ionic Liquids[a]

Ion	MW	T_m, T_g, T_{cl}[b]/K	Density at 298 K/g cm^{-3}	Viscosity at 298 K /mPa s	Ionic Conductivity at 298 K/ mS cm^{-1}
[C$_1$C$_1$im][(FH)$_{2.3}$F]	162	272	1.17	5.1	110
[C$_2$C$_1$im][(FH)$_{2.3}$F]	176	208 (148 g)	1.13	4.9	100
[C$_{allyl}$C$_1$im][(FH)$_{2.3}$F]	188	–	1.11	5.5	90
[C$_3$C$_1$im][(FH)$_{2.3}$F]	190	(152 g)	1.11	7.0	61
[C$_4$C$_1$im][(FH)$_{2.3}$F]	204	(154 g)	1.08	19.6	33
[C$_5$C$_1$im][(FH)$_{2.3}$F]	218	(158 g)	1.05	26.7	27
[C$_6$C$_1$im][(FH)$_{2.3}$F]	232	(157 g)	1.00	25.8	16
[C$_8$C$_1$im][(FH)$_{2.1}$F]	256	248 (288 cl)	–	–	–
[C$_{10}$C$_1$im][(FH)$_{2.0}$F]	285	278 (314 cl)	–	–	–
[C$_{12}$C$_1$im][(FH)$_{2.0}$F]	315	295 (324 cl)	–	–	–
[C$_{14}$C$_1$im][(FH)$_{2.0}$F]	341	307 (376 cl)	–	–	–
[C$_{16}$C$_1$im][(FH)$_{2.0}$F]	369	327 (424 cl)	–	–	–
[C$_{18}$C$_1$im][(FH)$_{2.0}$F]	398	340 (463 cl)	–	–	–
[C$_2$C$_1$pyrr][(FH)$_{2.3}$F]	179	–	1.07	9.9	74.6
[C$_{allyl}$C$_1$pyrr][(FH)$_{2.3}$F]	191	–	1.05	8.5	78
[C$_3$C$_1$pyrr][(FH)$_{2.3}$F]	193	–	1.05	11.2	58.1
[C$_4$C$_1$pyrr][(FH)$_{2.3}$F]	207	–	1.04	14.5	35.9

TABLE 5.1 Selected Physical Properties of Fluorohydrogenate Ionic Liquids[a]—cont'd

Ion	MW	T_m, T_g, T_{cl}[b]/K	Density at 298 K/g cm^{-3}	Viscosity at 298 K /mPa s	Ionic Conductivity at 298 K/ mS cm^{-1}
[C$_6$C$_1$pyrr] [(FH)$_{2.3}$F]			0.993	18.0	23.7
[C$_{1O1}$C$_1$pyrr] [(FH)$_{2.2}$F]	193	<153	1.13	7.9	74.7
[C$_2$C$_1$pip] [(FH)$_{2.3}$F]	193	217, 237	1.07	24.2	37.2
[C$_{allyl}$C$_1$pip] [(FH)$_{2.3}$F]	205	–	1.06	16.7	63
[C$_3$C$_1$pip] [(FH)$_{2.3}$F]	207	(164 g)	1.06	33.0	23.9
[C$_4$C$_1$pip] [(FH)$_{2.3}$F]	221	(162 g)	1.04	37.1	12.3
[C$_2$pyr] [(FH)$_{2.3}$F]	173	(165 g)	1.13	10.1	71
[C$_{allyl}$pyr] [(FH)$_{2.3}$F]	185	203	1.12	5.3	82
[C$_3$pyr] [(FH)$_{2.3}$F]	187	–	1.11	12.3	48
[C$_4$pyr] [(FH)$_{2.3}$F]	201	–	1.09	29.7	25
[C$_{allyl}$C$_1$mor] [(FH)$_{2.3}$F]	207	–	1.16	34.6	35
[P$_{4441}$] [(FH)$_{2.3}$F]	282	249 (169 g)	0.969	36	6.0
[P$_{4444}$] [(FH)$_{2.3}$F]	324	239, 255	0.954	47	3.7
[P$_{4448}$] [(FH)$_{2.3}$F]	381	(174 g)	0.934	74	1.5
[P$_{222(1O1)}$] [(FH)$_{2.1}$F]	224	206, 239	1.07	44	11
[P$_{2225}$] [(FH)$_{2.1}$F]	250	218 (166 g)	0.999	28	12.4

Continued

TABLE 5.1 Selected Physical Properties of Fluorohydrogenate Ionic Liquids[a]—cont'd

Ion	MW	T_m, T_g, T_{cl}[b]/K	Density at 298 K/g cm^{-3}	Viscosity at 298 K /mPa s	Ionic Conductivity at 298 K/ mS cm^{-1}
[P$_{2228}$][(FH)$_{2.1}$F]	292	227, 246	0.968	47	6.1
[S$_{111}$][(FH)$_{1.9}$F]	134	242	1.18	7.8	131
[S$_{112}$][(FH)$_{2.0}$F]	150	227	1.14	8.2	111
[S$_{122}$][(FH)$_{2.0}$F]	164	–	1.11	8.9	91
[S$_{222}$][(FH)$_{2.0}$F]	178	217	1.09	8.3	83

[a]These data were reported in Refs. [34–42,60,89]. See Figure 9.2 for abbreviation of the cations. The number after "C" in the cation name denotes the number of carbon atoms in the alkyl chain and "allyl" denotes the allyl group. The four numbers after "P" and "S" for the tetraalkylphosphonium and trialkylsulfonium cations denote the carbon numbers in the alkyl chains. The symbol "1O1" denotes the methoxymethyl group.
[b]T_m, T_g, and T_{cl} denote melting temperature, glass-transition temperature, and clearing temperature. T_g and T_{cl} are listed with "g" and "cl", respectively.

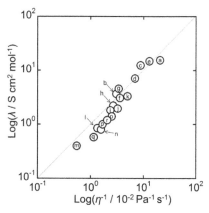

FIGURE 5.4 Walden plot (molar ionic conductivity (λ) against reciprocal viscosity (η)) for selected ionic liquids ((a) [C$_2$C$_1$im][(FH)$_{2.3}$F], (b) [C$_4$C$_1$im][(FH)$_{2.3}$F], (c) [C$_2$C$_1$pyrr][(FH)$_{2.3}$F], (d) [C$_4$C$_1$pyrr][(FH)$_{2.3}$F], (e) [S$_{111}$][(FH)$_{1.9}$F], (f) [P$_{2225}$][(FH)$_{2.1}$F], (g) [C$_4$pyr][(FH)$_{2.3}$F]) (h) [C$_2$C$_1$im][BF$_4$], (i) [C$_2$C$_1$im][SO$_3$CF$_3$], (j) [C$_2$C$_1$im][N(SO$_2$CF$_3$)$_2$], (k) [C$_2$C$_1$im][N(SO$_2$F)$_2$], (l) [C$_4$C$_1$im][BF$_4$], (m) [C$_4$C$_1$im][PF$_6$], (n) [C$_4$C$_1$im][SO$_3$CF$_3$], (o) [C$_4$C$_1$im][N(SO$_2$CF$_3$)$_2$], (p) [C$_4$pyrr][N(SO$_2$CF$_3$)$_2$], (q) [P$_{2225}$][N(SO$_2$CF$_3$)$_2$], and (r) [C$_4$pyr][N(SO$_2$CF$_3$)$_2$]) [34,36,39,41,42,47–52]. See Figure 5.3 for the abbreviations of the cations.

based on alkylimidazolium, alkylpyrrolidinium, and alkylpiperidinium generally exhibit high ionic conductivities (e.g., 110 mS cm^{-1} for [C$_1$C$_1$im][(FH)$_{2.3}$F] (C$_1$C$_1$im$^+$ = 1,3-dimethylimidazolium), 100 mS cm^{-1} for [C$_2$C$_1$im][(FH)$_{2.3}$F], and 75 mS cm^{-1} for [C$_2$C$_1$pyrr][(FH)$_{2.3}$F] (C$_2$C$_1$pyrr = N-ethyl-N-methylpyrrolidinium)), noncyclic tetraalkylphosphonium fluorohydrogenate ionic liquids show high viscosities and therefore low ionic conductivities (e.g., 6.0 mS cm^{-1} for [P$_{4441}$][(FH)$_{2.3}$F] (P$_{4441}$ = tributylmethylphosphonium)).

The diffusivity of ions in fluorohydrogenate ionic liquids was investigated by the pulsed gradient spin-echo nuclear magnetic resonance (NMR) technique [53]. Although the diffusion coefficient of (FH)$_n$F$^-$ in [C$_2$C$_1$im][(FH)$_{2.3}$F] (4.1 × 10^{-6} cm^2 s^{-1}) is larger than that of C$_2$C$_1$im$^+$ (3.0 × 10^{-6} cm^2 s^{-1}), it is just 1.3 times and clearly denies a possible special conduction mechanism such as the ion hopping mechanism. Another important point here is the higher diffusion coefficient of C$_2$C$_1$im$^+$ in [C$_2$C$_1$im][(FH)$_{2.3}$F] than in other ionic liquids (e.g., 0.5 × 10^{-6} cm^2 s^{-1} for C$_2$C$_1$im$^+$ in [C$_2$C$_1$im][BF$_4$]) [50], which suggests that the cation in [C$_2$C$_1$im][(FH)$_{2.3}$F] diffuses faster than that in [C$_2$C$_1$im][BF$_4$]. Consequently, the high ionic conductivity of fluorohydrogenate ionic liquids results from their low viscosity that enhances the diffusion of the component ions.

By the method proposed by Tsuzuki et al. [54], ion–ion interactions in fluorohydrogenate ionic liquids were investigated using the crystal structures of three different salts, [C$_2$C$_1$im][FHF] for FHF$^-$ [17], [C$_1$C$_1$im][(FH)$_2$F] for (FH)$_2$F$^-$ [12], and [C$_1$C$_1$im][(FH)$_3$F] for (FH)$_3$F$^-$ [12] (Figure 5.5). The [C$_2$C$_1$im][FHF] crystal structure is characterized by its layered structure, where the imidazolium ring of C$_2$C$_1$im$^+$ and the entire FHF$^-$ are located in the same layer and linked through relatively strong hydrogen bond. The ion configuration within the layer is closely related to those of [C$_2$C$_1$im][NO$_2$] [55] and [C$_2$C$_1$im][OCN] [56], whereas the stacking mode perpendicular to the layer is different; the cation–cation stacking is observed in [C$_2$C$_1$im][FHF], whereas the cation–anion stacking is found in [C$_2$C$_1$im][NO$_2$] and [C$_2$C$_1$im][OCN]. The cations and anions in [C$_1$C$_1$im][(FH)$_2$F] and [C$_1$C$_1$im][(FH)$_3$F] form one-dimensional columns as is observed in crystal structures of some imidazolium-based salts (e.g., [C$_4$C$_1$im] Cl [57], [C$_2$C$_1$im][FHF] [17], and [C$_1$C$_1$im][SO$_3$CH$_3$] [58]). The interaction energies between the cation and anion in these crystal structures were evaluated by quantum mechanical calculations at the MP2 level and the interaction energies were plotted against the distance between the two ions. The plots showed linearity, which indicated that the interaction energies were mainly dominated by electrostatic interactions as in the cases of other imidazolium salts, and the low viscosity and high conductivity of fluorohydrogenate ILs are derived from their dynamic properties. It is suggested that the HF unit in the anion of fluorohydrogenate ionic liquids is exchanged between the anions. Such a HF unit is considered to weaken the cation–anion interactions as a dielectric spacer and produces smaller anionic diffusion species.

Figure 5.6 shows cyclic voltammograms of a glass-like carbon electrode in (a) [C$_2$C$_1$pyrr][(FH)$_{2.3}$F], (b) [C$_2$C$_1$im][(FH)$_{2.3}$F], and (c) [C$_2$C$_1$im] [BF$_4$] [36,39,59]. For [C$_2$C$_1$im][(FH)$_{2.3}$F], the cathodic and anodic limits on

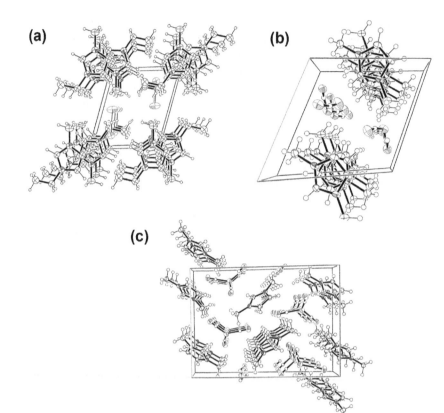

FIGURE 5.5 Crystal packings of (a) [C_2C_1im][FHF], (b) [C_1C_1im][$(FH)_2$F], and (c) [C_1C_1im][$(FH)_3$F] [12,17].

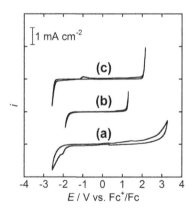

FIGURE 5.6 Cyclic voltammograms of a glass-like carbon in (a) [C_2C_1pyrr][$(FH)_{2.3}$F], (b) [C_2C_1im][$(FH)_{2.3}$F], and (c) [C_2C_1im][BF_4] [36,39].

a glass-like carbon electrode are around −2 and 1 V vs. Fc+/Fc, resulting in an electrochemical window of 3 V. This is smaller than the electrochemical windows of typical ionic liquids such as [C_2C_1im][BF_4] (~4.5 V). The cathodic limit potential of [C_2C_1im][$(FH)_{2.3}F$] is positively shifted significantly on a Pt electrode due to the small overpotential for hydrogen gas evolution from $(FH)_nF^-$. Replacement of C_2C_1im$^+$ with C_2C_1pyrr$^+$ increases the electrochemical window on a glass-like carbon, especially to the anodic direction, which results from the high oxidative resistance of pyrrolidinium cation. In the cases of aliphatic cation such as alkylpyrrolidinium, phosphonium, and sulfonium cations, fluorohydrogenate ionic liquids give electrochemical windows over 5 V on a glass-like carbon.

Introduction of fluoroalkyl groups to the imidazolium cation was attempted to give further hydrophobicity to fluorohydrogenate ionic liquids [60]. A series of 1-methyl-3-polyfluoroalkylimidazolium fluorohydrogenate salts ([($C_xF_{2(x-2)+1}$)C_1im][$(FH)_{2.0}F$], "$C_xF_{2(x-2)+1}$" is the polyfluoroalkyl chain of which the two carbon atoms from the root are not fluorinated, e.g., "C_3F_3" denotes 3,3,3-trifluoropropyl group) were synthesized and characterized. The three salts with relatively short side chains have melting temperatures below room temperature: [(C_3F_3)C_1im][$(FH)_{1.7}F$] (melting temperature: 274 K), [(C_4F_5)C_1im][$(FH)_{1.7}F$] (glass-transition temperature: 186 K), and [(C_6F_9)C_1im][$(FH)_{1.8}F$] (melting temperature: 276 K). The other salts such as [(C_8F_{13})C_1im][$(FH)_{2.0}F$] and [($C_{10}F_{17}$)C_1im][$(FH)_{2.0}F$] are in the crystal phase at room temperature. Although addition of water to 1-methyl-3-polyfluoroalkylimidazolium fluorohydrogenate salts first resulted in phase separation, the two phases slowly mixed with each other. As the number of fluorine atoms increases, the density and viscosity increases and the ionic conductivity and electrochemical stability decreases. Liquid crystal phase was observed for the salts with long chains ([(C_8F_{13})C_1im][$(FH)_{2.0}F$] and [($C_{10}F_{17}$)C_1im][$(FH)_{2.0}F$]) according to polarized optical microscopy and X-ray diffraction analysis.

There are several interesting applications of fluorohydrogenate ionic liquids as electrolytes in electrochemical devices by making full use of their high ionic conductivities. Applications as electrolytes for polymer electrolyte fuel cells (PEFCs) were investigated, targeting operation in the intermediate temperature range without humidification [61–63]. Fuel cells operating above 373 K are attracting attention due to the improved electrode reaction kinetics at elevated temperatures. Current PEFCs using perfluorosulfonate polymer electrolyte such as Nafion require humidification for operation because water is essential for proton conduction in the electrolyte. Moreover, the degradation of the electrolyte occurs above 373 K. Since the fuel cell generates heat by itself, cooling and humidification equipment are inevitable for the current PEFC to avoid drying up and degradation of the electrolyte which increase the volume and weight of the system. Ionic liquids with high thermal stability are interesting candidates as electrolytes for a fuel cell operating at higher temperatures without humidification. Fundamental studies on the polarization behavior of the H_2 and O_2

electrodes in fluorohydrogenates revealed that this fuel cell certainly works and $(FH)_nF^-$ anions are able to carry proton from anode to cathode (not via proton hopping). Figure 5.7 shows schematic drawing of the principle of operation of a fuel cell using fluorohydrogenate ionic liquids as a nonvolatile electrolyte. The following equations are the reactions of the fuel cell using the $[C_2C_1pyrr][(FH)_{1.7}F]$ electrolyte which contains $(FH)_2F^-$ and FHF^- in it:

$$\text{Cathode}: 1/2O_2 + 4(FH)_2F^- + 2e^- \rightarrow H_2O + 6(FH)F^- \qquad (5.4)$$

$$\text{Anode}: H_2 + 6FHF^- \rightarrow 4(FH)_2F^- + 2e^- \qquad (5.5)$$

$$\text{Total}: 1/2O_2 + H_2 \rightarrow H_2O \qquad (5.6)$$

Fuel cell operation using the fluorohydrogenate ionic liquid–polymer composite electrolyte membrane was also performed [62,63]. Such a composite is prepared by selecting an appropriate monomer including hydroxyethylmethacrylate (HEMA) [64]. High ionic conductivities are observed for fluorohydrogenate ionic liquid–polymer composite materials (e.g., 8.7 at 298 K and 81.9 mS cm^{-1} at 393 K for the composite of $[C_2C_1pyrr][(FH)_{2.3}F]$ and poly(HEMA) in the ratio of 9:1). The maximum power density observed for the $[C_2C_1pyrr][(FH)_{2.3}F]$-poly(HEMA) composite electrolyte is 32 mW cm^{-2} at 323 K [62].

Electrochemical capacitors are attractive energy storage devices for energy regeneration. High power density and long lifetime are the beneficial points of electrochemical capacitors compared to secondary batteries [65–67]. Electrode materials with a high surface area such as the activated carbon are required for double-layer capacitors, whereas the capacitance is usually nonlinear to the specific surface area because of the effect of the pore structure. Redox capacitors also work in the same way but the electric charge is stored based on

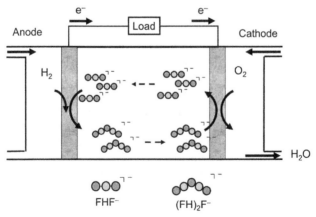

FIGURE 5.7 Schematic drawing of the fuel cell using fluorohydrogenate ionic liquids as electrolytes [61].

Faradaic reactions. Minimization of internal resistance is an important factor to design electrochemical capacitors because operation at high power densities is expected for them and high ionic conductivities of fluorohydrogenate ionic liquids are attractive for this purpose. Figure 5.8 shows voltage dependence of capacitance for electrochemical capacitors using activated carbon electrodes and the three nonaqueous electrolytes, $[C_2C_1im][(FH)_{2.3}F]$, $[C_2C_1im][BF_4]$, and 1 M $[N(C_2H_5)_4][BF_4]$/PC (PC = propylene carbonate) [40]. Voltage dependence of capacitance for $[C_2C_1im][(FH)_{2.3}F]$ is significantly larger than those for $[C_2C_1im][BF_4]$ and 1 M $[N(C_2H_5)_4][BF_4]$/PC. The capacitance value obtained for $[C_2C_1im][(FH)_{2.3}F]$ (162 F g^{-1} at 2.5 V) is four times larger than that for $[C_2C_1im][BF_4]$ (42 F g^{-1} at 3.0 V). Although the change in cationic structure (length of alkyl side chain, the change of the main frame from imidazolium to pyrrolidnium, and the introduction of alkoxy group) results in different capacitance values, the basic voltage dependences of capacitances are similar to each other. The origins of such unusual behavior were investigated by separating the positive and negative electrode behavior using a three-electrode cell [68]. The most probable redox reaction at the positive electrode is the oxidation of carbon accompanied by absorption of $(FH)_nF^-$ or F^- into the activated carbon species, and the reverse process. The redox reaction at the negative electrode involves the adsorption of atomic hydrogen generated from $(FH)_nF^-$ and the reverse process, although the redox reaction involving surface functional groups cannot be ruled out.

Fluorohydrogenate ionic liquids work as a Lewis base when they are reacted with a fluoride (or oxidefluoride) which acts as a stronger Lewis acid than HF to liberate HF from $(FH)_nF^-$ [69–73]. The following is the example of this reaction:

$$[C_2C_1im][(FH)_{2.3}F] + BF_3 \rightarrow [C_2C_1im][BF_4] + 2.3HF \quad (5.7)$$

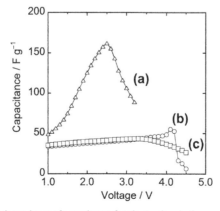

FIGURE 5.8 Voltage dependence of capacitance for electrochemical capacitors using (a) $[C_2C_1im]$ $[(FH)_{2.3}F]$, (b) $[C_2C_1im][BF_4]$, and (c) 1 M $[N(C_2H_5)_4][BF_4]$/PC [40].

This reaction gives a variety of fluorocomplex salts with high purity (e.g., $C_2C_1im^+$ salts of BF_4^-, PF_6^-, AsF_6^-, NbF_6^-, TaF_6^-, UF_6^-, WF_7^-, VOF_4^-, $MoOF_5^-$, and WOF_5^-) owing to the trace amounts of water and halide in the starting fluorohydrogenate ionic liquids. The byproduct HF is easily removed under vacuum at elevated temperatures.

Several organic reactions in fluorohydrogenate ionic liquids are known, including halofluorination of alkenes and ring-opening fluorination of epoxides [74–76]. Some difficulties in the use of previously known fluorinating reagents such as hydrogen fluoride are avoided by using fluorohydrogenate ionic liquids. For example, 1-fluoro-2-iodo-1-phenylpropane is prepared from (E)-1-phenylpropene with $[C_2C_1im][(FH)_{2.3}F]$ and N-iodosuccinimide in CH_2Cl_2 (85% yield) [74]. Ring-opening fluorination of styrene oxide with $[C_2C_1im][(FH)_{2.3}F]$ in CH_2Cl_2 gives 2-fluoro-2-phenyl-ethanol (72% yield) [76]. A mild condition, that is, slow HF transfer from the anion to the reactant, is considered to be a key step of this reaction because a rapid reaction does not give fluorohydrin.

It is noteworthy that the largest alkali metal cation, cesium cation, forms a melt with a low HF dissociation pressure around $Cs[(FH)_{2.45}F]$ where the phase diagram has a eutectic point [77,78]. Application of this fluorohydrogenate salt as an electrolytic bath for elemental fluorine generation was attempted in old days [79], though this electrolytic cell was not industrially applied due to the high cost of the salt.

5.4 FLUOROHYDROGENATE IONIC LIQUID CRYSTALS

Ionic liquid crystals can be considered as substances that possess the properties of both ionic liquids and liquid crystals, being composed of ionic species and exhibiting a liquid crystal mesophase in a certain temperature range [80]. Although various alkylammonium cations are known to form ionic liquid crystals, alkylimidazolium salts have been most widely studied. Examples of alkylimidazolium-based ionic liquid crystals are chloride [81], bromide [82], tetrachlorometallate (MCL_4^-, M = Co and Ni) [81], hexafluorophosphate [83], tetrafluoroborate [84], and trifluoromethanesulfonate salts [82]. In these materials, long alkyl chains on the imidazolium cation interact with each other and usually form smectic liquid crystal phases. Ionic liquid crystal phases based on 1-alkyl-3-methylimidazolium cations, $C_xC_1im^+$, are generally observed with alkyl chain longer than dodecyl and their temperature range increases with increasing alkyl chain length. Ionic liquid crystals attract attention as anisotropic ion conductors owing to their anisotropic structures [85–88].

Fluorohydrogenate ionic liquid crystals were first reported in 2010 [89]. Differential scanning calorimetric curves of $[C_xC_1im][(FH)_2F]$ (x = 10, 12, 14, 16, and 18) show typical shapes for salts with a mesophase; a large melting peak is followed by a small clearing peak during the heating scan. The clearing temperature shows a large dependence on alkyl chain length compared with the melting temperature, leading to the wider temperature range of liquid crystal

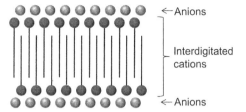

FIGURE 5.9 Schematic drawing of the structure for a smectic A_2-type ionic liquid crystal.

mesophase for the cation with a longer alkyl chain (e.g., 283.1 K for $[C_{10}C_1im]$ $[(FH)_2F]$ and 399.6 K for $[C_{18}C_1im][(FH)_2F]$). This is mainly due to the increase in van der Waals interactions between alkyl chains.

Polarized optical microscopy is often used to see texture of liquid crystal phase. Each type of liquid crystal has its own texture which is useful to judge the structure of the liquid crystal phase. Usually, C_xC_1im-based ionic liquid crystals have smectic-type structures and $[C_xC_1im][(FH)_2F]$ salts are not exceptional. The texture of $[C_xC_1im][(FH)_2F]$ is characterized by spontaneous formation of smooth fanlike or focal conic structures without the broken fanlike texture which is indicative of the smectic C mesophase, suggesting the formation of smectic A mesophase. Schematic drawing of the smectic A_2-type ionic liquid crystal is shown in Figure 5.9. X-ray diffraction patterns of the $[C_xC_1im][(FH)_2F]$ salts have peaks in the low-angle region for both the crystal phase and liquid crystal mesophase, indicating formation of layered structures. The layer spacings of the $[C_xC_1im][(FH)_2F]$ crystal phases are smaller than those of the liquid crystal phases. The temperature dependence of the layer spacing for the $[C_{12}C_1im]$ $[(FH)_2F]$ liquid crystal was investigated in detail and the peak position in the X-ray diffraction pattern shifts to the high angle with an increase in temperature. In general, the tilted angle of the layered structure decreases with increasing temperature for smectic C mesophases, leading to an increase of the layer spacing. On the other hand, the bilayer structure of smectic A mesophases interdigitates more deeply with increasing temperature due to the increase of thermal mobility of the alkyl chains, leading to the decrease of the layer spacing. Thus, the behavior of the $[C_{12}C_1im][(FH)_2F]$ liquid crystal agrees with that of smectic A mesophase. Although the same type of ionic liquid crystal was observed for N-alkyl-N-methylpyrrolidinium fluorohydrogenate salts ($[C_xC_1Pyr][(FH)_2F]$), alkyl chain longer than the tetradecyl group ($x > 14$ in $C_xC_1Pyr^+$) is required to form liquid crystal phase [90]. The difference in interaction between the polar head group with the anion causes the difference in thermal behavior between the $C_xC_1im^+$ and $C_xC_1pyrr^+$ salts.

Ionic conductivities parallel ($\sigma_{//}$) and perpendicular ($\sigma_{\perp}$) to the smectic layer were measured using orientation controlled cells (ITO electrode cell and comb-shaped electrode cell) for $[C_xC_1im][(FH)_2F]$ ($x = 10$, 12, 14, and 16) [89]. The results show clear anisotropy in ionic conductivity for $x = 12$, 14, and 16, as shown in Figure 5.10, whereas $[C_{10}C_1Im][(FH)_2F]$ does not exhibit significant

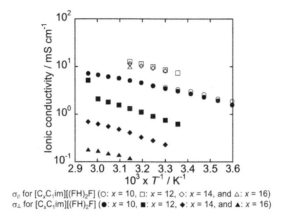

FIGURE 5.10 Temperature dependence of anisotropic ionic conductivity (($\sigma_\perp$) and ($\sigma_{//}$)) for [$C_xC_1$im][(FH)$_2$F] (x = 10, 12, 14, and 16) [89].

anisotropy probably because of the weak interaction between the decyl groups. The $\sigma_{//}$ value is roughly 10 times higher than the $\sigma_\perp$ value (e.g., $\sigma_{//}$ = 7.37 and $\sigma_\perp$ = 0.62 mS cm^{-1} at 298 K for [$C_{12}C_1$im][(FH)$_2$F]), and the ions are considered to move faster within the layer than perpendicular to the layer. The dominant charge carrier in these ion-conductive layers is thought to be (FH)$_2$F$^-$, since the cation is much larger than the anion and linked to the adjacent cation by van der Waals interactions.

Effects of the HF composition, n ($1.0 \le n \le 2.3$), on the formation of ionic liquid crystal was investigated in combination with $C_{12}C_1$im$^+$ [91]. The phase diagram of [$C_{12}C_1$Im][(FH)$_n$F] (n vs transition temperature) suggests that [$C_{12}C_1$Im][(FH)$_n$F] is a mixed crystal system that has a boundary around n = 1.9. For all the compositions, a liquid crystal mesophase with a smectic A interdigitated bilayer structure is observed. With increasing n, the temperature range of the mesophase decreases and the layer spacing of the smectic structure decreases. The $\sigma_{//}$ and $\sigma_\perp$ values increase with increasing n, whereas the anisotropy of the ionic conductivities ($\sigma_{//}/\sigma_\perp$) is independent of n, because the thickness of the insulating sheet formed by the dodecyl group remains nearly unchanged.

5.5 FLUOROHYDROGENATE IONIC PLASTIC CRYSTALS

Ionic plastic crystals are sometimes observed as intermediate phase in the temperature range between ionic liquids and crystals [92–94]. The constituting ions in ionic plastic crystals are highly disordered and sometimes rotating [95,96]. Recent reports suggest ionic plastic crystals are interesting candidates of safe solid-state ion conductors since such disordering in structure enhances diffusion of the ions. Ionic plastic crystals themselves have intrinsic unique properties such as low volatility, low flammability, and high electrochemical stabilities. Relatively symmetric organic cations such as tetraalkylammonium

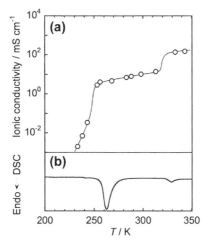

FIGURE 5.11 (a) Ionic conductivity and (b) differential scanning calorimetric curve of [C_1C_1pyrr][(FH)$_2$F] [99].

and dialkylpyrrolidinium tend to form ionic plastic crystals [92–94]. Addition of another salt sometimes results in an increase in ionic conductivity. For example, the N-ethyl-N-methylpyrrolidinium bis(trifluoromethylsulfonyl) amide ([C_2C_1pyrr][TFSA]) was studied as a solid-state electrolyte for lithium secondary batteries. Although the ionic conductivity of this ionic plastic crystal is 10^{-4} mS cm^{-1} at 333 K in the neat form, it increases up to 2×10^{-1} mS cm^{-1} at 333 K by the addition of a lithium salt (5 mol% Li[TFSA]) [97,98]

Fluorohydrogenate salts based on some alkylammonium and phosphonium cations form ionic plastic crystals. N,N-dimethylpyrrolidinium and N-ethyl-N-methylpyrrolidinium fluorohydrogenate salts ([C_1C_1pyrr][(FH)$_2$F] and [C_2C_1pyrr][(FH)$_2$F]) have ionic plastic crystal phases in the temperature ranges of 258–325 and 236–303 K, respectively [99]. The differential scanning calorimetric curve for [C_1C_1pyrr][(FH)$_2$F] is shown in Figure 5.11. The entropy changes of melting for these two salts are extremely small (4.1 J K^{-1} mol^{-1} for [C_1C_1pyrr][(FH)$_2$F] and 2.0 J K^{-1} mol^{-1} for [C_2C_1pyrr][(FH)$_2$F]), reflecting their soft textures and highly disordered structures.

Although ions in ionic plastic crystals are orientationally disordered, they are located in a certain position in a lattice. Thus, their crystal structures can be analyzed by X-ray diffraction. X-ray diffraction patterns of the crystal and plastic crystal phases for [C_1C_1pyrr][(FH)$_2$F] are shown in Figure 5.12 [99]. The ionic plastic crystal phases of [C_1C_1pyrr][(FH)$_2$F] and [C_2C_1pyrr][(FH)$_2$F] give three diffraction peaks which are indexed with the indices, 111, 200, and 220, of NaCl-type structures ($a = 9.90$ Å for [C_1C_1pyrr][(FH)$_2$F] and $a = 10.18$ Å for [C_2C_1pyrr][(FH)$_2$F]). Three-dimensional rotation of the constituent ions results in such a lattice with high symmetry. Another fluorohydrogenate ionic plastic crystal, [P_{2222}][(FH)$_2$F] (P_{2222}^+ = tetraethylphosphonium), has a hexagonal

120 Advanced Fluoride-Based Materials for Energy Conversion

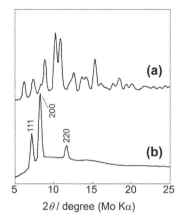

FIGURE 5.12 X-ray diffraction patterns of (a) crystal phase at 203 K and (b) plastic crystal phase at 298 K of [C$_1$C$_1$pyrr][(FH)$_2$F]. The indices in (b) are the ones for the NaCl-type structure [99].

lattice of an inverse NiAs-type structure with $a = 8.02$ Å and $c = 11.84$ Å, where the larger P$_{2222}^+$ are hexagonally close packed and the smaller (FH)$_2$F$^-$ occupies the octahedral interstitial site [34].

As in the cases of fluorohydrogenate ionic liquids mentioned above, fluorohydrogenate ionic plastic crystals also exhibit high ionic conductivities even without addition of another salt. Arrhenius plot of ionic conductivity for [C$_1$C$_1$pyrr][(FH)$_2$F] is shown in Figure 5.11. Ionic conductivities of these ionic plastic crystal phases range from 10^0 to 10^1 mS cm^{-1} (e.g., 10.3 mS cm^{-1} at 298 K for [C$_1$C$_1$pyrr][(FH)$_2$F] and 14.4 mS cm^{-1} at 288 K for [C$_2$C$_1$pyrr][(FH)$_2$F]) and are as high as those of popular ionic liquids (e.g., 14 mS cm^{-1} at 298 K for [C$_2$C$_1$im][BF$_4$] in the liquid state [50]). Pulsed-field gradient spin-echo NMR spectroscopy revealed that only (FH)$_n$F$^-$ moves in the ionic plastic crystal phase as a charge carrier with a diffusion coefficient of ~10^{-7} cm^2 s^{-1}. The self-diffusion coefficient of the cation in the ionic plastic crystal phase of [C$_1$C$_1$pyrr][(FH)$_2$F] is too small to measure, although the cation in [C$_2$C$_1$pyrr][(FH)$_2$F] has a slight mobility below 303 K.

The [C$_1$C$_1$pyrr][(FH)$_2$F] ionic plastic crystal has a reasonably large electrochemical window of nearly 4 V. Its application as a solid-state electrolyte in electrochemical capacitors using activated carbon electrodes was attempted [100]. Such electrochemical capacitors have an advantage in safety issues because the use of solid-state electrolytes leads to suppression of electrolyte leakage. Although a decrease in capacitance is observed in the first 20 cycles, this electrochemical capacitor stably operates more than 300 cycles. The capacitances for the positive and negative electrodes in [C$_1$C$_1$pyrr][(FH)$_2$F] ionic plastic crystals are 263 and 221 F g^{-1}, respectively, at the 300th cycle when the cell is charged up to the voltage of 2.5 V with a current density of 238 mA g^{-1}. These values are significantly higher than those observed for usual ionic liquid electrolytes.

REFERENCES

[1] D.T. Meshri, in: T. Nakajima, A. Tressaud, B. Žemva (Eds.), Chap. 18 in Advanced Inorganic Fluorides, Elsevier, Lausanne, 2000.
[2] H. Groult, J. Fluorine Chem. 119 (2003) 173–189.
[3] S.P. Vavalides, R.E. Cable, W.K. Henderson, C.A. Powell, Ind. Eng. Chem. 50 (1958) 178–180.
[4] G.H. Cady, J. Am, Chem. Soc. 56 (1934) 1431–1434.
[5] R.M. Bozorth, J. Am. Chem. Soc. 45 (1923) 2128–2132.
[6] B.A. Coyle, S.L. W., J.A. Ibers, J. Solid State Chem. 1 (1970) 386–393.
[7] G.A. Olah, J.G. Shih, G.K.S. Prakash, J. Fluorine Chem. 33 (1986) 377–396.
[8] N. Yoneda, Tetrahedron 47 (1991) 5329–5365.
[9] J.E. Huheey, E.A. Keiter, P.L. Keiter, R.L, Inorganic Chemistry: Principles of Structure and Reactivity, Chap. 8, Harper & Row, New York, 1993.
[10] W.D. Chandler, K.E. Johnson, J.L.E. Campbell, Inorg. Chem. 34 (1995) 4943–4949.
[11] J.H. Clark, J. Emsley, R.E. Overill, J. Chem. Soc. Dalton Trans. (1981) 1219–1222.
[12] T. Enomoto, Y. Nakamori, K. Matsumoto, R. Hagiwara, J. Phys. Chem. C 115 (2011) 4324–4332.
[13] Y. Shodai, S. Kohara, Y. Ohishi, M. Inaba, A. Tasaka, J. Phys. Chem. A 108 (2004) 1127–1132.
[14] T. Bunic, M. Tramsek, E. Goreshnik, B. Žemva, J. Solid State Chem. 181 (2008) 2318–2324.
[15] M. Gerken, J.P. Mack, G.J. Schrobilgen, R.J. Suontamo, J. Fluorine Chem. 125 (2004) 1663–1670.
[16] R. Kruh, K. Fuwa, T.E. McEver, J. Am. Chem. Soc. 78 (1956) 4256–4258.
[17] K. Matsumoto, T. Tsuda, R. Hagiwara, Y. Ito, Solid State Sci. 4 (2002) 23–26.
[18] T.R. Mcdonald, Acta Cryst., A 25 (1969) 374–375.
[19] B.L. Mcgaw, J.A. Ibers, J. Chem. Phys. 39 (1963) 2677–2684.
[20] N.S. Ostlund, L.W. Ballenger, J. Am. Chem. Soc. 97 (1975) 1237–1238.
[21] S.W. Peterson, H.A. Levy, J. Chem. Phys. 20 (1952) 704–707.
[22] M.T. Rogers, L.A. Helmholz, J. Am. Chem. Soc. 62 (1940) 1533–1536.
[23] M.A. Beno, G.S. Blackman, J.M. Williams, K. Bechgaard, Inorg. Chem. 21 (1982) 3860–3862.
[24] D. Boenigk, D. Mootz, J. Am. Chem. Soc. 110 (1988) 2135–2139.
[25] J.D. Forrester, D.H. Templeton, A. Zalkin, M.E. Senko, Acta Cryst 16 (1963) 58–62.
[26] D. Mootz, D. Boenigk, J. Am. Chem. Soc. 108 (1986) 6634–6636.
[27] D. Mootz, D. Boenigk, Z. Anorg, Allg. Chem. 544 (1987) 159–166.
[28] D. Wiechert, D. Mootz, R. Franz, G. Siegemund, Chem. Eur. J. 4 (1998) 1043–1047.
[29] A.R. Mahjoub, D. Leopold, K. Seppelt, Eur. J. Solid State Inorg. Chem. 29 (1992) 635–647.
[30] D. Mootz, W. Poll, Z. Naturforsch B 39 (1984) 290–297.
[31] D. Mootz, W. Poll, Z. Naturforsch B 39 (1984) 1300–1305.
[32] D. Mootz, W. Poll, Z. Anorg. Allg. Chem. 484 (1982) 158–164.
[33] R. Hagiwara, T. Hirashige, T. Tsuda, Y. Ito, J. Fluorine Chem. 99 (1999) 1–3.
[34] T. Enomoto, S. Kanematsu, K. Tsunashima, K. Matsumoto, R. Hagiwara, Phys. Chem. Chem. Phys. 13 (2011) 12536–12544.
[35] R. Hagiwara, T. Hirashige, T. Tsuda, Y. Ito, J. Electrochem. Soc. 149 (2002) D1–D6.
[36] R. Hagiwara, K. Matsumoto, Y. Nakamori, T. Tsuda, Y. Ito, H. Matsumoto, K. Momota, J. Electrochem. Soc. 150 (2003) D195–D199.
[37] S. Kanematsu, K. Matsumoto, R. Hagiwara, Electrochem. Commun. 11 (2009) 1312–1315.

[38] K. Matsumoto, R. Hagiwara, Electrochemistry 73 (2005) 730–732.
[39] K. Matsumoto, R. Hagiwara, Y. Ito, Electrochem. Solid-State Lett. 7 (2004) E41–E44.
[40] A. Senda, K. Matsumoto, T. Nohira, R. Hagiwara, J. Power Sources 195 (2010) 4414–4417.
[41] R. Taniki, K. Matsumoto, R. Hagiwara, Electrochem. Solid-State Lett. 15 (2012) F13–F15.
[42] M. Yamagata, S. Konno, K. Matsumoto, R. Hagiwara, Electrochem. Solid-State Lett. 12 (2009) F9–F12.
[43] R. Hagiwara, Y. Nakamori, K. Matsumoto, Y. Ito, J. Phys. Chem. B 109 (2005) 5445–5449.
[44] K. Izutsu, Electrochemistry in Nonaqueous Solutions, 2nd Revised and Enlarged ed., Wiley-VCH, 2009.
[45] P. Walden, Organic solutions- and ionisation means. III. Chapter: internal friction and its connection with conductivity, Z. Phys. Chem-Stoch Ve 55 (1906) 207–249.
[46] W. Xu, E.I. Cooper, C.A. Angell, J. Phys. Chem. B 107 (2003) 6170–6178.
[47] E.I. Cooper, E.J.M. Sullivan, in: Proceedings of the 8th International Symposium on Molten Salts, vol. 92-16, The Electrochemical Society, Pennington, NJ, 1992, pp. 386–396.
[48] J. Fuller, R.T. Carlin, R.A. Osteryoung, J. Electrochem. Soc. 144 (1997) 3881–3886.
[49] M. Ishikawa, T. Sugimoto, M. Kikuta, E. Ishiko, M. Kono, J. Power Sources 162 (2006) 658–662.
[50] A. Noda, K. Hayamizu, M. Watanabe, J. Phys. Chem. B 105 (2001) 4603–4610.
[51] H. Tokuda, S. Tsuzuki, M.A.B.H. Susan, K. Hayamizu, M. Watanabe, J. Phys. Chem. B 110 (2006).
[52] K. Tsunashima, M. Sugiya, Electrochem. Commun. 9 (2007) 2353–2358.
[53] Y. Saito, K. Hirai, K. Matsumoto, R. Hagiwara, Y. Minamizaki, J. Phys. Chem. B 109 (2005) 2942–2948.
[54] S. Tsuzuki, H. Tokuda, M. Mikami, Phys. Chem. Chem. Phys. 9 (2007) 4780–4784.
[55] J.S. Wilkes, M.J. Zaworotko, J. Chem. Soc. Chem. Commun. (1992) 965–967.
[56] J. Janikowski, C. Forsyth, D.R. MacFarlane, J.M. Pringle, J. Mater. Chem. 21 (2011) 19219–19225.
[57] S. Saha, S. Hayashi, A. Kobayashi, H. Hamaguchi, Chem. Lett. 32 (2003) 740–741.
[58] J.D. Holbrey, W.M. Reichert, R.P. Swatloski, G.A. Broker, W.R. Pitner, K.R. Seddon, R.D. Rogers, Green Chem. 4 (2002) 407–413.
[59] J. Ohtsuki, K. Matsumoto, R. Hagiwara, Electrochemistry 77 (2009) 624–626.
[60] R. Taniki, N. Kenmochi, K. Matsumoto, R. Hagiwara, J. Fluorine Chem. 149 (2013) 112–118.
[61] R. Hagiwara, T. Nohira, K. Matsumoto, Y. Tamba, Electrochem. Solid-State Lett. 8 (2005) A231–A233.
[62] P. Kiatkittikul, T. Nohira, R. Hagiwara, J. Power Sources 220 (2012) 10–14.
[63] J.S. Lee, T. Nohira, R. Hagiwara, J. Power Sources 171 (2007) 535–539.
[64] T. Tsuda, T. Nohira, Y. Nakamori, K. Matsumoto, R. Hagiwara, Y. Ito, Solid State Ionics 149 (2002) 295–298.
[65] A. Chu, P. Braatz, J. Power Sources 112 (2002) 236–246.
[66] Y. Gogotsi, P. Simon, Science 334 (2011) 917–918.
[67] P. Simon, Y. Gogotsi, Nat. Mater. 7 (2008) 845–854.
[68] R. Taniki, K. Matsumoto, T. Nohira, R. Hagiwara, J. Electrochem. Soc. 160 (2013) A734–A738.
[69] T. Kanatani, K. Matsumoto, R. Hagiwara, Eur. Inorg. Chem. (2010) 1049–1055.
[70] T. Kanatani, K. Matsumoto, R. Hagiwara, Chem. Lett. 38 (2009) 714–715.
[71] T. Kanatani, R. Ueno, K. Matsumoto, T. Nohira, R. Hagiwara, J. Fluorine Chem. 130 (2009) 979–984.

[72] K. Matsumoto, R. Hagiwara, J. Fluorine Chem. 126 (2005) 1095–1100.
[73] K. Matsumoto, R. Hagiwara, R. Yoshida, Y. Ito, Z. Mazej, P. Benkic, B. Žemva, O. Tamada, H. Yoshino, S. Matsubara, Dalton Trans. (2004) 144–149.
[74] H. Yoshino, S. Matsubara, K. Oshima, K. Matsumoto, R. Hagiwara, Y. Ito, J. Fluorine Chem. 125 (2004) 455–458.
[75] H. Yoshino, K. Matsumoto, R. Hagiwara, Y. Ito, K. Oshima, S. Matsubara, J. Fluorine Chem. 127 (2006) 29–35.
[76] H. Yoshino, K. Nomura, S. Matsubara, K. Oshima, K. Matsumoto, R. Hagiwara, Y. Ito, J. Fluorine Chem. 125 (2004) 1127–1129.
[77] K. Matsumoto, J. Ohtsuki, R. Hagiwara, S. Matsubara, J. Fluorine Chem. 127 (2006) 1339–1343.
[78] R.V. Winsor, G.H. Cady, J. Am. Chem. Soc. 70 (1948) 1500–1502.
[79] F.C. Mathers, P.T. Stroup, Trans. Am. Electrochem. Soc. 66 (1934) 245–252.
[80] K. Binnemans, Chem. Rev. 105 (2005) 4148–4204.
[81] C.J. Bowlas, D.W. Bruce, K.R. Seddon, Chem. Commun. (1996) 1625–1626.
[82] A.E. Bradley, C. Hardacre, J.D. Holbrey, S. Johnston, S.E.J. McMath, Chem. Mater. 14 (2002) 629–635.
[83] C.M. Gordon, J.D. Holbrey, A.R. Kennedy, K.R. Seddon, J. Mater. Chem. 8 (1998) 2627–2636.
[84] J.D. Holbrey, K.R. Seddon, J. Chem. Soc. Dalton Trans. (1999) 2133–2139.
[85] T. Mukai, M. Yoshio, T. Kato, M. Yoshizawa, H. Ohno, Chem. Commun. (2005) 1333–1335.
[86] M. Yoshio, T. Kagata, K. Hoshino, T. Mukai, H. Ohno, T. Kato, J. Am. Chem. Soc. 128 (2006) 5570–5577.
[87] M. Yoshio, T. Mukai, K. Kanie, M. Yoshizawa, H. Ohno, T. Kato, Adv. Mater. 14 (2002) 351–354.
[88] M. Yoshio, T. Mukai, H. Ohno, T. Kato, J. Am. Chem. Soc. 126 (2004) 994–995.
[89] F. Xu, K. Matsumoto, R. Hagiwara, Chem. Eur. J. 16 (2010) 12970–12976.
[90] F. Xu, S. Matsubara, K. Matsumoto, R. Hagiwara, J. Fluorine Chem. 135 (2012) 344–349.
[91] F. Xu, K. Matsumoto, R. Hagiwara, J. Phys. Chem. B 116 (2012) 10106–10112.
[92] D.R. Macfarlane, M. Forsyth, P.C. Howlett, J.M. Pringle, J. Sun, G. Annat, W. Neil, Acc. Chem. Res. 40 (2007) 1165–1173.
[93] J.M. Pringle, Phys. Chem. Chem. Phys. 15 (2013) 1339–1351.
[94] J.M. Pringle, P.C. Howlett, D.R. MacFarlane, M. Forsyth, J. Mater. Chem. 20 (2010) 2056–2062.
[95] H. Ishida, S. Kashino, R. Ikeda, Z. Naturforsch A 55 (2000) 765–768.
[96] L.Y. Jin, K.M. Nairn, C.M. Forsyth, A.J. Seeber, D.R. MacFarlane, P.C. Howlett, M. Forsyth, J.M. Pringle, J. Am. Chem. Soc. 134 (2012) 9688–9697.
[97] M. Forsyth, J. Huang, D.R. MacFarlane, J. Mater. Chem. 10 (2000) 2259–2265.
[98] D.R. MacFarlane, J.H. Huang, M. Forsyth, Nature 402 (1999) 792–794.
[99] R. Taniki, K. Matsumoto, R. Hagiwara, K. Hachiya, T. Morinaga, T. Sato, J. Phys. Chem. B 117 (2013) 955–960.
[100] R. Taniki, K. Matsumoto, T. Nohira, R. Hagiwara, J. Power Sources 245 (2014) 758–763.

Chapter 6

Novel Fluorinated Solvents and Additives for Lithium-Ion Batteries

T. Böttcher,[1] N. Kalinovich,[1] O. Kazakova,[1] M. Ponomarenko,[1] K. Vlasov,[1] M. Winter[2] and G.-V. Röschenthaler[1]

[1]*School of Engineering and Science, Jacobs University GmbH, Bremen, Germany;* [2]*MEET—Münster Electrochemical Energy Technology, Westfälische Wilhelms-Universität, Münster, Germany*

Chapter Outline

- 6.1 Introduction 125
- 6.2 Lithium Conductive Salts: Polyfluorinated Lithium Sulfonates 125
- 6.3 Solvents and Cosolvents for Electrolyte Systems 132
  - 6.3.1 Alkyl Tetrafluoro-2-(Alkoxy) Propionates as Cosolvents for Propylene-Carbonate-Based Electrolytes 132
  - 6.3.2 Derivatives of Hexafluorbutandiol as Cosolvents and SEI Additives 135
  - 6.3.3 Fluorinated 1,2,4-Thiadiazianan-3,5-dion-1,1-dioxides Useful SEI Additives in PC 138
- 6.4 Overcharge Protecting Agents: PF_5–Carbene Adducts 139
- References 143

6.1 INTRODUCTION

Fluorinated components [1] are indispensable for rechargeable lithium-ion battery electrolyte systems, in order to improve, e.g., high- and low-temperature performance, oxidative stability, inflammability and safety, the ability to form solid electrolyte interphases (SEIs) at the anode and cathode. Here synthetic concepts for new fluorinated electrolyte salts, ionic and neutral additives, and solvents are described.

6.2 LITHIUM CONDUCTIVE SALTS: POLYFLUORINATED LITHIUM SULFONATES

Future applications for lithium-ion batteries require high operative temperature range. The currently most used lithium salt in lithium-ion batteries is $LiPF_6$.

Despite its commercial success, its chemical behavior limits the thermal stability of the electrolyte [2]. In recent years, it was intensively tried to establish alternative salts for $LiPF_6$ as the common conductive salt. Therefore, especially lithium sulfonates with perfluorinated organic rests were tested.

Sulfonates with a $–CF_2CF_2SO_3Li$ unit have been synthesized and zwitterionic functionalized imidazolium ionic liquids with the $–CF_2CFHOCF_2CF_2SO_3Li$ group were also obtained [3]. Other salts of fluorinated sulfonic acids became available, i.e., perfluoroalkyl sulfonic acids $CF_3(CF_2)_nSO_3Li$ [4] (n = 1–9), bis(perfluoroalkyl) sulfonic acids [1,5] $(CF_2)_n(SO_2OLi)_2$ (n = 1–4) and $O(CF_2CF_2SO_2OLi)_2$ [1] with high conductivity at 100 °C, $(CF_3)_2NCF_2CF_2SO_3Li$ [4f,6] and $F_3CF_2OCF_2CF_2SO_3Li$ [7], as well as $CF_3OCF_2CF_2SO_3Na$ [8] and $CF_3OCFHCF_2SO_3K$ [9] but not the corresponding lithium salts. Also, aryl-containing lithium perfluorosulfonates were synthesized [10].

Two new salts were investigated, namely, lithium-1,1,2,2,-tetrafluoro-2-methoxy-ethanesufonate 3 [11] and lithium-1,2,2,2-tetrafluoro-1-(trifluoromethoxy)-ethanesulfonate 5 [12]. The electrochemical studies showed that these compounds do not to generate HF in the presence of moisture; have good lithium-ion conductivity, a high electrochemical stability, and good SEI formation properties; and can be used in a wide temperature range. An efficient and safe method for their synthesis in a multigram scale was developed [13].

Fluorosulfonyltetrafluoroethylate 1 was synthesized from 2,2-difluoro-(1-fluorosulfonyl)acetyl fluoride [14], and Et_3Nx3HF (Scheme 6.1), precursor for lithium-1,1,2,2-tetrafluoro-2-methoxy-ethanesulfonate 3 (Scheme 6.2).

When generated in an aprotic solvent, such as CH_2Cl_2, CH_3CN, or diglyme, not isolated, and immediately reacted with dimethyl sulfate or methyl triflate [15] the highest yield of tetrafluoroethylether 2 of 74% is provided in CH_3CN at 50 °C. The conversion of 2 into the corresponding lithium sulfonate 3 was carried out in methanol at 0 °C with two equivalents of LiOH (Scheme 6.2).

Electrochemical studies [11,12] showed that electrolytes containing lithium-1,1,2,2-tetrafluoro-2-methoxy-ethanesulfonate 3 did not form hydrogen fluoride even at 95 °C (Figure 6.1). Thermal aging experiments for an ethylene carbonate

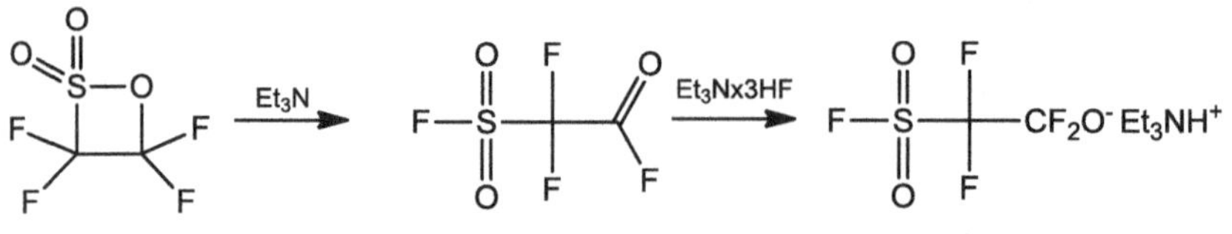

SCHEME 6.1 Fluoridation of 2,2-difluoro-(1-fluorosulfonyl)acetyl fluoride.

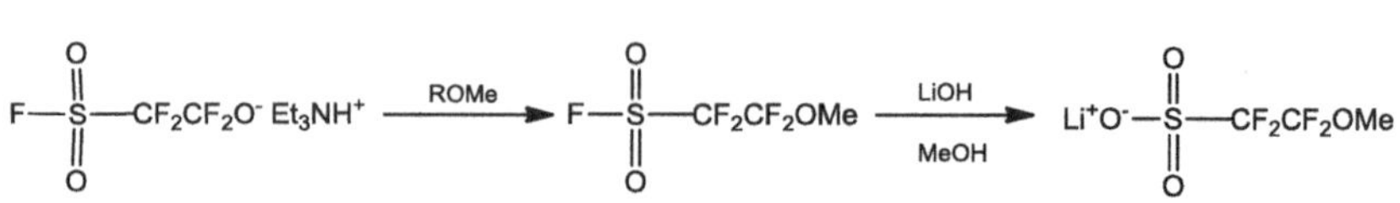

SCHEME 6.2 Synthesis of lithium-1,1,2,2-tetrafluoro-2-methoxy-ethanesulfonate 3.

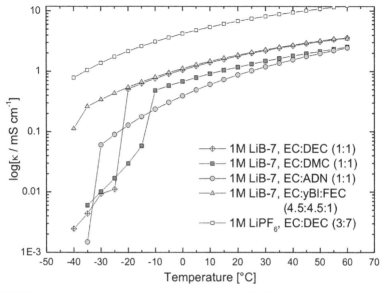

FIGURE 6.1 Conductivity measurements for **3** (marked as LiB-7). Solvents: ethylene carbonate (EC), 2 diethyl carbonate (DEC), dimethyl carbonate (DMC), adiponitrile (AND), γ-butyrolactone (γ-Bl), fluoroethylene carbonate (FEC). *Taken from Ref. [11].*

(EC) solution disclosed that even after 2 weeks at 95 °C, no HF could be detected in contrast to LiPF$_6$-based electrolytes, opening a broader temperature window for the application as a conductive salt. In a mixture of EC and diethyl carbonate (DEC), **3** shows an anodic stability up to 5.6 V rendering **3** for the usage for high-volt cathode materials. No aluminum corrosion at the cathode side was observed, but there was formation of a protection layer, just as in the case of LiPF$_6$.

The electrochemical stability limits can be obtained from Figure 6.2. The lithium salt **3** showed an acceptable stability, reaching from 0.2 V up to 5.6 V vs Li/Li$^+$. This fulfills the demands for a stable electrolyte. Although 1 M lithium 2-methoxy-1,1,2,2-tetrafluoro-ethanesulfonate **3** EC:DEC (1:1) electrolyte obtained higher anodic stabilities than the standard LiPF$_6$ electrolyte (see also Figure 6.3), the extra stability gained is not significantly higher (1 M LiPF$_6$, EC:DEC (3:7) anodic limit vs Li/Li$^+$ is 5.9 V).

The EC:DEC electrolyte shows the highest and the EC: dimethyl carbonate (DMC) the lowest electrochemical stability window proving that lithium-2-methoxy-1,1,2,2-tetrafluoro-ethanesulfonate **3** has sufficient electrochemical stability for all electrochemical applications in the usual carbonate solvents.

Even electrolytes with high electrochemical stabilities may not be suitable for application in a commercial lithium-ion battery, especially in case of salts containing sulfur, e.g., lithium bis(trifluoromethylsulfonyl)imide (LiTFSI) [16], because of the aluminum current collector corrosion at the positive electrode shown in Figure 6.3 at a typical cell voltage of 4.5 V, caused by the LiTFSI electrolyte. Thus, no corrosion could be detected for LiPF$_6$ and

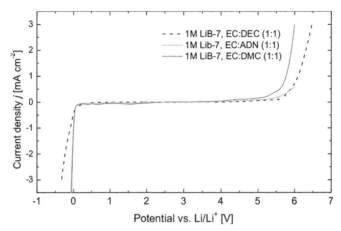

FIGURE 6.2 Electrochemical stability windows of three 1 M lithium-2-methoxy-1,1,2,2-tetrafluoro-ethanesulfonate **3** (marked as LiB-7) binary mixtures. *Taken from Ref. [11].*

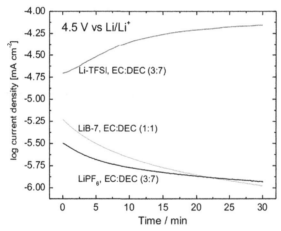

FIGURE 6.3 Corrosion test on aluminum foil at 4.5 V. LiTFSI shows a corrosion current, while LiPF$_6$ and lithium-2-methoxy-1,1,2,2-tetrafluoro-ethanesulfonate **3** (marked as LiB-7) do not corrode the aluminum surface. *Taken from Ref. [11].*

lithium-2-methoxy-1,1,2,2-tetrafluoro-ethanesulfonate **3** (marked as LiB-7) at this positive electrode potential.

The respective electrolyte solutions were tested in a lithium-ion cell, using overdimensioned nickel cobalt manganese (NCM) positive electrodes. The results were compared with the experimental data of a commercial 1 M LiPF$_6$ EC:DEC (3:7) electrolyte.

The lithium-2-methoxy-1,1,2,2-tetrafluoro-ethanesulfonate **3** (marked as LiB-7) electrolyte reaches the same capacities as the standard electrolyte (Figure 6.4). Electrolytes with lithium-2-methoxy-1,1,2,2-tetrafluoro-ethanesulfonate **3** as

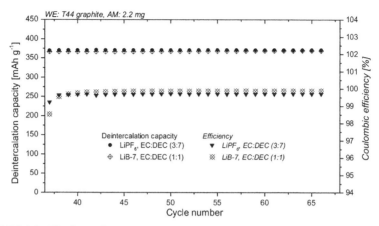

FIGURE 6.4 The figure displays 30 standard cycles following the D-rate test and the recovery cycles. *Taken from Ref. [11].*

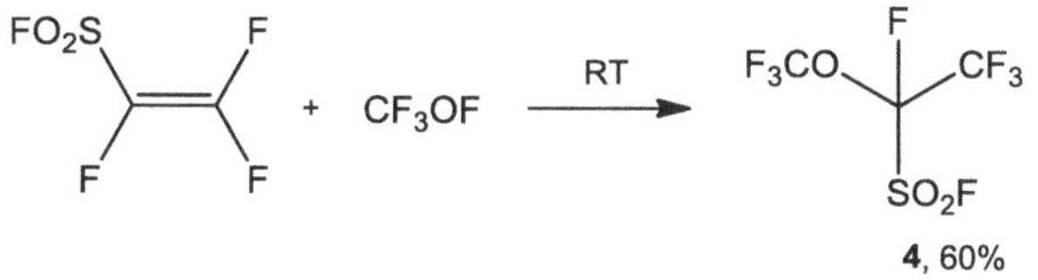

SCHEME 6.3 Synthesis of 1,2,2,2-tetrafluoro-1-(trifluoromethoxy)ethanesulfonyl fluoride **4** using the Navarrini [17] method.

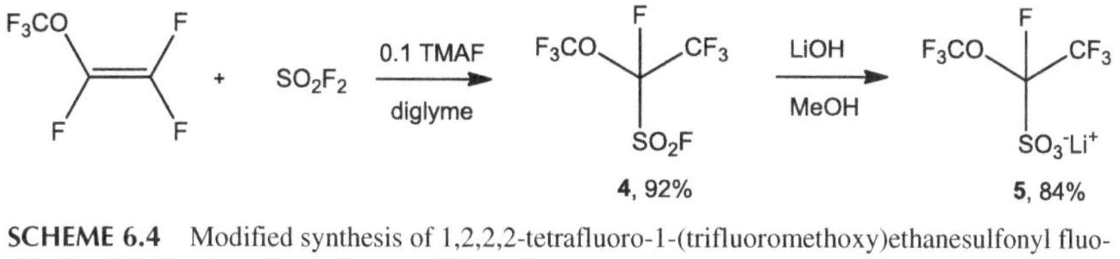

SCHEME 6.4 Modified synthesis of 1,2,2,2-tetrafluoro-1-(trifluoromethoxy)ethanesulfonyl fluoride **4** and further transformation to sulfonate **5**.

conducting salt possess a slightly lower conductivity and oxidation stability as compared to the standard electrolyte salt LiPF$_6$, but an excellent cycle stability and no HF formation after thermal aging at 95 °C.

In the case of lithium-1,2,2,2-tetrafluoro-1-(trifluoromethoxy)-ethanesulfonate **5**, a novel synthetic pathway for the precursor 1,2,2,2-tetrafluoro-1-(trifluoromethoxy)ethanesulfonyl fluoride **4** was developed. Earlier, it had been synthesized from 1,2,2-trifluoroethenesulfonyl fluoride and CF$_3$OF [17], (Scheme 6.3) having the difficulties to handle the powerful oxidant CF$_3$OF.

A straightforward route was reaction of industrially available 1,1,2-trifluoro-2-(trifluoromethoxy) ethylene with sulfuryl fluoride in an aprotic solvent at 50 °C (Scheme 6.4) furnishing **4** followed by reaction with 2 eq LiOH in methanol to finally give the perfluorinated lithium sulfonate **5** (Scheme 6.4).

The conductivity of some tested lithium-1,2,2,2-tetrafluoro-1-(trifluoromethoxy)-ethanesulfonate **5** electrolyte solutions are given in Figure 6.5. These electrolytes included 1 M lithium-1,2,2,2-tetrafluoro-1-(trifluoromethoxy)-ethanesulfonate **5** (marked as LiB-8) solutions with two binary EC:DEC and EC:γ-BL (γ-butyrolactone)(1:1 ratios) solvent mixtures and a ternary EC:γ-BL: fluoroethylene carbonate (FEC) (4.5:4.5:1) solvent mixture compared to the standard 1 M LiPF$_6$ EC:DEC (3:7) electrolyte.

Compared to the lithium-2-methoxy-1,1,2,2-tetrafluoro-ethanesulfonate **3** EC:DEC (1:1) electrolyte, the 1 M lithium-1,2,2,2-tetrafluoro-1-(trifluoromethoxy)-ethanesulfonate **5** (marked as LiB-8) EC:DEC (1:1) electrolyte reached lower conductivities. Two tested electrolytes based on γ-BL obtained nearly the same conductivity at temperatures higher than −10 °C. At lower temperatures the lithium-1,2,2,2-tetrafluoro-1-(trifluoromethoxy)-ethanesulfonate **5** EC:γ-BL:FEC (4.5:4.5:1) electrolyte obtained better conductivities [18].

An overlay of the 1 M lithium-1,2,2,2-tetrafluoro-1-(trifluoromethoxy)-ethanesulfonate **5** EC:DEC (1:1) and EC:γ-BL:FEC (4.5:4.5:1) electrolytes benchmarked against the LiPF$_6$ EC:DEC (3:7) standard electrolyte is displayed Figure 6.6.

The 1 M lithium-1,2,2,2-tetrafluoro-1-(trifluoromethoxy)-ethanesulfonate **5** EC:DEC (1:1) obtained the highest anodic stability, reaching 5.8 V vs Li/Li$^+$. The standard electrolyte was stable up to 5.4 V vs Li/Li$^+$. The FEC containing the LiB-8 electrolyte showed early reactions, starting before 4 V vs Li/Li$^+$ and stabilizing into a current density plateau.

The formation cycles and the subsequent 20 standard cycles are presented in Figure 6.7 displaying the discharge capacities and the corresponding

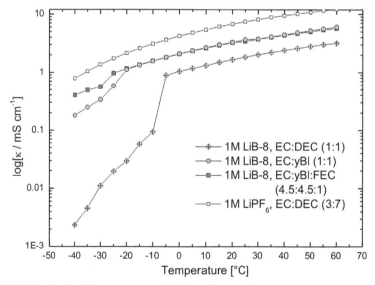

FIGURE 6.5 Conductivity measurements for **5** (marked as LiB-8). Solvents: ethylene carbonate (EC), diethyl carbonate (DEC), γ-butyrolactone (γ-Bl), fluoroethylene carbonate (FEC). *Taken from Ref. [12].*

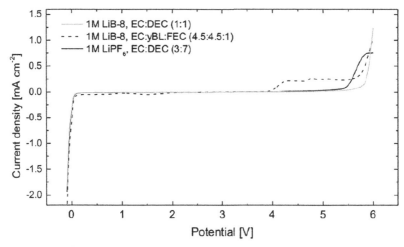

FIGURE 6.6 Electrochemical stability windows of the standard electrolyte and of 1 M **5** in EC:DEC and EC:γ-BL:FEC solvent mixtures. The experiments were conducted on a platinum WE (working electrode) (Ø 1 mm), in oversized lithium metal CE (counter electrode) and a lithium metal reference. *Taken from Ref. [12].*

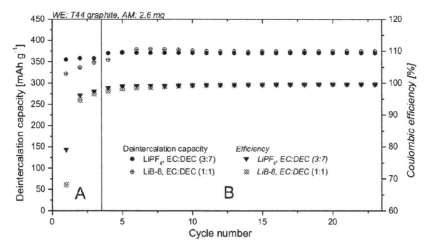

FIGURE 6.7 The constant current/potentiostatic cycling test of the 1 M lithium-1,2,2,2-tetrafluoro-1-(trifluoromethoxy)-ethanesulfonate **5** EC:DEC (1:1) electrolyte and the standard electrolyte (1 M LiPF$_6$, EC:DEC (3:7)). *Taken from Ref. [12].*

efficiencies. Part A shows the three formation cycles and part B the subsequent 30 1C standard cycles. The **5** electrolyte obtained lower capacities and efficiencies within the first cycles. After the formation step, both electrolytes reached the theoretical capacity of the graphite electrode.

Lithium-2-methoxy-1,1,2,2-tetrafluoro-ethanesulfonate **3** and lithium-1,2,2,2-tetrafluoro-1-(trifluoromethoxy)-ethanesulfonate **5** would provide a way to replace LiPF$_6$ as the supporting electrolyte in lithium-ion batteries.

6.3 SOLVENTS AND COSOLVENTS FOR ELECTROLYTE SYSTEMS

6.3.1 Alkyl Tetrafluoro-2-(Alkoxy)Propionates as Cosolvents for Propylene-Carbonate-Based Electrolytes

EC is found to be indispensable in the electrolyte system since it is able to form stable SEI on graphitic anodes, but having poor low-temperature performance [19]. To extend the liquid range of the electrolyte, linear carbonates are usually added, which are, however, highly flammable. Therefore, their use involves safety hazards, particularly for cells with high capacity. An alternative to the state-of-the art electrolyte solvents, namely, propylene carbonate (PC) exhibits low melting point (m.p.) (−55 °C), high boiling point (b.p.) (240 °C), and high flash point (132 °C) combined with high conductivities and good salt dissociation due to its high permittivity (64.92 at 25 °C). On the other hand, PC cannot form an SEI on graphite electrodes, which causes exfoliation of the graphite anode by solvent cointercalation [20]. To overcome this problem different additives have been proposed, such as vinylene carbonate (VC) [21], FEC [18], vinyl compounds [22], sulfite [23] and isocyanate derivatives [24], aromatic esters [25], phenyl tris-2-methoxydiethoxy silane [25b], 2-phenylimidazole [25c], 2-cyanofurane [25d], lithium bis(oxalato)borate [26], *N*-vinyl-2-pyrrolidone [27], 1,3-propane sulfone [28], and cyclic sulfates [26b]. Upon addition of these additives, lithium-ion cells based on graphite anodes can be cycled in PC-based electrolytes without exfoliation. But, there are only a few reports regarding the use of pure PC with additives in combination with graphite anodes. Most of the results reported, in fact, were obtained using mixtures of PC with EC and/or carbonates. Jeong et al. reported in 2001 [21b] on the use of PC with VC as additive in combination with graphite electrodes, using $LiClO_4$, which is not applied in commercial lithium-ion batteries [29].

Methyl tetrafluoro-2-(methoxy) propionate **6a** (MTFMP) successfully suppresses graphite exfoliation in PC-based electrolytes with $LiPF_6$ and is considered a new cosolvent for PC.

MTFMP, **6a** was first described in a patent in 2001 as a solvent for lithium-ion battery electrolytes [30] and synthesized according to the literature [31]. Analogously, ethyl **6b** and isopropyl **6c** derivatives were also produced (Scheme 6.5). Except the report of the solubility of lithium bis(trifluoromethanesulfonyl)imide

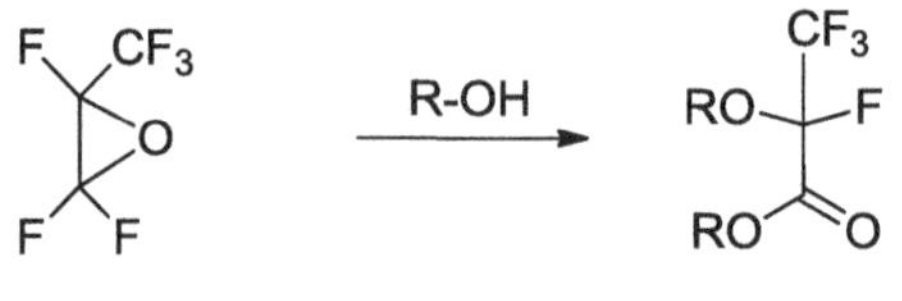

6a: R = Me, 85%
6b: R = Et, 80%
6c: R = *i*Pr, 82%

SCHEME 6.5 Syntheses of alkyl tetrafluoro-2-(alkoxy) propionates **6a–c**.

(LiTFSI) in MTFMP no further physical or electrochemical properties of MTFMP as electrolyte component were reported.

The use of MTFMP **6a** as cosolvent for PC renders the formation of an effective SEI on graphite electrodes, making the use of PC possible. Electrolytes with different ratios of PC and MTFMP **6a** and LiPF$_6$ were studied due to their SEI film-forming ability on graphite anodes. The content of MTFMP **6a** was kept as low as possible. But 10 wt% (5.6 mol%) of MTFMP **6a** was necessary to achieve an effective SEI (see Figure 6.8). Tests without MTFMP **6a** were not performed due to graphite exfoliation. To evaluate the performance of the new electrolyte with respect to a conventional electrolyte, the comparison was extended to 1 M LiPF$_6$ in EC:DEC (3:7 wt%) electrolyte.

The temperature-dependent conductivity of the three electrolytes (see Figure 6.8) was measured in the range of −40 °C and 60 °C. In Figure 6.9, it can be seen that the standard electrolyte 1 M LiPF$_6$ in EC:DEC (3:7 wt%) shows the highest conductivity up to 45 °C, and the two PC-based electrolytes show nearly the same conductivity. Remarkably, above 45 °C and 55 °C the conductivities of 1 M LiPF$_6$ in PC and PC:MTFMP **6a** (9:1 wt%), respectively, are higher than that of 1 M LiPF$_6$ in EC:DEC (3:7 wt%) [33].

To verify the effect of MTFMP **6a** in PC-based electrolytes in terms of formation of an effective SEI, cells with a graphite working electrode were cycled in PC-based electrolytes with and without MTFMP **6a**. For comparison, the same test was also performed with the standard electrolyte 1 M LiPF$_6$ in EC:DEC (3:7 wt%). To avoid limitation of cyclability due to dendritic growth, oversized NCM cathodes were used as lithium source instead of lithium metal foil. In Figure 6.10, the first cycle of the cells with the different electrolytes is shown. The standard electrolyte shows just a small plateau at around 0.8 V

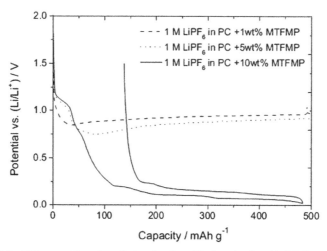

FIGURE 6.8 Voltage profiles of the first cycle of graphite-based cells with PC electrolytes containing different amounts of MTFMP **6a**. *Taken from Ref. [33].*

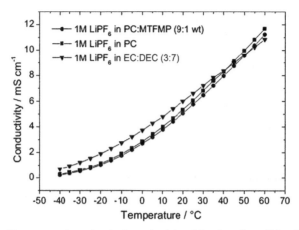

FIGURE 6.9 Temperature-dependent ionic conductivity of the electrolytes. *Taken from Ref. [33].*

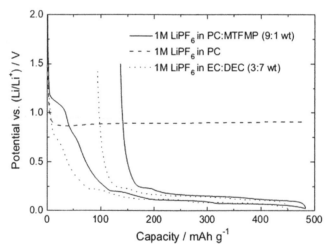

FIGURE 6.10 Voltage profiles of the first cycle of graphite-based cells with different electrolytes. *Taken from Ref. [33].*

versus Li/Li+, which is known to be associated with the formation of an SEI via the decomposition of EC. The formation of the SEI enables cycling of the graphite anode without exfoliation. In contrast, when the PC-based electrolyte is used a continuous electrochemical reaction at 0.9 V versus Li/Li+ takes place, which is associated with the intercalation of PC-solvated lithium ions leading to graphite exfoliation [32]. Due to the exfoliation, the graphite particles are destroyed and, consequently, cycling of the cell is not possible. If 10 wt% MTFMP **6a** is added to the PC-based electrolyte, however, the first faradaic reaction occurs already at 1.2 V versus Li/Li+, being associated with the decomposition of MTFMP [33].

MTFMP forms an effective SEI preventing the intercalation of the solvated lithium ions and thus, exfoliation of the graphite particles. Additionally, the cell with 1 M LiPF$_6$ in PC:MTFMP (9:1 wt%) displays a high coulombic efficiency of 72% in the first cycle, which is 7% lower than that of the cell containing the standard electrolyte (79%). The efficiency of the second and third cycles is 93% and 95%, respectively, for the PC:MTFMP **6a** mixture and 97% and 99%, respectively, for the EC:DEC mixture [33].

The results showed that the addition of 10 wt% of MTFMP **6a** fully prevents exfoliation thus enabling the use of PC-based electrolytes in combination with graphite electrodes. The formation of an effective SEI is observed during the initial three cycles. Long-term cycling tests proved the stability of this SEI layer as the capacity retention was 99% after 300 cycles. In addition, the rate capability of this new electrolyte showed extremely promising results. These results support the feasibility of using MTFMP **6a** as cosolvent for PC in graphite-based lithium-ion batteries [33].

6.3.2 Derivatives of Hexafluorbutandiol as Cosolvents and SEI Additives

Electrolyte additives are known to be an effective and economic approach to improving the stability of electrode surface films [18]. In the past decade, many organic and inorganic compounds have been identified as effective electrolyte additives: examples are VC [21], vinyl ethylene carbonate [34], ethylene sulfite [35], and FEC [18]. Fluorine substitution of organic compounds improves their oxidation stability [36]. Nevertheless, the fluorine substitution simultaneously increases reduction potentials of organic compounds, i.e., causes electrochemical decomposition at higher potentials than those for organic solvents such as EC, PC, and DEC [37]. If electrochemical reduction of organofluorine compounds continues without forming protective surface film (SEI) on carbon anode, irreversible capacity highly increases. However, if decomposed products quickly form SEI on carbon electrode, such fluorine compounds can be used as nonflammable solvents for lithium-ion batteries [38].

The anode material of the commercially available lithium-ion batteries is currently based mostly on graphite. For graphite electrodes, however, a pronounced exfoliation is observed, especially when using PC-containing electrolytes, which leads to destruction of the electrode [20a].

Many fluorinated cyclic carbonates have been synthesized and investigated as additives for lithium-ion batteries [38,39]. But organic solvents such as EC are still very flammable and have high vapor pressure [2]. Further improvement in the safety of the electrolytes was reported by Nakajima et al. [38], namely, solvent mixtures containing cyclic carbonates and 2-oxo-4-trifluoromethyl-1,3-dioxolane. *cis*-4,5-Bis(trifluoromethyl)-1,3-dioxolane-2-one **7** (di-CF$_3$-EC) can prevent the exfoliation of the graphite electrode. The cell at the same time exhibits a high cycling stability. *cis*-4,5-Bis(trifluoromethyl)-1,3-dioxolane-2-one **7** also has high thermal stability, and a high b.p. and a low m.p. In particular, *cis*-4,5-bis(trifluoromethyl)-1,3-dioxolane-2-one **7** is nonflammable, and has a flash point of over 122 °C, allowing its easy handling as an additive or a solvent for electrolytes and a significant improvement in the operational reliability. Moreover, *cis*-4,5-bis(trifluoromethyl)-1,3-dioxolane-2-one **7** shows a small irreversible capacity loss in the formation of a cell, and a high oxidation stability of 5.5 V versus lithium [40].

cis-4,5-Bis(trifluoromethyl)-1,3-dioxolane-2-one **7** is a derivative of hexafluorbutandiol **6** (Scheme 6.6). *meso*-1,1,1,4,4,4-Hexafluoro-2,3-butanediol was synthesized according to the literature [41].

cis-4,5-Bis(trifluoromethyl)-1,3-dioxolane-2-one **7** was accessible by reacting phosgene with *meso*-1,1,1,4,4,4-hexafluoro-2,3-butanediol in the presence of a base [40] (Scheme 6.7).

The conductivity was determined of an electrolyte containing 1 M $LiPF_6$ in a mixture of PC and *cis*-4,5-bis(trifluoromethyl)-1,3-dioxolane-2-one **7** at a temperature range from −10 °C to +60 °C (see Figure 6.11).

The variation of conductivity of the electrolyte 1 M $LiPF_6$ in a 1:1 mixture (wt%) of PC and **7** (di-CF_3-EC/PC 1:1.) in the temperature range of −10 °C until +60 °C is shown in Figure 6.11, revealing good conductivity at 25 °C.

A higher irreversible capacity of the 1 M $LiPF_6$ di-CF_3-EC/PC (1:1) electrolyte can also be noticed in Figure 6.12. The figure plots the voltage versus capacity of the first cycle using the electrolyte 1 M $LiPF_6$ in a mixture of 50 wt% PC and 50 wt% *cis*-4,5-bis(trifluoromethyl)-1,3-dioxolane-2-one **7** (di-CF_3-EC/PC 1:1) compared to a 1 M solution of $LiPF_6$ in a mixture of 30 wt% EC and 70 wt% DEC (EC:DEC, 3:7) and PC. The decomposition and SEI formation of the di-CF_3-EC/PC 1:1 electrolyte occurred at approx. 1.2 V versus Li/Li⁺. There is thus no evidence of exfoliation by PC or *cis*-4,5-bis(trifluoromethyl)-1,3-dioxolan-2-one **7**. In addition, the electrolyte with 72 wt% showed a very good efficiency in the first cycle [40].

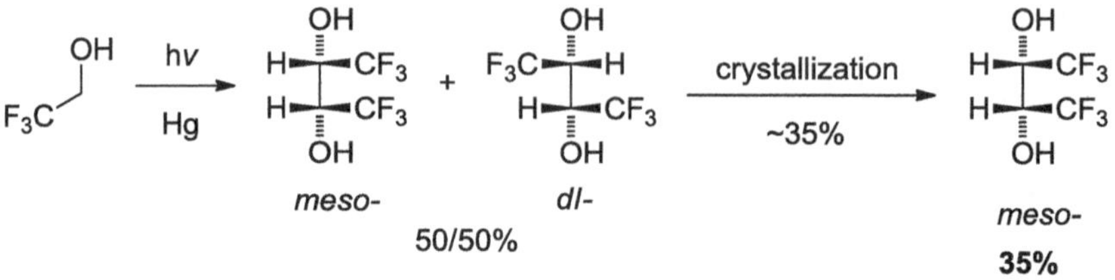

SCHEME 6.6 Synthesis of *meso*-1,1,1,4,4,4-hexafluoro-2,3-butanediol.

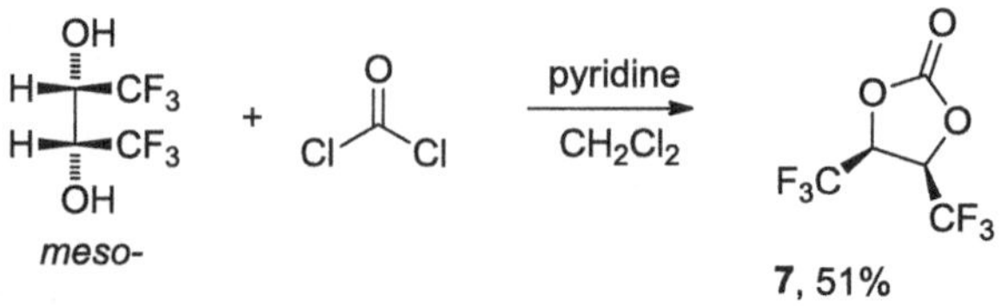

SCHEME 6.7 Synthesis of *cis*-4,5-bis(trifluoromethyl)-1,3-dioxolan-2-one **7**.

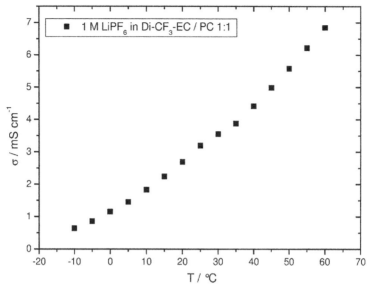

FIGURE 6.11 Conductivity measurements for *cis*-4,5-bis(trifluoromethyl)-1,3-dioxolane-2-one **7**. *Taken from Ref. [40].*

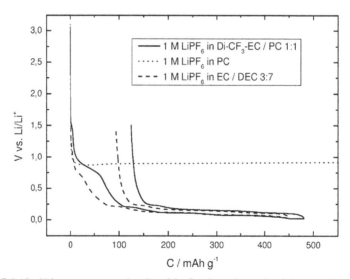

FIGURE 6.12 Voltage versus capacity plot of the first formation cycle of the 1 M LiPF$_6$ di-CF$_3$-EC/PC (1:1) electrolyte and standard electrolytes 1 M LiPF$_6$ EC:DEC (3:7) and 1 M LiPF$_6$ PC. *Taken from Ref. [40].*

For comparison, during SEI formation through the electrolyte containing EC and DEC, the decomposition of the EC is 0.8 V. On the other hand, the representation of the potential profile of the first cycle using the electrolyte 1 M LiPF$_6$ in PC shows the exfoliation of the graphite electrode at 0.9 V.

An electrolyte containing cis-4,5-bis(trifluoromethyl)-1,3-dioxolan-2-one **7** and PC prevents exfoliation of the graphite anode and showed excellent discharge capacity.

6.3.3 Fluorinated 1,2,4-Thiadiazianan-3,5-dion-1,1-dioxides Useful SEI Additives in PC

Nonfluorinated 1,2,4-thiadiazianan-3,5-dion-1,1-dioxide derivatives are already known as cosolvents for Li-ion batteries due to their low volatility, high flash point, and stability [42]. New fluorinated cyclic sulfonamides were synthesized from $FO_2SCF_2C(O)F$ and $FO_2SCF(CF_3)C(O)F$ with dimethyl urea [43] (Scheme 6.8).

A cell with an electrolyte of 1 M $LiPF_6$ and 6-difluoro-2,4-dimethyl-1,2,4-thiadiazianan-3,5-dion-1,1-dioxide **8a** (5 wt%) in PC was prepared and investigated. Exfoliation of graphite was completely prevented. Three stages of lithium intercalation into the graphite were found, as well as an irreversible capacity loss of 200 mAh g^{-1} and a 63% efficiency in the first cycle (Figure 6.13).

Furthermore, 6-difluoro-2,4-dimethyl-1,2,4-thiadiazianan-3,5-dion-1,1-dioxide **8a** has a positive impact on the formation of an SEI layer on the negative electrode, increasing the thermal stability, decreasing the capacity loss, and aging [43].

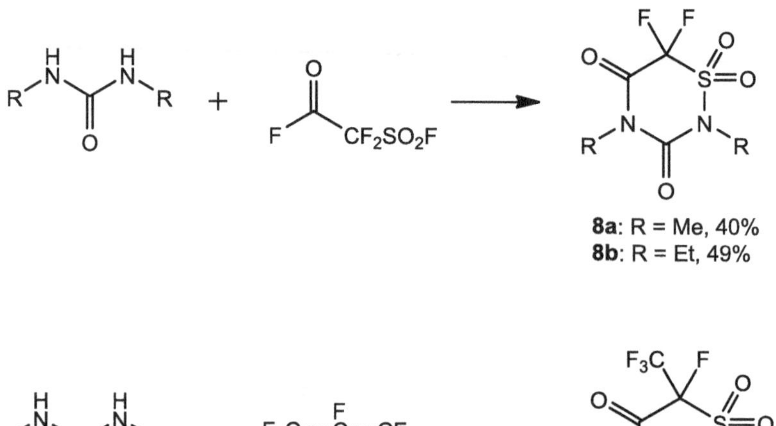

8a: R = Me, 40%
8b: R = Et, 49%

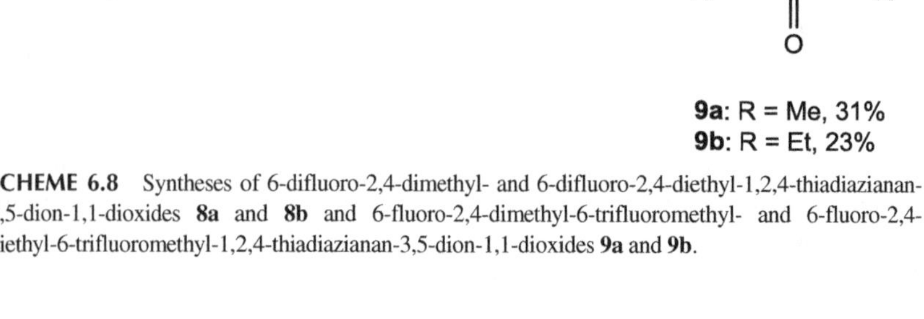

9a: R = Me, 31%
9b: R = Et, 23%

SCHEME 6.8 Syntheses of 6-difluoro-2,4-dimethyl- and 6-difluoro-2,4-diethyl-1,2,4-thiadiazianan-3,5-dion-1,1-dioxides **8a** and **8b** and 6-fluoro-2,4-dimethyl-6-trifluoromethyl- and 6-fluoro-2,4-diethyl-6-trifluoromethyl-1,2,4-thiadiazianan-3,5-dion-1,1-dioxides **9a** and **9b**.

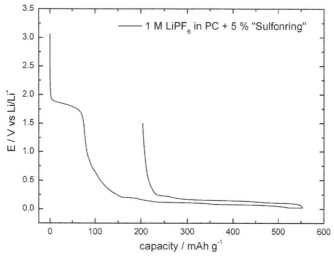

FIGURE 6.13 Voltage versus capacity plot of the first formation cycle of 1 M LiPF$_6$ with 5 wt% of 6-difluoro-2,4-dimethyl-1,2,4-thiadiazianan-3,5-dion-1,1-dioxide **8a** (marked as Sulfonring). *Taken from Ref. [43].*

6.4 OVERCHARGE PROTECTING AGENTS: PF$_5$–CARBENE ADDUCTS

Cell safety can be enhanced by an overcharge electrolyte additive given to the base electrolyte in order to stop the charging process at a defined potential. Overcharge additives should be active under overcharge conditions only, and not take part in any charge-consuming reaction in the normal operation voltage window.

Overcharge additives can decompose irreversibly at the cathode at a defined overcharge potential. This means any overcharge and also the regular charge/discharge reactions are blocked [44]. These additives are frequently called "shutdown additives." The state-of-the-art shutdown additive is biphenyl which decomposes and forms a polymer layer on the cathode, but there is evidence that it influences the cycling behavior in the normal operation voltage window [45]. Other examples for decomposition additives are aromatic compounds such as cyclohexylbenzene, isopropylbenzene and toluene [46], bromobenzyl isocyanate [24]; and vinylene compounds [47]. Investigating other reactive species suitable as overcharge additives, we identified an *N*-heterocyclic diamino carbene (NHC)–PF$_5$ complex, where carbenes are the Lewis bases, potentially reactive. With a Lewis acid partner, here PF$_5$, the reactivity can be possibly moderated in a way that the carbene part of the adduct becomes reactive only after decomposition of the adduct. This decomposition can be probably initiated electrochemically by oxidation [48].

The oxidative addition of 2,2-difluoro-1,3-dimethylimidazolidine and bis(dimethylamino)difluoromethane to PF_3, Cl_2PPh, and Cl_2PMe gave the *N*-heterocyclic diaminocarbenes-PF_4R (NHC^{Me})-PF_4R (R = F, Ph, Me), and acyclic diaminocarbenes-PF_4R (ADC-PF_4R, R = F, Ph, Me), as depicted in Scheme 6.9 [49]. All reactions proceeded quantitatively.

The adduct NHC^{Me}-PF_5 **10b** was studied as a possible overcharge additive for Li-ion batteries [50]. Figure 6.14(a) compares the electrochemical stability window of a $LiPF_6$/EC/DMC electrolyte with the NHC-PF_5 **10b** additive with an electrolyte without the overcharge additive at a platinum working electrode. At higher potentials, in the anodic region, the behavior of the two electrolytes is different. Without additive, the current flow increases at c. 5.5 V vs Li/Li$^+$, indicating electrolyte oxidation. With the NHC-PF_5 **10b** additive a strong oxidation reaction starts already at 4.6 V versus Li/Li$^+$, passes a maximum at 4.9 V versus Li/Li$^+$, decreases up to a potential of 5.3 V versus Li/Li$^+$, and then finally increases similarly to the electrolyte without additive at slightly higher potentials. Obviously, the first oxidation process can be assigned to the reaction of the additive and the second one to the oxidation of the base electrolyte. In the cathodic region a similar behavior can be seen for both electrolytes. Lithium plating starts at c. 0 V versus Li/Li$^+$ and no other reduction reactions, for instance, SEI [20a,32,51] formation or reduction of the additive in solution, are visible. The latter indicates that the adduct does not act as a redox shuttle, having no significant influence on the discharge (=reduction) reaction.

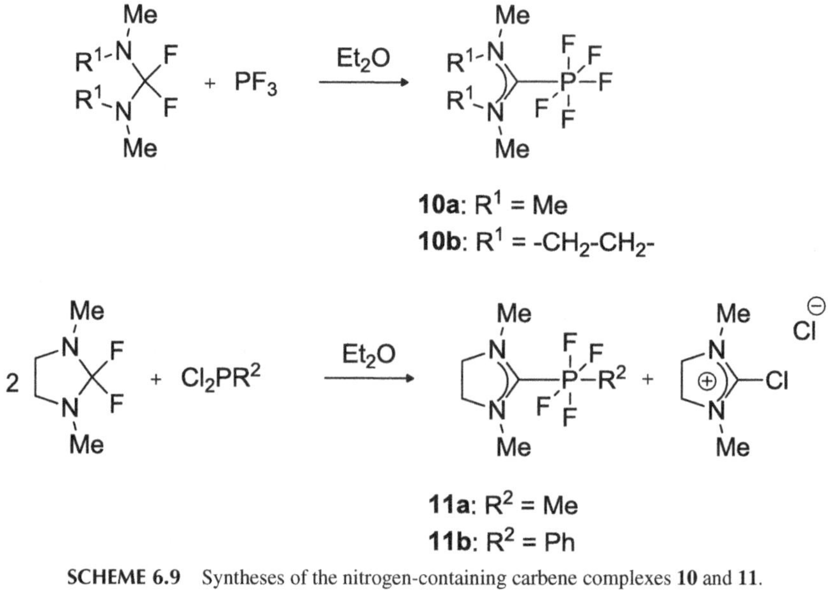

10a: R^1 = Me
10b: R^1 = -CH_2-CH_2-

11a: R^2 = Me
11b: R^2 = Ph

SCHEME 6.9 Syntheses of the nitrogen-containing carbene complexes **10** and **11**.

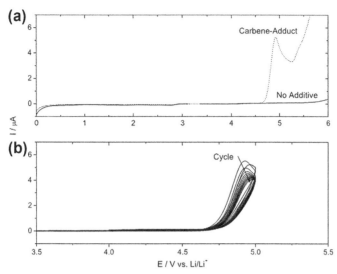

FIGURE 6.14 (a) Stability window in a potential range from −0.025 to 6.0 V versus Li/Li⁺ of 1 M LiPF$_6$/EC/DMC and 0.9 M LiPF$_6$/0.1 M NHC-PF$_5$/EC/DMC (dotted). (b) Cyclic voltammetry in a potential range from 3.0 to 5.0 V versus Li/Li⁺ of 0.9 M LiPF$_6$/0.1 M NHC-PF$_5$/EC/DMC. Working electrode, Pt; counter electrode, Li; reference electrode, Li. Scan rate, 1 mV/s. *Taken from Ref. [52].*

The reversibility of the discovered reaction was checked by cyclic voltammetry on a platinum working electrode within the potential range of 3.0–5.0 V (Figure 6.14(b)). In the first sweep the decomposition starts at 4.6 V versus Li/Li⁺. In the second sweep the decomposition reaction has shifted to slightly higher values which may be explained with an increased resistance, for instance, by surface layer formation on the Pt consisting of decomposition products of the electrolyte. In the following cycles the oxidation reaction can be still observed but the onset of the reaction is shifted toward higher potential values sweep by sweep. It is intrinsic to this voltammetric experiment, that the working electrode is only for a too low amount of time within the potential region, where the oxidation of the electrolyte additive can take place [52].

This is different in constant current experiment. A state-of-the-art LiNi$_{1/3}$Co$_{1/3}$Mn$_{1/3}$O$_2$ (NCM) cathode material was combined with a lithium anode and reference electrode. Figure 6.15 shows the potential profiles of the constant current charge/discharge experiment in a LiPF$_6$/EC/DMC based on an electrolyte without (Figure 6.15(a)) and with (Figure 6.15(b)) additive. For both cells two cycles from 3.0 to 4.3 V versus Li/Li⁺ followed by one overcharge cycle with an increased cutoff potential of 4.8 V versus Li/Li⁺ were performed two times. In the last cycle the potential was increased to 4.95 V versus Li/Li⁺ [52].

In Figure 6.15(a) the cell without additive shows the typical voltage profile of NCM for the first two cycles. In the following overcharge process the NCM

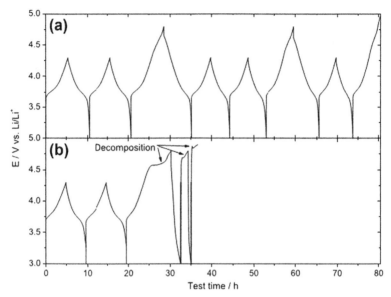

FIGURE 6.15 Constant current cycling of (a) 1 M LiPF$_6$/EC/DMC and (b) 0.9 M LiPF$_6$/0.1 M NHC-PF$_5$/EC/DMC in a half-cell arrangement. Working electrode, NCM; counter electrode, Li; reference electrode, Li. Two cycles in a potential range from 3.0 to 4.3 V versus Li/Li$^+$ with one cycle from 3.0 to 4.8 V versus Li/Li$^+$ in between and one cycle to 4.95 V versus Li/Li$^+$ at C/10, D/10. *Taken from Ref. [52].*

material gets delithiated to a high state of charge. Nevertheless, the NCM material recovers and performs the following cycles to 4.3, 4.8, and to 4.95 V versus Li/Li$^+$ in the last cycle. Consequently, the cathode material becomes highly delithiated without additive, which is always correlated with a decrease in safety. For the electrolyte with additive (Figure 6.15(b)), the cycling behavior is not influenced by the additive in the first two cycles (with a cutoff potential between 3.0 and 4.3 V vs Li/Li$^+$). In the third cycle, the cutoff potential is increased, and a long potential plateau starting at 4.57 V versus Li/Li$^+$ can be observed, which can be correlated to the continuous oxidation of the adduct additive. The increase in potential after the plateau can be attributed to electrolyte decomposition. The decomposition products of the adduct by oxidation obviously block the NCM cathode material from further lithium insertion and extraction, as in the following discharge step and in the following cycles almost no discharge capacity is observed. It can be concluded that under overcharge conditions, the adduct additive hinders the NCM electrode to undergo a fast potential increase, whereas without overcharge additive the potential shoots up fast. The effect of the adduct additive, however, is an irreversible one, as the cell is electrochemically inactive afterward. This indicates that a film of decomposition products has been irreversibly formed on the NCM electrode surface and assigns the adduct additive to the group of shutdown additives [52].

REFERENCES

[1] S.A. Ullrich, G.L. Gard, R.L. Nafshun, M.M. Lerner, J. Fluorine Chem. 79 (1996) 33–38.
[2] K. Xu, Chem. Rev. 104 (2004) 4303–4418.
[3] B.B. Hallac, O.E. Geiculescu, R.V. Rajagopal, S.E. Creager, D.D. DesMarteau, Electrochim. Acta 53 (2008) 5985–5991.
[4] [a] J.B. Chao, S. Aoyama, JP2005267857 (2005).
 [b] S.J. Hamrock, A.D. Fanta, W.M. Lamanna, B.J. Johnson, S.D. Boyd, H. Shimada, P.P. Pham, WO2001052341 (2001).
 [c] S. Iwatani, JP2011159496 (2011).
 [d] M. Kim, US20140065476 (2014).
 [e] F. Kita, H. Sakata, S. Sinomoto, A. Kawakami, H. Kamizori, T. Sonoda, H. Nagashima, J. Nie, N.V. Pavlenko, Y.L. Yagupolskii, J. Power Sources 90 (2000) 27–32
 [f] W.M. Lamanna, R.B. Loch, WO9930381 (1999).
 [g] S. Mochizuki, T. Kubota, JP 2008198409 (2008).
 [h] G. Nagasubramanian, D.H. Shen, S. Surampudi, Q. Wang, G.K.S. Prakash, Electrochim. Acta 40 (1995) 2277–2280
 [i] L. Sun, WO9706207 (1997).
[5] [a] V.R. Koch, L.A. Dominey, J.L. Goldman, M.E. Langmuir, J. Power Sources 20 (1987) 287–291.
 [b] W.M. Lamanna, A. Xiao, M.J. Triemert, P.P. Pham, WO2012170240 (2012).
 [c] A. Yano, Y. Sasaki, K. Kojima, F. Kita, US20120301784 (2012).
[6] W.M. Lamanna, L.J. Krause, G. Moore, S.J. Hamrock, US5514493 (1996).
[7] [a] M. Amereller, R. Schmitz, R. Schmitz, R.A. Mueller, M. Winter, S. Passerini, C. Schreiner, M. Kunze, WO 2013127864 (2013).
 [b] G. Sobotta, J. Pazderski, W. Skatschko, W. Lezjuk, O. Tagajew, EP 596485 (1994).
[8] A. Blakeney, O. Dimov, L. Ferreira, J. Hatfield, J. Kocab, J. Spaziano, US20020197558 (2002).
[9] A. Feiring, M. Harmer, C. Junk, F. Schadt, and Z. Schnepp, US20060276670 (2006).
[10] [a] R. Arvai, F. Toulgoat, M. Médébielle, B. Langlois, F. Alloin, C. Iojoiu, J.Y. Sanchez, J. Fluorine Chem. 129 (2008) 1029–1035.
 [b] E. Paillard, F. Toulgoat, C. Iojoiu, F. Alloin, J. Guindet, M. Medebielle, B. Langlois, J.Y. Sanchez, J. Fluorine Chem. 134 (2012) 72–76.
 [c] F. Toulgoat, B.R. Langlois, M. Médébielle, J.Y. Sanchez, J. Org. Chem. 72 (2007) 9046–9052.
[11] M. Winter, C. Schreiner, A. Lex-Balducci, R. Müller, R. Schmitz, R. Schmitz, N. Kalinovich, K. Vlasov, G.V. Röschenthaler, DE102011052156 (2011).
[12] M. Winter, C. Schreiner, A. Lex-Balducci, R. Müller, R. Schmitz, R. Schmitz, N. Kalinovich, K. Vlasov, G.V. Röschenthaler, EP2013/053949 (2013).
[13] N. Kalinovich, O. Kazakova, K. Vlasov, M. Winter, G.V. Röschenthaler, Org. Process Res. Dev., Submitted for publication.
[14] D.C. England, M.A. Dietrich, R.V. Lindsey, J. Am. Chem. Soc. 82 (1960) 6181–6188.
[15] P.N.D. Singh, R.F. Klima, S. Muthukrishnan, R.S. Murthy, J. Sankaranarayanan, H.M. Stahlecker, B. Patel, A.D. Gudmundsdóttir, Tetrahedron Lett. 46 (2005) 4213–4217.
[16] [a] L.J. Krause, W.M. Lamanna, J. Summerfield, M. Engle, G. Korba, R. Loch, R. Atanasoski, J. Power Sources 68 (1997) 320–325.
 [b] S.S. Zhang, T.R. Jow, J. Power Sources 109 (2002) 458–464.
[17] V. Montanari, W. Navarrini, A.M. Staccione, EP0466483 (1996).
[18] R. McMillan, H. Slegr, Z.X. Shu, W. Wang, J. Power Sources 81–82 (1999) 20–26.

[19] Y. Ein-Eli, S.R. Thomas, R. Chadha, T.J. Blakley, V.R. Koch, J. Electrochem. Soc. 144 (1997) 823–829.
[20] [a] M. Winter, Z. Phys. Chem 223 (2009) 1395–1406.
[b] M.R. Wagner, J.H. Albering, K.C. Moeller, J.O. Besenhard, M. Winter, Electrochem. Commun. 7 (2005) 947–952.
[21] [a] H. Buqa, A. Würsig, J. Vetter, M.E. Spahr, F. Krumeich, P. Novák, J. Power Sources 153 (2006) 385–390.
[b] S.-K. Jeong, M. Inaba, R. Mogi, Y. Iriyama, T. Abe, Z. Ogumi, Langmuir 17 (2001) 8281–8286.
[22] [a] H.J. Santner, C. Korepp, M. Winter, J.O. Besenhard, K.C. Möller, Anal. Bioanal. Chem. 379 (2004) 266–271.
[b] K. Abe, H. Yoshitake, T. Kitakura, T. Hattori, H. Wang, M. Yoshio, Electrochim. Acta 49 (2004) 4613–4622.
[c] J.-T. Lee, Y.-W. Lin, Y.-S. Jan, J. Power Sources 132 (2004) 244–248.
[23] [a] R. Chen, F. Wu, L. Li, Y. Guan, X. Qiu, S. Chen, Y. Li, S. Wu, J. Power Sources 172 (2007) 395–403.
[b] H. Ota, T. Akai, H. Namita, S. Yamaguchi, M. Nomura, J. Power Sources 119–121 (2003) 567–571.
[c] G.H. Wrodnigg, J.O. Besenhard, M. Winter, J. Electrochem. Soc. 146 (1999) 470–472.
[d] G.H. Wrodnigg, J.O. Besenhard, M. Winter, J. Power Sources 97–98 (2001) 592–594.
[e] G.H. Wrodnigg, T.M. Wrodnigg, J.O. Besenhard, M. Winter, Electrochem. Commun. 1 (1999) 148–150.
[24] [a] C. Korepp, W. Kern, E.A. Lanzer, P.R. Raimann, J.O. Besenhard, M. Yang, K.C. Möller, D.T. Shieh, M. Winter, J. Power Sources 174 (2007) 637–642.
[b] C. Korepp, W. Kern, E.A. Lanzer, P.R. Raimann, J.O. Besenhard, M.H. Yang, K.C. Möller, D.T. Shieh, M. Winter, J. Power Sources 174 (2007) 387–393.
[25] [a] J.-T. Lee, M.-S. Wu, F.-M. Wang, Y.-W. Lin, M.-Y. Bai, P.-C.J. Chiang, J. Electrochem. Soc. 152 (2005) A1837–A1843.
[b] Q. Xia, B. Wang, Y.P. Wu, H.J. Luo, S.Y. Zhao, T. van Ree, J. Power Sources 180 (2008) 602–606.
[c] B. Wang, Q.T. Qu, L.C. Yang, Q. Xia, Y.P. Wu, D.L. Zhou, X.J. Gu, T. van Ree, J. Power Sources 189 (2009) 757–760.
[d] C. Korepp, H.J. Santner, T. Fujii, M. Ue, J.O. Besenhard, K.C. Möller, M. Winter, J. Power Sources 158 (2006) 578–582.
[26] [a] K. Xu, S. Zhang, T.R. Jow, Electrochem. Solid-State Lett. 8 (2005) A365–A368.
[b] A. Sano, S. Maruyama, J. Power Sources 192 (2009) 714–718.
[27] B. Wang, H.P. Zhang, L.C. Yang, Q.T. Qu, Y.P. Wu, C.L. Gan, D.L. Zhou, Electrochem. Commun. 10 (2008) 1571–1574.
[28] G. Park, H. Nakamura, Y. Lee, M. Yoshio, J. Power Sources 189 (2009) 602–606.
[29] G.H. Newman, R.W. Francis, L.H. Gaines, B.M.L. Rao, J. Electrochem. Soc. 127 (1980) 2025–2027.
[30] O. Boese, M. Rieland, D. Seffer, W. Kalbreyer, WO0178183 (2001).
[31] B. Dolenský, J. Kvíčala, J. Paleček, O. Paleta, J. Fluorine Chem. 115 (2002) 67–74.
[32] M. Winter, W.K. Appel, B. Evers, T. Hodal, K.-C. Möller, I. Schneider, M. Wachtler, M.R. Wagner, G.H. Wrodnigg, J.O. Besenhard, Monatsh. Chem. 132 (2001) 473–486.
[33] [a] R. Schmitz, R. Schmitz, R. Müller, O. Kazakova, N. Kalinovich, G.V. Röschenthaler, M. Winter, S. Passerini, A. Lex-Balducci, J. Power Sources 205 (2012) 408–413.
[b] M. Winter, G.V. Röschenthaler, R. Schmitz, N. Kalinovich, S. Passerini, A. Lex-Balducci, R. Müller, S. Nowak, M. Kunze, E. Kraemer, O. Kazakova, WO2012084066 (2012).

[34] Y. Hu, W. Kong, H. Li, X. Huang, L. Chen, Electrochem. Commun. 6 (2004) 126–131.
[35] H. Sano, H. Sakaebe, H.J. Matsumoto, Electrochem. Soc. 158 (2011) A316–A321.
[36] T. Nakajima, K. Dan, M. Koh, T. Ino, T. Shimizu, J. Fluorine Chem. 111 (2001) 167–174.
[37] T. Achiha, T. Nakajima, Y. Ohzawa, M. Koh, A. Yamauchi, M. Kagawa, H. Aoyama, J. Electrochem. Soc. 157 (2010) A707–A712.
[38] Y. Matsuda, T. Nakajima, Y. Ohzawa, M. Koh, A. Yamauchi, M. Kagawa, H. Aoyama, J. Fluorine Chem. 132 (2011) 1174–1181.
[39] [a] M. Bomkamp, J. Olschimke, J. Eicher, WO2011048053 (2011).
[b] Y. Fujiwara, N. Saino, G. Ko, JP2010126477 (2010).
[c] M. Kobayashi, K. Muraishi, Y. Fujiwara, G. Ko, JP2010024216 (2010).
[d] M. Koh, M. Kagawa, A. Yamauchi, US2011118485 (2011).
[e] H. Sakata, M. Koh, A. Yamauchi, H. Nakazawa, T. Sanagi, A. Nakazono, Y. Adachi, K. Sawaki, A. Tani, M. Tomita, JP2012216389 (2012).
[f] H. Sakata, M. Koh, A. Yamauchi, H. Nakazawa, T. Sanagi, A. Nakazono, Y. Adachi, K. Sawaki, A. Tani, M. Tomita, JP2012216387 (2012). (g) X.J. Wang, H.S. Lee, H. Li, X.Q. Yang, X. Huang, J. Electrochem. Commun. 12 (2010) 386–389(h)X. Zhang, Z. Yang, H. Lu, G. Wu, L. Zhou, WO2014026432 (2014).
[i] Y. Zhu, M.D. Casselman, Y. Li, A. Wei, D.P. Abraham, J. Power Sources 246 (2014) 184–191.
[40] R. Schmitz, R.A. Müller, S. Passerini, R.W. Schmitz, J. Kasnatscheew, M. Winter, T. Schedlbauer, C. Schreiner, G.V. Röschenthaler, N. Kalinovich, DE102012104567 (2012).
[41] S.H. Brown, R.H. Crabtree, J. Am. Chem. Soc. 111 (1989) 2935–2946.
[42] S.B. Bae, DE1020090053469 (2009).
[43] N. Kalinovich, G.V. Röschenthaler, T. Schedlbauer, J. Kasnatscheew, K. Vlasov, R. Schmitz, R.W. Schmitz, R.A. Mueller, M. Winter, S. Passerini, WO2013189481 (2013).
[44] L. Xiao, X. Ai, Y. Cao, H. Yang, Electrochim. Acta 49 (2004) 4189–4196.
[45] S. Tobishima, Y. Ogino, Y. Watanabe, J. Appl. Electrochem. 33 (2003) 143–150.
[46] N. Iwayasu, H. Honbou, T. Horiba, J. Power Sources 196 (2011) 3881–3886.
[47] H.J. Santner, K.C. Möller, J. Ivančo, M.G. Ramsey, F.P. Netzer, S. Yamaguchi, J.O. Besenhard, M. Winter, J. Power Sources 119–121 (2003) 368–372.
[48] [a] K.C. Möller, T. Hodal, W.K. Appel, M. Winter, J.O. Besenhard, J. Power Sources 97–98 (2001) 595–597.
[b] E. Krämer, R. Schmitz, S. Passerini, M. Winter, C. Schreiner, Electrochem. Commun. 16 (2012) 41–43.
[49] [a] T. Böttcher, O. Shyshkov, M. Bremer, B.S. Bassil, G.V. Röschenthaler, Organometallics 31 (2011) 1278–1280.
[b] T. Böttcher, S. Steinhauer, N. Allefeld, B. Hoge, B. Neumann, H.G. Stammler, B.S. Bassil, M. Winter, N.W. Mitzel, G.V. Röschenthaler, Dalton Trans. 43 (2014) 2979–2987.
[50] C. Dippel, A. Lex-Balducci, M. Winter, M. Kunze, R. Schmitz, R.A. Müller, S. Passerini, N. Kalinovich, G.V. Röschenthaler, T. Böttcher, DE102011055028 (2013).
[51] H. Schranzhofer, J. Bugajski, H.J. Santner, C. Korepp, K.C. Möller, J.O. Besenhard, M. Winter, W. Sitte, J. Power Sources 153 (2006) 391–395.
[52] C. Dippel, R. Schmitz, R. Müller, T. Böttcher, M. Kunze, A. Lex-Balducci, G.V. Röschenthaler, S. Passerini, M.J. Winter, Electrochem. Soc. 159 (2012) A1587–A1590.

Chapter 7

Safety Improvement of Lithium Ion Battery by Organofluorine Compounds

Tsuyoshi Nakajima
Department of Applied Chemistry, Aichi Institute of Technology, Yakusa, Toyota, Japan

Chapter Outline

7.1 Introduction 147
7.2 Organofluorine Compounds 148
7.3 Differential Scanning Calorimetry Study on the Thermal Stability of Fluorine Compound-Mixed Electrolyte Solutions 151
    7.3.1 Reactions of Metallic Lithium (Li) or Lithium-Intercalated Graphite (LiC_6) with Cyclic and Linear Carbonate Solvents 151
    7.3.2 Reactions of Metallic Lithium (Li) with Organofluorine Compounds and Those of Lithium-Intercalated Graphite (LiC_6) with Fluorine Compound-Mixed Electrolyte Solutions 154
    7.3.3 Influence of Molecular Structures of Fluorine Compounds on the Reactivity with Li 161
7.4 Electrochemical Oxidation Stability of Fluorine Compound-Mixed Electrolyte Solutions 161
7.5 Charge/Discharge Behavior of Natural Graphite Electrodes in Fluorine Compound-Mixed Electrolyte Solutions 165
7.6 Conclusions 171
References 171

7.1 INTRODUCTION

Lithium ion batteries have a possibility of firing and/or exploding at high temperatures, by short circuit formation, by overcharging, and so on since they use flammable organic solvents. Mixtures of cyclic carbonates, such as ethylene carbonate (EC) and propylene carbonate (PC), with high dielectric constants and linear carbonates, such as dimethyl carbonate (DMC), ethyl methyl carbonate (EMC), and diethyl carbonate (DEC), with low viscosities are usually used as solvents for lithium ion batteries. Cyclic carbonates with high dielectric constants enable the dissolution of inorganic electrolytes, such as $LiPF_6$, and linear carbonates with low viscosities make the diffusion of Li^+ ions easy. High thermal and oxidation stability is one of the most important issues for the practical applications of lithium ion batteries. Recently, lithium ion batteries with high charge and discharge rates are required for their application to hybrid and electric vehicles. High oxidation stability of electrolyte solutions is particularly requested for the use of 5 V class cathodes. Several different types of compounds have been examined to improve the thermal and oxidation stability of electrolyte solutions. Phosphorus compounds, such as phosphates, usually show good flame retardant properties [1–32]. Mixing of ionic liquids is also useful to increase the oxidation stability of electrolyte solutions [33–37]. In addition to these compounds, organofluorine compounds are new candidates to improve the thermal and oxidation stability of lithium ion batteries [38–49].

7.2 ORGANOFLUORINE COMPOUNDS

It is expected that fluorine substitution of organic compounds enhances the stability to oxidation because it decreases highest occupied molecular orbital (HOMO) levels of organic compounds. Partially fluorinated organic compounds shown in Figure 7.1 have been examined in the study [42–46]. Their HOMO and lowest unoccupied molecular orbital (LUMO) energies are given in Table 7.1, in which compounds, A~J are fluorine compounds used in the study, and A-H~J-H are the same type of compounds where fluorine is replaced with hydrogen [42,43]. The fluorine compounds are mainly cyclic and linear fluorocarbonates and -ethers, and are miscible with solvent mixtures consisting of cyclic and linear carbonates. The miscibility of highly fluorinated organic compounds with EC and PC having high dielectric constants is low. Further, highly fluorinated organic compounds reduce the solubility of inorganic electrolyte such as $LiPF_6$. Therefore, partially fluorinated carbonates, ethers, and so on are preferable as new solvents with high oxidation stability. However, the decrease in LUMO levels by fluorine substitution also suggests the increase in reduction potentials of organic compounds, that is, it is inferred that organofluorine compounds shown in Figure 7.1 are electrochemically reduced at the higher potentials than those for EC, PC, DMC, EMC, and DEC. If electrochemically decomposed products contribute to the formation of a protective surface film (solid electrolyte interphase: SEI) on a graphite anode, organofluorine compounds are quite useful solvents. It is therefore necessary to check charge/discharge characteristics of graphite in the low potential region in organofluorine compound-mixed electrolyte solutions.

(A)

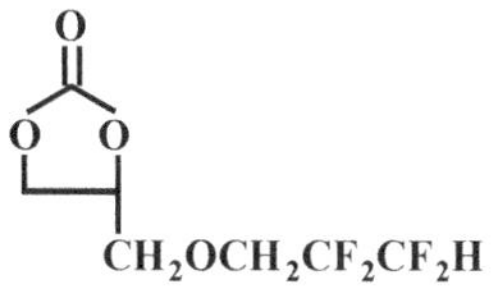

[structure: 1,3-dioxolan-2-one with CH$_2$OCH$_2$CF$_2$CF$_2$H substituent]

4-(2,2,3,3-tetrafluoro-propoxymethyl)-[1,3]dioxolan-2-one

(B)

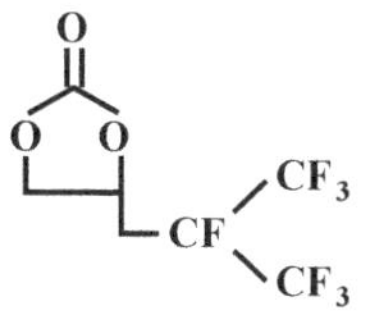

[structure: 1,3-dioxolan-2-one with CF(CF$_3$)$_2$ substituent]

4-(2,3,3,3-tetrafluoro-2-trifluoromethyl-propyl)-[1,3]dioxolan-2-one

(C)

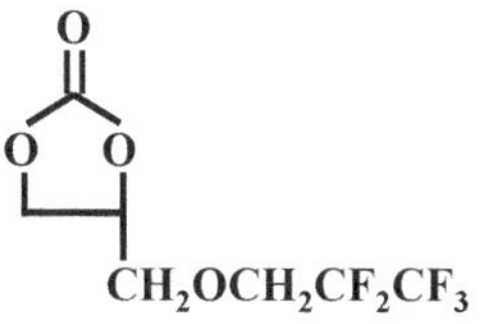

[structure: 1,3-dioxolan-2-one with CH$_2$OCH$_2$CF$_2$CF$_3$ substituent]

4-(2,2,3,3,3-pentafluoro-propoxymethyl)-[1,3]dioxolan-2-one

(D)

HCF$_2$CF$_2$CH$_2$OCOOCH$_2$CF$_2$CF$_2$H

bis(2,2,3,3-tetrafluoro-propyl) carbonate

(E)

CF$_3$CH$_2$OCOOCH$_2$CF$_3$

bis(2,2,2-trifluoroethyl) carbonate

(F)

CF$_3$CH$_2$OCOOCH$_3$

2,2,2-trifluoroethyl methyl carbonate

FIGURE 7.1 Organofluorine compounds used in the study.

(G)

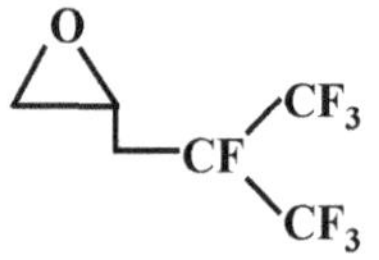

[structure: oxirane ring with -CF(CF₃)(CF₃) substituent]

2-(2,3,3,3-tetrafluoro-2-trifluoromethyl-propyl)-1-oxitrane

(H)

HCF₂CF₂CH₂OCF₂CF₂H

3-(1,1,2,2-tetrafluoroethoxy)-1,1,2,2-tetrafluoropropane

(I)

HCF₂CF₂CH₂OCF₂CFClH

3-(2-chloro-1,1,2-trifluoroethoxy)-1,1,2,2-tetrafluoropropane

(J)

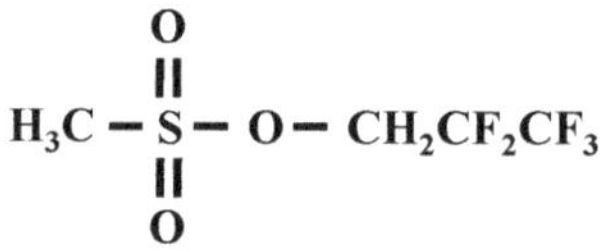

$$H_3C - \underset{\underset{O}{\|}}{\overset{\overset{O}{\|}}{S}} - O - CH_2CF_2CF_3$$

2,2,3,3,3-pentafluoropropyl methanesulfonate

FIGURE 7.1 — Continued

TABLE 7.1 HOMO and LUMO Energies of Fluorine Compounds and the Same Type Compounds Without Fluorine, Calculated by Spartan '06 Semiempirical Method Using AM1

Compound	HOMO (kJmol^{-1})	LUMO (kJmol^{-1})
A-H	−1068.5	100.4
A	−1143.3	64.2
B-H	−1122.1	111.3
B	−1183.6	−1.3
C-H	−1067.9	120.3
C	−1148.3	−19.1
D-H	−1100.2	100.6
D	−1151.1	15.5
E-H	−1099.4	132.4
E	−1207.0	2.0

TABLE 7.1 HOMO and LUMO Energies of Fluorine Compounds and the Same Type Compounds Without Fluorine, Calculated by Spartan '06 Semiempirical Method Using AM1—cont'd

Compound	HOMO (kJmol^{-1})	LUMO (kJmol^{-1})
F-H	−1105.0	133.5
F	−1165.0	60.2
G-H	−1073.2	257.7
G	−1140.7	−8.3
H-H	−1002.5	285.2
H	−1154.4	23.0
I-H	−1031.7	122.9
I	−1143.2	−12.4
J-H	−1095.6	−67.5
J	−1163.3	−126.0

A~J: fluorine compounds shown in Figure 7.1.
A-H~J-H: same-type compounds in which F is replaced with H.

7.3 DIFFERENTIAL SCANNING CALORIMETRY STUDY ON THE THERMAL STABILITY OF FLUORINE COMPOUND-MIXED ELECTROLYTE SOLUTIONS

7.3.1 Reactions of Metallic Lithium (Li) or Lithium-Intercalated Graphite (LiC$_6$) with Cyclic and Linear Carbonate Solvents

Lithium-intercalated graphite (LiC$_6$) decomposes above approximately 100 °C, releasing fresh Li, which then reacts with organic solvents of the lithium ion battery. Figure 7.2 shows differential scanning calorimetry (DSC) profiles for the reactions of metallic Li with EC and PC [50]. Endothermic peaks observed at 38 and 182 °C indicate the melting of EC and metallic Li, respectively. No exothermic peak was observed below 182 °C probably because the surface oxide film on metallic Li prevents the reactions of Li with EC and PC. As soon as metallic Li melted, the reaction of Li with EC started, giving an exothermic peak at 217 °C. On the other hand, PC is more stable than EC. The reaction of Li with PC started at 234 °C, giving an exothermic peak at 257 °C. Figure 7.3 shows the reactions of LiC$_6$ with EC and PC [50]. EC reacted with Li released from graphite, giving an exothermic peak at 174 °C while PC did not react up to 300 °C.

Lithium alkyl dicarbonate (CH$_2$OCO$_2$Li)$_2$ is the main product of the reaction of EC with LiC$_6$ [51–56]. The formation of (CH$_2$OCO$_2$Li)$_2$ may consist

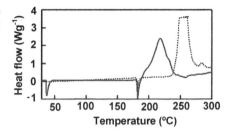

FIGURE 7.2 DSC curves for mixtures of metallic Li and EC or PC [50]. Li/EC: ———, Li/PC: ··········.

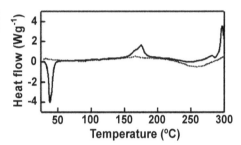

FIGURE 7.3 DSC curves for $Li_{0.99}C_6$/EC and $Li_{0.96}C_6$/PC mixtures [50]. $Li_{0.99}C_6$/EC: ———, $Li_{0.96}C_6$/PC: ··········.

of two-step reactions: radical formation reaction (7.1) and the subsequent dimerization reaction of two radical species (7.2). The result of the DSC study suggests that lithium alkyl dicarbonate $(CH_2OC(O)O^-Li^+)_2$ is easily formed from EC.

$$\text{(EC)} + Li \rightarrow [\cdot CH_2CH_2OC(O)O^-Li^+] \quad (7.1)$$

$$2[\cdot CH_2CH_2OC(O)O^-Li^+] \rightarrow (CH_2OC(O)O^-Li^+)_2 + CH_2=CH_2 \quad (7.2)$$

$$\text{(PC)} + Li \rightleftarrows [CH_3\dot{C}HCH_2OC(O)O^-Li^+] \quad (7.3)$$

$$[CH_3\dot{C}HCH_2OC(O)O^-Li^+] + Li \rightarrow Li_2CO_3 + CH_2=CH_2CH_3 \quad (7.4)$$

However, the reaction of PC with Li or LiC_6 does not proceed smoothly as shown in Figures 7.2 and 7.3, respectively. Table 7.2 shows that the C–O bond in –O–CH(CH$_3$)– group of PC is slightly longer than that for the –O–CH$_2$– group, that is, more ionic in the –O–CH(CH$_3$)– group (1.451 nm) than in the –O–CH$_2$– group (1.441 nm), which is consistent with the fact that the electrostatic charge of the oxygen atom is lower in the –O–CH(CH$_3$)– group

TABLE 7.2 Bond Lengths and Electrostatic Charges of Oxygen and Carbon of Propylene Carbonate (PC) [50]

C–O bond	O=C<	–O–CH$_2$–	–O–CH(CH$_3$)–
C–O bond length (nm)	1.216	1.441	1.451
Electrostatic charge of O	−0.504	−0.411	−0.445
Electrostatic charge of C	0.908	−0.096	+0.200

TABLE 7.3 Bond Lengths and Electrostatic Charges of Oxygen and Carbon of Ethylene Carbonate (EC) [50]

C–O Bond	O=C<	–O–CH$_2$–
C–O bond length (nm)	1.215	1.442
Electrostatic charge of O	−0.514	−0.414
Electrostatic charge of C	0.939	−0.066

(E: −0.445) than in the –O–CH$_2$– group (E: −0.411), and that of carbon atom bonded to oxygen is higher in the –O–CH(CH$_3$)– group (E: +0.200) than in the –O–CH$_2$– group (E: −0.096) (Table 7.3: data of EC for comparison) [50]. This suggests that the longer C–O bond in the –O–CH(CH$_3$)– group is more easily broken than that in –O–CH$_2$– group by the reaction with Li. However, a combination reaction of two radicals similar to Eqn (7.2) may not easily proceed due to the presence of a methyl group. This may be the reason why PC is more stable than EC against Li and LiC$_6$. PC did not react with Li when Li$_{0.96}$C$_6$ was used (Figure 7.3), but reacted with Li at a higher temperature than for EC when metallic Li was used because the amount of metallic Li used was much larger than that of Li released from LiC$_6$. The radical species formed in Eqn (7.3) may further react with Li if the amount of Li is enough. A possible reaction is the formation of Li$_2$CO$_3$ as given in Eqn (7.4) as previously reported [56]. The difficulty in the formation of lithium alkyl carbonate may be the main reason why PC is electrochemically decomposed on graphite without the formation of SEI.

On the other hand, DEC, EMC, and DMC provided similar DSC curves to each other, giving exothermic peaks by the reactions with deintercalated Li at 130–135 °C lower than 174 °C for EC (Figure 7.4) [50]. Exothermic peaks >250 °C may be due to the thermal decomposition of organic solvents and SEI. In the case of linear carbonates, lithium alkyl carbonates may be formed by one-step reactions as shown in Eqns (7.5)–(7.7). This would make the reactions of DEC, EMC, and DMC with Li easier than those of PC and EC.

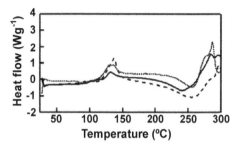

FIGURE 7.4 DSC curves for $Li_{0.97}C_6$/DEC, $Li_{0.95}C_6$/EMC and $Li_{0.85}C_6$/DMC mixtures [50]. $Li_{0.97}C_6$/DEC: ——, $Li_{0.95}C_6$/EMC: - - - - -, $Li_{0.85}C_6$/DMC: ••••••••••.

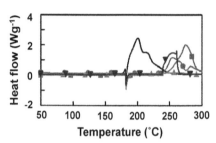

FIGURE 7.5 DSC curves for mixtures of metallic Li and organic solvent or organofluorine compound [46]. ——: EC/DMC(1:1 vol), —●—: **A**, —■—: **B**, —▼—: **G**, —▲—: **J**.

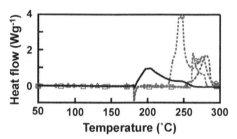

FIGURE 7.6 DSC curves for mixtures of metallic Li and organic solvent or organofluorine compound [45]. ——: EC/DMC, ---◇---: **C**, - - -▲- - -: **H**, ----□----: **I**.

$$CH_3CH_2OC(O)OCH_2CH_3 + Li \rightarrow CH_3CH_2OC(O)O^-Li^+ + [\cdot CH_2CH_3] \quad (7.5)$$

$$CH_3OC(O)OCH_2CH_3 + Li \rightarrow CH_3OC(O)O^-Li^+ + [\cdot CH_2CH_3] \quad (7.6)$$

$$CH_3OC(O)OCH_3 + Li \rightarrow CH_3OC(O)O^-Li^+ + [\cdot CH_3] \quad (7.7)$$

7.3.2 Reactions of Metallic Lithium (Li) with Organofluorine Compounds and Those of Lithium-Intercalated Graphite (LiC_6) with Fluorine Compound- Mixed Electrolyte Solutions

DSC curves for the reactions of metallic Li with organofluorine compounds are shown in Figures 7.5 and 7.6 [45,46]. No exothermic reaction is observed below the melting point of Li, 180 °C, because Li surface is protected by a thin oxide

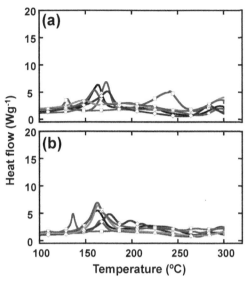

FIGURE 7.7 DSC curves for lithiated graphite ($Li_{0.92-0.98}C_6$) with the SEI film [44]. (a) $Li_{0.94-0.98}C_6$ prepared in 0.78 molL^{-1} LiClO$_4$-EC/DEC/PC (1:1:1 vol) or EC/DEC/PC/(**A, B, D, E**, or **F**) (1:1:1:0.83 vol, 21.7 vol%). (b) $Li_{0.92-0.95}C_6$ prepared in 0.67 molL^{-1} LiClO$_4$-EC/DEC/PC (1:1:1 vol) or EC/DEC/PC/(**A, B, D, E**, or **F**) (1:1:1:1.5 vol, 33.3 vol%). ⎯⎯⎯: EC/DEC/PC, ⎯○⎯: EC/DEC/PC/**A**, ⎯□⎯: EC/DEC/PC/**B**, ⎯◇⎯: EC/DEC/PC/**D**, ⎯▽⎯: EC/DEC/PC/**E**, ⎯△⎯: EC/DEC/PC/**F**.

film. Exothermic reactions started at 238 °C for fluorine compounds **A**, **B**, and **G**, and at 226 °C for **C**. Those for **H** and **I** started at slightly higher temperatures, 254 and 261 °C, respectively, and their exothermic peaks were found between 246 and 283 °C. It is interesting that pentafluoropropyl methanesulfonate, **J** is very stable, giving no exothermic peak up to 300 °C, that is, **J** neither reacts with Li nor thermally decomposes below 300 °C. Thus, fluorine compounds used in the study have a higher stability than that of the solvent mixture, EC/DMC against metallic Li.

Figure 7.7 shows DSC curves obtained for lithiated graphite samples with SEI films without the electrolyte solution [44]. Lithiated graphite samples prepared in 0.78 and 0.67 molL^{-1} LiClO$_4$-EC/DEC/PC gave weak exothermic peaks at 164 and 162 °C. These exothermic peaks would be due to the reaction of Li deintercalated from LiC$_6$ with SEI because lithiated graphite is decomposed by a temperature increase to approximately 200 °C [52,57–61]. It was reported that LiC$_6$ decomposes to LiC$_{12}$ by a temperature increase to 120 °C and Li-intercalated graphite completely decomposes at around 200 °C [52]. The reactions of deintercalated Li with the main components of SEI were also reported: $2Li + (CH_2OCO_2Li)_2 \rightarrow 2Li_2CO_3 + C_2H_4$ [52]. For lithiated graphite samples, $Li_{0.94-0.98}C_6$ and $Li_{0.92-0.95}C_6$ prepared in fluorine compound-mixed electrolytes, no significant difference is found from the DSC curves obtained without fluorine compounds.

Figure 7.8 shows DSC curves obtained for mixtures of lithiated graphite and electrolyte solution [44]. Mixtures of lithiated graphite and 0.90, 0.78, or 0.67 molL^{-1} LiClO$_4$-EC/DEC/PC yielded three exothermic peaks at 148–153 °C (medium peaks), 194–203 °C (weak peaks), and 284–288 °C (strong peaks). The medium and weak peaks at 148–153 °C and 194–203 °C, respectively, would be due to the reactions of Li released from LiC$_6$ with SEI and electrolyte solutions.

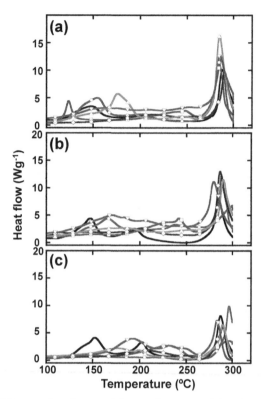

FIGURE 7.8 DSC curves for mixtures of fluorocarbonate-containing electrolyte solution and lithiated graphite ($Li_{0.92-0.98}C_6$) [44]. (a) $0.90\,molL^{-1}$ $LiClO_4$-EC/DEC/PC (1:1:1 vol) or EC/DEC/PC/(**A, B, D, E,** or **F**) (1:1:1:0.33 vol, 10.0 vol%) and $Li_{0.96-0.98}C_6$. (b) $0.78\,molL^{-1}$ $LiClO_4$-EC/DEC/PC (1:1:1 vol) or EC/DEC/PC/(**A, B, D, E,** or **F**) (1:1:1:0.83 vol, 21.7 vol%) and $Li_{0.96-0.97}C_6$. (c) $0.67\,molL^{-1}$ $LiClO_4$-EC/DEC/PC (1:1:1 vol) or EC/DEC/PC/(**A, B, D, E,** or **F**) (1:1:1:1.5 vol, 33.3 vol%) and $Li_{0.92-0.96}C_6$. ———: EC/DEC/PC, —○—: EC/DEC/PC/A, —□—: EC/DEC/PC/B, —◇—: E/DEC/PC/D, —▽—: EC/DEC/PC/E, —△—: EC/DEC/PC/F.

The medium peaks at 148–153 °C may be caused by the decomposition of stage 1 LiC_6, and weak peaks at 194–203 °C are probably due to the decomposition of high stage lithiated graphite [52]. The strong peaks observed at 284–288 °C would be due to the thermal decomposition of electrolyte solutions. Reactions of deintercalated Li with SEI and electrolyte solutions are influenced by fluorocarbonates and their concentrations. With an increase in the concentration of fluorocarbonates, exothermic peaks at around 150 and 200 °C disappeared or shifted to higher temperatures. When fluorocarbonates were mixed with $1\,molL^{-1}$ $LiClO_4$-EC/DEC/PC by 10.0 vol% (Figure 7.8(a)), fluorocarbonate **D** gave the highest effect. When the amount of mixed fluorocarbonate was increased to 21.7 vol%, thermal stability was improved. Figure 7.8(c) indicates DSC curves obtained when fluorocarbonates were mixed with $1\,molL^{-1}$ $LiClO_4$-EC/DEC/PC by 33.3 vol%. Under this condition, the reaction of deintercalated Li with SEI and electrolyte solution was more pronouncedly suppressed for the solutions containing fluorocarbonates **A**, **B**, **D**, and **F**. Even in the case of the fluorocarbonate, **E**-containing solution, the exothermic peak shifted to 191 °C. Thus, five fluorocabonates improve the thermal stability of electrolyte solutions. The amounts of fluorocarbonates needed to suppress the reaction of deintercalated Li with SEI and electrolyte solution are 10.0 vol% for fluorocarbonate **D**, 21.7 vol% for fluorocarbonates **B** and **F**, and 33.3 vol% for fluorocarbonates **A** and **E**.

Figure 7.9 shows the DSC curves obtained when 1 molL^{-1} LiPF$_6$-EC/DMC and 1 molL^{-1} LiPF$_6$-EC/DMC/(**A**, **B**, **G**, or **J**) (1:1:1 vol.) were used [46].

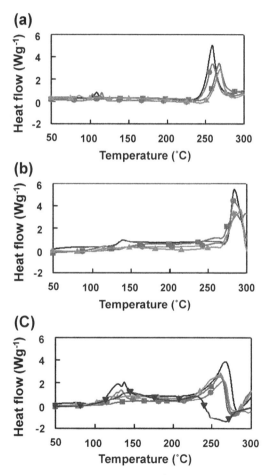

FIGURE 7.9 DSC curves for mixtures of 1 molL^{-1} LiPF$_6$-EC/DMC (1:1 vol) or 1 molL^{-1} LiPF$_6$-EC/DMC/(**A**, **B**, or **J**) (1:1:1 vol) and delithiated graphite with SEI film (a), only lithiated graphite samples (Li$_{0.90-0.98}$C$_6$) with SEI film (b), and mixtures of 1 molL^{-1} LiPF$_6$-EC/DMC (1:1 vol) or 1 molL^{-1} LiPF$_6$-EC/DMC/(**A**, **B**, **G**, or **J**) (1:1:1 vol) and lithiated graphite (Li$_{0.90-0.98}$C$_6$) with SEI film (c) [46]. ────: EC/DMC, ─●─: EC/DMC/A, ─■─: EC/DMC/B, ─▼─: EC/DMC/G, ─▲─: EC/DMC/J.

Mixtures of delithiated graphite and electrolyte solution with or without the fluorine compound gave weak exothermic peaks at around 100 °C and strong ones between 258 and 268 °C (Figure 7.9(a)). The weak peaks at around 100 °C are attributed to the reactions of the electrolyte solution with PF_5 generated by the dissociation of $LiPF_6$, and strong ones at 258–268 °C are mainly due to the thermal decomposition of electrolyte solutions [58–61]. Exothermic heats are smaller in fluorine compound-mixed solutions than 1 molL^{-1} $LiPF_6$-EC/DMC. Figure 7.9(b) exhibits the DSC curves of only lithiated graphite samples prepared in 1 molL^{-1} $LiPF_6$-EC/DMC and 1 molL^{-1} $LiPF_6$-EC/DMC/(**A**, **B**, or **J**) (1:1:1 vol). LiC_6 prepared in 1 molL^{-1} $LiPF_6$-EC/DMC provided a weak peak at 140 °C and a strong one at 283 °C. The former is caused by the reaction of Li released from graphite with SEI, and the latter is due to the thermal decomposition of SEI [58–61]. However, the $Li_{0.90-0.98}C_6$ samples prepared in 1 molL^{-1} $LiPF_6$-EC/DMC/(**A**, **B**, or **J**) gave no peaks at around 140 °C and the exothermic peaks due to decomposition of SEI are also weaker. Figure 7.9(c) shows the DSC profiles for mixtures of lithiated graphite and 1 molL^{-1} $LiPF_6$-EC/DMC or 1 molL^{-1} $LiPF_6$-EC/DMC/(**A**, **B**, **G**, or **J**) (1:1:1 vol). The effect of mixing of fluorine compounds is found when they are mixed by 33 vol%. The exothermic peaks between 100 and 150 °C due to the reactions with $Li_{0.90-0.98}C_6$ and those between 230 and 300 °C by the thermal decomposition of electrolyte solutions and SEIs are both weakened in the cases of fluorine compound-mixed electrolyte solutions. Endothermic peaks are mixed between 240 and 300 °C in Figure 7.9(c), which is attributed to the exfoliation of graphite by final decomposition of Li-intercalated graphite [57]. Among four fluorine compounds **A**, **B**, **G**, and **J**, exothermic peak is the weakest in **G**-mixed solution (Figure 7.9(c)). The exothermic peaks between 100 and 150 °C are weaker than those in Figure 7.8. When $LiPF_6$ is used as an electrolyte, LiF is deposited on the graphite electrode, which would suppress the reactions of Li with SEI and organic solvents.

The influence of LiF deposition on a graphite electrode is clearly found in fluoroether-mixed electrolyte solutions as shown in Figures 7.10 and 7.11 [45]. Figure 7.10 shows DSC data obtained by using an $LiClO_4$-containing electrolyte solution. In Figure 7.10(a), exothermic peaks indicating the thermal decomposition of SEI on delithiated graphite and electrolyte solution were found at 286 °C and above 300 °C. In Figure 7.10(b) obtained for only lithiated graphites ($Li_{0.91-0.99}C_6$) without electrolyte solutions, exothermic peaks were located at 164–169 °C for all samples prepared in the electrolyte solutions with and without fluoroethers **H** and **I**, indicating the reactions of Li released from graphite with SEI. Mixtures of electrolyte solution and lithiated graphite ($Li_{0.91-0.99}C_6$) provided two exothermic peaks at 153–162 °C and 280 °C (Figure 7.10(c)). The peaks at 153–162 °C and 280 °C would be due to reactions of deintercalated Li with SEI and electrolyte solution, and thermal decomposition of SEI and electrolyte solution, respectively. No effect of fluoroethers **H** and **I** is observed for the reactions with deintercalated Li. This may be because fluoroethers mainly interact with hydrocarbon groups of EC,

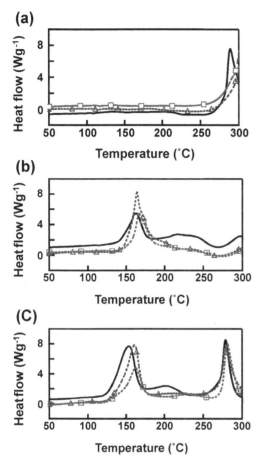

FIGURE 7.10 DSC curves for mixtures of 0.67 molL^{-1} LiClO$_4$-EC/DEC/PC (1:1:1 vol) or 0.67 molL^{-1} LiClO$_4$-EC/DEC/PC/(**H** or **I**) (1:1:1:1.5 vol) and delithiated graphite with an SEI film (a), only lithiated graphite (Li$_{0.91-0.99}$C$_6$) (b), and 0.67 molL^{-1} LiClO$_4$-EC/DEC/PC (1:1:1 vol) or 0.67 molL^{-1} LiClO$_4$-EC/DEC/PC/(**H** or **I**) (1:1:1:1.5 vol) and lithiated graphite (Li$_{0.91-0.99}$C$_6$) (c) [45]. ———: EC/DEC/PC, ---▲---: EC/DEC/PC/**H**, ----□----: EC/DEC/PC/**I**.

PC, and DEC, and therefore, organic solvents EC, PC, and DEC preferentially react with Li. In LiPF$_6$-containing electrolyte solutions, DSC measurements give somewhat different results as shown in Figure 7.11. In Figure 7.11(a) obtained for mixtures of delithiated graphite and electrolyte solution, the peaks at 85–103 °C and 244–264 °C may be due to the reaction of PF$_5$ generated by the dissociation of LiPF$_6$, and thermal decomposition of SEI and electrolyte solutions, respectively [52,57–61]. A large difference from Figure 7.10(b) and (c) is observed in Figure 7.11(b) and (c) obtained for only lithiated graphite samples (Li$_{0.90-0.98}$C$_6$), and mixtures of lithiated graphite (Li$_{0.90-0.98}$C$_6$) and electrolyte solution, respectively. In Figure 7.11(b), strong exothermic peaks were not observed at 150–160 °C, but DSC curves only slightly shifted to the exothermic side in the range of 120–260 °C, which suggests that the reaction of deintercalated Li with SEI is suppressed by LiF contained in SEI. X-Ray photoelectron spectroscopy spectra obtained for used electrodes show the

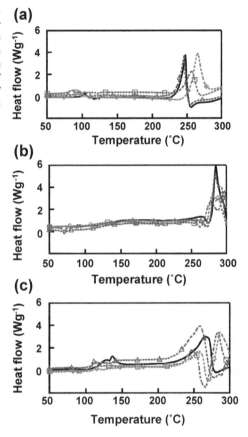

FIGURE 7.11 DSC curves for mixtures of 1 molL^{-1} LiPF$_6$-EC/DMC (1:1 vol) or 1 molL^{-1} LiPF$_6$-EC/DMC/(**C, H,** or **I**) (1:1:1 vol) and delithiated graphite with an SEI film (a), only lithiated graphite (Li$_{0.90-0.98}$C$_6$) (b), and 1 molL^{-1} LiPF$_6$-EC/DMC (1:1 vol) or 1 molL^{-1} LiPF$_6$-EC/DMC/(**C, H,** or **I**) (1:1:1 vol) and lithiated graphite (Li$_{0.90-0.98}$C$_6$) (c) [45]. ———: EC/DMC, ···◇···: EC/DMC/**C**, ---△---: EC/DMC/**H**, ----☐----: EC/DMC/**I**.

presence of LiF that would have been formed by the dissociation of LiPF$_6$ and reaction of resulting PF$_5$ with Li. The exothermic peaks at 275–297 °C would be attributed to the thermal decomposition of SEI [52,57–61]. Mixtures of electrolyte solution and lithiated graphite (Li$_{0.90-0.98}$C$_6$) gave weak exothermic peaks at 111–137 °C and strong ones at 250 °C or at higher temperatures (Figure 7.11(c)). Among the peaks observed at 111–137 °C, 1 molL^{-1} LiPF$_6$-EC/DMC/**C** (1:1:1 vol) containing the fluorocarbonate gave the weakest peak, showing the highest effect to prevent the reactions with Li deintercalated from graphite. It is inferred that fluorocarbonate **C** has a (–O)$_2$C = O group and a larger dielectric constant than that of fluoroethers **H** and **I**; therefore, miscibility of fluorocarbonate **C** with EC is better than that of fluoroethers at the molecular level. Further, fluoroethers would mainly interact with hydrocarbon groups of EC and DMC as mentioned above. These may be the reasons why fluorocarbonate **C** has a better effect on the reaction with deintercalated Li than fluoroethers **H** and **I**. The exothermic peaks between 250 and 300 °C are due to the decomposition of SEI and electrolyte solutions [52,57–61]. However, an endothermic peak due to exfoliation of graphite caused by the final decomposition of Li-intercalated graphite is overlapped in the same temperature range [52]. Therefore, the exothermic peaks are clearly separated into two peaks.

7.3.3 Influence of Molecular Structures of Fluorine Compounds on the Reactivity with Li

Reaction of fluorine compounds with Li and electrochemical reduction are both reductive reactions. As shown later in Section 7.5, fluorine compounds used in the study are electrochemically reduced at higher potentials (1.9–2.7 V vs Li/Li$^+$) [42,43] than those of EC, PC, DMC, and DEC (1.3–1.6 V) [56,62]. However, the reactivity of fluorine compounds with metallic Li and LiC$_6$ is low as shown in Section 7.3.2. It seems that the low reactivity of fluorine compounds with metallic Li and LiC$_6$ is due to the difficulty in the formation of organic Li salts or slow reaction rates for the formation of Li salts as in the case of PC. Linear fluoroethers **H** and **I** are composed of –CH$_2$–O–CF$_2$– in which O is bonded with CH$_2$ and CF$_2$. Therefore, the direct contact of Li atom with O of –CH$_2$–O–CF$_2$– is not easy, that is, reactivity of the linear fluoroethers with Li is low. Reduction currents for linear fluoroethers are also small [43]. In cyclic fluorocarbonates **A**, **B**, and **C**, the $\underline{C}$–$\underline{O}$ bond lengths of –$\underline{O}$–$\underline{C}$H(CH$_2$R$_f$)– (R$_f$: ligand composed of C, H, F, and O atoms) are longer than those of –$\underline{O}$–$\underline{C}$H$_2$– and O=$\underline{C}$(–$\underline{O}$)$_2$– (–$\underline{O}$–$\underline{C}$H(CH$_2$R$_f$)–: 1.446 nm; –$\underline{O}$–$\underline{C}$H$_2$–: 1.441–1.443 nm; O=$\underline{C}$(–$\underline{O}$)$_2$–: 1.368–1.373 nm). This suggests that the $\underline{C}$–$\underline{O}$ bonds of –$\underline{O}$–$\underline{C}$H(CH$_2$R$_f$)– are broken by the reaction with Li. However, the formation of organic Li salts may not proceed smoothly because of the long R$_f$ ligands of **A**, **B**, and **C** similarly to the case of PC shown in Section 7.3.1. In the case of linear fluorocarbonates **D**, **E**, and **F**, Li salts such as CHF$_2$CF$_2$CH$_2$OCO$_2$Li, CF$_3$CH$_2$OCO$_2$Li, and CF$_3$CH$_2$OCO$_2$Li, respectively, may be more easily formed than those of cyclic fluorocarbonates having long R$_f$ ligands because the reactions with Li are similar to those of DEC, EMC, and DMC. Nevertheless, the reactivity of **D**, **E**, and **F** with Li is less than those of DEC, EMC, and DMC as shown in Figure 7.8.

7.4 ELECTROCHEMICAL OXIDATION STABILITY OF FLUORINE COMPOUND-MIXED ELECTROLYTE SOLUTIONS

Oxidation current measurements were performed using a Pt wire or a glassy carbon electrode. Figure 7.12 shows oxidation currents measured in 0.90, 0.78, and 0.67 mol L^{-1} LiClO$_4$-EC/DEC/PC (1:1:1 vol) and -EC/DEC/PC/(**A**, **B**, **D**, **E**, or **F**) (1:1:1:0.33, 0.83, or 1.5 in vol, or 10.0, 21.7, or 33.3 vol%, respectively) using a Pt wire electrode [44]. Small oxidation currents were observed from 5.8 V, increasing after 6.0 V versus Li/Li$^+$ in the EC/DEC/PC electrolytes. Small currents below 5.8 V were not detected because the Pt wire electrode had a small

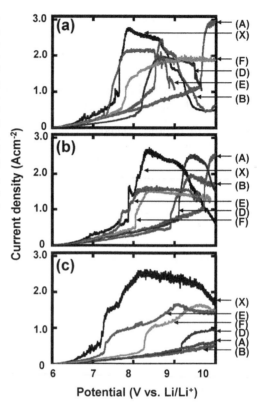

FIGURE 7.12 Linear sweep voltammograms for a Pt wire electrode in fluorocarbonate-containing electrolyte solutions [44]. (a): 0.90 molL^{-1} LiClO$_4$-EC/DEC/PC (1:1:1 vol) and EC/DEC/PC/(**A**, **B**, **D**, **E**, or **F**) (1:1:1:0.33 vol, 10.0 vol%). (b): 0.78 molL^{-1} LiClO$_4$-EC/DEC/PC (1:1:1 vol) and EC/DEC/PC/(**A**, **B**, **D**, **E**, or **F**) (1:1:1:0.83 vol, 21.7 vol%). (c): 0.67 molL^{-1} LiClO$_4$-EC/DEC/PC (1:1:1 vol) and EC/DEC/PC/(**A**, **B**, **D**, **E**, or **F**) (1:1:1:1.5 vol, 33.3 vol%). X: EC/DEC/PC, A: EC/DEC/PC/**A**, B: EC/DEC/PC/**B**, D: EC/DEC/PC/**D**, E: EC/DEC/PC/**E**, F: EC/DEC/PC/**F**.

surface area. Oxidation currents are significantly reduced with an increase in the amounts of mixed fluorocarbonates. In the solutions containing fluorocarbonates **A**, **B**, and **D** by 21.7 and 33.3 vol%, oxidation currents were much lower than those in EC/DEC/PC and electrolytes containing fluorocarbonates **E** and **F**. Mixing by 21.7 vol% is enough for fluorocarbonates **A**, **B**, and **D** to reduce oxidation currents while 33.3 vol% mixing is necessary for fluorocarbonates **E** and **F**.

Figure 7.13 shows the oxidation currents measured using a glassy carbon electrode in 1 molL^{-1} LiPF$_6$-EC/DMC (1:1 vol) and 1 molL^{-1} LiPF$_6$-EC/DMC/(**A**, **B**, or **J**) (1:1:1 vol), and in 1 molL^{-1} LiPF$_6$-EC/DEC (1:1 vol) and 0.67 molL^{-1} LiPF$_6$-EC/DEC/(**A**, **B**, **D**, or **J**) (1:1:1 vol) [46]. A Pt electrode cannot be used in an LiPF$_6$-containing solution because it is lost by the reaction with fluorine above approximately 7.5 V. Oxidation currents have current peaks, and no current is observed after that. This phenomenon would be the "anode effect", which is observed in the electrolytic production of F$_2$ gas using a carbonaceous anode. Once the anode effect occurs, the anode potential suddenly increases, and simultaneously current falls down to zero. Glassy carbon would be fluorinated by the oxidative decomposition of PF$_6^-$ anion at

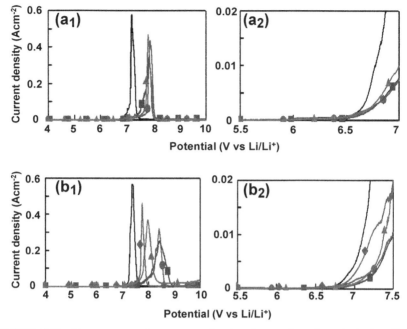

FIGURE 7.13 Linear sweep voltammograms for the glassy carbon electrode in 1 molL^{-1} LiPF$_6$-EC/DMC (1:1 vol) and 1 molL^{-1} LiPF$_6$-EC/DMC/(**A**, **B**, or **J**) (1:1:1 vol) (a$_1$ and a$_2$), and in 1 molL^{-1} LiPF$_6$-EC/DEC (1:1 vol) and 0.67 molL^{-1} LiPF$_6$-EC/DEC/(**A**, **B**, **D**, or **J**) (1:1:1 vol) (b$_1$ and b$_2$) [46]. ———: EC/DMC and EC/DEC, —●—: EC/DMC/**A** and EC/DEC/**A**, —■—: EC/DMC/**B** and EC/DEC/**B**, —◆—: EC/DEC/**D**, —▲—: EC/DMC/**J** and EC/DEC/**J**.

a high potential, yielding hydrophobic carbon–fluorine (CF) film. The wettability of the electrolyte solution with glassy carbon is extremely reduced by the formation of a hydrophobic CF film, which leads to the occurrence of the anode effect. Oxidation current peaks are found at the higher potentials in fluorine compound-mixed solutions than in original 1 molL^{-1} LiPF$_6$-EC/DMC and -EC/DEC. As shown in Figure 7.13(a$_2$) and (b$_2$), oxidation currents started to increase at 6.5 and 6.7 V relative to Li/Li$^+$, respectively. Overpotential of the carbon anode is usually large in fluoride-containing solutions due to the formation of the CF film. Oxidation currents are clearly reduced in both 1 molL^{-1} LiPF$_6$-EC/DMC/(**A**, **B**, or **J**) (1:1:1 vol) and 0.67 molL^{-1} LiPF$_6$-EC/DEC/(**A**, **B**, **D**, or **J**) (1:1:1 vol), compared with those in original solutions without fluorine compounds. Among four fluorine compounds **A**, **B**, and **J** give slightly better results than **D**.

Figure 7.14 shows electrochemical oxidation currents measured in 0.90, 0.78, and 0.67 molL^{-1} LiClO$_4$-EC/DEC/PC (1:1:1 vol) and -EC/DEC/PC/(**H** or **I**) (1:1:1:0.33, 0.83, or 1.5 in vol, or 10.0, 21.7, or 33.3 vol%, respectively) using a Pt wire electrode [45]. Small oxidation currents started to flow from 5.8 V versus Li/Li$^+$, gradually increasing with electrode potential. In the

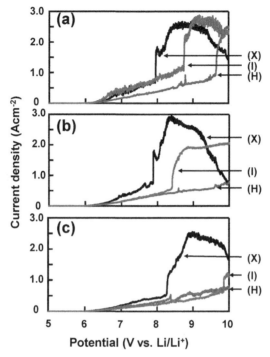

FIGURE 7.14 Linear sweep voltammograms for the Pt wire electrode in fluoroether-mixed electrolyte solutions [45]. (a) 0.90 molL^{-1} LiClO$_4$-EC/DEC/PC (1:1:1 vol) and 0.90 molL^{-1} LiClO$_4$-EC/DEC/PC/(**H** or **I**) (1:1:1:0.33 vol), (b) 0.78 molL^{-1} LiClO$_4$-EC/DEC/PC (1:1:1 vol) and 0.78 molL^{-1} LiClO$_4$-EC/DEC/PC/(**H** or **I**) (1:1:1:0.83 vol), (c) 0.67 molL^{-1} LiClO$_4$-EC/DEC/PC (1:1:1 vol) and 0.67 molL^{-1} LiClO$_4$-EC/DEC/PC/(**H** or **I**) (1:1:1:1.5 vol). X: EC/DEC/PC, H: EC/DEC/PC/**H**, I: EC/DEC/PC/**I**.

electrolyte solutions without fluoroethers, oxidation currents increased at approximately 8 V, indicating the vigorous decomposition of the solvents. However, oxidation currents are largely reduced by the mixing of fluoroethers, decreasing with increasing fluoroether for both **H** and **I**. In LiPF$_6$-containing electrolyte solutions, large overpotentials are observed, that is, the oxidation currents started to flow at 6.5 V higher than 5.8 V in LiClO$_4$-containing solutions as shown in Figure 7.15 [45]. The result shows that oxidation currents are smaller in fluoroether-mixed solutions than in 1 molL^{-1} LiPF$_6$-EC/DMC. The decrease in oxidation currents would be attributed to the high stability of fluorine compounds to electrochemical oxidation. The decrease in the oxidation currents would be due to the reduction of electrode surface areas of the Pt wire and glassy carbon electrodes by adsorption of stable fluorine compounds.

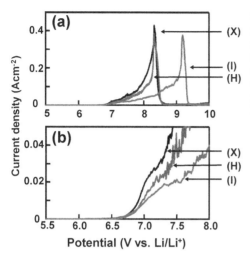

FIGURE 7.15 Linear sweep voltammograms for the glassy carbon electrode in fluoroether-mixed electrolyte solutions [45]. (a) 1 molL^{-1} LiPF$_6$-EC/DMC (1:1 vol) and 1 molL^{-1} LiPF$_6$-EC/DMC/(**H** or **I**) (1:1:1 vol), and (b) magnified figure between 5.5 and 8.0V. X: EC/DMC, H: EC/DMC/H, I: EC/DMC/I.

7.5 CHARGE/DISCHARGE BEHAVIOR OF NATURAL GRAPHITE ELECTRODES IN FLUORINE COMPOUND-MIXED ELECTROLYTE SOLUTIONS

Charge/discharge characteristics of natural graphite electrodes were investigated because fluorine compounds are electrochemically reduced at higher potentials than EC, PC, DMC, and DEC. Electrochemical reduction of fluorine compounds used in the study starts between 1.9 and 2.7V versus Li/Li$^+$ [42,43]. These potentials are higher than the reduction potentials of EC (1.4V), PC (1.0–1.6V), DMC (1.3V), and DEC (1.3V) [56,62]. EC-based solvents should be used for high crystalline graphite such as natural graphite for the smooth formation of SEI on the electrode. Many fluorine compounds can be used for EC/DMC or EC/DEC electrolytes because EC easily forms SEI on natural graphite electrodes [42,44–46]. First coulombic efficiencies obtained in fluorocarbonate-mixed electrolyte solutions are nearly the same as or slightly higher than those obtained in EC/DMC and EC/DEC electrolytes without fluorine compounds. If decomposed products of fluorine compounds facilitate SEI formation on graphite electrode, PC-containing electrolytes with low melting points can be also used. Figure 7.16 shows the first charge/discharge curves obtained using a natural graphite electrode (NG15 μm) in 0.90, 0.78, and 0.67 molL^{-1} LiClO$_4$-EC/DEC/PC (1:1:1 vol) and -EC/DEC/PC/(**A**, **B**, **D**, **E**, or **F**) (1:1:1:0.33, 0.83 or 1.5 in vol, or 10.0, 21.7, or 33.3 vol%, respectively) electrolytes as functions of the concentration of fluorocarbonate and current density [44]. The potential plateaus at 0.8V versus Li/Li$^+$ indicate the reductive decomposition of PC. In EC/DEC/PC electrolyte without the fluorine compound, the potential plateau was prolonged with decreasing

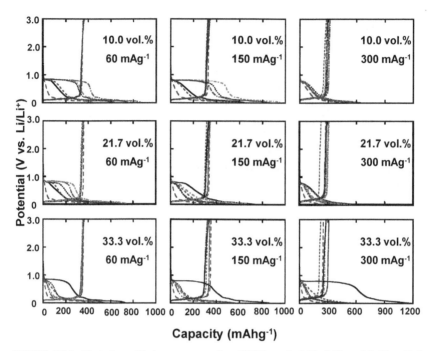

FIGURE 7.16 First charge/discharge curves of an NG15 μm in 0.90, 0.78, and 0.67 molL^{-1} LiClO$_4$-EC/DEC/PC (1:1:1 vol) and 0.90, 0.78, and 0.67 molL^{-1} LiClO$_4$-EC/DEC/PC/(**A, B, D, E,** or **F** 1:1:1:0.33, 0.83 or 1.5 vol, 10.0, 21.7 or 33.3 vol%, respectively) as functions of the concentration of fluorocarbonate and current density [44]. ⎯⎯⎯: EC/DEC/PC, ⎯ ⎯ ⎯ ⎯: EC/DEC/PC/A, ⎯ ⎯ ⎯ ⎯ ⎯: EC/DEC/PC/B, ▪▪▪▪▪▪▪▪▪▪▪: EC/DEC/PC/D, ⎯ ▪ ▪ ⎯ ▪ ▪ ⎯: EC/DEC/PC/E, ⎯ ▪ ⎯ ▪ ⎯ ▪: EC/DEC/PC/F.

concentrations of LiClO$_4$ and increasing current densities. According to these changes, the first columbic efficiency in the EC/DEC/PC electrolyte also decreased. On the other hand, in the fluorocarbonate-mixed solutions, the potential plateau was shortened with increasing concentrations of fluorocarbonate and current density. The difference in the EC/DEC/PC electrolytes with and without fluorine compound was clearly observed when fluorocarbonate was mixed by 33.3 vol%. First coulombic efficiency in EC/DEC/PC/ (**A, B, D, E,** or **F**) electrolyte increased, that is, irreversible capacity decreased with increasing concentrations of fluorocarbonate and current density. Fluorocarbonate **B** is the best among five fluorocarbonates, giving much higher first coulombic efficiencies, that is, lower irreversible capacities than others. For other fluorocarbonates except for **B**, much higher first coulombic efficiencies than those in the EC/DEC/PC electrolyte were also obtained by mixing of fluorocarbonates by 33.3 vol%. Charge capacities are nearly the same as for each other in the electrolyte solutions with and without fluorocarbonates at 60 mAg^{-1}.

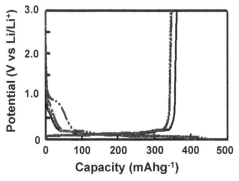

FIGURE 7.17 First charge/discharge curves of an NG15 μm in 1 molL^{-1} LiPF$_6$-EC/DMC (1:1:1 vol) and 1 molL^{-1} LiPF$_6$-EC/DMC/(**A**, **B**, **G**, or **J**) (1:1:1 vol) at 60 mAg^{-1} [46]. ⎯⎯: EC/DMC, ⋯⋯⋯⋯⋯: EC/DMC/**A**, ⎯ · ⎯ : EC/DMC/**B**, ⎯ · · ⎯ · : EC/DMC/**G**, ⎯ ⎯ ⎯: EC/DMC/**J**.

Figure 7.17 shows charge/discharge potential curves at the first cycle in 1 molL^{-1} LiPF$_6$-EC/DMC (1:1:1 vol) and 1 molL^{-1} LiPF$_6$-EC/DMC/(**A**, **B**, **G**, or **J**) (1:1:1 vol) at 60 mAg^{-1} [46]. Electrode potentials quickly decreased except that in 1 molL^{-1} LiPF$_6$-EC/DMC/**G** (1:1:1 vol), in which a short potential plateau indicating the reduction of **G** was observed. First coulombic efficiencies obtained in fluorine compound-mixed electrolyte solutions are similar to those obtained in 1 molL^{-1} LiPF$_6$-EC/DMC (1:1:1 vol) without fluorine compounds except that for 1 molL^{-1} LiPF$_6$-EC/DMC/**G** (1:1:1 vol). Contribution of **G** to SEI formation is slightly lower than others. The results indicate that fluorine compounds **A**, **B**, and **J** can be used for 1 molL^{-1} LiPF$_6$-EC/DMC (1:1:1 vol). The effect of mixing of fluorine compounds was also examined using PC-containing electrolytes. Figure 7.18 shows charge/discharge potential curves at the first cycle in 1 molL^{-1} LiPF$_6$-EC/EMC/PC/(**A**, **B**, or **J**) (1:1:1:0.33 or 1:1:1:1.5 vol) [46]. The potential plateau observed in 1 molL^{-1} LiPF$_6$-EC/EMC/PC (1:1:1 vol) indicates the electrochemical reduction of PC. However, the potential plateau almost disappeared in fluorine compound-mixed solutions. Table 7.4 gives electrochemical data obtained in PC-containing electrolytes. First coulombic efficiencies were largely increased by mixing of fluorine compounds without any decrease in capacities [46]. First coulombic efficiencies were in the range of 69–74% when **A**, **B**, and **J** were mixed by 10 vol%, but it reached 81% when **J** was mixed by 33 vol%. The increments of first coulombic efficiencies were in the range of 16–20% in 1 molL^{-1} LiPF$_6$-EC/EMC/PC/(**A**, **B**, or **J**) (1:1:1:0.33 vol), and 28% in 1 molL^{-1} LiPF$_6$-EC/EMC/PC/**J** (1:1:1:1.5 vol) at 60 mAg^{-1}. This means that fluorine compounds **A**, **B**, and **J** effectively facilitate SEI formation in PC-containing electrolyte solutions.

Figure 7.19 shows charge/discharge potential curves at the first cycle in 0.90, 0.78, and 0.67 molL^{-1} LiClO$_4$-EC/DEC/PC with or without fluoroether [45]. In all cases, potential plateaus at 0.8 V versus Li/Li$^+$ indicating the electrochemical reduction of PC are reduced by mixing of fluoroethers. This trend becomes more distinct with increasing fluoroethers. Coulombic efficiencies are

TABLE 7.4 Charge/Discharge Capacities and Coulombic Efficiencies of NG15 μm in 1 molL^{-1} LiPF$_6$-EC/EMC/PC (1:1:1 vol), 1 molL^{-1} LiPF$_6$-EC/EMC/PC/(A, B or J) (1:1:1:0.33 vol), and 1 molL^{-1} LiPF$_6$-EC/EMC/PC/J (1:1:1:1.5 vol) at 60 mAg^{-1} [46]

Electrolyte Solution	Cycle Number	Discharge Capacity (mAhg^{-1})	Charge Capacity (mAhg^{-1})	Coulombic Efficiency (%)
EC/EMC/PC (1:1:1:0.33)	1st	630	337	53.5
	10th	337	333	98.6
EC/EMC/PC/**A** (1:1:1:0.33)	1st	476	326	69.3
	10th	315	311	98.7
EC/EMC/PC/**B** (1:1:1:0.33)	1st	456	335	73.5
	10th	313	309	98.7
EC/EMC/PC/**J** (1:1:1:0.33)	1st	486	342	70.3
	10th	335	331	98.8
EC/EMC/PC/**J** (1:1:1:1.5)	1st	419	340	81.2
	10th	327	324	99.2

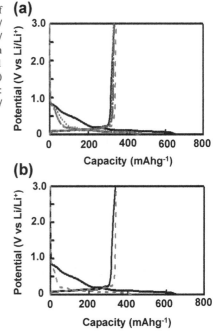

FIGURE 7.18 First charge/discharge curves of an NG15 μm at 60 mAg^{-1} in 1 molL^{-1} LiPF$_6$-EC/EMC/PC (1:1:1 vol) and 1 molL^{-1} LiPF$_6$-EC/EMC/PC/(**A**, **B**, or **J**) (1:1:1:0.33 vol) (a), and in 1 molL^{-1} LiPF$_6$-EC/EMC/PC (1:1:1 vol) and 1 molL^{-1} LiPF$_6$-EC/EMC/PC/**J** (1:1:1:1.5 vol) (b) [46]. ——: EC/EMC/PC, ··········: EC/EMC/PC/**A**, — · —: EC/EMC/PC/**B**, — — —: EC/EMC/PC/**J**.

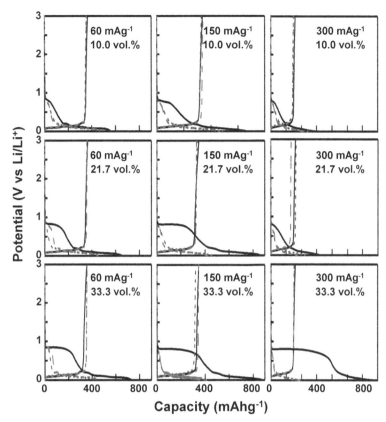

FIGURE 7.19 Charge/discharge potential curves of an NG15 μm at the first cycle, obtained in 0.90, 0.78, and 0.67 molL^{-1} LiClO$_4$-EC/DEC/PC (1:1:1 vol), and 0.90, 0.78, and 0.67 molL^{-1} LiClO$_4$-EC/DEC/PC/(**H** or **I**) (1:1:1:0.33, 0.83, or 1.5 vol) at current densities of 60, 150, and 300 mAg^{-1} [45]. ———: EC/DEC/PC, — — — —: EC/DEC/PC/**H**, ═ ═ ═ ═: EC/DEC/PC/**I**.

shown in Figure 7.20 as a function of cycle number [45]. Mixing of fluoroethers largely increases first coulombic efficiencies, which indicates that fluoroethers effectively facilitate SEI formation on natural graphite powder because electrochemical reduction of fluoroethers **H** and **I** starts at 2.1 and 2.3V versus Li/Li$^+$, respectively [43], higher than 1.3–1.6V for PC, EC, and DEC [56,62]. The increments in first coulombic efficiencies by mixing of fluoroethers were approximately 10–30%, 20–40%, and 10–50% at 60, 150, and 300 mAg^{-1}, respectively. The charge capacities obtained in fluoroether-mixed solutions are nearly the same as those in original electrolyte solutions without fluorine compounds at 60 mAg^{-1}; however, they slightly decrease at higher current densities.

In LiPF$_6$-containing electrolyte solutions, the first coulombic efficiencies are high in most of the cases. Figure 7.21 shows charge/discharge curves at the first cycle obtained in 1 molL^{-1} LiPF$_6$-EC/EMC/PC (1:1:1 vol) and 1 molL^{-1} LiPF$_6$-EC/EMC/PC/(**H** or **I**) (1:1:1:1.5 vol) [45]. In the PC-containing electrolytes, SEI

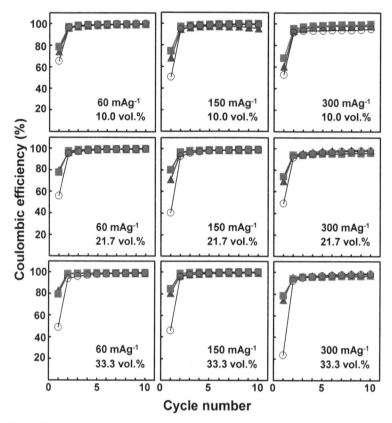

FIGURE 7.20 Coulombic efficiencies for an NG15 µm as a function of cycle number, obtained in 0.90, 0.78, and 0.67 molL^{-1} LiClO$_4$-EC/DEC/PC (1:1:1 vol), and 0.90, 0.78, and 0.67 molL^{-1} LiClO$_4$-EC/DEC/PC/(**H** or **I**) (1:1:1:0.33, 0.83, or 1.5 vol) at current densities of 60, 150, and 300 mAg^{-1} [45]. ─○─: EC/DEC/PC, ─▲─: EC/DEC/PC/**H**, ─■─: EC/DEC/PC/**I**.

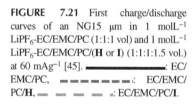
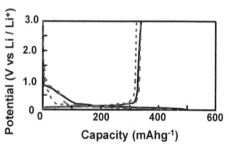

FIGURE 7.21 First charge/discharge curves of an NG15 µm in 1 molL^{-1} LiPF$_6$-EC/EMC/PC (1:1:1 vol) and 1 molL^{-1} LiPF$_6$-EC/EMC/PC/(**H** or **I**) (1:1:1:1.5 vol.) at 60 mAg^{-1} [45]. ─────: EC/EMC/PC, ━ ━ ━ ━: EC/EMC/PC/**H**, ── ── ──: EC/EMC/PC/**I**.

formation is faster in fluoroether-mixed solutions than in the original one. First coulombic efficiencies obtained in fluoroethers **H**- and **I**-mixed solutions were 78 and 74%, respectively, >68% in the original solution. The results indicate that fluoroethers **H** and **I** also facilitate SEI formation in PC-containing electrolytes.

7.6 CONCLUSIONS

Fluorine substitution of organic compounds lowers HOMO and LUMO levels, increasing their oxidation stability and reduction potentials. Organofluorine compounds such as fluorocarbonates and fluoroethers used in the study are excellent solvents to improve the safety of lithium ion batteries. Mixing of fluorine compounds decreases the reactivity with metallic Li and LiC_6, and increases the electrochemical oxidation stability of electrolyte solutions, that is, it highly improves the thermal and oxidation stability of electrolyte solutions. Mixing of fluorine compounds facilitates SEI formation on graphite electrode, giving high coulombic efficiencies at the first cycle in PC-containing electrolytes.

REFERENCES

[1] X. Wang, E. Yasukawa, S. Kasuya, J. Electrochem. Soc. 148 (2001) A1058.
[2] X. Wang, E. Yasukawa, S. Kasuya, J. Electrochem. Soc. 148 (2001) A1066.
[3] K. Xu, M.S. Ding, S. Zhang, J.L. Allen, T.R. Jow, J. Electrochem. Soc. 149 (2002) A622.
[4] X.L. Yao, S. Xie, C.H. Chn, Q.S. Wang, J.H. Sun, Y.L. Li, S.X. Lu, J. Power Sources 144 (2005) 170.
[5] X. Wang, C. Yamada, H. Naito, G. Segami, K. Kibe, J. Electrochem. Soc. 153 (2006) A135.
[6] Y.E. Hyung, D.R. Vissers, K. Amine, J. Power Sources 119–121 (2003) 383.
[7] E.-G. Shim, T.-H. Nam, J.-G. Kim, H.-S. Kim, S.-I. Moon, J. Power Sources 172 (2007) 919.
[8] T.-H. Nam, E.-G. Shim, J.-G. Kim, H.-S. Kim, S.-I. Moon, J. Electrochem. Soc. 154 (2007) A957.
[9] D.H. Doughty, E.P. Roth, C.C. Crafts, G. Nagasubramanian, G. Henriksen, K. Amine, J. Power Sources 146 (2005) 116.
[10] K. Xu, S. Zhang, J.L. Allen, T.R. Jow, J. Electrochem. Soc. 149 (2002) A1079.
[11] K. Xu, M.S. Ding, S. Zhang, J.L. Allen, T.R. Jow, J. Electrochem. Soc. 150 (2003) A161.
[12] K. Xu, S. Zhang, J.L. Allen, T.R. Jow, J. Electrochem. Soc. 150 (2003) A170.
[13] S.S. Zhang, K. Xu, T.R. Jow, J. Power Sources 113 (2003) 166.
[14] D. Zhou, W. Li, C. Tan, X. Zuo, Y. Huang, J. Power Sources 184 (2008) 589.
[15] Q. Wang, J. Sun, X. Yao, C. Chen, Electrochem. Solid-State Lett. 8 (2005) A467.
[16] Q. Wang, J. Sun, C. Chen, J. Power Sources 162 (2006) 1363.
[17] E.-G. Shim, T.-H. Nam, J.-G. Kim, H.-S. Kim, S.-I. Moon, J. Power Sources 175 (2008) 533.
[18] T.-H. Nam, E.-G. Shim, J.-G. Kim, H.-S. Kim, S.-I. Moon, J. Power Sources 180 (2008) 561.
[19] S. Izquierdo-Gonzales, W. Li, B.L. Lucht, J. Power Sources 135 (2004) 291.
[20] H.F. Xiang, Q.Y. Jin, C.H. Chen, X.W. Ge, S. Guo, J.H. Sun, J. Power Sources 174 (2007) 335.
[21] Y.-B. He, Q. Liu, Z.-Y. Tang, Y.-H. Chen, Q.-S. Song, Electrochim. Acta 52 (2007) 3534.
[22] H.F. Xiang, Q.Y. Jin, R. Wang, C.H. Chen, X.W. Ge, J. Power Sources 179 (2008) 351.
[23] H.F. Xiang, Q.Y. Jin, R. Wang, C.H. Chen, X.W. Ge, J. Power Sources 195 (2010) 335.
[24] B.S. Lalia, T. Fujita, N. Yoshimoto, M. Egashira, M. Morita, J. Power Sources 186 (2009) 211.
[25] B.S. Lalia, N. Yoshimoto, M. Egashira, M. Morita, J. Power Sources 195 (2010) 7426.
[26] L. Wu, Z. Song, L. Liu, X. Guo, L. Kong, H. Zhan, Y. Zhou, Z. Li, J. Power Sources 188 (2009) 570.
[27] E.-G. Shim, T.-H. Nam, J.-G. Kim, H.-S. Kim, S.I. Moon, Electrochim. Acta 54 (2009) 2276.
[28] B.S. Lalia, N. Yoshimoto, M. Egashira, M. Morita, J. Power Sources 195 (2010) 7426.
[29] S. Dalavi, M. Xu, B. Ravdel, L. Zhou, B.L. Lucht, J. Electrochem. Soc. 157 (2010) A1113.
[30] J.-A. Choi, Y.-K. Sun, E.-G. Shim, B. Scrosati, D.W. Kim, Electrochim. Acta 56 (2011) 10179.

[31] J. Hu, Z. Jin, H. Zhong, H. Zhan, Y. Zhou, Z. Li, J. Power Sources 197 (2012) 297.
[32] B. Wu, F. Pei, Y. Wu, R. Mao, X. Ai, H. Yang, Y. Cao, J. Power Sources 227 (2013) 106.
[33] T. Sato, T. Maruo, S. Marukane, K. Takagi, J. Power Sources 138 (2004) 253.
[34] A. Guerfi, M. Dontigny, P. Charest, M. Petitclerc, M. Lagacé, A. Vijh, K. Zaghib, J. Power Sources 195 (2010) 845.
[35] C. Arbizzani, G. Gabrielli, M. Mastragostino, J. Power Sources 196 (2011) 4801.
[36] L. Lombardo, S. Brutti, M.A. Navarra, S. Panero, P. Reale, J. Power Sources 227 (2013) 8.
[37] J. Xiang, F. Wu, R. Chen, L. Li, H. Yu, J. Power Sources 233 (2013) 115.
[38] K.A. Smith, M.C. Smart, G.K.S. Prakash, B.V. Ratnakumar, ECS Trans. 11 (2008) 91.
[39] S. Chen, Z. Wang, H. Zhao, H. Qiao, H. Luan, L. Chen, J. Power Sources 187 (2009) 229.
[40] K. Naoi, E. Iwama, N. Ogihara, Y. Nakamura, H. Segawa, Y. Ino, J. Electrochem. Soc. 156 (2009) A272.
[41] K. Naoi, E. Iwama, Y. Honda, F. Shimodate, J. Electrochem. Soc. 157 (2010) A190.
[42] T. Achiha, T. Nakajima, Y. Ohzawa, M. Koh, A. Yamauchi, M. Kagawa, H. Aoyama, J. Electrochem. Soc. 156 (2009) A483.
[43] T. Achiha, T. Nakajima, Y. Ohzawa, M. Koh, A. Yamauchi, M. Kagawa, H. Aoyama, J. Electrochem. Soc. 157 (2010) A707.
[44] Y. Matsuda, T. Nakajima, Y. Ohzawa, M. Koh, A. Yamauchi, M. Kagawa, H. Aoyama, J. Fluorine Chem. 132 (2011) 1174.
[45] N. Ohmi, T. Nakajima, Y. Ohzawa, M. Koh, A. Yamauchi, M. Kagawa, H. Aoyama, J. Power Sources 221 (2013) 6.
[46] D. Nishikawa, T. Nakajima, Y. Ohzawa, M. Koh, A. Yamauchi, M. Kagawa, H. Aoyama, J. Power Sources 243 (2013) 573.
[47] N. Nanbu, S. Watanabe, M. Takehara, M. Ue, Y. Sasaki, J. Electroanal. Chem. 625 (2009) 7.
[48] Y. Sasaki, Physical and electrochemical properties and application to lithium batteries of fluorinated organic solvents, in: T. Nakajima, H. Groult (Eds.), Fluorinated Materials for Energy Conversion, Elsevier, Oxford, 2005, pp. 285–304.
[49] K. Sato, L. Zhao, S. Okada, J. Yamaki, J. Power Sources 196 (2011) 5617.
[50] T. Nakajima, Y. Hirobayashi, Y. Takayanagi, Y. Ohzawa, J. Power Sources 243 (2013) 581.
[51] D. Aurbach, B. Markovsky, A. Shechter, Y. Ein-Eli, J. Electrochem. Soc. 143 (1996) 3809.
[52] O. Haik, S. Ganin, G. Gershinsky, E. Zinigrad, B. Markovsky, D. Aurbach, I. Halalay, J. Electrochem. Soc. 158 (2011) A913.
[53] S.-H. Kang, D.P. Abraham, A. Xiao, B.L. Lucht, J. Power Sources 175 (2008) 526.
[54] X. Zhang, R. Kostecki, T. Richardson, J.K. Pugh, P.N. Ross Jr., J. Electrochem. Soc. 148 (2001) A1341.
[55] Y. Wang, S. Nakamura, M. Ue, P.B. Balbuena, J. Am. Chem. Soc. 123 (2001) 11708.
[56] J.M. Vollmer, L.A. Curtiss, D.R. Vissers, K. Amine, J. Electrochem. Soc. 151 (2004) A178.
[57] E. Zinigrad, L. Larush-Asraf, J.S. Gnanaraj, H.E. Gottlieb, M. Sprecher, D. Aurbach, J. Power Sources 146 (2005) 176.
[58] J. Yamaki, Thermally stable fluoro-organic solvents for lithium ion battery, in: T. Nakajima, H. Groult (Eds.), Fluorinated Materials for Energy Conversion, Elsevier, Oxford, 2005, pp. 267–284.
[59] I. Watanabe, T. Doi, J. Yamaki, Y.Y. Lin, G.T.-K. Fey, J. Power Sources 176 (2008) 347.
[60] T. Doi, L. Zhao, M. Zhou, S. Okada, J. Yamaki, J. Power Sources 185 (2008) 1380.
[61] A. Sano, M. Kurihara, T. Abe, Z. Ogumi, J. Electrochem. Soc. 156 (2009) A682.
[62] X. Zhang, R. Kostecki, T. Richardson, J.K. Pugh, P.N. Ross Jr., J. Electrochem. Soc. 148 (2001) A1341.

Chapter 8

Artificial SEI for Lithium-Ion Battery Anodes: Impact of Fluorinated and Nonfluorinated Additives

D. Lemordant,[1] W. Zhang,[2] F. Ghamouss,[1] D. Farhat,[1] A. Darwiche,[3] L. Monconduit,[3] R. Dedryvère,[2] H. Martinez[2], S. Cadra[4] and B. Lestriez[5]

[1]PCM2E, Université F. Rabelais, Parc de Grandmont, Tours, France; [2]ICG-AIME, Université Montpellier 2, Montpellier, France; [3]IPREM-ECP CNRS UMR 5254, Pau, France; [4]CEA/DAM, Le Ripault, Monts, France; [5]Institut des Matériaux Jean Rouxel (IMN), CNRS UMR 6502, Université de Nantes, Nantes, France

Chapter Outline

8.1 Introduction	174
8.1.1 Li-Ion Principle	174
8.1.2 SEI Formation	175
8.1.3 SEI Structure and Role	178
8.1.4 SEI Components at the Electrode Surface	178
8.1.5 SEI Components Dissolved in the Electrolyte	180
8.2 Application to TiSnSb Anodes	181
8.2.1 Electrochemical Properties of TiSnSb Electrode	183
8.2.2 Surface Analysis Results	186
8.2.2.1 XPS Results (First Cycle)	186
8.2.2.2 EIS Results	192
8.2.2.3 Scanning Electron Microscopic Analysis	193
8.2.3 Discussion	194
8.2.3.1 At the first Cycle	194
8.2.3.2 At the 20th Cycle	197
8.2.4 Experimental Section	198
8.2.4.1 Electrochemical Tests	198
8.2.4.2 X-ray Photoelectron Spectroscopy	199
8.2.4.3 Scanning Electron Microscopy	199
8.2.4.4 Electrochemical Impedance Spectroscopy	200
8.3 Conclusion	200
Acknowledgment	200
References	201

8.1 INTRODUCTION

Rechargeable Li-ion batteries are now widespread for various portable applications due to their high energy densities, high calendar lifetime, high cycling ability, and low self-discharge [1]. This means that this technology is mature but the increasing demand for stored energy of electronic devices, like smartphones, requires further progress in energy density. Moreover, the use of this technology in electrical vehicles or even hybrid plug-in vehicles implies that security problem must be resolved in parallel with the increase in energy density of the batteries.

8.1.1 Li-Ion Principle

Li-ion batteries are secondary (rechargeable) batteries consisting of an anode material which is able to host the lithium ions dissolved in the electrolyte like graphite or others carbonaceous materials and a cathode material made up of lithium-liberating compounds, like lithiated transition metal oxides (LMO) with M = Co, Ni, Mn, or phosphate ($LiFePO_4$). In order to obtain high energy densities, a high voltage is desirable for the complete cell which means that a low-voltage anode (near 0 V vs Li^+/Li) and a high-voltage cathode (over 4 V) must be associated. For insertion materials during the charge and discharge processes, lithium ions are inserted or extracted from the interstitial space between atomic layers within the active materials of the battery. Hence, Li ions are transferred between the anode and cathode through the electrolyte which usually is a mixture of cyclic and acyclic alkyl carbonates (AC) in which is dissolved a lithium salt, which is mostly $LiPF_6$. The reason for use of $LiPF_6$ for this purpose originates from many reasons: high electrochemical stability toward oxidation and reduction, high degree of dissociation in AC mixtures, high conductivity, and high passivation power toward aluminum collectors. The main drawback of $LiPF_6$ is its lack of thermodynamic stability when the temperature is raised over 50–60 °C, leading to the harmful PF_5 and LiF formation. Since neither the anode nor the cathode materials essentially change during the charge and discharge processes, the operation is safer than that of a lithium metal battery.

As the conventional graphite electrode is limited by its theoretical capacity of 372 mAh g^{-1} [2], further improvement in term of capacity will require new materials. One of the possible ways for that is to focus on Sb- [3–7] and Sn-based [8–11] conversion materials or silicon- and Si-based materials which exhibit storage capacities over 500 mAh g^{-1}. The conversion equation involved is given in Eqn (8.1), where M designs Sn or Sb:

$$M + xLi^+ + xe^- \leftrightarrow Li_x M \quad (8.1)$$

For instance, the Li-rich $Li_{22}Sn_5$ phase is able to theoretically accommodate 22 Li for 5 Sn which leads to a theoretical capacity of 994 mAh g^{-1}. In the case of Sb, Li_3Sb gives a theoretical capacity of 660 mAh g^{-1} [12]. Silicon in Si-based material forms various alloys with lithium as expressed in Eqn (8.1) and Li-rich alloys may contain as much as 4.4 Li per Si ($Li_{22}Si_5$) corresponding to a theoretical specific capacity of 4400 mA/g. All Si and Sn compounds exhibit working potential (E^-) close to that of graphite and less than 0.5 V vs Li^+/Li. This means that AC solvents are not electrochemically stable when these materials are lithiated as their reduction potential is in the range 0.8–0.7 V vs lithium. Nevertheless, the reduction of the electrolyte may stop owing to the formation of a passive layer at the surface of the active material, which is known as the solid electrolyte interphase (SEI). If the SEI is Li^+ conductive, Li ions are continuously inserted in the host material or make alloy with the anode material when the potential is driven below the solvent reduction potential. A similar phenomenon may occur when high-voltage cathodes ($E^+=4.5–5.0$ V) are employed like lithium manganese nickel oxide (LMNO) spinels. These compounds have a working voltage which is of the order of or over the oxidation potential of cyclic ACs, but the formation of a stable SEI often does not occur. The scheme reported in Figure 8.1 summarizes the conditions in which a "natural" SEI is spontaneously formed as a function of the electrolyte electrochemical window ($EW=EW^+-EW^-$) and the electrodes' working potentials:

- Anodic side: $E^- < EW^-$
- Cathodic side: $E^+ > EW^+$

8.1.2 SEI Formation

Reduction of AC in the presence of $LiPF_6$, occurs at a potential below 1 V vs Li^+/Li, but all ACs are not equivalent, even if they are reduced in the same potential range. As an example, ethylene carbonate (EC) is the best SEI builder and propylene carbonate (PC) is the worst on graphite anodes. PC does not provide a stable SEI layer, mostly as the products of its reduction dissolve easily in the electrolyte. Some

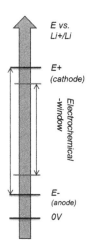

FIGURE 8.1 Electrochemical window of an electrolyte and SEI formation.

acyclic ACs are good SEI builders like dimethyl carbonate (DMC), methyl propyl carbonate (MPC) [13], or methyl isopropyl carbonate (MiPC) [13]. On the contrary, diethyl carbonate is not a good SEI builder. In mixtures of cyclic and acyclic AC, 20% of EC is often considered as sufficient to obtain a good SEI on graphite.

When non-SEI-forming solvents are employed, e.g., PC, it is, nevertheless, possible to cycle graphite if an SEI-forming additive is introduced in the electrolyte. This type of additive is always added to the electrolyte in small quantities, i.e., 0.3–5% (by mass). SEI-forming additives may have very different structures but often they are five-membered oxygen-containing heterocycles additives (–O–C(=X)–O–) where X=O, S, or a vinyl group like vinylene carbonate (VC) or acrylic acid nitrile [14]. Fluorine (F) or cyano groups (CN) are common substituents for hydrogen atoms on cyclic or acyclic compounds like fluoroethylene carbonate (FEC) or difluoroethylene carbonate (F_2EC). The main reason is that all these additives exhibit a kind of instability like ring strain or reactivity like vinyl group. Ring strain results from a combination of angle strain, conformational strain, and transannular strain and this occurs in five-membered rings. When the potential is driven to a low value, the ring will open to form radicals which are able to react in a complicated manner to give anions (carbonate, semicarbonates, dicarbonates, etc.) or to polymerize. Moreover, the presence of Lewis base groups (–O–, C=O, –O–C=O, –S–, C≡N) in the additive's structure is useful for the transport of lithium ions throughout the SEI layer. The introduction of fluorine atoms in the additive formulas will decrease the lowest unoccupied molecular orbital (LUMO) energy and hence make the reduction easier, i.e., at higher potential than the bulk electrolyte (alkyl carbonates).

In Figure 8.2 is reported, as an example, the galvanostatic cycling of a graphite anode in EC:DMC (1:1)+$LiPF_6$ (1M). Reduction of graphite surface groups occurs

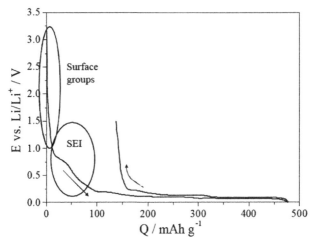

FIGURE 8.2 Galvanostatic cycling measurements at 10 mAh g^{-1} in 1 M LiPF6, EC:DMC (1:1 by wt.) electrolyte for KS6-based electrodes (first cycle).

at the first cycle between 3 and 1V followed by the reduction of the electrolyte beginning at 0.8V. At lower potentials (<0.25V) and when the SEI is formed, insertion of Li ions in graphite occurs. In order to avoid the deposition of metallic lithium the scanning potential is stopped at 0.02V. On the reverse scan deinsertion occurs between 0.02 and 0.3V and the reversible capacity can be measured. Often, more than one cycle is required to achieve the SEI formation and to stabilize it. Then the irreversible capacity (lost) is the sum of the capacities lost during these cycles.

In Figure 8.3 is reported the galvanostatic cycling of a Li/graphite half-cell using $PC + LiPF_6$ (1M) as electrolyte with or without the addition of SEI-forming additives: VC, FEC, and F_2EC. In the absence of any additive, the reduction of the electrolyte occurs continuously at 0.8V vs Li^+/Li. Cointercalation of solvent molecules happens in the graphite plane interspace and their reduction generates propylene gas and CO_2 as major components which are able to exfoliate the graphite structure. In the presence of SEI-forming additives, the pattern is completely different and the shape of the curves looks like those reported in Figure 8.2. The coulombic efficiency (CE%) is defined as:

$$CE\% = 100 \cdot Q(\text{de-lithiation})/Q(\text{lithiation})$$

where Q is the charge in coulomb corresponding to Li insertion or deinsertion for the successive cycles. The graphite reversible capacity Q_{rev} corresponding to the curves reported in Figure 8.2 have been determined and the results are reported in Table 8.1.

Results reported in Table 8.1 show that these SEI-forming additives are all efficient as they permit to cycle graphite in PC-based electrolytes. Nevertheless, their

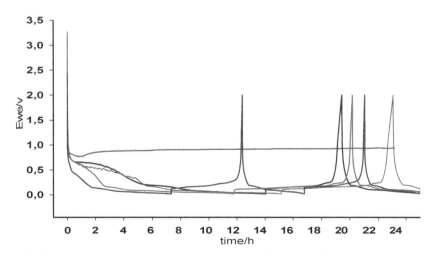

FIGURE 8.3 Galvanostatic cycling of a Li/graphite half-cell using 1M LiPF6 in electrolytes containing PC (LiPF6, 1M) without (red) or with VC (2%, blue), FEC (2%, green), F_2EC (2%, pink). Cycling in Novolyte LP100 is given as a reference (purple). *(For interpretation of the references to color in this figure legend, the reader is referred to the online version of this book.)*

TABLE 8.1 Coulombic Efficiency (CE in %), Reversible Capacity (Q_{rev}). and Approximate Reduction Potential of VC, FEC and F_2EC Additives for a Graphite Anode Cycling in PC + LiPF$_6$ (1M) Electrolyte: All Values Are Means Over 12 Cycles and s Is the Standard Deviation.

Additive	None	VC (2%)	FEC (2%)	F_2EC (2%)
E_{red} (V vs Li+/Li)	0.8	1.2	1.4	1.7
CE%	0	93 (σ=2.0)	91 (σ=1.5)	>98 (σ=1.1)
Q_{rev} (mAh/g)	0	228 (σ=6)	311 (σ=10)	324 (σ=7)

performances in term of CE and reversible capacity are not equivalent. The highest CE and reversible capacity are obtained with F_2EC additive which is reduced at the highest potential and hence has the lowest LUMO energy. FEC leads to a larger reversible capacity than VC with a similar CE value.

8.1.3 SEI Structure and Role

The essential properties of a functional SEI (natural or artificial) are:

1. A stable interphase with negligible dissolution and deposition over the cycle life of the electrode.
2. Good mechanical properties: high storage modulus and great elasticity in order to accommodate active material volume expansion during lithiation.
3. A constant thickness and a low electrical impedance.
4. A high conductivity for unsolvated lithium ion.
5. A barrier to solvent molecules.

All these properties are better verified by Li$^+$ conducting polymer membranes than a mineral layer. As a matter of fact, the analysis of the SEI formed on graphite and others active materials shows that the composition is clearly organomineral composed of a polymeric phase [15,16]. The schematic view of an SEI is displayed in Figure 8.4(a) for a flat material where mineral inclusions are embedded in a polymeric phase. The polymeric phase is Li$^+$ conducting owing to numerous oxygenated groups (or cyano groups) in the polymer chain. The mineral inclusion, mainly nanosized, is also able to exchange lithium ions like Li$_2$CO$_3$, lithium semicarbonates, lithium dicarbonates, and LiF ($Li_nF_m^{(n-m)+}$).

As graphite is an anisotropic material, it is probable that the structure of the SEI is not the same on the edges and on the planes of the grapheme sheets. In Figure 8.4(b), a schematic view of the SEI formed on graphite is reported.

8.1.4 SEI Components at the Electrode Surface

The SEI formed on graphite electrodes in AC mixtures containing LiPF$_6$ has been extensively studied by means of X-ray photoelectron spectroscopy (XPS)

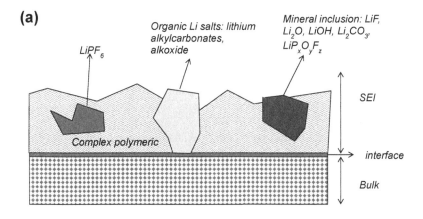
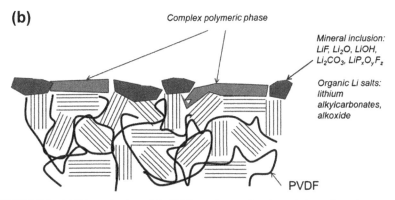

FIGURE 8.4 (a) Schematic view of the SEI formed over a flat electrode in an electrolyte containing a mixture of ACs and LiPF$_6$ and (b) Schematic view of the SEI formed over a graphite electrode in an electrolyte containing a mixture of ACs and LiPF$_6$.

analysis. Aurbach et al. confirmed via Fourier transform infrared spectroscopy that the SEI is composed of mainly (CH$_2$OCO$_2$Li)$_2$, Li$_2$CO$_3$, LiF, etc. in EC-based electrolyte [3], whereas ROCO$_2$Li is the main species of SEI in PC-based electrolyte [4].

The SEI composition is also affected by the presence of SEI-forming additives. As an example, when fluorinated additives are used, the amount of LiF in the SEI is increased. This is the case of FEC [17,18] and probably also F$_2$EC. The mechanisms leading to the formation of LiF as a component of the SEI are known: the first originates from the instability of LiPF$_6$ which undergoes dissociation into LiF and PF$_5$ relatively easily:

$$LiPF_6 \rightarrow PF_5 + LiF$$

The strong Lewis acid PF$_5$ binds to the basic sites of solvent molecules (O–C(=O)–O) groups in AC. In the presence of fluorinated additives like FEC, the following processes [16] may occur at the electrode surface:

1. FEC is reduced through the opening of the five-membered ring leading to the formation of lithium poly(vinyl carbonate), LiF, and some dimers,
2. The FEC-derived lithium poly(vinyl carbonate) enhances the stability of the SEI film. This reduction mechanism opens a new path to explore new electrolyte additives that can improve the cycling ability of many anode materials.

The benefit expected from SEI builders are:

- A high CE (>99%): no side reactions (oxidation/reduction of the electrolyte components),
- A stable film (increase in cycle number and calendar life),
- A stable electrode capacity: no dissolution of active material (transition metals like Mn, Ni, Co,…), and
- A high conductivity (low ohmic drop, high charge transfer rate).

8.1.5 SEI Components Dissolved in the Electrolyte

The electrolyte has been removed from a Li/Gr cell after cycling and analyzed by gas chromatography (GC)–mass spectrometry (MS). From the GC–MS spectra, some peaks (designed as A, B, C, and D) have been selected and the corresponding retention times and molecular masses have been reported in Table 8.2.

The mass spectra of A, B, C, and D and their fragments are reported in Figure 8.5.

The following is the analysis of selected species A, B, C, and D from extracted-ion chromatogram

1. A: This compound presents signals at m/z=29, 45 (base peak), 59, 74, 89, 117, and 127 Da. Signals below m/z ≤ 117 Da may correspond to oligoethylene glycol, whereas the signal at 127 Da is specific for the dimethyl phosphate fragment. A plausible structure for A is: 2-(2-ethoxyethoxy)ethyl dimethyl phosphate as reported in Figure 8.6(a).
2. B: This compound presents a base peak at m/z=45 and signals at m/z=15 (weak), 29(weak), 59, 89(weak), and 91(weak) Da. Signals at 91, 59, 45, and 29 Da correspond well to an AC fragment and the peak at m/z=89 to an

TABLE 8.2 Selected Compounds from CG-MS Spectra

Label	Retention time (min)	(m/z) Da^{-1}
A	9.15	242
B	10.16	192
C	11.97	250
D	13.04	>103

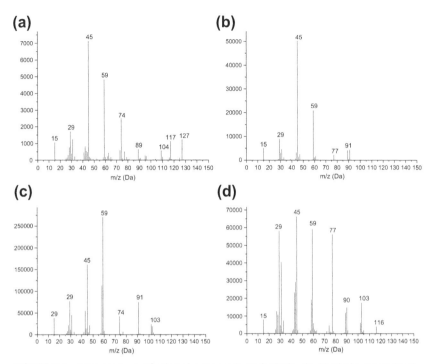

FIGURE 8.5 Mass spectrum of selected compounds (a) 9.15 min, (b) 10.16 min, (c) 11.97 min, and (d) 13.04 min.

oligoethylene glycol (diglyme). Hence a plausible structure for B is: ethyl (2-(2-methoxyethoxy)ethyl) carbonate.

3. C: It has been found mostly on the electrodes and separator, probably as it is more viscous. The signals classified by decreasing intensity order are found at m/z = 59, 45, 29, 91, 74, 29, and 103 Da. This mass spectrum matches well with a glycol dicarbonate compound as the diethyl diglycol carbonate.
4. D is localized as C mostly on the electrodes and separator. The signals classified by decreasing intensity order are found at m/z = 45, 59, 29, 77, 103, 90, 89, 102 and 116 Da. This spectrum which is not very different from that of a dimethyl diglycol carbonate, matches better with an oligocarbonate glycol.

Hence dissolved SEI compounds are essentially low-molecular-weight ethylene glycol and ethylene glycol carbonate derivatives. This means that higher molecular weight compounds, such as polyethylene oxide (PEO), polycarbonate, and polyether carbonate, remain in the SEI, playing the role of a binder and a functional membrane.

8.2 APPLICATION TO TiSnSb ANODES

TiSnSb has been developed as a negative electrode material for Li-ion batteries as this conversion material can reversibly take up as much as 6.5 lithium per formula

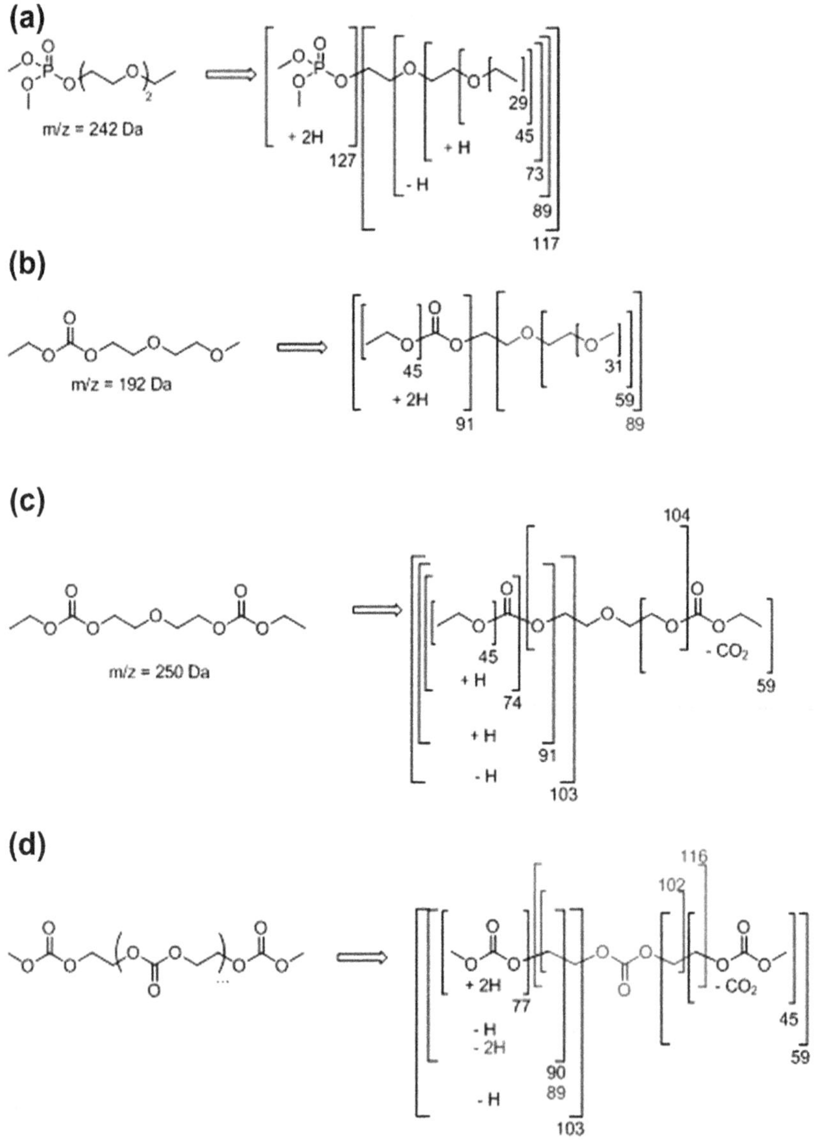

FIGURE 8.6 Structure proposed for compounds A,B,C and D according to their respectively mass spectrum (a), (b), (c) and (d).

unit leading to a theoretical capacity of 580 mAh g^{-1} with noteworthy high rate capabilities [19,20]. The conversion reaction may be summarized by Eqn (8.2):

$$TiSnSb + 6.5Li = Ti + Li_3Sb + 0.5Li_7Sn_2 \qquad (8.2)$$

However, like other conversion and alloying materials TiSnSb suffers from drastic volume change during lithium insertion/extraction. As an example, the lattice of the tin antimony alloy (SnSb) expands to 137% when it accommodates 6.5 lithium atoms. The expansion of the material causes high mechanical strains which results in the mechanical disintegration of the active material particles and finally to fast capacity fading. A way to improve the cyclability of SnSb alloys has been to introduce transition metals such as titanium in order to create a buffering matrix [21]. During charge, when Sn or Sb reacts with Li, Ti has a buffering effect which helps to increase the mechanical resistance to the volume expansion, and to retain the initial capacity of the electrode material. In the following, the 1C rate will be defined by the insertion of 1 Li in 1 h which corresponds approximately to a specific current of 96 mA h^{-1}.

Since the formation of SEI layer plays a crucial role in the cycling ability of the electrode, it is very important to understand the mechanisms which lead to its formation as well as the composition of this layer and the most convenient way to improve the quality of the SEI layer is to use SEI-forming additives such as vinylene carbonate (VC) and fluorinated EC. Using LiCoO$_2$/graphite batteries, it has been demonstrated [22] that the radical polymerization leads to poly(VC) as the main VC-derived product contributing to the formation of the surface film. VC has also been applied to improve the cycle life of other systems such as Si thin films [23–25], Si-based composite electrodes [26], and Si nanowire (SiNW) electrodes [27]. Besides VC, FEC has also been proposed as an SEI improver, as it is reduced before EC and others ACs and contributes to form a more stable SEI. Long-chain flexible polycarbonates could be the major surface film component in FEC and VC-containing electrolyte solutions [27]. This polymeric surface film may explain the ability of the electrode to accommodate large volume expansions as is the case for Si. FEC also promotes longer cycle life in the case of sodium batteries using Sb anode [28] and hard carbon anode [29]. The addition of FEC and VC favors the precipitation of more stable degradation products limiting further EC decomposition and strongly improving the electrochemical performances [26]. Recently, VC- and FEC-containing electrolytes have also been reported to significantly enhance the cycling ability of conversion material like TiSnSb used as negative electrode in Li-ion batteries [30].

In this study, XPS and electrochemical impedance spectroscopy (EIS) were carried out to carefully investigate the physical properties and chemical composition of the SEI layer formed at the TiSnSb electrode surface in the presence or absence of SEI-forming additives.

8.2.1 Electrochemical Properties of TiSnSb Electrode

Cyclic voltammetry was performed to highlight the irreversible processes occurring during the first reduction before the conversion reaction. Cyclic voltammetry (CV) curves, reported in Figure 8.7, show that the electrochemical reduction of the electrolyte occurs in the 0.5–2 V potential range. In this range, three reduction processes are identified. The first peak located at 1.15 V (Figure 8.7(a)) could be assigned to the irreversible reduction of VC and formation of AC derivatives and polyalkyl carbonates on the electrode surface. Further reduction of cyclic carbonates, EC and PC, is expected to occur at lower potential, i.e., below 0.7 V. The small peaks at 0.78 and 0.66 V can be assigned to the reduction of Sn, Ti, and Sb oxides at the surface of the electrode as these peaks are always visible even when the electrolyte formulation is changed. The first CV cycle after addition of FEC to the electrolyte is reported in Figure 8.7(b). From the inset of Figure 8.7(b), one can see that the first reduction peak is shifted to a higher potential. In that case, the formation of the SEI layer is expected to be more efficient since it could be formed at higher potential and thus can be achieved well before the reduction of the bulk electrolyte. Furthermore, the conversion peak at a low potential of 0.125 V is thinner and more pronounced when FEC is used. So we can expect from these observations that the conversion reaction is faster when FEC is used as the SEI is less resistive (see further in the text).

TiSnSb electrode undergoes a low-voltage discharge and a large irreversible capacity during the first cycle even in the presence of VC as additive, as shown in Figure 8.8(a) where the galvanostatic curves obtained at the first charge/discharge cycle are displayed. When FEC is added to the electrolyte, as shown in Figure 8.8(b), the irreversible capacity is reduced but the discharge profile indicates that the global shape of the galvanostatic curve remains unchanged, which reveals that the electrochemical processes are unmodified by the choice of the additive, i.e., a rapid decrease of the potential down to a plateau at 0.5 V and then to a second plateau at 0.25 V. The lengths of these plateaus correspond, respectively, to the lithiation of Sb and Sn. In Figure 8.8(a), the galvanostatic curve indicates that the reduction of the standard electrolyte solution is equivalent to the insertion of 6.9 Li per formula unit, which is more than the 6.5 Li corresponding to Li_3Sb and Li_7Sn_2 phase formation and, when FEC is added to the electrolyte, 5.5 Li per formula unit are inserted in the active material. During

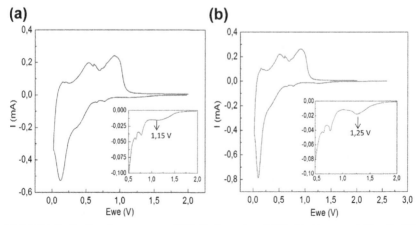

FIGURE 8.7 Cyclic voltammograms of TiSnSb electrode at $0.025\,mVs^{-1}$ between 0.02 and 2.0 V, (a) without FEC, (b) with FEC.

the reverse scan, 5.1 Li and 4.3 Li per formula unit can be removed from the active material, respectively, in the absence and presence of FEC. As a conclusion, the addition of FEC induced both a reduction of the irreversible capacity (from 1.8 Li to 1.2 Li) and of the reversible capacity at the first cycle (5.1 Li to 4.3 Li). As FEC is reduced before VC and other components of the electrolyte, the SEI composition must contain more reduction products coming from FEC than VC and the smaller irreversible capacity indicates that FEC is an efficient SEI-forming additive.

Galvanostatic curves obtained at the 20th cycle exhibit shape similar to that obtained at the first cycle. Measurements of the capacity retention during the first 20 cycles in the standard electrolyte and in the presence or absence of FEC are displayed in Figure 8.9. The irreversible capacity becomes negligible for both electrolytes at the second or third cycle which means that the CE approaches 100% during the following cycles. Nevertheless, the cell using the FEC-containing electrolyte displays better cycling performances as the specific capacity is steady at 450 mAh g^{-1} from the second to the 20th cycle, while it continuously fades, down

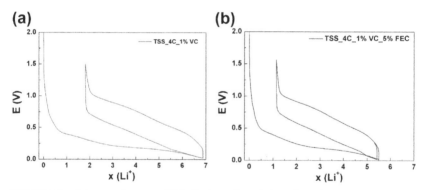

FIGURE 8.8 Galvanostatic curve of TiSnSb electrode cycled at 4C rate between 0.02 and 1.5V, (a) the first discharge/charge without FEC, (b) the first discharge/charge with FEC.

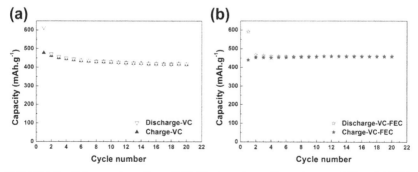

FIGURE 8.9 Galvanostatic cycling of TiSnSb electrode at 4C rate between 0.02 and 1.5V in the standard electrolyte at the 20th discharge/charge cycle: capacities curves vs cycle number: (a) without FEC, (b) with FEC.

to 415 mAh g^{-1} at the 20th cycle in the standard electrolyte. This is another proof that FEC helps to form a more stable SEI layer at the electrode surface.

8.2.2 Surface Analysis Results

8.2.2.1 XPS Results (First Cycle)

Table 8.3 shows the XPS quantification data of 1s peaks for C, O, Li, and F and 3d core peaks for Sn. Although titanium and antimony are active elements in the TiSnSb electrode, XPS analysis at the electrode surface reveals that these two species are not observed after cycling. Concerning the absence of Ti, it is assumed that lithiated Li_7Sn_2 and Li_3Sb phases, formed upon Li insertion at the surface of the material particle, dissimulate Ti which is buried in the bulk. The signals of Sn and Sb species are hence more easily detected than that of the nonelectrochemically active Ti within the analysis depth of XPS. This assumption is supported by the detection of Ti species in Ti 2p core peak spectra (not given here) after mechanical removal (blade-scratching) of the top layer (<5 nm) of the electrode surface. On the other hand, Sb 3d core peak spectra share the same range of binding energies as that of oxygen, leading to an overlap of Sb and O peaks. From a quantitative point of view, the amount of Sb species at the surface is relatively low as compared to that of oxygen. Therefore, O 1s spectra will be considered as pure O species in the following. As a result, Sn species are the only really distinguishable component of the active material.

The XPS Sn 3d core peak spectra are recorded at the end of the first discharge and first charge. Binding energies of 484.7 and 493.2 eV are assigned to metallic tin and binding energies of 486.0 and 494.5 eV are assigned to tin oxide. The corresponding quantification data are presented in Table 8.3. Without FEC, there is no tin species detected in the analysis depth of XPS at the end of the first discharge. In case of a charged electrode, Sn species could be detected (metallic Sn is 0.02%, Sn oxide is 0.07%). On the other hand, with the addition of FEC, a few amount of tin species can be observed both at the discharged and the charged state. It is worth noticing that the quantitative analysis indicates that the amount of Sn species is higher at the charged state than the discharged state in both cases with or without the addition of FEC.

The XPS C 1s core peak spectra, displayed in Figure 8.10, provides valuable information regarding the SEI nature when jointly analyzed with the O 1s (displayed in Figure 8.11) and Li 1s core peaks. The F 1s and P 2p core peaks data are reported in Table 8.3 with the corresponding binding energies. The component with a binding energy of 285.0 eV is assigned to CH_x environment, which is attributed to hydrocarbon contamination (always detected at the extreme surface) and to carbon atoms of organic species bound to carbon or hydrogen atoms only. The component observed at 286.5 eV can be assigned to carbon atoms bound to one oxygen atom (C–O), while the component at 289.0 eV corresponds to carbon atoms bound to two oxygen atoms (O=C–O). The component observed at 290.2 eV is the characteristic of carbon bound to three oxygen atoms, which is

TABLE 8.3 XPS Quantification Data at the End of the 1st Discharge (D) / Charge (C), without (DMC) / with the Addition of FEC (FEC). (The Abbreviation Begins with D or C Indicates When the XPS Analyses Were Carried Out, at the End of Discharge or Charge; 4C Points Out the Cycling Rate; Ends with FEC Means the Electrode Was Cycled Using the Standard Electrolyte with the Addition of FEC as the Second Electrolyte Additive.)

	D1-4C-DMC		C1-4C-DMC		D1-4C-FEC		C1-4C-FEC	
	B.E. (eV)	at. (%)	B.E. (eV)	at. (%)	B.E. (eV)	at. (%)	B.E. (eV)	at. (%)
C 1s	285.0	(7.7)	285.0	(10.1)	285.0	(16.6)	285.0	(19.9)
	286.6	(0.9)	286.6	(4.2)	286.4	(3.7)	286.4	(13.8)
	289.0	(1.0)	289.0	(2.3)	289.2	(2.1)	288.9	(4.8)
	290.2	(13.4)	290.2	(11.5)	290.6	(4.0)	290.5	(3.9)
O 1s	528.6	(1.5)	–	–	529.3	(1.8)	529.5	(2.0)
	–	–	531.7	(46.8)	–	–	531.8	(17.7)
	532.0	(47.2)	–	–	532.1	(33.5)	–	–
	–	–	–	–	533.1	(5.7)	533.0	(11.0)
Li 1s	55.4	(28.1)	55.2	(24.1)	55.7	(30.7)	56.0	(19.0)

Continued

TABLE 8.3—cont'd

	D1-4C-DMC		C1-4C-DMC		D1-4C-FEC		C1-4C-FEC	
	B.E. (eV)	at. (%)	B.E. (eV)	at. (%)	B.E. (eV)	at. (%)	B.E. (eV)	at. (%)
F 1s	–	–	684.8	(0.8)	685.6	(1.8)	685.4	(7.0)
	687.5	(0.3)	–	–	–	–	–	–
P 2p	N.D.		N.D.		N.D.		N.D.	
Sn 3d	N.D.		484.9	(0.01)	484.7	(0.01)	485.3	(0.02)
			486.8	(0.04)	486.0	(0.01)	486.5	(0.03)
			492.6	(0.01)	493.2	(0.01)	493.6	(0.01)
			494.3	(0.03)	494.5	(0.01)	494.7	(0.02)

B.E., Binding energy; N.D., Not detected.

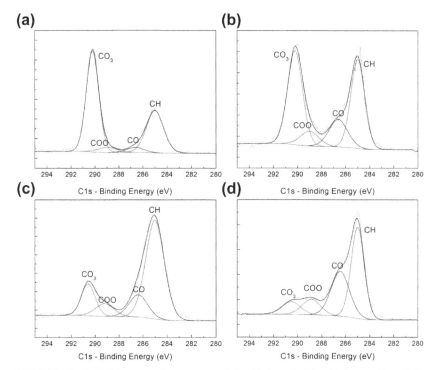

FIGURE 8.10 XPS C 1s core peak spectra recorded at (a) the end of first discharge without FEC, (b) the end of first charge without FEC, (c) the end of first discharge with FEC, (d) the end of first charge with FEC.

typical of carbonate-like species (–CO_3) that could be Li_2CO_3 or AC $ROCO_2Li$. In the presence of FEC, the amount of CO_3-like carbon decreases from 13.4% to 4.0% at the end of discharge, and from 11.5% to 3.9% at the end of charge. Charged electrodes indicate less amount of CO_3 component in cases both with and without FEC. The amount of C–O component increases in the presence of FEC from 0.9% to 3.7% at the discharged state, and from 4.2% to 13.8% at the charged state.

Without FEC, the O 1s core peaks (Figure 8.11) at the end of discharge or charge present a broad peak, located around 531.8 eV. These components could be attributed either to Li_2CO_3 (532.2 eV) or LiOH (531.5 eV). Note that the small-intensity component detected at the end of discharge, at 528.7 eV, is assigned to Li_2O. With FEC, the broad peak (33.5% of all the detected atoms) observed in the discharged electrode is located at a binding energy of 532.1 eV. The amount of Li_2O is 1.8%, higher than 1.5% in the electrode without FEC. On the other hand, in the charged electrode, Li_2CO_3 is still the major compound around 17.7%, although much lower than that of the electrodes cycled without FEC. Note that the total amount of O species decreased from 41.0% to 30.8% after the charge process. The amount of Li_2O is 2.0% compared to 0% in the

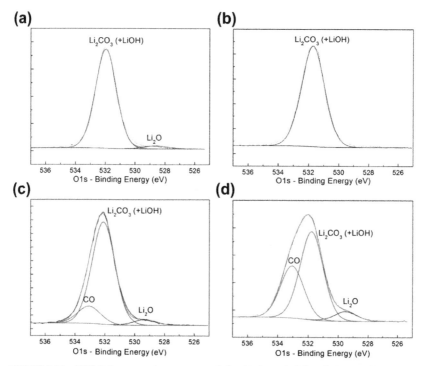

FIGURE 8.11 XPS O 1s core peak spectra recorded at (a) the end of first discharge without FEC, (b) the end of first charge without FEC, (c) the end of first discharge with FEC, (d) the end of first charge with FEC.

electrode cycled without FEC. The peak located at 533.0 eV (5.7% and 11.0%, respectively, at the end of the discharge and charge) can be assigned to a CO environment, and will be examined into more details in the discussion.

The Li 1s core peak spectra (55.2 eV) represents multiple components Li_2CO_3 and LiOH when TiSnSb electrode is cycled without FEC. The total amount of Li species is 28.1% and 24.1% for discharged and charged electrodes, respectively. With FEC, the Li 1s core peaks are located at binding energies of 55.7 and 56.0 eV for discharged and charged electrodes, respectively. The total amount of Li species decreased from 30.7% in the discharged electrode to 19.0% in the charged electrode.

The F 1s core peaks (Table 8.3) are characterized by a main component at 685.0 eV, which could be assigned to LiF, and by a very weak component at 687.5 eV attributed to $LiPF_6$ (only observed for discharge electrode without FEC). In the presence of FEC, the amount of LiF increases from 1.8% (discharge) to 7.0% at the end of charge. In general, charged electrodes exhibit a higher amount of LiF at their surface in both cases (i.e., with or without FEC).

In Table 8.4 the XPS quantification data of C 1s, O 1s, Li 1s, F 1s, and Sn 3d core peaks at the 20th cycle are displayed. As for the first cycle, the elements

TABLE 8.4 XPS Quantification Data after the 20th Discharge (D) or Charge (C), without (DMC) or with the Addition of FEC (FEC)

	D20-4C-DMC		C20-4C-DMC		D20-4C-FEC		C20-4C-FEC	
	B.E. (eV)	at. (%)	B.E. (eV)	at. (%)	B.E. (eV)	at. (%)	B.E. (eV)	at. (%)
C 1s	285.0	(23.1)	285.0	(29.2)	285.0	(16.1)	285.0	(24.0)
	286.5	(6.6)	286.6	(11.2)	286.7	(2.3)	286.6	(11.9)
	289.0	(1.4)	288.4	(2.4)	288.8	(1.6)	288.7	(4.0)
	290.4	(1.6)	290.6	(3.1)	290.2	(10.6)	290.5	(5.3)
O 1s	528.7	(5.6)	528.7	(1.4)	528.7	(1.8)	529.1	(1.0)
	531.5	(28.9)	531.5	(24.6)	531.9	(40.1)	531.5	(11.6)
	–	–	532.8	(4.8)	–	–	532.7	(17.4)
	–	–	534.0	(3.6)	–	–	534.3	(1.4)
Li 1s	55.4	(30.9)	55.8	(17.7)	55.5	(26.9)	56.0	(18.0)
F 1s	685.4	(1.2)	684.9	(1.3)	685.0	(0.5)	685.5	(3.9)
	687.6	(0.3)	687.1	(0.9)	686.9	(0.1)	687.9	(0.3)
P 2p	133.5	(0.3)	134.0	(0.5)	N.D.		133.8	(0.7)
	136.9	(0.1)	136.6	(0.1)			–	–
Sn 3d	N.D.		N.D.		N.D.		484.9	(0.01)
							486.8	(0.02)
							492.7	(0.01)
							494.2	(0.01)

N.D., Not detected.

present in the active material are difficult to observe. Only Sn species (metallic Sn of 0.02%, Sn oxide of 0.03%) are observed at the charged state and in the presence of FEC.

The XPS C 1s core peak spectra recorded after the 20th discharge/charge presents, as in the first cycle, the components with binding energies of 285.0, 286.5, 289.0, and 290.2 eV which correspond to CH_x, C–O, O=C–O, and –CO_3 species, respectively.

In the presence of FEC, the amount of –CO_3 increases from 1.6% to 10.6% at the end of discharge and from 3.1% to 5.3% at the end of charge according to the quantification data given in Table 8.4.

Without FEC and at the end of the discharge, a broad peak observed in the XPS O 1s core peak spectra at 531.5 eV could be mainly attributed to LiOH. The peak located at low binding energy (528.7 eV) is assigned to Li_2O (5.6%). In the charged electrode, the O1s peak is composed of multiple components located at 528.7, 531.5, 532.8, and 534.0 eV. These components are, respectively, attributed to Li_2O, LiOH, C–O, and POF_x environments.

With FEC, in the discharged electrode, a broad peak located at 531.9 eV represents Li_2CO_3 as the major component (40.1%). This result is in agreement with the corresponding C 1s spectrum, where CO_3 is 10.6% with FEC (and 1.6% without FEC). The amount of Li_2O is 1.8%, lower than that of 5.6% in the electrode cycled without FEC.

In the charged electrode, the broad peak is composed of multiple components, where LiOH is 11.6%, CO environment (532.7 eV) is 17.4%, and a possible component of POF_x (534.3 eV) is 1.4%. The amount of Li_2O (529.1 eV) is 1.0% compared to 1.4% in the electrode without FEC. The total amount of O species decreases from 41.9% to 31.4% after charge process as already observed during first cycle in the presence of FEC (from 41.0% to 30.8%).

A small amount of LiF (F 1s: 685 eV) is detected. In the presence of FEC, it increases from 0.5% to 3.9% between discharged and charged states. Note also that a small amount of $LiPF_6$ salt is observed for all electrodes analyzed at the 20th cycle. In general, charged electrodes contain a higher total amount of F species in cases both with and without FEC.

8.2.2.2 EIS Results

In order to evaluate the effect of FEC, EIS spectrum has been recorded during the first cycle and displayed in Figure 8.12. The EIS spectrum reported in Figure 8.12(a) is relative to the end of first charge at 1.6 V. In the high frequency (HF) domain ($\approx 10^4$ Hz) the resistance of interfaces, which includes the SEI layer, is lower when FEC is added to the electrolyte. EIS spectrum recorded at 0.5 V is presented in Figure 8.12(b), where the depressed shape of the semicircle indicates the contribution from both interface resistance and charge transfer, and at lower frequencies, the straight line at 45° is assigned to a Warburg element that represents Li diffusion through the TiSnSb electrode. It should be noted that the resistance attributed to charge transfer in the medium frequency (MF) domain at 142 Hz drastically

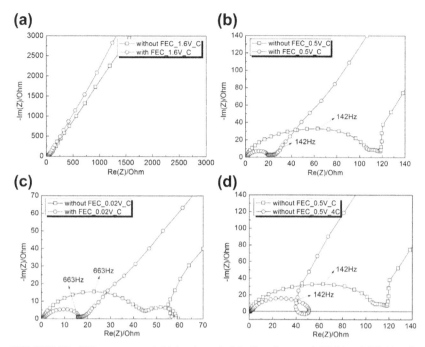

FIGURE 8.12 EIS spectra recorded (a) at the end of the first charge at 1.6 V, (b) at 0.5 V, (c) at the end of the first discharge at 0.02 V, (d) at 0.5 V between different cycling rate of 4C and C.

decreases from 60 to 20 Ω with the addition of FEC to the electrolyte. At the end of the first discharge (0.02 V) an inductive loop, as seen in Figure 8.12(c), in the low frequency (LF) range is clearly observed when using the standard electrolyte. With the addition of FEC, the inductive loop disappears, and the interfaces and charge transfer resistances decrease. In addition, the effect of the cycling rate is displayed in Figure 8.12(d), where the inductive loop is clearly observed at the cycling rate of 4C but not at C. This means that the active material is not at equilibrium when a high speed of 4C is applied.

8.2.2.3 Scanning Electron Microscopic Analysis

The surface morphology of pristine TiSnSb electrode is presented in Figure 8.13. Scanning electron microscopic (SEM) images illustrate that TiSnSb microparticles are evenly distributed at the electrode surface (Figure 8.13(a)) and in the bulk material based on the cross-sectional view (Figure 8.13(b)). Under a higher magnification, electrode displays a texture that is favorable to material volume change during lithiation as the net formed by vapor-grown carbon fibers (VGCF) where the TiSnSb microparticles get entangled is also tightened by the carboxymethyl cellulose (CMC) binder (Figure 8.13(c)). The corresponding backscattered electron image (Figure 8.13(d)) indicates that the TiSnSb microparticles (bright area) and carbon fibers (dark area) are clearly distinguishable.

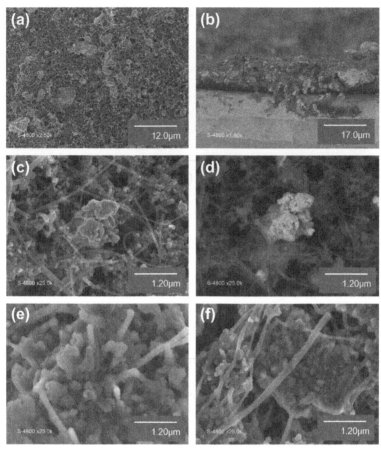

FIGURE 8.13 SEM images of pristine TiSnSb electrode, (a) top view in the scale of 12.0 μm, (b) cross-sectional view in the scale of 16.7 μm, (c) top view in the scale of 1.20 μm, (d) backscattered electron image of Figure 8.10(c), (e) at the end of the 20th discharge, and (f) at the end of the 20th charge.

The surface morphology of TiSnSb electrode cycled after 20th discharge is presented in Figure 8.13(e), where carbon fibers and the TiSnSb microparticles are clearly covered by a surface layer. At the end of 20th charge this surface layer disappears (Figure 8.13(f)).

8.2.3 Discussion

8.2.3.1 At the first Cycle

The discussion will be focused first on the thickness of the SEI layer during the first cycle. Without the addition of FEC, no tin species are detected in the analysis depth of XPS (5 nm) at the end of discharge, which illustrates that the SEI layer is at least thicker than 5 nm. However, Sn species are detectable from the electrode at the charged state, around 0.1%. Since signals of Sn are given by the active material that is covered by the SEI, this means that the SEI layer is thinner at the charged state. From the end of discharge to the end of charge, XPS analysis suggests that the SEI layer partially dissolves during the charge process.

With the addition of FEC, the Sn 3d spectrum of the discharged electrode presents a small amount of tin species, which indicates that the SEI thickness is close to the analysis depth. As Sn species are detected in the electrode cycled with FEC, we may conclude that the presence of FEC results in a thinner SEI layer compared to those electrodes cycled with the standard electrolyte which contains only VC as additive. The conclusion of thinner SEI formed in FEC-containing electrolyte was also made in a previous work on SiNW electrodes [27]. Partial dissolution of the SEI layer during the charge process occurs, in the presence or absence FEC. In addition, the presence of LiF in the electrode cycled with FEC may further prove a thinner SEI layer.

EIS results demonstrate that when FEC is added the HF and MF semicircles decrease in size, which indicates lower interface and charge transfer resistances. This correlates well with the presence of a thinner SEI layer as indicated by XPS analysis.

In addition, the inductive loop recorded in the EIS spectra is not observed in the HF but only in the LF range, which has also been reported in other studies [30,31]. This phenomenon usually exists under the conditions of rough electrodes or electrosorption of corrosion process [32,33] fuel cells [34] and electrodeposition systems [35]. J. Song et al. [36] suggests that this inductive loop corresponds to adsorption during lithium intercalation into carbon. Thus the appearance of inductive loop is only due to the presence of an SEI layer which is modified by the electrolyte additives and the cycling rate. Having accepted this premise, the formation of an inductive loop could be due to the adsorption of ions in a porous, thick, and imperfect SEI. Nevertheless, another possible mechanism can be proposed: current flow occurring between particles having different state of charge [37], as illustrated by Figure 8.14. When lithium ions are been inserted into TiSnSb electrode during discharge, lithium-rich and lithium-deficient regions in the electrode may be isolated by the SEI film (due to imbalanced electronic continuity) and, in this case, a concentration cell is built between as the lithium content in the antimony is not the same in all particles. The direction of current flow within this concentration cell is opposite to that of lithium insertion, so this current flow will generate a field that opposes the field due to lithium ion insertion. Such situation accomplishes the requirements to form an inductive loop.

As inductive loops become smaller in size when FEC is added, this could show that the quality of the SEI is improved by the FEC additive, and indicate a more homogeneous SEI layer no matter which mechanism leading to the inductive loop is operating. In addition, as shown in Figure 8.12(d), the magnitude of the inductive loop depends on the cycling rate. The inductive loop is larger when the electrode is cycled at a relatively high regime of 4C, and it almost disappears

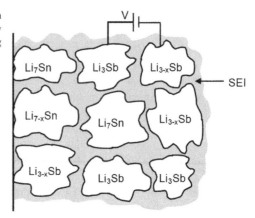

FIGURE 8.14 Schematic presentation of the mechanism where a current flow is occurring between particles having different state of charge.

when the cycling rate is decreased to 1C. This is in agreement with the schematic presentation of Figure 8.14 where high cycling rates could then promote the formation of alloys with different state of charge (lithiation). Indeed, lithiation is more favorable at lower cycling rate even if it is improved by the addition of FEC.

The chemical nature of the SEI layer will now be discussed. Without the addition of FEC, in C 1s core peak spectra, the presence of Li_2CO_3 results in a peak at 290.2 eV corresponding to $-CO_3$ species, while $ROCO_2Li$ species presents two peaks of equal intensity at 290.2 and 286.5 eV due to the same number of carbon atoms present in CO and CO_3 environment. Since the C 1s core peak spectra displayed in Figure 8.10 does not support the presence of $ROCO_2Li$, Li_2CO_3 could be considered as the major component among these two species. Moreover, from a quantitative point of view, the relative proportion of C to O in Li_2CO_3 equals 1:3 and the excess amount of the species in the corresponding O 1s spectra may indicate the formation of LiOH (531.5 eV). The origin of LiOH can be assigned to water contamination, which comes from CMC binder present in the electrode formulation. It is worth noticing that the decomposition of LiOH phase into Li_2O and H_2O under the X-ray beam in ultrahigh vacuum has often been observed. Therefore, Li_2CO_3 is the major component accompanying LiOH at the extreme surface of the SEI layer in the absence of FEC.

On the other hand, in the presence of FEC, the increase in C–O and the decrease in CO_3 species observed in the C 1s quantification data indicates that there may be a composition change at the extreme surface of the SEI layer. This occurs especially at the end of discharge, where data indicates a relative proportion for these two components of 3.7: 4.0, which suggests that $ROCO_2Li$ instead of Li_2CO_3 is now the major carbonate species. In the corresponding O 1s spectrum, the broad peak observed at a binding energy of 532.1 eV can be assigned to a mixture of oxygen atoms present in the environment of Li_2CO_3 (532.2 eV), LiOH, and also C-O (533.0 eV) which can also be assigned to $ROCO_2Li$. In the charged state, in spite of $ROCO_2Li$ that presents the C–O environment, the excess amount of CO species given in the quantification data of C 1s may be assigned to other degradation products like ROLi or PEO oligomers ($-CH_2-CH_2-O-)_n$ [38] for which carbon atoms are in a one-oxygen environment. The presence of LiF (55.9 eV) is confirmed by its corresponding quantification data presented in Table 8.3 (1.8% for discharged electrode, 7.0% for charged electrode), and in the Li 1s spectra of the discharged electrode, the broad peak is also attributed to: Li_2CO_3, LiOH, $ROCO_2Li$, and LiF. Therefore, $ROCO_2Li$ and LiF together with LiOH are the major components in the presence of FEC.

Hence, Li_2CO_3 is suggested as the major component of the SEI layer when using standard electrolyte, while $ROCO_2Li$ and LiF are detected when using FEC-containing electrolyte. This result reveals that the addition of FEC modifies the chemical composition of the very near surface (~5 nm) of the SEI layer.

8.2.3.2 At the 20th Cycle

As indicated in the discussion of the first cycle, after the partial dissolution during charge process, there is 0.1% Sn species in total observed at the end of first charge. However, after 20 cycles without the addition of FEC, none of the Sn 3d spectra indicate Sn species at their extreme surface, which demonstrates that the SEI layer grows thicker than the analysis depth of XPS (5 nm) upon cycling.

With the addition of FEC, Sn species are present at the 20th discharge and this amount has to be compared to the 0.04% Sn species detected at the end of the first discharge. This leads to the conclusion that we made for the electrodes cycled without FEC: SEI layer grows thicker than the analysis depth of XPS upon cycling. Nevertheless, at the charged state, the presence of Sn species implies that the SEI thickness is again close to the analysis depth after the partial dissolution during the charge process. This confirms that the addition of FEC results in a thinner SEI for the first cycle as well as the 20th cycle.

Table 8.4 also indicates a significant decrease of CO_3 species from the first discharge (13.4%) to the 20th discharge (1.6%) when FEC is not added. Meanwhile in O 1s spectra, no CO environment related to $ROCO_2Li$ but instead a broad peak (28.9%) is observed at a binding energy of 531.5 eV, which can be mainly assigned to LiOH. Thus, at the end of the 20th discharge, the major component at the outermost surface of the SEI layer is no longer $-CO_3$ species but LiOH, which could still be attributed to the water contained in the CMC binder. As LiOH undergoes a transformation to Li_2O when it is exposed to X-ray beam, this may also explain a rather high amount of Li_2O (5.6%) detected in the O 1s spectra. At the charged state, although the amount of CO_3 species in C 1s spectrum is a little higher (3.1%) than that of the discharged state, the broad peak in the corresponding O 1s spectrum again indicates the mixture of Li_2CO_3 and LiOH, where LiOH is still the major component. As a result, LiOH together with Li_2CO_3 is the main species present at the surface of the electrode when using electrolyte without FEC.

On the other hand, in the presence of FEC, the amount of CO_3 in C 1s increases from the first discharge (4.0%) to the 20th discharge (10.6%). The broad peak (40.1%) in O 1s spectrum gives no evidence of CO environment associated with $ROCO_2Li$ but confirms $-CO_3$ species, which makes Li_2CO_3 the major carbonate-related species at the surface. In addition, the broad peak (26.9%) found at 55.6 eV in Li 1s spectra implies a mixture of LiOH, Li_2CO_3, and LiF (~1.0%).

At the charged state, the $-CO_3$ species is detected in the C 1s spectrum together with the peak observed at 532.7 eV in O 1s spectrum, which can be assigned to CO environment in $ROCO_2Li$, suggesting the presence of $ROCO_2Li$ instead of Li_2CO_3. In addition, the existence of LiOH is confirmed by the detection of the 531.5-eV peak in O 1s spectrum. Furthermore, the quantification data of F 1s implies that LiF (3.9% in F 1s) could be considered as a major component at the surface. Thus, $ROCO_2Li$ and LiF accompanied by LiOH are suggested as the major components at the surface of the charged electrode with the addition of FEC.

There are a some more details to be noticed about LiF. The amount of LiF detected in the charged electrodes which is always higher than that of the discharged electrodes may support the hypothesis of a partial dissolution of the SEI layer during the charge process, as degradation product from this salt should be the first layer deposited at the electrode surface. Moreover, the decreased amount of LiF after 20 cycles compared to that of the first cycle supports the idea that growth of the SEI layer occurs upon cycling as already concluded from the discussion of Sn 3d spectra.

Impedance spectra exhibiting smaller capacitive semicircles indicate that the SEI layer and the charge transfer have a lower resistance meaning that a thinner or/and more conductive SEI layer is formed. The presence of Sn species in the XPS spectra indicates that the thickness reduction of the SEI is, at least partially, due to the dissolution process which occurs when FEC is added. Moreover, SEM images show that, in the presence of FEC, a more homogeneous SEI layer (composed mainly of Li_2CO_3) is observed at the 20th cycle which helps to explain the differences in nature of its outermost surface (Figure 8.13). Table 8.5 summarizes the main components present in the SEI as a function of the cycle number, the state of charge, and the nature of the additives in the electrolyte.

8.2.4 Experimental Section

8.2.4.1 Electrochemical Tests

TiSnSb electrodes were prepared by mixing TiSnSb microparticles, carbon conductive additives (VGCF, carbon black), and CMC binder as described in previous works [19,20,39]. TiSnSb electrodes were cycled against lithium using a standard electrolyte: EC/PC/3DMC (1M $LiPF_6$)+VC (1% v:v). FEC was used as the second additive (5% v:v) designed to improve the cycling performance of the active material. Swagelok cells were assembled in a glove box using the composite electrode as the positive electrode, microporous Celgard membrane and Whatman glass fiber paper filled with the electrolyte solution as the separator, and metallic lithium as the negative electrode. Cells

TABLE 8.5 Main Components in the SEI at the First and the 20th Cycle and as a Function of the State of Charge

SEI Composition (Main Components)	Discharged Standard Electrolyte	Charged Standard Electrolyte	Discharged FEC-Added Electrolyte	Charged FEC-Added Electrolyte
First cycle	Li$_2$CO$_3$ (+LiOH)	Li$_2$CO$_3$ (+LiOH)	ROCO$_2$Li + LiF + LiOH	ROCO$_2$Li + LiF + LiOH
20th cycle	LiOH (+Li$_2$CO$_3$)	LiOH (+Li$_2$CO$_3$)	Li$_2$CO$_3$	ROCO$_2$Li + LiF + LiOH

were cycled at 20 °C using a VMP system (Bio-Logic, South Africa) in galvanostatic mode from 0.02 to 1.5 V vs Li$^+$/Li at 4C rate (i.e., 1 Li in 1/4 h). Cyclic voltammetry was conducted at a scan rate of 25 μV s^{-1} from the open-circuit voltage (OCV) to 20 mV vs Li$^+$/Li and then to a set potential of 1.6 V vs Li/Li$^+$.

8.2.4.2 X-ray Photoelectron Spectroscopy

XPS measurements were carried out with a Thermo Scientific K-Alpha X-ray photoelectron spectrometer, using a focused monochromatized Al Kα radiation (hv = 1486.6 eV). The XPS spectrometer was directly connected through a glove box under argon atmosphere, in order to avoid moisture/air exposure of the samples. For the Ag 3d5/2 line the full width at half maximum was 0.50 eV under the recording conditions. The X-ray spot size was 400 μm. Peaks were recorded with a constant pass energy of 20 eV. The pressure in the analysis chamber was less than 2 × 10^{-7} Pa. Short acquisition time spectra were recorded at the beginning and end of each experiment to check that the samples did not suffer from degradation during the measurements. The binding energy scale was calibrated from the hydrocarbon contamination using the C 1s peak at 285.0 eV. Core peaks were analyzed using a nonlinear Shirley-type background [31]. The peak positions and areas were optimized by a weighted least-squares fitting method using 70% Gaussian and 30% Lorentzian line shapes. Quantification was performed on the basis of Scofield's relative sensitivity factors [30]. TiSnSb electrodes were thoroughly rinsed with pure DMC and dried before XPS measurements; it is assumed that there was no trace of LiPF$_6$ salt and solvents left at the electrode surface during these measurements. For each electrode sample, several XPS analyses were performed at different positions to make the results statistically reliable.

8.2.4.3 Scanning Electron Microscopy

The morphology and the texture of the electrodes were examined by a scanning electron microscope (JEOL 7600).

8.2.4.4 Electrochemical Impedance Spectroscopy

Three-electrode Swagelok-type cell, with lithium foils as counter and reference electrodes, was assembled in an argon-filled glove box. Microporous Celgard membranes and Whatman paper filters, filled with the electrolyte solution, were used as separator. Impedance experiments were performed using a multichannel galvanostat–potentiostat (VMP-Bio-Logic, South Africa) piloted by an EC Lab V10.20 interface. Impedance spectra were recorded at 1C rate at the end of discharge (0.02 V) of each cycle. The state of charge of the electrode was controlled galvanostatically at the cycling rate of C. The impedance of the cell was measured in the potentiostatic mode (potentio-electrochemical impedance spectroscopy: PEIS) in the frequency range from 1 to 2 mHz with an AC amplitude of 10 mV.

8.3 CONCLUSION

As suggested in previous works on graphite and Sn- and Si-based electrodes, VC and FEC-containing electrolytes are able to reduce the cumulative capacity losses as they favor the precipitation of more stable degradation products from the electrolyte and limit its further decomposition. One reason that can explain why the SEI is more stable in the presence of VC and especially FEC is that these additives are able to form a polymeric network at a higher potential than the reduction potential of electrolyte components. A small fraction of the liquid electrolyte will be embedded in this polymeric network and reduced at lower potential to give all mineral and organic products generally observed in the SEI by XPS experiments. "Moreover, the Li_2CO_3 species observed at the 20th cycle in the discharged state of the electrode cycled in FEC-containing electrolyte are though to originate from the reduction of the remaining solvent molecules embedded in the polymeric network"

XPS study confirmed that partial dissolution of the SEI layer happens during cycling and that change in composition of its extreme surface occurs in the presence of FEC. This is the reason why low-molecular-weight SEI components dissolved in the electrolyte can be identified by GC-MS analysis. Ionic lithium compounds embedded in the SEI like Li_2CO_3, $ROCO_2Li$, and LiF may play an important role in the transport of Li^+ ions throughout this protective layer which do not permit easy solvent crossing. Thus, improvement of the cycling ability of anode material in the presence of fluorinated additives like FEC or F_2EC is due to a decrease in the interfacial and charge transfer resistances as indicated by EIS experiments. As a conclusion, FEC (or F_2EC) associated or not with VC can be regarded as the universal SEI builder electrolyte additive.

ACKNOWLEDGMENT

Financial support by the ANR (*Projet Blanc* ICARES) and EU-FP7 program (Hi-C) is gratefully acknowledged.

REFERENCES

[1] P.G. Bruce, B. Scrosati, J.-M. Tarascon, Angew. Chem. Int. Ed. 47 (2008) 2980–2946.
[2] O. Yamamoto, N. Imanishi, Y. Takeda, H. Kashiwagi, J. Power Sources 54 (1995) 72–75.
[3] L. Monconduit, J.C. Jumas, R. Alcántara, J.L. Tirado, C. Pérez Vicente, J. Power Sources 107 (2002) 74–79.
[4] L.M.L. Fransson, J.T. Vaughey, K. Edstrom, M.M. Thackeray, J. Electrochem. Soc. 150 (2003) A86–A91.
[5] J. Xie, X. Zhao, G. Cao, Y. Zhong, M. Zhao, J. Electroanal, Chem. 542 (2003) 1–6.
[6] J. Xie, G. Cao, Y. Zhong, X.B. Zhao, J. Electroanal, Chem. 568 (2004) 323–327.
[7] R.J. Alcantara, F. Fernandez-Madrigal, P.L. Lavela, J. Tirado, J.C. Jumas, J. Olivier-Fourcade, J. Mater. Chem. 9 (1999) 2517–2521.
[8] J. Wolfenstine, S. Campos, D. Foster, J. Read, W.K. Behl, J. Power Sources 109 (2002) 230–233.
[9] Q.F. Dong, C.Z. Wu, M.G. Jin, Z.C. Huang, M.S. Zheng, J.K. You, Z.G. Lin, Solid State Ionics 167 (2004) 49–54.
[10] L. Fang, B.V.R. Chowdari, J. Power Sources, 97–98 (2001) 181–184.
[11] K.K.D. Ehinon, S. Naille, R. Dedryvère, P.-E. Lippens, J.-C. Jumas, D. Gonbeau, Chem. Mater 20 (2008) 5388–5398.
[12] M.M. Thackeray, J.T. Vaughey, C.S. Johnson, A.J. Kropf, R. Benedek, L.M.L. Fransson, K. Edstrom, J. Power Sources 113 (2003) 124–130.
[13] I. Geoffroy, A. Chagnes, B. Carré, D. Lemordant, P. Biensan, S. Herreyre, J. Power Sources 112 (2002) 191–198.
[14] H.J. Santner, K.-C. Möller, J. Ivančo, M.G. Ramsey, F.P. Netzer, S. Yamaguchi, J.O. Besenhard, M. Winter, J. Power Sources, 119–121 (2003) 368–372.
[15] S.-D. Xu, Q.-C. Zhuang, J. Wang, Y.-Q. Xu, Y.-B. Zhu, Int. J. Electrochem. Sci. 8 (2013) 8058–8076.
[16] X. Chen, X. Li, D. Mei, J. Feng, M.Y. Hu, J. Hu, M. Engelhard, J. Zheng, W. Xu, J. Xiao, J. Liu, J.G. Zhang, ChemSusChem 7 (2014) 549–554.
[17] W. Zhang, H. Martinez, R. Dedryvère, F. Ghamouss, D. Lemordant, A. Darwiche, L. Monconduit, abstract #639, 225th ECS Meeting, Orlando, FL, May 11-15, 2014.
[18] N.-S. Choi, K.H. Yew, K.Y. Lee, M. Sung, H. Kim, S.-S. Kim, J. Power Sources 161 (2006) 1254–1259.
[19] M.T. Sougrati, J. Fullenwarth, A. Debenedetti, B. Fraisse, J.C. Jumas, L. Monconduit, J. Mater, Chem. 21 (2011) 10069–10076.
[20] C. Marino, A. Darwiche, N. Dupré, H.A. Wilhelm, B. Lestriez, H. Martinez, R. Dedryvère, W. Zhang, F. Ghamouss, D. Lemordant, L. Monconduit, J. Phys. Chem. C 117 (2013) 19302–19313.
[21] M. Winter, J.O. Besenhard, Electrochimica Acta 45 (1999) 31–50.
[22] L. El Ouatani, R. Dedryvère, C. Siret, P. Biensan, S. Reynaud, P. Iratçabal, D. Gonbeau, J. Electrochem. Soc. 156 (2009) A103–A113.
[23] M. Ulldemolins, F. Le Cras, B. Pecquenard, V.P. Phan, L. Martin, H. Martinez, J. Power Sources 206 (2012) 245–252.
[24] N.-S. Choi, Y. Lee, S. Kim, S.-C. Shin, Y.-M. Kang, J. Power Sources 195 (2010) 2368–2371.
[25] L. Chen, K. Wang, X. Xie, J. Xie, J. Power Sources 174 (2007) 538–543.
[26] D. Mazouzi, N. Delpuech, Y. Oumellal, M. Gauthier, M. Cerbelaud, J. Gaubicher, N. Dupré, P. Moreau, D. Guyomard, L. Roué, B. Lestriez, J. Power Sources 220 (2012) 180–184.
[27] V. Etacheri, O. Haik, Y. Goffer, G.A. Roberts, I.C. Stefan, R. Fasching, D. Aurbach, Langmuir 28 (2012) 965–976.

[28] J. Qian, Y. Chen, L. Wu, Y. Cao, X. Ai, H. Yang, Chem. Commun. 48 (2012) 7070–7072.
[29] S. Komaba, T. Ishikawa, N. Yabuuchi, W. Murata, A. Ito, Y. Ohsawa, ACS Appl. Mater. Interfaces 3 (2011) 4165–4168.
[30] E. Karden, S. Buller, R.W. DE Doncker, J. Power Sources 85 (2000) 72–78.
[31] G. Nagasubramanian, J. Power Sources 87 (2000) 226–229.
[32] N.A. Hampson, S.A.G.R. Karunathilaka, R. Leek, J. Appl. Electrochem. 10 (1980) 3–11.
[33] D.A. Harrington, B.E. Conway, Electrochim. Acta 32 (1987) 1703–1721.
[34] J.T. Muller, P.M. Urban, W.F. Holderich, J. Power Sources 84 (1999) 157–160.
[35] R. Wiart, Electrochim. Acta 35 (1990) 1587–1593.
[36] J.Y. Song, H.H. Lee, Y.Y. Wang, C.C. Wan, J. Power Sources 111 (2002) 255–260.
[37] Q.-C. Zhuang, X.-Y. Qiu, S.-D. Xu, Y.-H. Qiang, S.-G. Sun, Diagnosis of electrochemical impedance spectroscopy in lithium-Ion batteries, in: I. Belharouak (Ed.), Lithium Ion Batteries - New Developments, 2012. ISBN: 978-953-51-0077-5, InTech.
[38] R. Dedryvère, H. Martinez, S. Leroy, D. Lemordant, F. Bonhomme, P. Biensan, D. Gonbeau, J. Power Sources 174 (2007) 462–468.
[39] H.A. Wilhelm, C. Marino, A. Darwiche, L. Monconduit, B. Lestriez, Electrochem. Commun. 24 (2012) 89–92.

Chapter 9

Surface Modification of Carbon Anodes for Lithium Ion Batteries by Fluorine and Chlorine

Tsuyoshi Nakajima
Department of Applied Chemistry, Aichi Institute of Technology, Yakusa, Toyota, Japan

Chapter Outline

- 9.1 Introduction 203
- 9.2 Effect of Surface Fluorination and Chlorination of Natural Graphite Samples 204
  - 9.2.1 Surface Fluorination of Natural Graphite Samples with Small Surface Areas 204
  - 9.2.2 Surface Fluorination of Natural Graphite Samples with Large Surface Areas 208
  - 9.2.3 Surface Chlorination of Natural Graphite Samples with Large Surface Areas 210
- 9.3 Effect of Surface Fluorination of Petroleum Cokes 214
- 9.4 Conclusions 219
- References 221

9.1 INTRODUCTION

Synthetic and natural graphites are usually used as anode materials of lithium ion batteries. The theoretical capacity of graphite is limited to be 372 mAhg^{-1} corresponding to LiC$_6$. In addition, ethylene carbonate (EC)-based solvents should be used for graphite anodes for the quick formation of surface film (solid electrolyte interphase or SEI) by the decomposition of a small amount of solvents. However, lithium ion batteries using EC-based solvents have a disadvantage on the low temperature operation because EC has a high melting point of 36 °C. In order to improve the low temperature operation, it is preferable to use propylene carbonate (PC) with a low melting point, −55 °C as a cosolvent. However, it is difficult to use PC for natural graphite with high crystallinity because electrochemical reduction continues on graphite surface without the formation of SEI, which gives rise to a large irreversible capacity.

Natural graphite powder is prepared by the mechanical pulverization of large particles. Therefore many lattice defects would exist at the surface, working as active sites for the electrochemical reduction of the solvents.

Various methods of surface modification have been applied to improve electrochemical properties of carbonaceous anodes of lithium ion batteries. They are carbon coating [1–14], metal or metal oxide coating [15–25], surface oxidation [26–33], surface fluorination and chlorination [34–52], and polymer, Si or Sb coating [53–60]. These methods of surface modification improve the electrochemical properties of carbon anodes. Among them, surface fluorination using fluorinating gases such as F_2, ClF_3, and NF_3 and plasma fluorination improve charge/discharge characteristics of natural and synthetic graphites. Due to the small dissociation energy of F_2 (155 kJmol^{-1}) and the highest electronegativity of fluorine, F_2 has high reactivity with other simple substances and compounds. Even light fluorination causes surface structure change of graphite with the formation of covalent C–F bonds. ClF_3 and NF_3 are strong fluorinating agents at high temperatures above ca. 200 °C. Plasma fluorination using CF_4 gas well modifies the surface structures of carbon materials at low temperatures. The surface fluorination of natural graphite powder samples with relatively small surface areas (<5 m^{2}g^{-1}) increases the capacities by increasing surface pore volumes [34–37]. On the other hand, the main effect of surface fluorination and chlorination of those having large surface areas (7–14 m^{2}g^{-1}) is the surface passivation by forming covalent C–F and C–Cl bonds at the surface, which suppresses electrochemical reduction of PC, increasing first coulombic efficiencies of natural graphite in PC-containing solvents [38–43]. The surface fluorination of graphitized petroleum cokes opens closed edge surface and enhances surface disorder, which facilitates the formation of SEI on graphitized petroleum cokes, leading to increase in first coulombic efficiencies (decrease in irreversible capacities) [44–50]. Thus, the effect of surface modification on electrochemical properties is different depending on the surface structures of carbon materials. The present chapter summarizes the results of surface modification of natural graphite and petroleum coke samples by fluorination and chlorination.

9.2 EFFECT OF SURFACE FLUORINATION AND CHLORINATION OF NATURAL GRAPHITE SAMPLES

9.2.1 Surface Fluorination of Natural Graphite Samples with Small Surface Areas

The surface modification of natural graphite samples with average particle sizes of 7, 25, and 40 µm (abbreviated to NG7 µm, NG25 µm, and NG40 µm) and their surface areas of 4.8, 3.7, and 2.9 m^{2}g^{-1}, respectively, was performed with F_2 gas of 3 × 10^4 Pa for 2 min and plasma fluorination using CF_4 [34–37]. Fluorine contents obtained by elemental analysis are less than 1 at% except the samples fluorinated by F_2 at high temperatures (Table 9.1) [34–37].

Surface fluorine concentrations obtained by X-ray Photoelectron Spectroscopy (XPS) are about 10 at% or less except those fluorinated by F_2 at high temperatures (Table 9.2) [34–37]. Plasma-fluorinated samples have slightly lower fluorine contents than those fluorinated by F_2. Fluorination of carbon materials by F_2 is an electrophilic reaction to form C–F covalent bonds, accompanying C–C bond rupture [61–63]. On the other hand, plasma fluorination is a radical reaction accompanying surface etching, which takes place releasing CF_4, COF_2, and so on. However, the fluorine contents are only slightly lower than those of the samples fluorinated by F_2 because plasma fluorination was made at a low temperature, 90 °C. Surface oxygen concentrations are decreased by fluorination. Surface fluorination significantly increases surface areas (Table 9.3) [35–37] and mesopore volumes [35]. Figure 9.1 shows the change in surface pore volumes by fluorination. Surface mesopores with diameters of 2–3 nm are particularly increased. Surface disorder evaluated by peak intensity ratios of Raman shifts of carbon materials is also increased by surface fluorination, which well coincides with increase in the surface areas

TABLE 9.1 Fluorine Contents in Surface-Fluorinated Natural Graphite Samples Obtained by Elemental Analysis

Fluorinated Graphite	Fluorine Content (at%)		
	NG7 μm	NG25 μm	NG40 μm
Fluorinated by F_2			
150–300 °C	0.3–0.4	0.3–0.4	0.2–0.4
350–500 °C	0.6–2.2	0.6–4.7	1.7–4.2
Plasma-fluorinated at 90 °C	0	0.3	0.3

TABLE 9.2 Surface Fluorine Concentrations of Surface-Fluorinated Natural Graphite Samples Obtained by XPS

Fluorinated Graphite	Surface Fluorine Concentration (at%)		
	NG7 μm	NG25 μm	NG40 μm
Fluorinated by F_2			
150–300 °C	6.0–10.4	4.5–6.9	3.6–12.0
350–500 °C	11.3–14.6	8.1–28.9	14.3–33.1
Plasma-fluorinated at 90 °C	6.7–8.8	7.1–11.5	3.3

TABLE 9.3 Surface Areas of Original and Surface-Fluorinated Natural Graphite Samples Obtained by BET Method

Graphite Sample	Surface Area (m^2g^{-1})		
	NG7 μm	NG25 μm	NG40 μm
Original	4.8	3.7	2.9
Fluorinated by F_2			
150 °C	5.2	3.5	–
250 °C	7.7	5.2	4.9
350 °C	8.5	6.1	5.0
Plasma-fluorinated at 90 °C	7.4	4.7	3.7

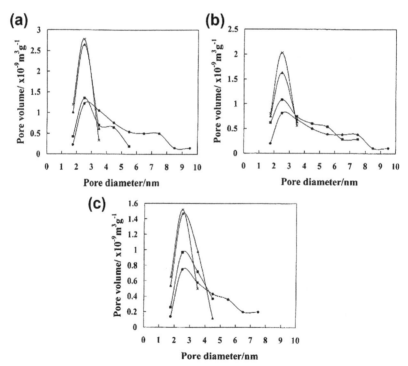

FIGURE 9.1 Pore volume distributions of natural graphite samples fluorinated by F_2 (Ref. [35]). (a): NG7 μm, (b): NG25 μm, (c): NG40 μm. ●: original, ■: fluorinated at 150 °C, ▲: fluorinated at 250 °C, ✕: fluorinated at 350 °C.

and mesopore volumes [34–37]. Figure 9.2 is an example of surface-fluorinated natural graphite. Fluorinated layers with high disorder are observed in the surface region. Thickness of the fluorinated layers is found to be several nanometers. Figure 9.3 shows an example of charge/discharge data, i.e., charge capacities of surface-fluorinated NG25 μm samples (corresponding to discharge capacities of practical lithium ion battery because counter electrode is metallic Li) as a function cycle number [35,37]. High capacities slightly exceeding the theoretical value of 372 mAhg^{-1} were observed in many cases. The optimum fluorination temperatures are less than 350 °C. The same high capacities were also obtained for plasma-fluorinated samples [36,37]. These results may be due to the accommodation of Li clusters in increased surface mesopores with diameters of 2–3 nm.

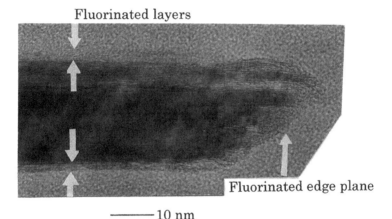

FIGURE 9.2 TEM image of NG40 μm fluorinated by F_2 at 300 °C.

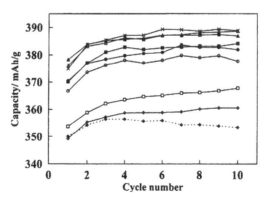

FIGURE 9.3 Charge capacities of NG25 μm samples fluorinated by F_2 in 1 molL^{-1} LiClO$_4$-EC/DEC (1:1 vol.) at 60 mAg^{-1} (Ref. [35]). ◆: NG25 μm, ■: 150 °C, ▲: 200 °C, ✕: 250 °C, △: 300 °C, ●: 350 °C, ○: 400 °C, □: 450 °C, ◇: 500 °C.

9.2.2 Surface Fluorination of Natural Graphite Samples with Large Surface Areas

Surface fluorination of natural graphite samples with average particle sizes of 5, 10, and 15 μm (abbreviated to NG5 μm, NG10 μm, and NG15 μm) and large surface areas of 13.9, 9.2, and 6.9 m^{2}g^{-1}, respectively, provides different surface structure changes and electrochemical properties [38–42]. Fluorination was conducted using F_2 (3×10^4 Pa) for 2 min, and ClF_3 and NF_3 (3×10^4 Pa) for 5 min. Surface areas are slightly increased by fluorination with F_2, however, reduced by the fluorination with ClF_3 and NF_3 and plasma treatment as given in Table 9.4 [39,40,42]. Mesopore volumes are nearly the same before and after fluorination by F_2, and highly reduced by the fluorination with ClF_3 and NF_3 and plasma treatment [39,40,42]. This may be the reason why charge capacities are not increased by the fluorination. Surface disorder slightly increases in all cases [38–42]. ClF_3 and NF_3 are strong fluorinating agents at higher temperatures than ca. 200 °C because they yield F atom and radical species such as NF_2 and ClF_2 though dissociation equilibrium, $ClF_3 \leftrightarrow ClF + F_2$ exists at high temperatures above 200 °C. Therefore the fluorination by NF_3 is a radical reaction, and radical reaction is also dominant in the case of ClF_3. Table 9.5 shows surface fluorine concentrations of fluorinated samples [39,40,42]. Surface fluorine

TABLE 9.4 Surface Areas and Total Mesopore Volumes of Original and Surface-Fluorinated Natural Graphite Samples

Graphite Sample	Surface Area (m^{2}g^{-1})			Mesopore Volume (cm^{3}g^{-1})		
	NG 5 μm	NG 10 μm	NG 15 μm	NG 5 μm	NG 10 μm	NG 15 μm
Original	13.9	9.2	6.9	0.047	0.035	0.026
200 °C (F_2)	16.2	9.9	7.3	0.049	0.036	0.028
300 °C (F_2)	19.3	10.8	7.7	0.059	0.038	0.029
200 °C (ClF_3)	11.7	7.3	6.3	0.031	0.036	0.030
300 °C (ClF_3)	11.2	6.5	6.2	0.030	0.037	0.031
200 °C (NF_3)	13.0	8.2	7.0	0.047	0.033	0.026
300 °C (NF_3)	11.1	6.6	4.8	0.045	0.032	0.023
Plasma-fluorinated at 90 °C	12.3	7.2	5.9	0.025	0.018	0.015

concentrations of the samples fluorinated by F_2 and plasma treatment are higher than those given in Table 9.2. This may be because NG5 μm, NG10 μm, and NG15 μm have the larger surface areas than NG7 μm, NG25 μm, and NG40 μm in the previous section. However, the samples fluorinated by ClF_3 and NF_3 have much lower surface fluorine concentrations than those fluorinated by F_2 and plasma treatment. Surface areas are also reduced by the fluorination using ClF_3 and NF_3. This means that strong surface etching of graphite occurs in these cases. Surface areas of plasma-fluorinated samples are also decreased by surface etching. This is different from the previous result for NG7 μm, NG25 μm, and NG40 μm. The reason may be that surface disorder of original NG5 μm, NG10 μm, and NG15 μm is higher than that of NG7 μm, NG25 μm, and NG40 μm. But surface fluorine concentrations of plasma-fluorinated NG5 μm, NG10 μm and NG15 μm are similar to the data obtained for those fluorinated by F_2 probably because the fluorination temperature is lower (90 °C) than those for ClF_3 and NF_3. Surface oxygen concentrations are reduced by the fluorination with F_2 and plasma fluorination. However, they are nearly the same or slightly increased in the case of ClF_3 and NF_3 probably because etched surface is oxidized when sample is taken out from the reactor into air.

Figures 9.4 shows first charge/discharge potential curves, charge capacities and coulombic efficiencies of natural graphite samples fluorinated by F_2, obtained in 1 molL^{-1} LiClO$_4$-EC/diethyl carbonate (DEC)/PC (1:1:1 vol.) at 150 mAg^{-1} [41]. Charge capacities are nearly the same before and after surface fluorination probably because surface mesopores with diameters of 2–3 nm are not changed or reduced by fluorination [39,40,42]. Important improvement by the fluorination is increase in first coulombic efficiencies, i.e., decrease in irreversible capacities of NG10 μm and NG15 μm in PC-containing solvents. The

TABLE 9.5 Surface Fluorine Concentrations of Natural Graphite Samples Fluorinated by F_2, ClF_3, NF_3 and Plasma Fluorination

Graphite Sample	Surface Fluorine Concentration (at%)		
	NG5 μm	NG10 μm	NG15 μm
200 °C (F_2)	11.6	15.3	11.1
300 °C (F_2)	19.0	17.7	20.5
200 °C (ClF_3)	0.0	0.0	2.5
300 °C (ClF_3)	0.0	11.5	12.0
200 °C (NF_3)	0.0	0.0	0.0
300 °C (NF_3)	1.4	1.2	1.4
Plasma-fluorinated at 90 °C	17.3	14.8	15.1

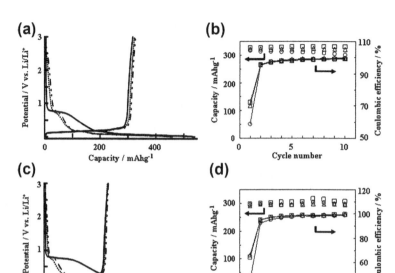

FIGURE 9.4 First charge/discharge curves, charge capacities and coulombic efficiencies for original and surface-fluorinated natural graphite samples at 150 mAg^{-1} in 1 molL^{-1} LiClO$_4$-EC/DEC/PC (1:1:1 vol.) (Ref. [41]). (a) and (b): NG10 μm, (c) and (d): NG15 μm. ———: original, ··········: fluorinated by F$_2$ at 200 °C, — ·· —: fluorinated by F$_2$ at 300 °C. ○: original, △: fluorinated by F$_2$ at 200 °C, □: fluorinated by F$_2$ at 300 °C.

potential plateaus at 0.8 V indicating electrochemical reduction of PC are significantly reduced in surface-fluorinated samples. First coulombic efficiencies are summarized in Table 9.6 [39,40,42]. NG5 μm having a large surface area has high first coulombic efficiencies both with and without surface fluorination except those highly fluorinated by F$_2$ at 300 °C probably because actual current densities are lower than in other cases. The increase in first coulombic efficiencies of NG10 μm and NG15 μm is consistent with surface fluorine concentrations given in Table 9.5. Natural graphite powder is prepared by the mechanical pulverization of large particles, which would create a large number of lattice defects at the surface working as active sites for electrochemical reduction of the solvents. Fluorination may reduce the surface active sites, i.e., cause surface passivation by the formation of covalent C–F bonds at graphite surface. This would effectively suppress the electrochemical decomposition of PC on natural graphite surface.

9.2.3 Surface Chlorination of Natural Graphite Samples with Large Surface Areas

Cl$_2$ gas has lower reactivity than F$_2$ because of its larger dissociation energy (239 kJmol^{-1}) and lower electronegativity of chlorine. Therefore, it is expected that chlorination of natural graphite powder by Cl$_2$ gas yields covalent C–Cl

TABLE 9.6 First Coulombic Efficiencies of Original and Surface-Fluorinated Natural Graphite Samples in 1 mol L^{-1} LiClO$_4$-EC/DEC/PC (1:1:1 vol.)

Graphite Sample	First Coulombic Efficiency (%)					
	NG5 μm		NG10 μm		NG15 μm	
	60 mAg^{-1}	150 mAg^{-1}	60 mAg^{-1}	150 mAg^{-1}	60 mAg^{-1}	150 mAg^{-1}
Original	81.8	78.7	66.2	58.7	51.8	43.7
200 °C (F$_2$)	78.4	79.2	78.0	70.3	72.3	66.0
300 °C (F$_2$)	53.9	53.9	76.1	71.9	75.1	64.0
200 °C (ClF$_3$)	80.5	81.7	64.4	69.7	56.4	64.8
300 °C (ClF$_3$)	79.8	81.4	75.5	71.1	72.3	68.9
200 °C (NF$_3$)	81.5	81.4	71.4	65.5	54.8	47.6
300 °C (NF$_3$)	82.4	80.6	72.2	66.3	54.9	50.9
Plasma-fluorinated at 90 °C	—	79.6	—	68.5	—	63.0

bonds at the surface of graphite without change of surface structure. Small amounts of surface chlorine are detected for NG10 μm and NG15 μm chlorinated by Cl_2 as shown in Table 9.7 [43]. The surface-chlorinated samples are stable in air and under high vacuum during XPS measurement. The chlorination of active carbon by Cl_2 at 550–600 °C gives hydrophobic chlorinated material [64]. These mean that high temperature reaction of Cl_2 with carbon materials gives covalent C–Cl bonds. However, only the surface is chlorinated in the case of graphite with high crystallinity. It is also known that chlorine is not intercalated into graphite. To obtain 0.5 at% or higher surface chlorine concentrations, chlorination conditions such as $1 \times 10^5 \, Pa \, Cl_2$, 400 °C and 10–30 min chlorination are necessary as given in Table 9.7. On the other hand, surface oxygen concentrations slightly increased compared with those of original NG10 μm and NG15 μm. When Cl_2 gas is introduced into Ni reactor, it reacts with adsorbed water molecules, yielding unstable HClO ($Cl_2 + H_2O \rightarrow HClO + HCl$). The complete removal of adsorbed water is normally difficult because the temperature in the vicinity of flange of the reactor is lower than the chlorination temperatures. The reactions of surface lattice defects with HClO and $HClO + Cl_2$ may give >C=O and –C=O(Cl) groups, respectively (>C + HClO → >C=O + HCl and $-C + HClO + 1/2Cl_2 \rightarrow -C=O(Cl) + HCl$). This would be the reason why surface oxygen concentrations are increased by chlorination. The d_{002} values of graphite samples and surface disorder are not changed by chlorination [43].

Figure 9.5 shows charge/discharge potential curves at first cycle, obtained in 1 molL^{-1} LiClO$_4$-EC/DEC/PC (1:1:1 vol.) at 60 mAg^{-1} [43]. The potential plateaus at 0.8 V versus Li/Li$^+$ indicate reductive decomposition of PC. First coulombic efficiencies are low for the samples, A–B and a–d chlorinated under mild conditions. This may be due to that unstable HClO simultaneously yields surface lattice defects by C–C bond breaking. First coulombic efficiency for NG10 μm increased from 58.9% to 81.0% with increasing surface chlorine as given in Table 9.8 [43]. Particularly samples D, E, and F with relatively larger surface chlorine concentrations of 0.5–1.4 at% gave high first coulombic efficiencies. The samples e and f with surface chlorine concentrations of 0.7 and 2.3 at%, respectively, exhibited high first coulombic efficiencies of 82.4%. First coulombic efficiencies thus increase with increasing surface chlorine and also oxygen concentrations. This result suggests that surface lattice defects acting as active sites for PC decomposition are reduced by the formation of covalent C–Cl bonds (–CCl$_3$ and >CCl$_2$ groups) and also C=O bonds (>C=O and –C=O(Cl) groups).

On the other hand, the results are somewhat different in 1 molL^{-1} LiPF$_6$-EC/EMC/PC (1:1:1 vol.) [43]. First coulombic efficiencies for original NG10 μm and NG15 μm were both higher in 1 molL^{-1} LiPF$_6$-EC/EMC/PC (1:1:1 vol.) than in 1 molL^{-1} LiClO$_4$-EC/DEC/PC (1:1:1 vol.), being 75.0% and 63.8% at 60 mAg^{-1}, and 66.4% and 58.3% at 300 mAg^{-1}, respectively. This is probably because a small amount of LiF generated by the reaction of LiPF$_6$ with Li is deposited on graphite to facilitate SEI formation. The increments of first

TABLE 9.7 Surface Composition of Chlorinated Natural Graphite Samples (Ref. [43])

Sample	Chlorination			NG10 μm				NG15 μm		
	Cl_2 (Pa)	Temp. (°C)	Time (min)	C (at%)	Cl	O		C	Cl	O
Original	–	–	–	92.8	–	7.2		93.3	–	6.7
A, a	3×10^4	200	3	86.3	0.1	13.6		90.4	0.2	9.4
B, b	3×10^4	300	3	89.8	0.1	10.1		88.5	0.2	11.3
C, c	1×10^5	300	10	90.4	0.2	9.3		90.0	0.3	9.7
D, d	1×10^5	400	10	88.6	0.5	10.9		90.4	0.3	9.3
E, e	1×10^5	400	20	91.1	0.9	8.0		90.6	0.7	8.7
F, f	1×10^5	400	30	90.1	1.4	8.5		85.8	2.3	11.9

A–F, Surface-chlorinated NG10 μm; a–f, Surface-chlorinated NG15 μm.

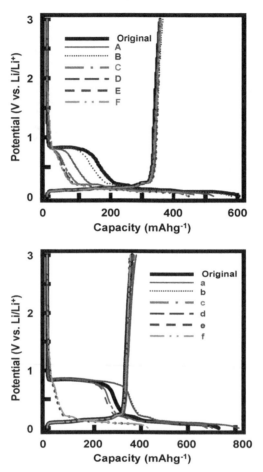

FIGURE 9.5 First charge/discharge potential curves of original and surface-chlorinated natural graphite samples at 60 mAg^{-1} in 1 molL^{-1} LiClO$_4$-EC/DEC/PC (1:1:1 vol.) (Ref. [43]). (Chlorination conditions of samples A–F (NG10 μm) and a–f (NG15 μm) are given in Table 9.7.)

coulombic efficiencies by chlorination reached ~15% and ~18% at 60 and 300 mAg^{-1}, respectively, for NG15 μm. In the case of NG10 μm, the increments of first coulombic efficiencies were ~8% even at 300 mAg^{-1}.

9.3 EFFECT OF SURFACE FLUORINATION OF PETROLEUM COKES

Petroleum coke samples are original one and those heat-treated at 1860, 2100, 2300, 2600, and 2800C (abbreviated to PC, PC1860, PC2100, PC2300, PC2600, and PC2800). First coulombic efficiency of petroleum coke varies depending on the heat-treatment temperatures as given in Table 9.9 [45,46,50]. PC1860

TABLE 9.8 First Coulombic Efficiencies for Original and Surface-chlorinated NG10 μm and NG15 μm Samples, Obtained in 1 mol L^{-1} LiClO$_4$-EC/DEC/PC (1:1:1 vol.) at 60 mAg^{-1}. (Chlorination Conditions of Samples A–F and a–f are Given in Table 9.7.)

Graphite Sample	First Coulombic Efficiency (%)	Graphite Sample	First Coulombic Efficiency (%)
NG10 μm	58.9	NG15 μm	49.5
A	68.2	a	45.6
B	62.4	b	44.7
C	75.4	c	52.7
D	79.5	d	53.3
E	79.8	e	82.4
F	81.0	f	82.4

TABLE 9.9 First Coulombic Efficiencies of Petroleum Cokes in 1 mol L^{-1} LiClO$_4$-EC/DEC (1:1 vol.) at 60 and 150 mAg^{-1}

Current Density (mAg^{-1})	Petroleum Coke Sample					
	PC	PC 1860	PC 2100	PC 2300	PC 2600	PC 2800
60	72.3	90.2	90.5	71.9	67.0	65.4
150	–	89.1	88.2	70.0	56.3	63.6

and PC2100 give high first coulombic efficiencies, 88–90%. However, they decrease with increasing heat-treatment temperature. Transmission electron microscopy (TEM) reveals that edge plane of heat-treated petroleum cokes is closed by carbon–carbon bond formation, which would make difficult the SEI formation and the insertion of Li$^+$ ion into petroleum cokes. The closed edge planes may be formed by the elimination of surface oxygen and subsequent carbon–carbon bond formation during the heat-treatment at high temperatures. Surface fluorination was performed by F$_2$, ClF$_3$, and NF$_3$ of 3×10^4 Pa for 2 min and plasma fluorination with CF$_4$. Figure 9.6 shows TEM images of surface-fluorinated PC2800 samples [47–49]. Top of closed edge plane is destroyed by fluorination. Tables 9.10 and 9.11 show surface concentrations of F, Cl and

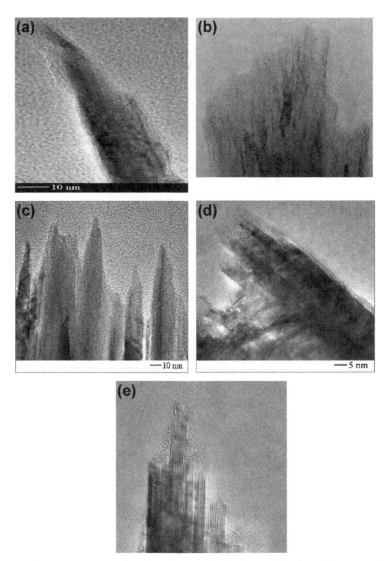

FIGURE 9.6 TEM images of surface-fluorinated PC2800. (a): fluorinated by F_2 at 400 °C for 2 min, (b): fluorinated by F_2 at 400 °C for 2 min, (c): plasma-fluorinated for 15 min at 90 °C, (d): plasma-fluorinated for 60 min at 90 °C, (e): fluorinated by NF_3 at 400 °C for 2 min.

N [44–50]. Surface fluorine concentrations are similar in the samples fluorinated by F_2 and plasma fluorination using CF_4. However, almost no fluorine is detected in those fluorinated by ClF_3 and NF_3, but trace amounts of Cl and N are detected. Fluorination with F_2 is an electrophilic reaction and those with ClF_3, NF_3 and plasma fluorination are radical reactions. Since plasma fluorination was made at low temperature of 90 °C, surface fluorine concentrations are

TABLE 9.10 Surface Fluorine Concentrations of Fluorinated Petroleum Cokes

Petroleum Coke	PC	PC 1860	PC 2100	PC 2300	PC 2600	PC 2800 (at%)
Fluorinated by F_2 at 300 °C	50.2	22.4	17.8	11.7	10.1	5.2
Plasma-fluorinated for 60 min at 90 °C	35.7	15.8	–	12.0	–	7.4

TABLE 9.11 Surface Concentrations of F, Cl and N of Petroleum Cokes Fluorinated by ClF_3 and NF_3

Sample	Fluorination	ClF_3		NF_3	
		F	Cl	F	N (at%)
PC	200 °C	0	1.0	0	0.5
	300 °C	0	1.4	2.6	0.4
PC1860	200 °C	0	0.6	0	0
	300 °C	0	0.9	0	0
PC2300	200 °C	0	1.1	0	0
	300 °C	0	0.4	0	0
	400 °C	0	0.5	0	0
	500 °C	0	1.1	0	0
PC2800	200 °C	0	1.1	0	0
	300 °C	0	0.3	0	0
	400 °C	0	0.7	0	0
	500 °C	0	1.0	0	0

TABLE 9.12 R Values (I_D/I_G) Obtained From Peak Intensity Ratios of Raman Shifts of Original and Surface-Fluorinated Petroleum Cokes

Petroleum Coke	PC	PC1860	PC2300	PC2800
Original	0.87	0.49	0.21	0.18
F_2				
300 °C	0.94	0.78	0.68	0.43
ClF_3				
200 °C	0.93	0.51	0.29	0.15
300 °C	0.93	0.49	0.26	0.16
400 °C	–	–	0.27	0.15
500 °C	–	–	0.26	0.16
NF_3				
200 °C	0.93	0.51	0.26	0.16
300 °C	0.78	0.50	0.26	0.16
400 °C	–	–	0.25	0.16
500 °C	–	–	0.25	0.15
Plasma-fluorinated at 90 °C	1.22	0.63	0.52	0.38

higher than in other two cases using ClF_3 and NF_3. These results are consistent with Table 9.5. Corresponding to this difference, surface disorder is increased by the fluorination using F_2 and plasma fluorination, however, decreased by the fluorination with ClF_3 and NF_3 (Table 9.12) [45,48–50]. Main effect by the fluorination with F_2 and plasma fluorination is increase of first coulombic efficiencies of graphitized petroleum cokes, PC2300, PC2600, and PC2800 by the opening of closed edge plane and subsequent increase in surface disorder facilitating SEI formation. Figure 9.7 shows charge/discharge potential curves of original and surface-fluorinated PC2800 at first cycle [45,50]. The potential plateau indicating the decomposition of organic solvents is largely reduced, i.e., SEI formation is facilitated by surface fluorination. Table 9.13 gives first coulombic efficiencies obtained at 60 and 150 mAg^{-1} in 1 molL^{-1} LiClO$_4$-EC/DEC (1:1 vol.) [44,46,49,50]. Large increase in first coulombic efficiencies is observed in the graphitized samples (PC2300, PC2600, and PC2800) fluorinated by F_2 at 300 °C and plasma treatment. In the case of fluorination by ClF_3

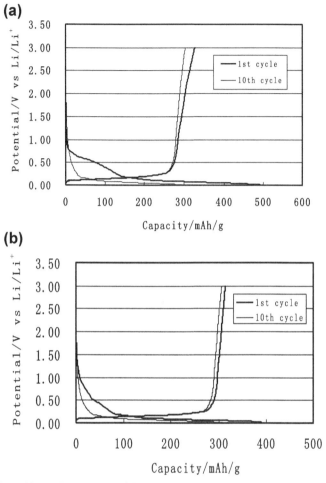

FIGURE 9.7 Charge/discharge potential curves of PC2800 (a) and those fluorinated by F_2 at 300 °C (b) at 60 mAg^{-1} in 1 molL^{-1} LiClO$_4$-EC/DEC (1:1 vol.) (Ref. [45]).

and NF$_3$, strong surface etching of petroleum cokes takes place. The main effect is different, i.e., increase in charge capacities [47,48,50].

9.4 CONCLUSIONS

Surface modification by fluorinating and chlorinating agents effectively improves electrochemical characteristics of carbon anodes for lithium ion battery. The effect of surface fluorination and chlorination is different depending on surface structures of carbon materials. Fluorination of natural graphite samples with small surface areas by F_2 gas and plasma treatment increases the mesopores with diameters of 2–3 nm, which leads to increase in charge capacities.

TABLE 9.13 First Coulombic Efficiencies (%) of Petroleum Cokes Fluorinated by F_2 and Plasma Fluorination in 1 mol L^{-1} LiClO$_4$-EC/DEC (1:1 vol.)

Sample	Current Density	PC	PC1860	PC2100	PC2300	PC2600	PC2800
Original	60 mAg^{-1}	72.3	90.2	90.5	71.9	67.0	65.4
200 °C (F_2)		68.6	87.9	91.7	72.4	66.7	60.5
300 °C (F_2)		47.4	83.3	87.1	84.1	79.1	83.6
Original	150 mAg^{-1}	—	89.1	88.2	70.0	56.3	63.6
200 °C (F_2)		—	88.0	89.8	76.5	60.3	59.6
300 °C (F_2)		—	81.8	84.6	83.3	82.1	79.5
Original	150 mAg^{-1}	—	89.1	—	70.0	—	63.6
Plasma-fluorinated at 90 °C		—	85.4	—	78.0	—	72.5

In the case of surface fluorination of other natural graphite samples with large surface areas by F_2 and plasma treatment, surface structure change is small and main effect is surface passivation of natural graphite surface by the formation of covalent C–F bonds, which brings about increase in first coulombic efficiencies, i.e., decrease in irreversible capacities in PC-containing solvents. The same effect, that is, increase in first coulombic efficiencies in PC-containing solvents, is also observed by surface chlorination using Cl_2 gas. Surface chlorination of the same natural graphite samples with large surface areas yields covalent C–Cl and C=O bonds at graphite surface without change of surface structures. High temperature treatment of petroleum coke causes the closure of edge planes by carbon–carbon bond formation with elimination of surface oxygen. This structure change brings about a decrease in first coulombic efficiencies, i.e., increase in irreversible capacities of graphitized petroleum cokes. The surface fluorination of graphitized petroleum cokes by F_2 and plasma treatment open the closed edges and increase the surface disorder, which provide increase in first coulombic efficiencies. On the other hand, the surface fluorination by ClF_3 and NF_3 causes surface etching without increase in the surface disorder of petroleum cokes. The main effect of surface fluorination by ClF_3 and NF_3 is increase of charge capacities.

REFERENCES

[1] H. Wang, M. Yoshio, J. Power Sources 93 (2001) 123.
[2] S. Soon, H. Kim, S.M. Oh, J. Power Sources 94 (2001) 68.
[3] M. Yoshio, H. Wang, K. Fukuda, Y. Hara, Y. Adachi, J. Electrochem. Soc. 147 (2000) 1245.
[4] H. Wang, M. Yoshio, T. Abe, Z. Ogumi, J. Electrochem. Soc. 149 (2002) A499.
[5] M. Yoshio, H. Wang, K. Fukuda, T. Umeno, N. Dimov, Z. Ogumi, J. Electrochem. Soc. 149 (2002) A1598.
[6] Y.-S. Han, J.-Y. Lee, Electrochim. Acta 48 (2003) 1073.
[7] Y. Ohzawa, M. Mitani, T. Suzuki, V. Gupta, T. Nakajima, J. Power Sources 122 (2003) 153.
[8] Y. Ohzawa, Y. Yamanaka, K. Naga, T. Nakajima, J. Power Sources 146 (2005) 125.
[9] Y. Ohzawa, T. Suzuki, T. Achiha, T. Nakajima, J. Phys. Chem. Solids 71 (2010) 654.
[10] H.-Y. Lee, J.-K. Baek, S.-M. Lee, H.-K. Park, K.-Y. Lee, M.-H. kim, J. Power Sources 128 (2004) 61.
[11] W.-H. Zhang, I. Fang, M. Yue, Z.-L. Yu, J. Power Sources 174 (2007) 766.
[12] J.-H. Lee, H.-Y. Lee, S.-M. Oh, S.-J. Lee, K.-Y. Lee, S.-M. Lee, J. Power Sources 166 (2007) 250.
[13] Y.-S. Park, H.J. Bang, S.-M. Oh, Y.-K. Sun, S.-M. Lee, J. Power Sources 190 (2009) 553.
[14] G. Park, N. Gunawardhana, H. Nakamura, Y.-S. Lee, M. Yoshio, J. Power Sources 196 (2011) 9820.
[15] R. Takagi, T. Okubo, K. Sekine, T. Takamura, Electrochemistry 65 (1997) 333.
[16] T. Takamura, K. Sumiya, J. Suzuki, C. Yamada, K. Sekine, J. Power Sources 81/82 (1999) 368.
[17] Y. Wu, C. Jiang, C. Wan, E. Tsuchida, Electrochem. Commun. 2 (2000) 626.
[18] S.-S. Kim, Y. Kadoma, H. Ikuta, Y. Uchimoto, M. Wakihara, Electrochem. Solid-State Lett. 4 (2001) A109.
[19] J.K. Lee, D.H. Ryu, J.B. Ju, Y.G. Shul, B.W. Cho, D. Park, J. Power Sources 107 (2002) 90.

[20] I.R.M. Kottegoda, Y. Kadoma, H. Ikuta, Y. Uchimoto, M. Wakihara, Electrochem. Solid-State Lett. 5 (2002) A275.
[21] I.R.M. Kottegoda, Y. Kadoma, H. Ikuta, Y. Uchimoto, M. Wakihara, J. Electrochem. Soc. 152 (2005) A1595.
[22] L.J. Fu, J. Gao, T. Zhang, Q. Cao, L.C. Yang, Y.P. Wu, R. Holze, J. Power Sources 171 (2007) 904.
[23] F. Nobili, S. Dsoke, M. Mancini, R. Tossici, R. Marassi, J. Power Sources 180 (2008) 845.
[24] M. Mancini, F. Nobili, S. Dsoke, F. D'Amico, R. Tossici, F. Croce, R. Marassi, J. Power Sources 190 (2009) 141.
[25] F. Nobili, M. Mancini, P.E. Stallworth, F. Croce, S.G. Greenbaum, R. Marassi, J. Power Sources 198 (2012) 243.
[26] E. Peled, C. Menachem, D. Bar-Tow, A. Melman, L4, J. Electrochem. Soc. 143 (1996).
[27] J.S. Xue, J.R. Dahn, J. Electrochem. Soc. 142 (1995) 3668.
[28] Y. Ein-Eli, V.R. Koch, J. Electrochem. Soc. 144 (1997) 2968.
[29] Y. Wu, C. Jiang, C. Wan, E. Tsuchida, J. Mater. Chem. 11 (2001) 1233.
[30] Y.P. Wu, C. Jiang, C. Wan, R. Holze, Electrochem. Commun. 4 (2002) 483.
[31] Y. Wu, C. Jiang, C. Wan, R. Holze, J. Power Sources 111 (2002) 329.
[32] Y.P. Wu, C. Jiang, C. Wan, R. Holze, J. Appl. Electrochem. 32 (2002) 1011.
[33] J. Shim, K.A. Striebel, J. Power Sources 164 (2007) 862.
[34] T. Nakajima, M. Koh, R.N. Singh, M. Shimada, Electrochim. Acta 44 (1999) 2879.
[35] V. Gupta, T. Nakajima, Y. Ohzawa, H. Iwata, J. Fluorine Chem. 112 (2001) 233.
[36] T. Nakajima, V. Gupta, Y. Ohzawa, M. Koh, R.N. Singh, A. Tressaud, E. Durand, J. Power Sources 104 (2002) 108.
[37] T. Nakajima, V. Gupta, Y. Ohzawa, H. Iwata, A. Tressaud, E. Durand, J. Fluorine Chem. 114 (2002) 209.
[38] K. Matsumoto, J. Li, Y. Ohzawa, T. Nakajima, Z. Mazej, B. Žemva, J. Fluorine Chem. 127 (2006) 1383.
[39] T. Achiha, T. Nakajima, Y. Ohzawa, J. Electrochem. Soc. 154 (2007) A827.
[40] T. Achiha, S. Shibata, T. Nakajima, Y. Ohzawa, A. Tressaud, E. Durand, J. Power Sources 171 (2007) 932.
[41] T. Nakajima, T. Achiha, Y. Ohzawa, A.M. Panich, A.I. Shames, J. Phys. Chem. Solids 69 (2008) 1292.
[42] X. Cheng, J. Li, T. Achiha, T. Nakajima, Y. Ohzawa, Z. Mazej, B. Žemva, J. Electrochem. Soc. 155 (2008) A405.
[43] S. Suzuki, Z. Mazej, B. Žemva, Y. Ohzawa, T. Nakajima, Acta Chim. Slov. 60 (2013) 513.
[44] T. Nakajima, J. Li, K. Naga, K. Yoneshima, T. Nakai, Y. Ohzawa, J. Power Sources 133 (2004) 243.
[45] J. Li, K. Naga, Y. Ohzawa, T. Nakajima, A.P. Shames, A.I. Panich, J. Fluorine Chem. 126 (2005) 265.
[46] J. Li, K. Naga, Y. Ohzawa, T. Nakajima, H. Iwata, J. Fluorine Chem. 126 (2005) 1028.
[47] K. Naga, T. Nakajima, Y. Ohzawa, B. Žemva, Z. Mazej, H. Groult, J. Electrochem. Soc. 154 (2007) A347.
[48] K. Naga, T. Nakajima, S. Aimura, Y. Ohzawa, B. Žemva, Z. Mazej, H. Groult, A. Yoshida, J. Power Sources 167 (2007) 192.
[49] T. Nakajima, S. Shibata, K. Naga, Y. Ohzawa, A. Tressaud, E. Durand, H. Groult, F. Warmont, J. Powerr Sources 168 (2007) 265.
[50] T. Nakajima, J. Fluorine Chem. 128 (2007) 277.
[51] H. Groult, T. Nakajima, L. Perrigaud, Y. Ohzawa, H. Yashiro, S. Komaba, N. Kumagai, J. Fluorine Chem. 126 (2005) 1111.

[52] H. Groult, T. Nakajima, A. Tressaud, S. Shibata, E. Durand, L. Perrigaud, F. Warmont, Electrochem. Solid-State Lett. 10 (2007) A212.
[53] S. Kuwabata, N. Tsumura, S. Goda, C.R. Martin, H. Yoneyama, J. Electrochem. Soc. 145 (1998) 1415.
[54] M. Gaberscek, M. Bele, J. Drofenik, R. Dominko, S. Pejovnik, Electrochem. Solid State Lett. 3 (2000) 171.
[55] J. Drofenik, M. Gaberscek, R. Dominko, M. Bele, S. Pejovnik, J. Power Sources 94 (2001) 97.
[56] M. Bele, M. Gaberscek, R. Dominko, J. Drofenik, K. Zupan, P. Komac, K. Kocevar, I. Musevic, S. Pejovnik, Carbon 40 (2002) 1117.
[57] M. Gaberscek, M. Bele, J. Drofenik, R. Dominko, S. Pejovnik, J. Power Sources 97/98 (2001) 67.
[58] B. Veeraraghavan, J. Paul, B. Haran, B. Popov, J. Power Sources 109 (2002) 377.
[59] M. Holzapfel, H. Buqa, F. Krumeich, P. Novak, F.M. Petrat, C. Veit, Electrochem. Solid-State Lett. 8 (2005) A516.
[60] C.-C. Chang, J. Power Sources 175 (2008) 874.
[61] S. Rozen, C. Gal, Terahedron Lett. 25 (1984) 449.
[62] S. Rozen, C. Gal, J. Fluorine Chem. 27 (1985) 143.
[63] M. Koh, H. Yumoto, H. Higashi, T. Nakajima, J. Fluorine Chem. 97 (1999) 239.
[64] Y. Kanaya, N. Watanabe, Denki Kagaku 42 (1974) 349 Japanese.

Chapter 10

Application of Polyvinylidene Fluoride Binders in Lithium-Ion Battery

Ramin Amin-Sanayei and Wensheng He
Arkema Inc, King of Prussia, PA, USA

Chapter Outline

10.1 Introduction	225	10.6 Electrode Preparation Method	231
10.2 Fluorine-Containing Binder	226	10.7 Peel Strength Measurement	232
10.3 Properties of Fluorinated Binder	228	10.8 Electrode Performance Test	233
10.4 Binder Swelling in Electrolyte Solvent	230	10.9 Fluorinated Waterborne Binders	234
10.5 Electrochemical Stability	230		

10.1 INTRODUCTION

Lithium-ion batteries (LIBs) are one of the most important technologies developed at the end of twentieth and into the twenty-first centuries, which can be deemed as the most impressive success story of modern electrochemistry. The world today cannot operate without LIBs because of their continued use in information technology and consumer electronics applications such as mobile phones, portable PCs, power tools, and other mobile equipments requiring power. In recent years, LIBs have shown great promise as power sources for the hybrid and full electric vehicles which in turn reduce demand for fossil fuel as well as emission of greenhouse gases.

The popularity of the LIB is due to its advantages over other secondary (or rechargeable) batteries:

- Lighter than other rechargeable batteries for a given charge capacity
- Lithium-ion chemistry delivers a high open-circuit voltage, which translates to higher power density at a given charge capacity

- Low self-discharge rate (about 1.5% per month) which provides longer shelf life
- Do not suffer from battery memory effect so it can drain to high depth of discharge without major side effect.

However, LIBs do need improvements to overcome:

- Insufficient cycle life, particularly in high-current applications
- Rising internal resistance with cycling and aging
- Safety concerns if overheated or overcharged
- Applications demanding more capacity from LIB

High-performance binders play a main role in alleviating some of the above concerns by providing better interconnectivity between electrode-forming particles such as electrochemically active particles, conductive carbon, electrolyte, and current collectors.

The development of new materials for LIB is the focus of research of prominent groups in the field of materials science throughout the world. Since this field is advancing rapidly and attracting an increasing number of researchers, it is important to provide current and timely updates for this constantly changing technology. One of the research and development topics is the enhancement of binder to match for ever demanding high voltage and high energy density LIBs. This chapter describes the key aspects of binder properties and the best mode of practice for fabrication of electrodes for LIB because battery properties could vary considerably when binders are combined with electrochemically active materials.

Both the anode and cathode are composed of electrochemically active materials coated onto current collectors (copper or aluminum). In the case of the anode, the active material is usually graphite and/or other carbonous matters. The cathode utilizes lithium metal oxides such as lithiated cobalt oxide ($LiCoO_2$), lithiated manganese oxide (Li_2MnO_2), lithium ion phosphate ($LiFePO_4$), lithiated manganese–nickel–cobalt oxide ($LiMnNiCoO_2$), and lithiated nickel–cobalt–aluminum oxide ($LiNiCoAlO_2$).

There are two main types of binders: fluorinated and nonfluorinated binders. Polyvinylidene fluoride (PVDF) resins are becoming the preferred cathode binder material because they provide higher resistance to oxidation than other binders, which is an important requirement for longevity and long calendar life. For anode, there is a significant use of nonfluorinated (aqueous) binders such as styrene-butadiene rubber (SBR) and polyacrylic acid. Although their resistance to oxidation is inferior to fluorinated binders, nonfluorinated binders have lower running costs and greater environmental compatibility than PVDF.

10.2 FLUORINE-CONTAINING BINDER

PVDF homopolymers and copolymers continue to gain success in the battery industry as binders for cathodes and anodes as well as battery separator in lithium-ion technology. The high electrochemical, thermal, and chemical stability of PVDF resins, as well as their ease of processing, yields unmatched performance compared to other polymeric binders in lithium-ion systems. The binder development takes into account the adhesive properties of the polymer, crystallinity, solvent uptake, solution behavior, and slurry preparation and processing.

The process for PVDF polymerization was first investigated by DuPont in the 1940s. In the 1960s, Pennwalt was the first to introduce to the market commercial production of PVDF under the trade name Kynar® PVDF for several applications. Today, Arkema owns the Kynar® PVDF trade name and as an industry leader, produces variety of PVDF-based polymers for various applications. In the 1990s, Sony was the first to use PVDF as the preferred binder because of its good electrochemical solubility, solvent resistance, and its ability to offer excellent performance with comparatively small amount.

PVDF is a linear partially fluorinate polymer containing 59.4 wt% fluorine, 3 wt% hydrogen, and balance is carbon. The high level of intrinsic crystallinity, typically near 60%, provides toughness and solvent-resistant properties. Incorporation of various fluorinated comonomers at low levels, typically less than 5 wt% to as much as 10 wt%, enhances flexibility and adhesion by reducing the crystallinity. However increasing comonomer can negatively impact the use at high temperature. PVDF is commercially produced via free radical polymerization either with emulsion or suspension processes.

The spatial arrangement of the alternating CH_2 and CF_2 groups along the polymer backbone creates a high dipole moment resulting in unique polarity, unusually high dielectric constant, complex polymorphism, and high piezoelectric and pyroelectric activity of the polymer. Typically, PVDF has four chain conformations:

1. Alpha: The alpha form is the most common form for PVDF and is the most thermodynamically stable. It prevails on coating and normal melt processing of structural parts.

2. Beta: The beta form develops under mechanical deformation of melt-processed materials, typically at temperatures close to its melting transition. This structure provides some unique piezo- and pyroelectric activity properties in PVDF.

3. Gamma: The gamma form arises infrequently, mostly by crystallization from solution.

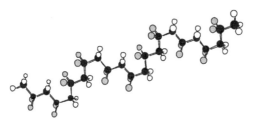

4. Delta: The delta form develops through a distortion of one of the other forms under high electric fields.

High dipole moment of PVDF facilitates its dissolution in highly polar solvents such as *N*-methyl-2-pyrrolidone (NMP), triethyl phosphate, dimethyl acetamide, while high fluorine content provides resistance to harsh acids and oxidization attacks, which makes the PVDF an ideal binder candidate for LIB.

10.3 PROPERTIES OF FLUORINATED BINDER

PVDF-based resins should be pure and free of any additives or ionic impurities that can interfere with electrochemical reactions. A few examples of Kynar® PVDF are listed in Table 10.1 that are specifically designed and optimized for use in LIB. Kynar® HSV is a high-molecular weight resin that provides improved adhesion with reduced loading level when utilized as a binder in electrode formulation. Kynar Powerflex® LBG, Kynar Flex® 2801 and 2751 are copolymers primarily used as a separator coating, which provide proper wetting and swelling properties as well as higher mechanical and electrochemical strength.

Pure PVDF-based polymers such as Kynar® and Kynar Flex® PVDF have exceptionally good electrochemical solubility because of high fluorine content. Moreover, when PVDF is produced via an emulsion polymerization process, which inherently yields a finer powder particle than a suspension polymerization process, the resulting PVDF may dissolve easier and faster.

The behavior of polymeric chains in solution is highly dependent on the nature of the polymer and the solvent. This behavior could range between compact swollen coils in a poor solvent to a highly extended chain in a good solvent such as NMP. The solution viscosity will increase as the polymer chains expand more in the solvent.

TABLE 10.1 Fluorine Content of Binders Used in Lithium-Ion Batteries

Grade	Melting Point (°C)	Melt Viscosity (kp)	Category	Fluorine Content
HSV-900	162–172	47–53	Homopolymer	59.4 wt%
HSV-800	162–172	44–50	Homopolymer	59.4 wt%
HSV-500	162–172	34–40	Homopolymer	59.4 wt%
761A	165–172	30–36	Homopolymer	59.4 wt%
761	165–172	23–29	Homopolymer	59.4 wt%
LBG	151-156	34–38	Copolymer	>59.4 wt%
2801	140–145	22–27	Copolymer	>60.0 wt%
2751	131–135	22–27	Copolymer	>60.0 wt%

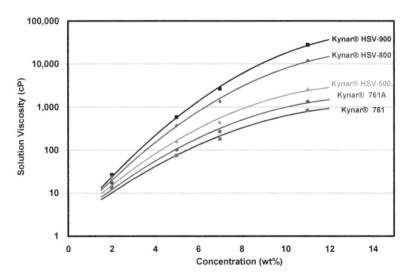

FIGURE 10.1 Solution viscosity of Kynar binders in N-methyl-2-pyrrolidone at 25 °C.

Dissolution of the polymer is a multistep process in which the solvent is first diffused into the amorphous region, creating a swollen polymer mass. In the second, but less thermodynamically favored step, the solvent penetrates the crystalline region resulting in total dissolution. The second step is aided by increasing the temperature of the solvent.

Molecular weight and concentration of PVDF have pronounced impact on its solution viscosity in NMP. Inspection of Figure 10.1 indicates that solution viscosity increased from 10 s of centipoises (cp) to about 50,000 cp when concentration increased from 2% to 12%. Also modest increase in molecular weight caused two orders of magnitude increase in solution viscosity at 12% concentration.

10.4 BINDER SWELLING IN ELECTROLYTE SOLVENT

Polymeric binders swell in electrolyte due to absorption of carbonate solvents into amorphous phase. Swelling with electrolyte is another important aspect of binder, which requires fine balance between allowing ions to pass though the binder facilitating better rate performance and battery cycle life. The amount of swelling can be estimated by soaking solid binder film in electrolyte solvent and measuring weight gain.

To accurately measure swelling, a solid film about 100 μm thick is needed. The film samples can be prepared by drying binder solution in NMP on a flat bottom surface, i.e., petri dish, at 120 °C using convection oven for several hours or preferably overnight. Then a sample of about 200 mg (~10 cm^2) should be soaked in the solvent mix of choice, generally EC/DMC/DEC 1/1/1, using a pressure tube in an oven at a fixed temperature (i.e., 60 °C) for a predetermined time. Weight gain should be immediately measured after taking the sample out of solvent and wiping it clean free of visible liquid.

The weight gain for Kynar® HSV-900 is around 20% at 60 °C which provides the right balance of ionic conductivity and cyclability.

10.5 ELECTROCHEMICAL STABILITY

Another important attribute for a LIB binder is its electrochemical stability, especially at higher voltage applications since modern LIBs hold more than twice as much energy by weight as the first commercial batteries sold by Sony in 1991. The higher energy was achieved by using higher voltage active materials composites as well as over charging batteries generally in excess of 4 V. The electrochemical stability test can be conducted using cyclic voltammetry (CV) technique with a coin cell configuration as shown in Figure 10.2.

The test electrode consists of carbon/binder with 30/70 ratio. The thickness of the sample electrode is about 10 μm, and diameter is ½ in (12.7 mm). Electrolyte is LP-57 from BASF. CV scans are from a Solartron 1287 Electrochemical working station between 2.5 and 5.0 V at 0.1 mV/s at room or elevated

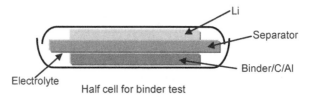

FIGURE 10.2 Schematic of coin cell configuration for binder electrochemical stability test.

temperatures. The CV scans showed that Kynar® PVDF is very stable up to 4.7 V where electrochemical activity of carbonate solvent starts.

10.6 ELECTRODE PREPARATION METHOD

Slurry processing method is the key factor impacting the dispersion of solid particles, quality of slurry, and electrode performance and cyclibility. In industrial or large-scale laboratory processes, shear mixers are generally used to dissolve PVDF in NMP at room or slightly elevated temperature. In the next stage, powder active materials are mixed using planetary mixer combined with homogenizer or dispersion disc blade. However, this process cannot be readily used for small-scale lab since minimal capacity is 500–1000 mL. Different mixing techniques can be adopted for preparation of small-scale slurry (about 50–100 mL), Arkema developed a process using a centrifugal planetary mixer made by Thinky. The flowchart in Figure 10.3 depicts the small-scale slurry process.

Step 1 is to disperse conductive additives in part of binder solution. Step 2 is to add active material and binder solution to the mixture in step 1, and mix the slurry at high solids level. Step 3 is to adjust slurry viscosity by mixing in small amount of NMP, and visually checking slurry viscosity. The slurry viscosity should be in the range of 3,000–15,000 mPa*s at ~10 s^{-1} shear rate.

Example of a NMC532 cathode slurry formulation with 3% conductive additive and 3% binder for laboratory scale (~16 mL) is shown in Table 10.2. The solids content in this slurry is 62%.

In small-scale laboratory setup, the electrode casting is usually achieved with an adjustable doctor blade and a variable-speed drawdown coater. Gap setting and the amount of slurry depend on blade design, slurry solids level, and

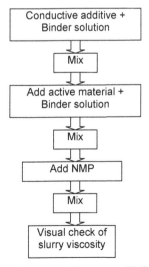

FIGURE 10.3 Laboratory slurry preparation process. NMP, *N*-methyl-2-pyrrolidone.

TABLE 10.2 Exemplary Cathode Slurry Formulation

Material	Grades	Weight (g)
Active material	Umicore NMC532, TX-10	18.8
Conductive additive	Denka black	0.6
Binder solution	Kynar® HSV-900, 8.0% in NMP	7.5
Solvent	NMP, ACS grade	5.2

NMP, N-methyl-2-pyrrolidone.

targeted mass loading. This gap setting should be determined experimentally to achieve desired dry thickness. The electrode is cast on the gloss finish side of the 14.5-μm-thick aluminum foil.

After casting, the electrodes are dried in a convection oven. Drying temperature can affect the distribution of binder in the electrodes and thus impacting the electrode performance and adhesion. In general, slower evaporation rate gives more uniform binder distribution. For NMP-based slurry, drying temperature in the range of 110–130 °C is recommended.

Calendering electrode is another important step for achieving higher energy and power density. The porosity of dry electrodes prior to calendering is usually around 50%, and it can be compacted to a porosity of ~30% without sacrifice of performance. For thick electrodes (>40 μm), calendering electrodes can significantly improve C-rate performance and cyclibility. Calendering can also substantially improve electrode adhesion.

The final step of electrode preparation is to thoroughly dry the electrodes in a vacuum oven overnight at 120 °C to remove trace moisture and solvent. The dried electrode should be stored in a dry box for battery assembly.

10.7 PEEL STRENGTH MEASUREMENT

Peel strength of dry electrode is often measured as an indicator of how well electrode active materials are adhering to current collector. The main property of binder is to provide interconnectivity and adhesion within electrode active materials and to current collectors. The peel strength tests can be conducted in 180° or 90° configurations using ASTM D903 standard. To measure the peel strength, 25.4-mm-wide electrodes were cut and fixed on support using double-sided 3M Scotch tape. Instron model 4201 universal testing machine with a 10 N load cell, having an unloaded grip weight of 3.7 N, was used to measure peel strength at peel-rates of 50 mm/min. Noteworthy that the peel-rate greatly affects the peel strength. For example, doubling the peel-rate yields 2–3x higher peel strength. Figure 10.4 indicates adhesion increased with viscosity of binder. Solution and melt viscosities of each binder are reported in Figure 10.1 and Table 10.1.

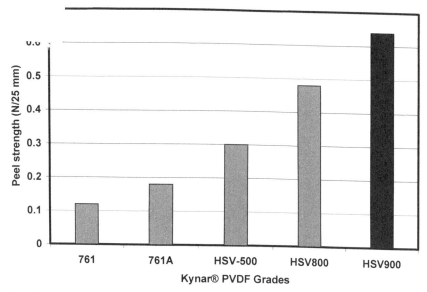

FIGURE 10.4 Kynar® PVDF binder peel strength, 3% Kynar binder in NMC (90° peel test, at 50 mm/min peel-rate, JIS 6854-1). PVDF, polyvinylidene fluoride.

10.8 ELECTRODE PERFORMANCE TEST

In small-scale laboratory setup, coin cells are used to screen electrode materials and evaluate their electrochemical performance. For electrode benchmark, standard electrodes were fabricated at Argonne National Lab (ANL). These electrodes are considered to be of high quality by the Department of Energy (DOE) of US. These electrodes are also recommended for global round robin test particularly throughout all institutes involving DOE-funded battery related projects.

A comparison of coin cell performance with Arkema and ANL benchmark electrodes are presented in Figure 10.5. ANL benchmark electrode uses NMC532 from Toda with composition of 90/5/5 ratio for active material, carbon, and binder. Arkema electrode used NMC532 from Umicore with Kynar® HSV-900 binder at 94/3/3 composition. The loading of both electrodes were as about 16 mg cm^{-2}.

As seen in Figure 10.5, Arkema electrodes showed the same performance as ANL benchmark in terms of initial capacity, C-rate performance, and cyclibility. As previously reported, the ANL benchmark electrode used a composition of 90/5/5 ratio for active material, carbon, and binder; whereas the Arkema electrodes utilizing Kynar® HSV 900 had a composition of 94/3/3 ratio. The Arkema electrodes utilizing less Kynar® binder and carbon performed slightly better at higher C-rates than the standard electrodes from ANL.

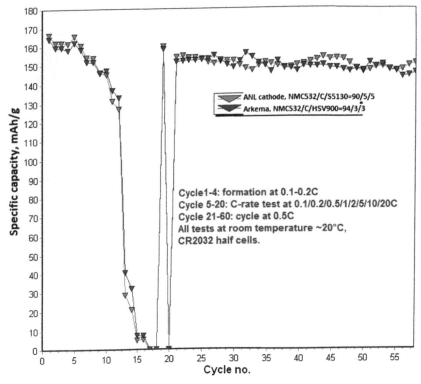

FIGURE 10.5 Electrode performance as function of C-rate.

10.9 FLUORINATED WATERBORNE BINDERS

Electrode slurries today are made with large amount of NMP. However, the NMP-based slurry presents safety, health, and environmental hazards that are not present in an aqueous system. Commonly, high-cost manufacturing controls are used to mitigate these risks. Recovering NMP requires large capital plus high operational cost and yet, it is not 100% effective.

Arkema has developed an advanced Kynar® PVDF waterborne binder (KWB) that significantly reduces volatile organic compounds (VOCs) in electrode slurry formulations by replacing most or all of NMP with water and does not compromise battery performance. KWB provides the desired properties of the intermediate and the final products. The design considerations include: (1) shelf life stability of the waterborne fluoropolymer dispersion, (2) stability of the slurry after admixing the powdery material, (3) an appropriate viscosity of slurry to facilitate good aqueous casting, (4) no foaming, and (5) superb interconnectivity within the electrode after drying.

Arkema has capitalized on its extensive experience and in-depth knowledge of waterborne PVDF to achieve the desired properties for the new binder. KWB

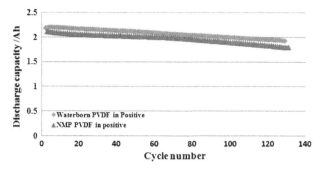

FIGURE 10.6 Cell capacity versus the cycle number for the cells for PVDF/NMP and for KWB positive at 60 °C. PVDF, polyvinylidene fluoride; NMP, *N*-methyl-2-pyrrolidone; KWB, Kynar® PVDF waterborne binder.

has been successfully used to fabricate both anodes and cathodes using water instead of NMP as the media for the slurry. The use of KWB can reduce VOC emissions, lessen worker exposure to the hazardous vapor of NMP, and decrease the amount of hazardous waste generated during electrode fabrication.

The 18,650 cells with KWB electrodes exhibit excellent Coulombic efficiency and capacity fade characteristics. The results were equivalent to standard cells made with solvent cast electrodes over 130 cycles without any notable differences of the cells in terms of cell capacity. Two sets of 18,650 cells of 2.0 Ah with $LiCoO_2$ cathodes, one with KWB and a second with a conventional PVDF binder in NMP, were cycled at room temperature at 0.5 C discharging/0.8 C charging. Similar tests were conducted on two additional sets of cells at 60 °C. The results in Figure 10.6 show that the cyclability of cells made with KWB cathodes is as good as if not better than the solvent cast electrode cells. The experiment at 60 °C is significant because the decay rate is generally faster at elevated temperature compared to room temperature. The results indicate that a cathode made with KWB can be equivalent to a solvent cast process for cell performance and calendar life.

Furthermore, 18,650 cells with KWB anodes and NMP solution cast cathode were tested and produced excellent Coulombic efficiency and capacity fade characteristics equivalent to if not better than standard battery made with SBR anode.

Chapter 11

Electrodeposition of Polypyrrole on CF_x Powders Used as Cathode in Primary Lithium Battery

Henri Groult,[1] Christian M. Julien,[1] Ahmed Bahloul,[2] Sandrine Leclerc,[1] Emmanuel Briot,[1] Ana-Gabriela Porras-Gutierrez[1] and Alain Mauger[1,3]
[1]*Sorbonne Universités, UPMC Univ., Laboratoire PHENIX, CNRS, Paris, France;* [2]*Laboratoire des Matériaux et Systèmes Électroniques, Centre Universitaire de Bordj Bou Arréridj, Bordj Bou Arréridj, Algeria;* [3]*UPMC Univ. Paris 06, Institut de Minéralogie et Physique de la Matière Condensée, Paris, France*

Chapter Outline
- 11.1 Introduction   237
- 11.2 Experimental Section   242
- 11.3 Results   243
  - 11.3.1 Preparation of PPy-CF_x Samples in Acetonitrile   243
  - 11.3.2 Physical–Chemical Characterizations of PPy-Coated Samples   246
  - 11.3.3 Electrochemical Studies in Primary Lithium Battery   251
- 11.4 Conclusions   259
- Acknowledgments   259
- References   259

11.1 INTRODUCTION

Since about 40 years ago, the use of graphite fluorides $(CF_x)_n$ (denoted CF_x henceforth) as positive electrodes in primary lithium batteries has been evidenced by Watanabe et al. [1–3] and such batteries have been commercially developed by the Matsushita Co. in the 1970s. Graphite fluorides are prepared from chemical reaction of fluorine with carbonaceous powders at high temperature [4]. A schematic evolution of the structure from graphite to graphite fluorides during the fluorination process was proposed recently by Nakajima's

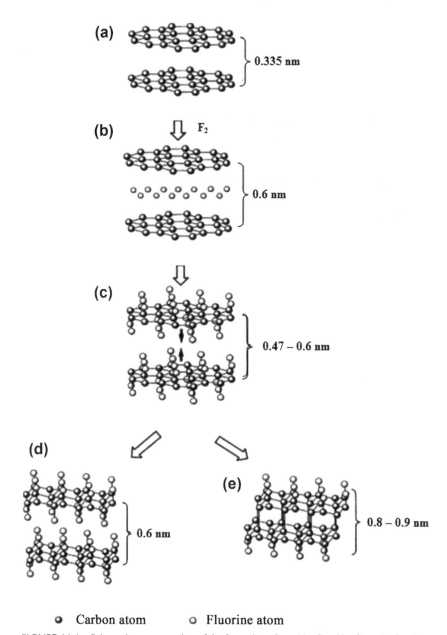

FIGURE 11.1 Schematic representation of the formation of graphite fluorides from the fluorination of graphite with F_2 [5].

group in Japan [5]. In this scheme (Figure 11.1), successive steps occur starting from graphite to the formation of intermediate C_xF phase with ionic C–F bonds and finally the occurrence of two kinds of graphite fluoride (CF and C_2F) with covalent C–F bonds.

Depending on the temperature, the concentration x in CF_x can vary between 0 and about 1.3. In fact, the graphite fluoride compounds are usually non-stoichiometric and the x value represents an average one. Thus, for $0.5 \leq x \leq 1$, the powder is composed of a mixture of C_2F and CF. Surface groups (CF, CF_2, and CF_3) are always present in varying amount regardless of the value of x [6–7]. The fluorination of graphite ($350 \leq T \leq 650$ °C) induces a dramatic change since the planar sp^2 hybridization in the pristine graphite turns to three-dimensional sp^3 hybridization in CF_x.

In spite of lower capacity and energy density, MnO_2 is most widely used as cathode materials in primary lithium battery due to its lower cost and higher rate capability. Graphite fluorides present some limitations: initial voltage delay, poor electrical conductivity, kinetics, limitation of the discharge rates ($\leq 2C$). Concerning the electrical conductivity, graphite fluorides exhibit very low value, which strongly decreases with an increase of fluorine content due to the covalency of the C–F bonds and the sp^3-type link of all carbon atoms. For example, the electrical conductivities of graphite and $CF_{0.465}$ are equal to 3×10^4 and 10^{-7} S cm^{-1}, respectively [4]. However, from a practical point of view, CF_x compounds present several advantages for practical uses in primary lithium battery such as high open-circuit voltage (OCV), flat discharge plateau, high energy density, possible uses in different technologies (bobbin, spiral, coin), large operating temperature range (from −60 to +130 °C), low self-discharge (<2% per year), thermal stability and safety. For instance, batteries with CF_x as positive materials are used for implantable medical devices [8–9] notably because of their flat discharge profiles, which allow predicting the end of the battery.

The reaction mechanism involved during the discharge process of the CF_x/primary lithium battery has been firstly proposed by Whittingham [10] and after widely studied in the literature. The overall discharge reaction can be written as follows:

$$CF_x + x\,Li \rightarrow C + x\,LiF. \tag{1}$$

The calculation of the theoretical OCV based on thermodynamic data gives values around 4.5 V versus Li^0/Li^+ depending on the CF_x composition [11–13], whereas experimental values about 3.1–3.4 V are usually observed. To explain such a difference, it is commonly admitted that an intermediate lithium–graphite intercalation compound (Li-GIC) with composition of Li_yCF_x and/or Li_yCF_x-S is firstly formed during the discharge process [2–3, 14–17]. In this compound, S is a solvent molecule co-intercalated with Li^+ upon discharge and the potential for the formation of this intermediate Li-GIC would be lower than that for direct reaction of lithium with CF_x. Then, the intermediate GIC decomposes into carbon and lithium fluoride through two processes of decomposition and desolvation [11].

Recently, Zhang et al. [18] have suggested that the Li/CF_x cell is discharged through a *"core–shell"* model as illustrated in Figure 11.2. In this model, the active CF_x phase is constantly remained in the core and completely consumed with discharge progress. The product shell is composed of Li-GIC intermediate, carbon and lithium fluoride and the amount of carbon and LiF depends on the

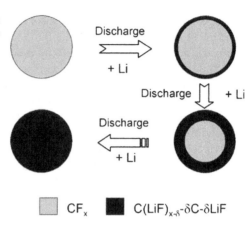

FIGURE 11.2 Schematic illustration of the "core–shell" model describing the discharge of a Li/CF$_x$ cell [18].

depth of discharge state. As mentioned by Zhang et al. [18], the outer product shell plays an important role in the cell performance such as discharge polarization, potential delay and heat generation. Finally, carbon, and LiF are produced at the end of the discharge reaction.

Because of their low electrical conductivity, the development of graphite fluorides is strongly limited even in the field of primary lithium battery (non-rechargeable) for which high power and high energy density (less than 1500 Wh kg^{-1}) are required. Thus, the maximum discharge rate is usually limited to 2C (full discharge in 30 min) for commercial graphite fluorides that are unable to deliver high power density. As reported above, a huge potential drop is usually observed at the beginning of the discharge, resulting in limited formation of Li-GIC around the CF$_x$ particles.

Several studies have been devoted to the enhancement of the electrochemical features of CF$_x$. Both physical and chemical modifications have been reported notably to suppress the initial voltage delay. The first attempt was proposed by Hagiwara et al. [19], who have performed a smooth decomposition of surface of graphite fluorides in a chlorine atmosphere in the range 300–400 °C. In such conditions, graphite fluorides decompose, i.e., the rupture of the C–F bond occurs onto the surface, giving rise to the formation of thin conductive graphite-like carbon around the particles. It results in a decrease of the initial overpotential without significant decrease in capacity. The chemical approaches were mainly focused on the partial and controlled reduction of CF$_x$ in presence of a reductive agent. In this process, the surface fluorine content was decreased to form subfluorinated carbon, which limits the initial voltage delay. Subfluorinated carbon nanofibers [20] or re-fluorinated graphite fluorides [21] have been proposed recently as alternative candidates for primary lithium battery. For the former, impressive performances were notably obtained in terms of power and energy density; indeed, such subfluorinated carbon nanofibers can support high discharge rate up to 6C delivering a maximum power density of 8057 W kg^{-1} associated with a high 1749 Wh kg^{-1} energy density achieved for an F/C ratio of 0.76. In contrast, at the maximum discharge 1C rate, power and energy density of 1370 W kg^{-1} and 2012 Wh kg^{-1}, respectively, are achieved with commercial carbon monofluoride (CF). Improvement of the electrochemical performance obtained with such fluorinated nanofibers has been explained by the presence of fluorinated and unfluorinated domains, which co-exist at the nanometer scale. In addition, TEM investigations have shown that only a surface fluorination of the nonfibers occurred, the core being nonfluorinated. In other hand, the nonfluorinated region (core of fibers) should act as an easy electron transport pathway within the graphite fluoride network.

Following the idea developed by Hagiwara et al. [19], we assume that better electrochemical performance should be obtained by modifying the surface properties, i.e., by enhancing the intrinsic electrical conductivity of the graphite fluorides. In this purpose, Zhang et al. [22] have reported improvements of the discharge performance of Li/CF$_x$ cell by thermal treatment of the CF$_x$ positive electrode material just below its decomposition temperature. In the same manner, Zhang et al. [23] have obtained similar enhancement by making a thin coating of the positive materials with carbon. Li et al. [24] have reported improvements of discharge performance of Li/CF$_x$ battery by using multi-walled carbon nanotubes as conductive additive, which enlarges the contact area of interphase and facilitates electron delivery. Recently, Reddy et al. have obtained remarkable results and succeeded in improving the energy density and power density of CF$_x$ by mechanical milling [25]; they achieved the power density of 9860 W kg^{-1} with a gravimetric energy density of 800 Wh kg^{-1} for the optimized material.

Among various alternative approaches to enhance the electrical conductivity of electrode materials, the electrodeposition of conducting polymers onto the surface of noble metals, graphite, oxide, etc. has been already achieved for instance in the field of lithium battery or supercapacitors. To be viable, the electropolymerization of the surface film must be easy, reproducible, and well-controlled, and performed if possible in a one-step procedure. Among available electroconducting polymers, much attention has been paid to the electrodeposition of polypyrrole (PPy) [26] onto MnO$_2$ [27] or Fe$_2$O$_3$ [28] to enhance their electrochemical performances when they are used as electrode materials in lithium battery. Few papers have been also published about the use of PPy film itself as electrode materials in lithium battery [29–30].

However, in the best of our knowledge, no attempts were achieved with graphite fluoride powders to modify the surface properties by the electropolymerization after our recent short communication [31]. Consequently, the main objective of the present chapter is to present an overview of the results obtained recently on the positive effect of the electrodeposition of PPy on commercial CF$_x$ powders. One of the most important aspects of this study is that such a surface modification allows a significant increase of the discharge currents, which can be applied to the electrodes of primary lithium battery. Consequently, it induces a drastic increase of the power density. These results will be described in details hereafter. Note that whatever the composition of the graphite fluorides used in this study, same tendencies have been observed. Therefore, results obtained with the different samples will be alternatively presented.

11.2 EXPERIMENTAL SECTION

A schematic representation of the cell used for the electrodeposition of large amount of PPy is presented in Figure 11.3. The crucible was filled with the CF_x powder and the electric contact is done with the glassy carbon crucible. This sample constitutes the working electrode and its potential was measured versus a saturated calomel electrode (SCE). A platinum wire was used as counter electrode. The supporting electrolyte was composed of 0.1 mol anhydride $LiClO_4$ dried at 110 °C in a vacuum dissolved into distilled acetonitrile (Acros Organics) containing 0.01 M of the pyrrole (Py) monomer. The electropolymerization of Py giving rise to PPy was studied at first by cyclic voltammetry (CV) in the potential range between −0.2 and 1.2 V versus SCE at scanning rate 50 mV s^{-1} and then by applying a constant positive potential to the electrode. In this work, commercial C_2F, $CF_{0.8}$, and CF powders were investigated.

The electrochemical lithium insertion reaction was studied in 1 mol L^{-1} $LiPF_6$ in EC:DMC (ethylene carbonate/dimethyl carbonate) solution as electrolyte (Merck LP30) at room temperature using Swagelok cells. The working electrode was composed of 70 wt% of active material, 20 wt% of acetylene black, and 10 wt% of PVDF. The counter and reference electrodes were metallic lithium foils. All potential values will be referred to this Li^0/Li^+ reference in the section devoted to the tests in primary lithium battery. The galvanostatic discharge curves as well as the electrochemical impedance spectroscopy (EIS) measurements (from 100 kHz to 1 mHz with ΔE = 10 mV) were performed using a potentiostat/galvanostat (VMP3 Bio-Logic).

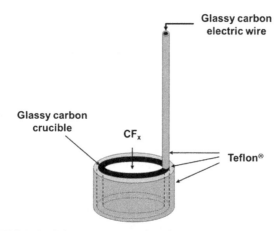

FIGURE 11.3 Schematic representation of the working electrode used for the electrodeposition of PPy on CF_x powder.

11.3 RESULTS

11.3.1 Preparation of PPy-CF$_x$ Samples in Acetonitrile

Organic conducting polymers such as PPy [32] can be synthesized from the appropriate monomer by electrochemical route [33]. In this case, they exhibit better conductivity than that obtained with PPy prepared by chemical route. Depending on doping level, they can act as metals, semiconductors, or insulators and thus can be used in a wide range of applications such as electric conductors, batteries, and electronic devices [34]. The reaction of polymerization occurs through the electrochemical oxidation of the Py monomer, which gives rise to PPy as described in Figure 11.4.

In this scheme, A^- is the counter-ion incorporated during polymerization. The polymer is produced during the oxidized state. The alternating occurrence of single and double bonds allows in conferring semiconductor characteristics to this polymer. There is significant overlap of neighboring π-bonds, and these electrons are therefore delocalized. Electrochemical polymerization provides several advantages compared to chemical method: the electroactive films are strongly adherent to the electrode surface and exhibit high electrical conductivity. In addition, by adjusting the operating conditions during the electrodeposition (potential, current, monomer content in solution, etc.), it is possible to control the mass of the electrodeposited polymers, i.e., its thickness.

In this study, the electrodeposition of Py giving rise to PPy was investigated by CV in 0.1 mol L^{-1} LiClO$_4$/CH$_3$CN electrolyte containing 0.01 mol of Py (Figure 11.5(a)). At first, blank curves were recorded in 0.1 mol L^{-1} LiClO$_4$/CH$_3$CN without Py to ensure that CF$_x$ powders are stable in the explored potential window and to verify that no side reactions occurred during the electropolymerization of PPy. Then, the experiments consist in performing the electrodeposition of PPy in 0.1 mol L^{-1} LiClO$_4$/CH$_3$CN electrolyte containing 0.01 mol of Py. The samples were referenced PPy-C$_2$F, PPy-CF$_{0.8}$, and PPy-CF after running 30 voltamograms cycled in 0.1 mol L^{-1} LiClO$_4$/CH$_3$CN electrolyte containing 0.01 mol of Py between −1.2 V and +0.2 V versus SCE. This potential was chosen from the analysis of voltamperometry curves because at higher voltages, the electrolyte should be oxidized and the polymer film fissured. The solution was deoxygenated by bubbling nitrogen and the cell was kept in a nitrogen environment throughout the synthesis. Results are as follows: first, an increase of the current density is observed due to beginning of the film formation process. After reaching a maximum, the current density decreases and then tends to a steady value. The film is formed more and more slowly due to a higher ohmic drop at the surface, the latter being larger versus time. After 20 min of electropolymerization, the coating has been achieved. Figure 11.5(b) shows the cyclic voltammograms

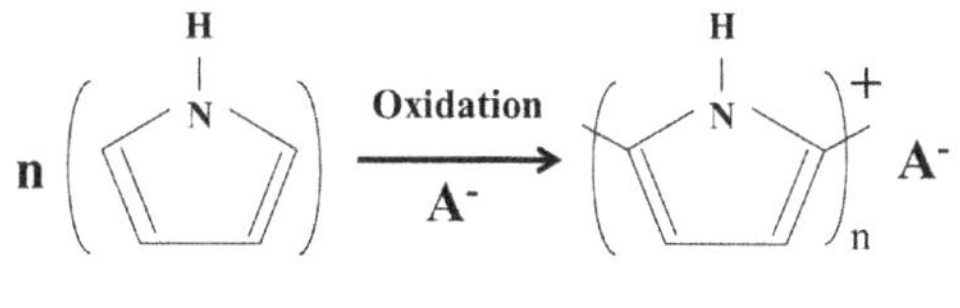

FIGURE 11.4 Electrochemical reaction for the synthesis of polypyrrole from pyrrole monomer.

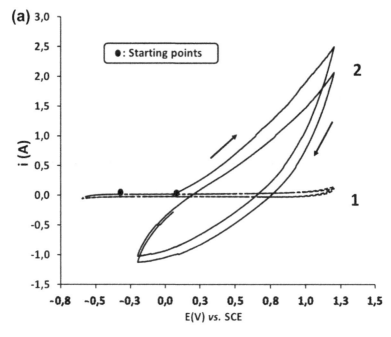

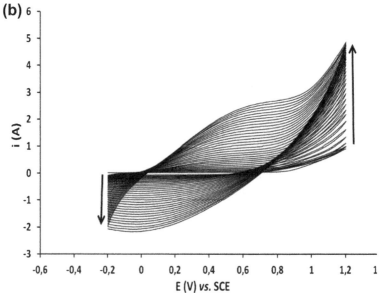

FIGURE 11.5 First cyclic voltammogram (v = 50 mV s^{-1}) recorded with CF in 0.1 mol L^{-1} LiClO4/CH$_3$CN electrolyte in absence (a) or in presence (b) of 0.01 mol of pyrrole.

(v = 50 mV s^{-1}) recorded with sample CF in 0.1 mol L^{-1} LiClO$_4$/CH$_3$CN without (curve 1) and with Py (curve 2) in the electrolyte. As a fact, in absence of Py (blank curve), the current was found to be very small confirming that the CF$_x$ is not reduced/oxidized in the chosen potential window (−0.2–1.3 V vs SCE).

The arrows in Figure 11.5(b) show the sense of evolution of the voltammograms just at the beginning of the deposition of PPy, up to the last cycle, which is the voltammogram for the CF–PPy-modified electrode. Again, the cyclic voltammograms profile did not reveal any reduction or oxidation peaks of CF except those relative to PPy. Thus, the anodic and the cathodic currents increase upon cycles indicating the growth of conducting polymer electrodeposited onto particles. Similar conclusions were deduced with C_2F and $CF_{0.8}$ samples in absence of monomer in the electrolyte.

To evaluate the impact of PPy electrodeposition, a series of experiments were carried out by EIS. First, impedance measurements were performed procedure in 0.1 mol L^{-1} $LiClO_4/CH_3CN$ electrolyte containing 0.01 mol of Py at OCV before and after the electrodeposition of PPy. Whatever the CF_x samples, the analysis of the impedance spectra gives rise to the same conclusions. For instance, the Nyquist diagrams obtained with CF and PPy-CF (Figures 11.6(a)–(b)) exhibit a semicircle in the high frequency region related to charge transfer. In the low frequency range, we observe a Warburg behavior with a slope related to the diffusion process of chlorate ions at the electrode-electrolyte interface. For PPy-modified graphite fluorides, the slope observed after the semicircle devoted to charge transfer is more pronounced (Figure 11.6(b)) than that observed with pristine graphite fluorides (Figure 11.6(a)). It means that the electrodeposition of PPy enhances the diffusion coefficient of Li^+ into fluorinated graphite. The comparison of the Nyquist plots shows that the presence of PPy induces a drastic decrease of the semicircle observed at medium frequency with PPy-$CF_{0.8}$, i.e., the charge transfer resistances were significantly decreased (10 times lower) owing to the presence of PPy. Interesting information has been also extracted from the expanded view of the high frequency region. The intercept between the semicircle and real axis at high frequency represents the bulk resistance (R_b) of the cell, including the resistance of the current collector, and the electrolyte resistance. A decrease of R_b values from about 45 Ω (before PPy electrodeposition) to about 26 Ω (after PPy electrodeposition) was observed, meaning that the electrodeposition of PPy enhanced significantly the electronic conductivity of CF_x materials by enlarging the electric contact

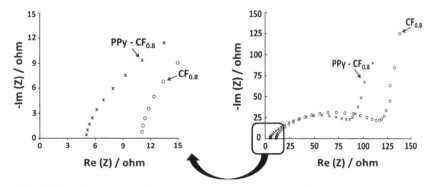

FIGURE 11.6 First and the following 20 cyclic voltammograms (v = 50 mV s^{-1}) recorded with CF in 0.1 mol L^{-1} $LiClO_4/CH_3CN$ electrolyte containing 0.01 mol of pyrrole.

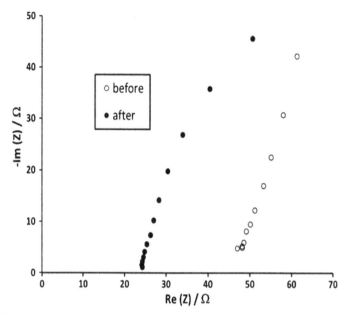

FIGURE 11.7 Impedance spectra of CF and CF-PPy recorded in 0.1 mol L^{-1} LiClO$_4$/CH$_3$CN electrolyte containing 0.01 mol of pyrrole.

grains, improving the diffusion phenomenon into the host lattice. Same tendency was also observed for others samples; in the case of CF$_{0.8}$, the resistance was found to be 45.3 Ω before electrodeposition of PPy and 27.3 Ω after. Finally, Figure 11.7 clearly shows that the highest frequencies are observed for PPy-CF$_{0.8}$, indicating a faster kinetics for PPy-modified samples.

11.3.2 Physical–Chemical Characterizations of PPy-Coated Samples

For comparison SEM and TEM investigations were done before and after PPy electrodeposition. Images in Figure 11.8 show that the surface morphology of PPy-CF$_x$ is slightly different from that of pristine CF$_x$. For a short electropolymerization time, uniform surface films have been deposited onto the CF$_x$ sample. The SEM (Figure 11.8(b)) and by TEM (Figure 11.8(c)) images of PPy-CF$_x$ have revealed typical surface nodular due the presence of small amount of PPy. The roughness is also slightly enhanced after deposition of PPy.

For increasing electropolymerization times, nanofibers with diameter between ~100 and 200 nm and length of about 0.5–1 μm have been observed by SEM (Figure 11.9(a)–(b)) and TEM (Figure 11.9(c). We remark that these fibers grew with the preferential direction parallel to the surface and gave rise to a 3D granular film.

As a result, it must be noted that a thin layer of PPy is enough to enhance the conduction between grains and to improve the performances of the raw materials. Moreover, even if a larger amount of PPy has been deposited onto the

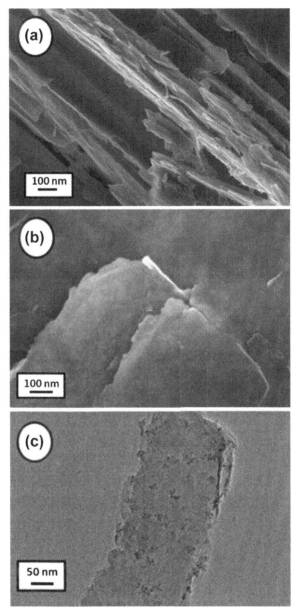

FIGURE 11.8 SEM (a,b) TEM (c) images of pristine C_2F (a) and PPy-C_2F (b,c).

surface of the graphite fluorides, it has been reported that PPy can be used in Li-ion battery as electroactive material [29–30]. Thus, PPy could also contribute to the total discharge capacity.

Adsorption–desorption analyses were also carried out mainly to access to the surface area and the pore size distribution. The samples were degassed at 250 °C

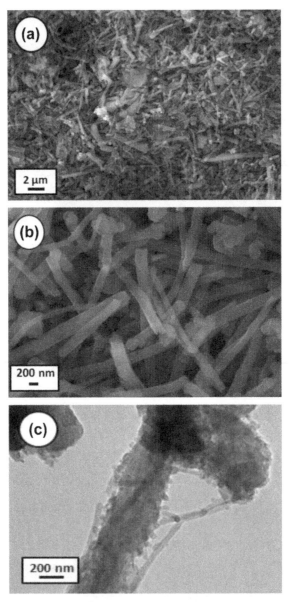

FIGURE 11.9 SEM (a,b) and TEM (c) images of pristine $CF_{0.8}$ (a) and PPy-$CF_{0.8}$ (b,c).

under vacuum during 15 h before each analysis and experiments were carried out under N_2 atmosphere. A typical adsorption isotherm obtained in the case of CF is presented in Figure 11.10(a). This isotherm shape is type I according to the IUPAC classification, with a small hysteresis loop due to the presence of mesoporosity (mesopores of width between 2 and 50 nm). Micropores (less than 2 nm) did not be evidenced whatever the CF_x samples. The carbon powder exhibits large porosity

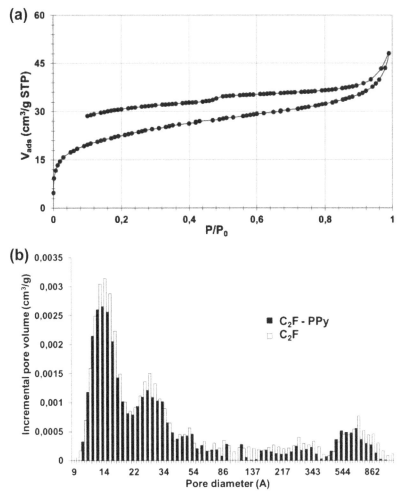

FIGURE 11.10 Adsorption–desorption isotherm (a) and pore volume distribution (b) obtained for C_2F before and after electrodeposition of PPy.

(meso and macroporosity) as presented in Figure 11.10(b). The presence of macroporosity is mainly attributed to the agglomeration of the carbon particles. The same tendency than that observed for the specific surface area was pointed out for the pore size distribution (Figure 11.10(b)). Thus, the coating procedure with PPy of the CF_x powders has only a small influence on the pore distribution and the elecytrodeposition of PPy does not affect deeply the adsorption properties of the CF_x sample. For the three tested samples, a reduction of about 10% of the BET (Brunauer, Emmett, and Teller) surface area has been pointed out. For instance, the BET surface areas of C_2F and PPy-C_2F were found to be 83 and 73 $m^2\ g^{-1}$, respectively. The largest value observed for $CF_{0.8}$ (127 $m^2\ g^{-1}$) may be explained by the presence of graphitic domains as revealed by X-ray diffraction (XRD) hereafter.

XRD measurements were performed in order to study the influence of the electrochemical deposition of PPy on the core of the CF_x. Typical XRD patterns of C_2F and $CF_{0.8}$ before and after PPy electrodeposition are given in Figure 11.11(a)–(b), respectively. It exhibits broad peaks corresponding to the fluorinated phase at around $2\theta \approx 14$, 30, 41, and 75° but does not reveal any peaks relative to the presence of PPy. Such broadness is typically observed with CF_x samples. For instance, the weak and broad peak relative to (001) diffraction lines of C_2F (Figure 11.11(a)) were observed at around $2\theta \approx 30°$ and 41°. In addition, the presence of a weak diffraction line at $2\theta \approx 41°$ for CF_x evidences the presence of intercalated phases. All CF_x samples tested in this study exhibit the

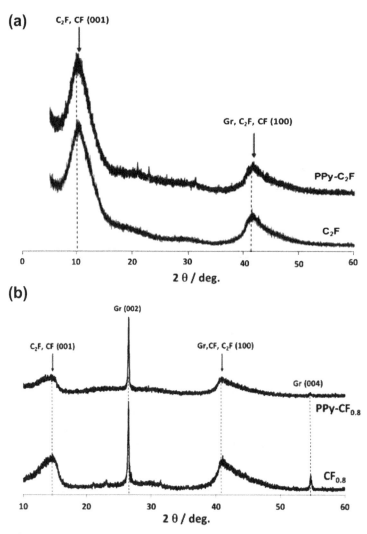

FIGURE 11.11 XRD patterns of (a) C_2F and PPy-C_2F and (b) $CF_{0.8}$ and PPy-$CF_{0.8}$.

same XRD patterns except $CF_{0.8}$, for which several diffraction lines belonging to graphite were observed (Figure 11.11(b)). Similarly to the case of subfluorinated carbon nanofibers proposed recently by Yazami and Hamwi [18], the presence of conducting graphitized domains, which co-exist at the nanometer scale with insulating fluorinated domains, should act as an easy electron transport pathway within the graphite fluoride and should enhance the electrochemical features of this material compared with those obtained with other selected CF_x powders.

11.3.3 Electrochemical Studies in Primary Lithium Battery

Let us consider now, the effect of PPy on the electrochemical performances of pristine and PPy-modified graphite fluorides when they are used as active materials in primary lithium battery. The experiments were performed in 1 mol L^{-1} $LiPF_6$-EC:DMC (1:1) electrolyte (LP30) provided by Merck. Figure 11.12 shows a typical discharge curve obtained with graphite fluoride. Several parameters can be extracted from this curve useful to compare the electrochemical performance of the different CF_x samples such as the OCV versus Li^+/Li^0 redox potential, the average discharge potential (E_d expressed in V versus Li^+/Li^0) measured at half-discharge capacities (Q expressed in mAh g^{-1}) for a potential cutoff of 1.5 V versus Li^+/Li^0; specific energy (E) and power (P) defined as $E_d \times Q$ (expressed in Wh kg^{-1}) and $E_d \times I$ (expressed in W kg^{-1}), respectively.

As mentioned above, the chronopotentiograms recorded with CF_x samples are characterized by an initial voltage delay due to the poor electrical conductivity of the CF_x. Thus, the voltage grows up at first drops rapidly from the OCV before and reaching a potential plateau. This flat potential profile is due to the two-phase nature of the discharge. Nevertheless, the voltage value

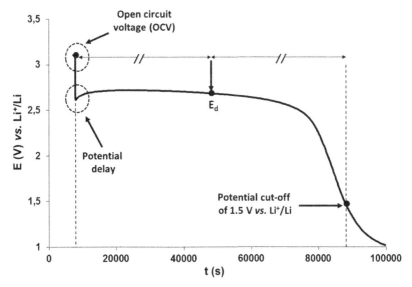

FIGURE 11.12 Typical discharge curve of CF_x.

corresponding to this plateau is much lower than the expected one due to the formation of an intermediate GIC such as Li_yCF_x and/or Li_yCF_x-S first formed during the discharge process.

In order to determine the kinetic parameters of the lithium insertion reaction, a galvanostatic intermittent titration method was used. At first, the cell is considered to be in thermodynamic equilibrium, which corresponds to a known stoichiometric composition and cell voltage. Then, a constant discharge current is applied at $t = 0$. During the current step, the potential of the cell decreases gradually. After a sufficient time, t, for reaching a particular value of the time (i.e., insertion ratio y), the current is interrupted. Then, the potential of the electrode gradually recovered toward the equilibrium value (OCV) due to the diffusion of active species from the surface to the bulk, i.e., the composition tends to become homogeneous by diffusion of mobile species. The cell voltage also tends gradually toward a new stationary value. A further galvanostatic pulse is applied to the cell, and so on. It gives rise to the variation of the OCV and the closed-circuit voltage (CCV) curves versus time versus depth of discharge.

For example, the OCV and CCV curves obtained with pristine C_2F are given in Figure 11.13. Whatever the samples, the OCV values obtained with PPy-CF_x are always greater than those of the pristine material. The OCV measured with CF and PPy-CF before the discharge procedure ($t = 0$) are 3.15 ± 0.05 V and 3.38 ± 0.02 V versus Li^+/Li^0, respectively, showing the beneficial effect of the polymer on the conductivity of the compounds.

Impedance measurements were carried out in 1 mol L^{-1} $LiPF_6$ in EC:DMC solution before discharge and at each point (Figure 11.15(a)) of the OCV curve. The analysis of the impedance spectra obtained with the different CF_x samples was similar. For instance, the Nyquist diagrams obtained with $CF_{0.8}$ and PPy-$CF_{0.8}$ (Figure 11.14) have shown the presence of a semicircle in the high frequency

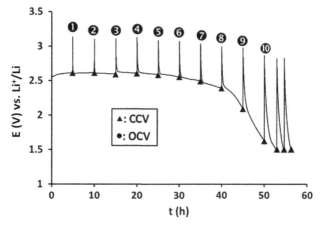

FIGURE 11.13 Evolution of the OCV and CCV recorded with C_2F in 1 mol L^{-1} LiPF6 in EC:DMC (1:1) at C/50 rate. Duration of each galvanostatic step was 5 h. Each galvanostatic step was numbered from 1 to 10.

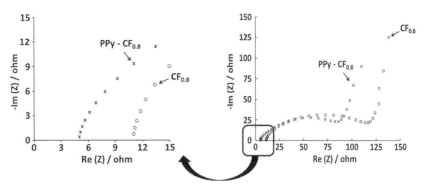

FIGURE 11.14 Nyquist diagrams recorded in 1 mol L^{-1} LiPF6 in EC:DMC electrolyte for CF$_{0.8}$ and PPy-CF$_{0.8}$ before the discharge procedure.

region related to charge transfer. At low frequencies, the spectra exhibit the well-known Warburg behavior with a slope segment related to the diffusion and migration process of Li$^+$ at the electrode–electrolyte interface. The very high frequency intercept (R_s) takes into account not only the electrolyte resistance but also the resistance of the electrodes in addition to the ohmic resistance of the battery. The medium frequency intersect allows determining the charge transfer resistance (R_{ct}). The low frequency range is related to the Warburg impedance. Typical Nyquist impedance spectra recorded at 50% of depth of discharge with CF$_{0.8}$ are given in Figure 11.15. The values of R_s and R_{ct} obtained at 0, 50, and 70% DOD are reported in Table 11.1. All the data were extracted from the fitting of the experimental curves. The experimental curves were perfectly fitted with a conventional Randles-type equivalent circuit which takes into account the charge transfer and the Warburg diffusion related to the diffusion of Li$^+$ into the host structure.

The comparison of the Nyquist plots has revealed that the presence of PPy induces a decrease of the semicircle observed at medium frequency, i.e., the charge transfer resistances were decreased owing to the presence of PPy. For example, R_{ct} decreased from ~125 Ω to about 71 Ω at 50% DOD with samples CF$_{0.8}$ and PPy-CF$_{0.8}$. It confirms the positive effect of PPy on the electrical conductivity of the CF$_x$ compounds. Interesting information have been also extracted from the expand view of the high frequency region (inset of Figure 11.14). The intercept between the semicircle and real axis at high frequency represents the bulk resistance (R_b) of the cell, including the resistance of the current collector, and the electrolyte resistance. As sown in the inset of Figure 11.14, a decrease of R_b values were observed after PPy electrodeposition (about 11 Ω before deposition of PPy to about 4 Ω after deposition in the case of sample CF) meaning that the electrodeposition of PPy enhanced the electronic conductivity of CF$_x$ materials by enlarging the contact area, improving the diffusion phenomenon into the host lattice.

The impedance spectra increased with increasing depth of discharge. This phenomenon is much more pronounced in the case of pristine CF$_x$. It is mainly due to the increase of the charge transfer resistance and also to the passivation of the electrodes. Nevertheless, this increase is moderated in the case of

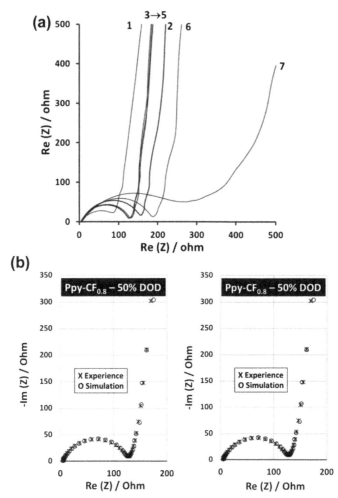

FIGURE 11.15 Nyquist diagrams recorded in 1 mol L^{-1} LiPF$_6$ in EC:DMC (1:1) for (a) PPy-CF$_{0.8}$ at different DOD numbered in the discharge curve of Figure 11.13 and (b) at 50% for CF$_{0.8}$ and PPy-CF$_{0.8}$.

PPy-modified CF$_x$. For example, R_{tc} values increased from about 115 to 338 Ω for CF at 0% and 70% DOD, respectively, whereas values obtained with PPy-CF were in the same order.

Discharges curves were recorded for various currents. A first set of experiments was done using stainless steel grid as current collector. For example, the discharge profiles of the CF$_{0.8}$ and PPy-CF$_{0.8}$ at various discharge rates are given in Figure 11.16(a) and (b), respectively.

The average discharge potential (E_d) measured at half-discharge capacity, the discharge capacities (Q), the specific energy (E), and the power (P) obtained for CF$_x$ and PPy-CF$_x$ are determined from the experimental curves. These values were calculated for a potential cutoff of 1.5 V versus Li$^+$/Li. Whatever the sample,

TABLE 11.1 Parameters Obtained From the Fit of Impedance Spectra (Nyquist Plots) for Different CF_x and PPy-CF_x Samples

DOD (%)	$CF_{0.8}$		PPy-$CF_{0.8}$		CF		PPy-CF	
	R_s (Ω)	R_{ct} (Ω)	R_s (Ω)	R_{ct} (Ω)	R_s (Ω)	R_{ct} (Ω)	R_s (Ω)	R_{ct} (Ω)
0	11.1	116.9	4.4	39.6	14	115.1	3.3	33.3
50	11.0	124.9	4.2	70.7	14.2	185.9	5	12.8
70	13	269.1	8.6	38.5	12.3	338.5	6.3	20.6

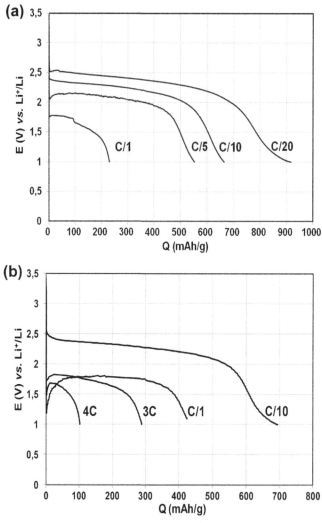

FIGURE 11.16 Discharge curves recorded in 1 mol L^{-1} LiPF6 in EC:DMC for (a) CF$_{0.80}$ and (b) PPy-CF$_{0.80}$, at various C-rates. Electrode substrate was stainless steel grid.

the voltage at first drops gradually from an OCV of 3.3 V to 2.3–1.6 V versus Li$^+$/Li depending on the current density, and then reaches a potential plateau due to the two-phase nature of the discharge. An increase in the discharge current caused a decrease in the average discharge voltages, i.e., the E_d values are strongly dependent on the discharge rates. The values of E_d and Q were not different at low current density (≤C/10 rate) for CF$_x$ and PPy-CF$_x$ indicating that the reaction is governed by the lithium ion diffusion mechanism. This difference is more significant for higher rates. Thus, E_d decreases from 2.37 to 1.73 V versus Li$^+$/Li at C/20- and 1C-rate, respectively, in the case of CF$_{0.8}$ and from 2.27 to 1.66 V versus Li$^+$/Li at C/10- and 4C-rate, respectively, in the case of PPy-CF$_{0.8}$. Almost

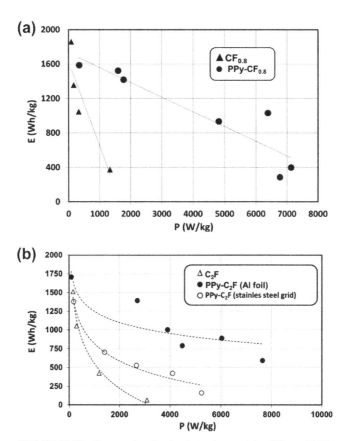

FIGURE 11.17 Ragone plots for the electrode materials of Figure 14.16.

99% of the theoretical capacity (788 mAh g^{-1}) is achieved at low discharge rates as for CF$_{0.80}$ at C/20. It is well known that the discharge of pristine CF$_x$ is not easy at rate higher than 1C. For higher discharge rates, only the PPy-CF$_x$ can be used; the best electrochemical performance at 4C-rate is obtained with PPy-CF$_{0.80}$. The Ragone plots giving the variation of the energy density versus the power density for CF$_{0.80}$ and PPy-CF$_{0.80}$ are presented in Figure 11.17(a).

For low discharge rates (≤C/10), high energy density and low power density values are observed, while the increasing current density results in a large decrease in the energy density due to the drop of the cell voltage and discharge capacity. However, for PPy-CF$_{0.80}$ powders, high current density (until 4C) can be applied and the average discharge potentials is higher. Consequently, the power density values reached with PPy-CF$_{0.80}$ are higher than in pristine CF$_{0.8}$ and a maximum value of 5235 W kg^{-1} has been achieved (discharge rate: 4C), whereas only 1340 W kg^{-1} was obtained with CF$_{0.80}$ (discharge rate: 1C).

A second set of experiments was done to improve the results already obtained. It consists in using an aluminum foil instead of a stainless steel grid as electrode substrate. In that case, the electrode material is deposited as a thin film which is

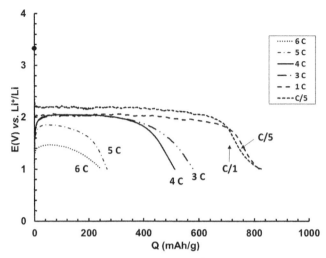

FIGURE 11.18 Discharge curves recorded in 1 mol L^{-1} LiPF6 in EC:DMC with (a) CF$_{0.80}$ and (b) PPy-CF$_{0.80}$, at various C-rates with Al foil as electrode substrate.

pressed to reach a uniform thickness. Significant improvements of the results were obtained. For example, the discharge profiles of the Li/PPy-CF$_{0.8}$ cells for various discharge rates are showed in Figure 11.18 and can be compared to the previous curves. As shown in this figure, higher discharge currents can be applied (until 6C rate). In addition, the potential plateau is rather the same until a discharge rate of 4C.

The Ragone plots giving the variation of the energy density versus the power density for CF$_{0.80}$ and PPy-CF$_{0.80}$ are presented in Figure 11.17(a). For low discharge rates (≤1C), high energy density and low power density values were observed, while the increasing current density results in a large decrease in the energy density due to the drop of the cell voltage and discharge capacity. However, for PPy-CF$_{0.80}$ or PPy-CF powders, high current density (until 6C) can be applied and the average discharge potentials were higher than those obtained without PPy and for a stainless steel grid as electrode substrate. Consequently, the power density values reached with PPy-CF$_{0.80}$ with an Al foil substrate were higher than with pristine CF$_{0.8}$ and a maximum value of 7130 W kg^{-1} has been achieved (discharge rate: 5C), whereas value of 5235 W kg^{-1} was obtained with a stainless steel grid substrate with PPy-CF$_{0.80}$ (discharge rate: 4C).

Same conclusions were obtained with other CF$_x$ compounds. For example, the effect of the nature of the substrate in the case of C$_2$F is reported in Figure 11.17(b). In that case, the power density values reached with PPy-C$_2$F were high and a maximum value of 7636 W kg^{-1} has been achieved (6C discharge rate) corresponding to an energy density value of about 600 Wh kg^{-1}.

To summarize, the presence of PPy allows enhancing the electronic conductivity, i.e., the particles connection. It means the electron pathway is enhanced and the polarization is decreased. The consequences are a significant increase of the energy and power densities. Even in presence of PPy, the discharge reaction is controlled by the diffusion of Li$^+$.

11.4 CONCLUSIONS

Electropolymerization of pyrrole (PPy) on graphite fluoride powders was performed in acetonitrile containing a small amount of Py. XRD measurements do not reveal any modification of the framework after deposition of PPy. Owing to this surface coating, the electrical conductivity between gains has been enhanced as revealed by impedance analysis. Pristine CF_x and PPy-CF_x have been tested as cathode materials in non-rechargeable Li battery. It has been shown that the electrochemical performances are significantly enhanced in the case of PPy-modified CF_x. A maximum of 2C-rate can be applied in the case of pristine C_2F of $CF_{0.8}$. In contrast, PPy-CF_x can support higher rates and the discharge process can be effective at more than 2C-rate. The use of Al foil as electrode substrate allowed optimizing the process and using this electrode configuration, and the maximum power density value was reached with PPy-C_2F of 7636 W kg^{-1} has been achieved at 6C discharge rate corresponding to an energy density value of about 600 Wh kg^{-1}.

ACKNOWLEDGMENTS

The authors express their gratitude to José Gomez (UPMC, Laboratoire Phenix, Paris, France) for his technical assistance.

REFERENCES

[1] N. Watanabe, M. Fukuda, US Patent 3 536 532, 1970.
[2] T. Nakajima, N. Watanabe, Graphite Fluorides and Carbon-Fluorine Compounds, CCR Press, Boca Raton, FL, 1991.
[3] T. Nakajima, Fluorine compounds as energy conversion materials, J. Fluorine Chem. 149 (2013) 104–111.
[4] N. Watanabe, T. Nakajima, H. Touhara, Chap. 2, in: Graphite Fluorides, vol. 8, Elsevier, Amsterdam, 1988, pp. 23–89.
[5] V. Gupta, T. Nakajima, Y. Ohzawa, B. Žemva, A study on the formation mechanism of graphite fluorides by Raman spectroscopy, J. Fluorine Chem. 1200 (2003) 143–150.
[6] V.N. Mitkin, I.P. Asanov, L.N. Mazalov, X-ray photoelectron- and auger-spectroscopic study of superstoichiometric fluorographite-like materials, J. Struct. Chem. 43 (2002) 843–855.
[7] P. Lam, R. Yazami, Physical characteristics and rate performance of $(CF_x)_n$ (0.33 < x < 0.66) in lithium batteries, J. Power Sources 153 (2006) 354–359.
[8] D.C. Bock, A.C. Marschilok, K.J. Takeuchi, E.S. Takeuchi, Batteries used to power implantable biomedical devices, Electrochim. Acta 84 (2012) 155–164.
[9] H.J. Zhu, M. Gravil, L. Feng, D. Karpinet, Li/CF_x medical battery development, ECS Trans. 11 (32) (2008) 11–17.
[10] M.S. Whittingham, Mechanism of reduction of the fluorographite cathode, J. Electrochem. Soc. 122 (1975) 526–527.
[11] J.L. Wood, R.B. Badachhape, R.J. Lagow, J.L. Margrave, J.L. Wood, R.B. Badachhape, R.J. Lagow, J.L. Margrave, J. Phys. Chem. 73 (1969) 3139–3142.
[12] N. Watanabe, Two types of graphite fluorides, $(CF)_n$ and $(C_2F)_n$, and discharge characteristics and mechanisms of electrodes of $(CF)_n$ and $(C_2F)_n$ in lithium batteries, Solid State Ionics 1 (1980) 87–110.
[13] H. Touhara, H. Fujimoto, N. Watanabe, A. Tressaud, Discharge reaction mechanism in graphite fluoride-lithium batteries, Solid State Ionics 14 (1984) 163–170.

[14] J. Read, E. Collins, B. Piekarski, S. Zhang, LiF formation and cathode swelling in the Li/CF$_x$ battery, J. Electrochem. Soc. 158 (2011) A504–A510.
[15] K. Guérin, M. Dubois, A. Hamwi, Electrochemical discharge mechanism of fluorinated graphite used as electrode in primary lithium batteries, J. Phys. Chem. Solids 67 (2006) 1173–1177.
[16] N. Watanabe, R. Hagiwara, T. Nakajima, H. Touhara, K. Ueno, Solvents effects on electrochemical characteristics of graphite fluoride-lithium batteries, Electrochim. Acta 27 (1982) 1615–1619.
[17] K. Ueno, N. Watanabe, T. Nakajima, Thermodynamic studies of discharge reaction of graphite fluoride-lithium battery, J. Fluorine Chem. 19 (1982) 323–332.
[18] S.S. Zhang, D. Foster, J. Wolfenstine, J. Read, J. Power Sources 187 (2009) 233–237.
[19] R. Hagiwara, T. Nakajima, K. Nogawa, N. Watanabe, Properties and initial discharges behavior of graphite fluorides decomposed under chlorine, J. Appl. Electrochem. 16 (1986) 223–228.
[20] R. Yazami, A. Hamwi, K. Guérin, Y. Ozawa, M. Dubois, J. Giraudet, F. Massin, Fluorinated carbon nanofibres for high energy and high power densities primary lithium batteries, Electrochem. Commun. 9 (2007) 1850–1855.
[21] J. Giraudet, C. Delabarre, K. Guérin, M. Dubois, F. Masin, A. Hamwi, Comparative performances for primary lithium batteries of some covalent and semi-covalent graphite fluorides, J. Power Sources 158 (2006) 1365–1372.
[22] S.S. Zhang, D. Foster, J. Read, Enhancement of discharge performance of Li/CF$_x$ cell by thermal treatment of CF$_x$ cathode material, J. Power Sources 188 (2009) 601–605.
[23] Q. Zhang, S. D'Astrog, P. Xiao, X. Zhang, L. Lu, Carbon-coated fluorinated graphite for high energy and high power densities primary lithium batteries, J. Power Sources 195 (2010) 2914–2917.
[24] Y. Li, Y. Chen, W. Feng, F. Ding, X. Liu, The improved discharge performance of Li/CF$_x$ batteries by using multi-walled carbon nanotubes as conductive additive, J. Power Sources 196 (2011) 2246–2250.
[25] M.A. Reddy, B. Breitung, M. Fichtner, Improving the energy density and power density of CF$_x$ by mechanical milling: a primary lithium battery electrode, ACS Appl. Mater. Interfaces 5 (2013) 11207–11211.
[26] T.V. Vernitskaya, O.N. Efimov, Polypyrrole: a conducting polymer; its synthesis, properties and applications, Russ. Chem. Rev. 66 (1997) 443–457.
[27] A. Bahloul, B. Nessark, N.-E. Chelali, H. Groult, A. Mauger, C.M. Julien, New composite cathode material for Zn//MnO$_2$ cells obtained by electro-deposition of polybithiophene on manganese dioxide particles, Solid State Ionics 204–205 (2011) 53–60.
[28] M. Mallouki, F. Tran-Van, C. Sarrazin, P. Simon, B. Daffos, A. De, C. Chevrot, J.-F. Fauvarque, Polypyrrole-Fe$_2$O$_3$ nanohybrid materials for electrochemical storage, J. Solid State Electrochem. 11 (2007) 398–406.
[29] S. Panero, P. Prosperi, F. Bonino, B. Scrosati, A. Corradini, M. Mastragostino, Characteristics of electrochemically synthesized polymer electrodes in lithium cells—III. Polypyrrole, Electrochim. Acta 32 (1987) 1007–1011.
[30] D. Naegele, R. Bittihn, Electrically conductive polymers as rechargeable battery electrodes, Solid State Ionics 28–30 (1988) 983–989.
[31] H. Groult, C.M. Julien, A. Bahloul, S. Leclerc, E. Briot, A. Mauger, Improvements of the electrochemical features of graphite fluorides in primary lithium battery by electrodeposition of polypyrrole, Electrochem. Commun. 13 (2011) 1074–1076.
[32] P. Audebert, P. Hapiot, Fast electrochemical studies of the polymerization mechanisms of pyrroles and thiophenes. Identification of the first steps, existence of 7r-dimers in solution, Synth. Met. 75 (1995) 95–102.
[33] C.M. Li, C.Q. Sun, W. Chen, L. Pan, Electrochemical thin film deposition of polypyrrole on different substrates, Surf. Coat. Technol. 198 (2005) 474–477.
[34] T.A. Skotheim, R.L. Elsenbaumer, J.R. Reynolds, Handbook of Conducting Polymers, second ed., Marcel Dekker Inc, 1998.

Chapter 12

New Nano-C–F Compounds for Nonrechargeable Lithium Batteries

K. Guérin and M. Dubois
Institute of Chemistry of Clermont-Ferrand, University of Blaise Pascal, Aubiere Cedex, France

Chapter Outline

12.1 Introduction	261	12.4 Increasing the Faradic Yield	277
12.2 Contribution of Nanomaterials to Enhance the Energy Density of a Primary Lithium Battery	263	12.5 Next-Generation Carbon Fluorides for Primary Lithium Batteries: Some Key Points	284
12.3 New Fluorination Ways to Increase the Power Density of a Primary Lithium Battery	271	References	285

12.1 INTRODUCTION

Primary lithium batteries are commonly used for many applications such as cameras, electrical locks, electronic counters, electronic measurement equipment, emergency power sources, memory backup, and spatial, military, and implantable medical devices. All these applications require power sources with high energy densities, good reliability, safety, and long life. One of the attractive chemistries is that provided with a fluorine-based cathode and more particularly carbon fluoride (denoted as CF_x). Generally, solid carbon fluorides are prepared by the direct reaction of fluorine gas with carbonaceous materials (conventional fluorination called direct fluorination). A temperature higher than 350 °C is needed if graphite or graphitized carbon materials (e.g., petroleum coke heat treated at 2800 °C) are used [1,2]. The higher the reaction temperature, the higher the fluorination level x ($x = F/C$, $0.5 < x < 1$) of compounds (then called $(CF)_n$) and the higher the C–F covalent character, where the carbon atoms are in sp^3 hybridization. Nevertheless, because of the decomposition reaction of CF_x (whatever the x value), the fluorination temperature must not be higher than 600 °C, at which the x value reaches 1 for graphite as the starting material. When amorphous or disordered (less graphitized) carbons (petroleum coke, carbon blacks, active carbon, etc.) [3,4] are used as starting materials, the fluorination level x sometimes exceeds one fluorine atom per carbon atom indicating the formation of CF_2 and CF_3 groups. The latter should be less active than C–F groups because the C–F bond energies in >CF_2 and –CF_3 are higher, and then the specific faradic capacity decreases. For such kind of disordered materials, different processes compete: (1) fluorination with C–F bond formation, (2) perfluorination with CF_2 and CF_3 formation on the sheet edges and defects, and even (3) decomposition, for which volatile carbon fluorides are evolved. Therefore, graphite (natural or artificial) and highly graphitized carbon (coke or others that are high temperature heat treated) used as starting materials are more suitable because of the easier control of the reaction.

Actually, in the Li/CF_x battery system, a high oxidation–reduction potential of the cathode reaction is combined with the low weight density of light C and F elements. The use of CF_x materials as the active cathode in primary lithium batteries was first demonstrated by Watanabe et al. [5], and then a few years later, the first Li/CF_1 batteries were commercialized by Matsushita Electric Co. in Japan [6]. Commercial Li/CF_x batteries consist of a coke-based cathode having an F/C molar ratio equal or slightly higher than unity. The main features of Li/CF_1 batteries are high energy density (up to 1560 Wh kg^{-1}), high average operating voltage (~2.4 V vs. Li$^+$/Li), long shelf life (>10 years at room temperature), stable operation and wide operating temperatures (–40 °C/170 °C) but low power density (~1400 W kg^{-1}), and low faradic yield (not >75% because of very high amounts of inactive CF_2 and CF_3 groups and dangling bonds that are considered as structural defects and hinder the ionic diffusion).

Up to now, the energy density of carbon fluorides is high mainly because of a high discharge capacity (near the theoretical one of 864 mAh kg^{-1} for an F/C = 1). To increase energy density, some special attention must be paid to increase the average discharge potential either (1) by decreasing the overvoltage due to the insulating behavior of covalent C–F bonding or (2) by developing new fluorination ways in order to decrease the C–F bonding strength.

This last strategy has been developed through the catalytic fluorination process, which allows us to enhance fluorine reactivity and therefore to use lower reaction temperatures. The lower the fluorination temperature, the lower is the C–F bonding energy. When minute amounts of a volatile fluoride, such as HF, AsF_5, IF_5, OsF_6, WF_6, and SbF_5 [7,8], were introduced into the fluorine atmosphere, fluorinated graphites were obtained from ambient temperatures up to 100 °C. All these compounds exhibit a fluorination level x = F/C <0.5, with C–F bonding either ionic (weak bonding energies, x < 0.25) or semiionic (or semicovalent) involving stronger bonding energies: 0.25 < x < 0.5, but less than that corresponding to covalent ones. Moreover, it has been shown that, whatever the fluorine content, sp^2 carbon hybridization (i.e., planarity of graphene layers) is maintained despite the fluorine intercalation reaction into graphite [8,9].

A high increase of discharge potential has been obtained [10,11] but, up to now, such materials are not industrialized because of both a low discharge capacity and some discharge capacity loss during aging. However, during long storage of the primary lithium battery, some residual intercalated species, which are always present in small amounts because of fluorination mechanisms, are removed from fluorographite interlayers and dramatically disturb electrochemical processes. So only covalent C–F bonding is convenient for primary lithium batteries.

As of today, the only available strategy to enhance the discharge voltage of a primary lithium battery made of CF_x is to use a high-temperature fluorination process in order to get a covalent C–F bonding. However, direct fluorination leads to high insulating carbon fluorides, so new high temperature fluorination methods are needed in order to increase the conductivity of the electrode material and to favor electronic diffusion. We will next describe three alternative fluorination strategies: subfluorination derived directly from direct fluorination, static fluorination, and controlled fluorination by decomposition of a solid fluorinating agent. The choice of the carbonaceous starting material for fluorination is also a key point, and the incoming of the carbonaceous nanomaterials appears as promising because the diffusion length may be decreased by structure control at the nanoscale level; shorter transport distances for both electrons and lithium/fluorine ions as well as a larger electrode/electrolyte contact area may be obtained [12]. This strategy is efficient for primary lithium batteries using fluorinated nanocarbons as electrodes because of their large surface area, novel size effects, significantly enhanced kinetics, and so on. Nanostructured electrode materials can not only increase the electroactivity of Li^+ ions but also improve the flow capacity (that is to say, to obtain high power densities) [13]. The present chapter focuses only on fluorinated (nano)carbons as a single phase. The other strategies using a mixture or hybrid materials will not be discussed although they may be of great interest [14–16]. The work of Groult at al [15] is a representative example of the possible improvement. Polypyrrole (PPy) electrodeposited onto graphite fluorides ($CF_{0.80}$) (in acetonitrile containing Pyrrole monomer) favors the electron flux into the electrode. The Li battery then delivered a power density of $5235\,kg^{-1}$ (with a 4C rate), whereas a maximum value of $1364\,W\,kg^{-1}$ (1C rate) was achieved with the untreated CF_x.

12.2 CONTRIBUTION OF NANOMATERIALS TO ENHANCE THE ENERGY DENSITY OF A PRIMARY LITHIUM BATTERY

One of the most interesting properties of fluorinated carbons lies in the nature of the interactions between the fluorine and carbon atoms, which can considerably vary. For the case of the fluorine adsorption on the surface of carbonaceous material, these interactions are very weak. On the other hand, a covalent, semiionic or ionic character can be obtained [17]. In particular, intermediate states are observed in compounds where fluorinated carbon atoms, with sp^3 hybridization, and nonfluorinated sp² ones coexist in the layers (hyperconjugation) [18]. The C–F bonding depends on the synthesis conditions; for covalent compounds, namely, graphite fluorides $(C_2F)_n$ and $(CF)_n$, prepared with molecular fluorine at 350 °C and 600 °C, respectively [19], the carbon skeleton consists of trans-linked cyclohexane chairs or cis-trans linked cyclohexane boats with sp³ bonding. In the case of fluorine–graphite intercalation compounds (C_xF) obtained at temperatures <100 °C, the planar configuration of graphite is partially preserved; the nature of the C–F bond evolves from ionic for low fluorine content to weak covalent for higher fluorine content. The carbon atoms are mainly in the sp² hybridization state. More recently, as discussed in the Introduction, fluorinated graphites were prepared using a room temperature synthesis in the presence of a gaseous mixture of fluorine, HF, and volatile fluorides (BF_3, IF_5, ClF_x, etc.) [20–24]. Another strategy may be developed.

The curvature of the carbon lattice and the resulting σ–π rehybridization act on both the chemical reactivity and the C–F bonding. Curvature results in strain in the carbon lattice as revealed by the pyramidalization and misalignment of the π-orbitals [25–27]; the pyramidalization angle θp in single walled carbon nanotubes, which quantitatively characterizes the σ–π rehybridization, is in the range of 0° for a pure sp² atom to 19.47° for a pure sp³ carbon atom, whereas it is equal to 11.6° for all the carbon atoms in fullerene. This parameter θp, which is inversely proportional to the diameter, has been used to evaluate the chemical reactivity of the nanotubes. Nanotubes are expected to be more reactive than flat graphene sheets, and the smaller the diameter, the higher the reactivity. In order to confirm this fact, fluorination of carbon lattices with different curvatures has been investigated. Single, double, and multiwalled carbon nanotubes (SWCNTs, DWCNTs, and MWCNTs, respectively) of different syntheses were chosen and compared to the two limit cases: on the one hand, spherical $C_{60}F_{48}$ fullerenes (the F/C ratio must be fixed at 48/60 to obtain spherical molecules) and planar covalent graphite fluorides. MWCNTs with a large average diameter (140 nm) are called carbon nanofibers (CNFs; provided by MER Corporation and denoted as CNFs). Two precautions have been used [28]: (1) the samples must exhibit a similar fluorine content, except for $C_{60}F_{48}$ and high-temperature graphite fluoride, in order to avoid different hyperconjugation and (2) only the outer nanotubes must be fluorinated to compare the curvature regarding the outer diameter d. So, the fluorine content F/C in the range 0.3–0.4 has been selected. Except for SWCNTs obtained by electroarc (EA), all the starting materials are commercial products from Unidym (HiPCO method), Helix for nanotubes, and MER Corporation for nanofibers. In agreement with producer data (Table 12.1), raw HiPCO SWCNTs exhibit a diameter distribution over a 1.4- to 0.7-nm range, in comparison with EA-SWCNTs for which d values are equal to 1.59 and 1.45 nm [29]. The diameter distribution of Helix SWCNTs is wider, and the nanotubes have higher diameters than do the HiPCO and EA nanotubes. The following qualitative classification with increasing diameter has been obtained at various Raman excitation wavelengths (457.9, 514.5, and

TABLE 12.1 Nanocarbon Synthesis Methods and Fluorination Conditions Using F_2 Gas; ^{13}C NMR and ^{19}F NMR Chemical Shifts and Wave numbers of the IR Vibration Mode

Starting materials	Producer of the raw material	Synthesis method	Purity	Average diameter[b] <Φ> (nm)	T_{F2} (°C)	F/C	Notation	$δ_{19F}$/ $CFCl_3$ (ppm)[d]	$δ_{13C}$/TMS (ppm)	$ν_{IR}$ (cm^{-1})[e]	$E_{1/2}$ or E_{C-F} (V)
SWCNTs	Helix	CVD[a] Co catalyzed	90%	~1.3	300	0.61	SWCNTs Helix	−180			
	HiPCO	CVD[a] CO	>95%	1.1	200	0.37[c]	SWCNTs HiPCO	−163	83.8	1127	2.7
	EA	Electroarc Ni-Y catalyzed	>95%	1.5	190	0.32	SWCNTs EA	−168	84.0	1188	
DWCNTs	Helix	CVD	90%	~4	300	0.42[c]		−178.4	84.9	1199	2.55
MWCNTs	Helix	CVD	95%	60±10	380	0.43[c]	MWCNTs	−183.8	85.3	1209	2.38
CNFs	MER	CVD	>90%	140±30	420	0.39[c]	CNFs	−190.0	87.0	1215	
C60	MER		99%	0.7	300	0.80[c] $C_{60}F_{48}$		−170.7 −162.9 −153.7 −145.4 −136.7	73 84	1214.6 1171.9 1140.3 1118.3 1060.7	2.16 2.54 3.02 3.45 3.63
Graphite UF4	Mersen		>99%		600	1	$(CF)_n$	−190	88.0	1215	2.4
Graphite UF4	Mersen		>99%		380	0.6	$(C_2F)_n$	−189	86.0	1215	

[a] Chemical vapor deposition (CVD).
[b] Producer data.
[c] From ^{13}C solid echo NMR (NMR experiments were performed at room temperature using a Tecmag spectrometer; working frequency ^{13}C of 73.4 MHz, respectively). The ^{13}C spectra were recorded using a solid echo sequence (two 5.5-μs π/2 pulses separated by 25 μs); this sequence allows the acquisition of the whole signal without loss due to the electronic dead time followed by a quantitative determination of the different contributions. ^{13}C Chemical shifts refer to TMS.
[d] Refer to $CFCl_3$.
[e] FTIR was carried out using a Thermo Nicolet 5700 in an attenuated total reflectance configuration.

1064.0 nm): HiPCO < EA < Helix. This method was extrapolated to DWCNTs. Inner and outer tube diameters were extracted, and predominant values of 0.9/1.3 and 1.6 nm were, respectively, obtained at 514.5 nm.

The fluorination temperature with the same duration (3 h, except for CNFs 16 h) must be increased for HiPCO SWCNTs (200 °C to reach $CF_{0.37}$), DWCNTs (300 °C, $CF_{0.42}$), MWCNTs (380 °C, $CF_{0.43}$), and CNFs (420 °C, $CF_{0.39}$). This confirms the first hypothesis about the higher reactivity with fluorine of curveted carbons. The next part is devoted to the effect of curvature on the C–F bonding.

The ^{13}C and ^{19}F NMR (nuclear magnetic resonance) chemical shifts as well as the wave number of the C–F vibration during Fourier transform infrared spectroscopy (FTIR) experiments are relevant indicators of C–F bonding [30]. The higher the IR wave number v_{IR}, the higher the covalence, the two limits being 1100 and 1220 cm^{-1} for weakened and pure covalence, respectively. ^{19}F Chemical shift δ_{19F} changes are in the range between −136 and −190 ppm/$CFCl_3$, and this range is larger than the ^{13}C values (82–88 ppm/tetramethylsilane (TMS) Table 12.1). When the diameter of the tubes progressively increases, from single (1.1 nm), double (4 nm), multiwalled tubes (60 nm) toward nanofibers (140 nm), the ^{19}F chemical shifts decrease, the values being −163.0, −178.4, −183.8, and −190.0 ppm, respectively. This highlights an increase of the covalent character of the C–F bonding. The data, regarding ^{19}F nuclei and also ^{13}C nuclei, of the fluorinated nanotubes are intermediate between those of two limits: planar fluorinated graphite (−190 ppm and 88 ppm in ^{19}F and ^{13}C NMR, respectively) and spherical highly fluorinated fullerenes $C_{60}F_{48}$ (−136.7 ppm for the highest ^{19}F value and 73 ppm for the lowest ^{13}C value). Several lines were observed for $C_{60}F_{48}$ during ^{19}F NMR and FTIR experiments. Several types of C–F bonds are then found in fluorinated fullerenes because of both different reaction sites (carbon can be involved in a pentagon or a hexagon) and steric hindrance [25].

Two reasons could explain why the C–F bonds in fluorinated fullerenes are significantly different from those for covalent graphite fluorides: (1) Fluorinated C_{60} are independent molecules in which the fluorine atoms stand outside of the fullerene cage. This could lead to steric hindrance between fluorine atoms. Pentagon and hexagons coexist in C_{60} leading to different neighborings. By analogy with graphite fluorides and fluorine-graphite intercalation compound (GIC), the hyperconjugation results in a weakening of the C–F covalence.

^{19}F and ^{13}C chemical shifts (measured by solid-state NMR) and wave number of the C–F vibration band (measured by IR spectroscopy) of the selected fluorinated nanotubes (single, double, and multiwalled), spherical fluorinated fullerene, and planar graphite fluorides unambiguously confirm the importance of the curvature of the carbon lattice. Both IR wave numbers and chemical shifts exhibit a progressive change as a function of the curvature indicating that the covalence is weakened when the curvature is increased. During the fluorination, the C–F bond formation requires the change of the carbon hybridization from sp^2 to sp^3. The curvature prevents the formation of pure sp^3 hybridization for the carbon atom, since it requires an important local strain. The residual sp² hybridized orbitals imply a weakening of the overlapping of the hybridized lobes of carbon and the fluorine atomic orbitals (Scheme 12.1). In other words, this results in a weakening of the C–F bond covalence. The change of the ^{19}F chemical shifts as a function of the diameter d (Figure 12.1) can be qualitatively explained by pyramidalization of the carbonaceous precursor. Because θp is related to σ–π rehybridization and evolves as 1/d, the changes are more important on δ_{19F} and then on the covalence for the lower diameter and nearly reach the saturation for nanofibers with a large diameter.

Fluorinated C_{60} exhibits a similar tendency but does not strictly follow the relationship observed for fluorinated nanotubes and graphites. An additional effect is expected to explain such a difference: steric hindrance between fluorine atoms for the case when the fluorination rate is high could require different bond lengths and then affects the C–F bonding. These different C–F bonds result in different properties when used as electrode material in primary lithium batteries. The NMR lines were narrower in ^{19}F experiments than in ^{13}C experiments thanks to the efficiency of the magic angle spinning (MAS). So, only the ^{19}F chemicals shifts will be discussed according to the electrochemical properties in primary lithium batteries. The galvanostatic mode with a similar current density was used; it was equal to $10\,\text{mA}\,\text{g}^{-1}$ to evaluate the discharge potential. Such a low current density is chosen in order to favor the electrochemical processes and to limit overpotential. Figure 12.2 displays the galvanostatic curves of the various samples (SWCNTs, MWCNTs, and CNFs).

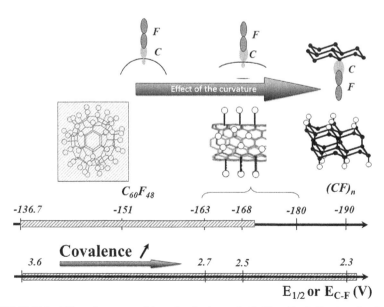

SCHEME 12.1 Effect of curvature of the carbon lattice on the C–F bonding in fluorinated carbons.

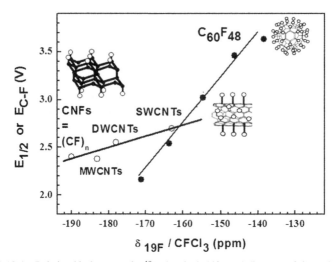

FIGURE 12.1 Relationship between the ^{19}F chemical shifts and diameter of the carbon lattice. NMR experiments were performed at room temperature using a Tecmag spectrometer (working frequency for ^{19}F of 282.2 MHz). For ^{19}F MAS spectra, a simple sequence was used with a single $\pi/2$ pulse duration of 5.5 μs. ^{19}F chemical shifts refer to CFCl$_3$.

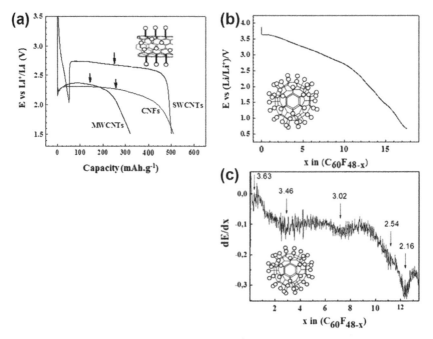

FIGURE 12.2 (a) Galvanostatic curves (10 mA g^{-1}) of the various fluorinated nanotubes (SWCNTs, MWCNTs, and CNFs). (b, c) Are related to C$_{60}$F$_{48}$. The arrows show the E$_{1/2}$ or E$_{C-F}$ potentials.

Whatever the number of walls, the fluorinated nanotubes exhibit a well-defined plateau (Figure 12.2(a)), related to reaction (1). The extracted potential at half of the total discharge capacity is easily measured on the plateau. On the contrary, for highly fluorinated fullerenes $C_{60}F_{48}$ (Figure 12.2(b)), no plateaus are observed and the derivative curves are necessary to underline the discharge potentials, which are called C–F decomposition potential (E_{C-F}). Five potentials are then obtained for $C_{60}F_{48}$. This result is in accordance with the previous studies on lithium batteries using a liquid or solid electrolyte [31–34]. The electrochemical curves during the first discharge exhibited a reduction wave located at higher potentials (from 1 to 1.5V) in comparison with nonfluorinated solid fullerene [35,36]. These preliminary studies revealed that the capacity is directly related to the initial rate of fluorination. As for graphite fluorides, the reduction mechanism of fluorinated fullerenes results from the progressive defluorination, associated with the formation of LiF, which is thermodynamically favored.

The initial potential drop for fluorinated SWCNTs may be related to overvoltage due to their insulating behavior. When the electroreduction starts, a few C–F bonds were converted in conductive carbon and LiF; the conduction of the electrode then slightly increased and the overvoltage decreased. Finally, the potential reaches the stabilized value that is taken into account for the comparison.

The ^{19}F NMR spectrum of fluorinated MWCNTs with $CF_{0.43}$ composition underlines a large amount of CF_2 groups [25], which could hinder the electrochemical processes. This high relative content of CF_2 may explain the low capacity for this sample.

Figure 12.3 underlines the nearly linear dependence observed for the discharge potential $E_{1/2}$ as a function of the ^{19}F NMR chemical shift, which is directly related to the covalence of the C–F bond and is then used to quantify the C–F bonding. The higher the curvature, the lower the diameter, and the higher the potential is. $E_{1/2}$ and δ_{19F} of covalent graphite fluorides are included in Figure 12.3 for comparison. The potential of fluorinated nanofibers and covalent graphite fluoride is close, meaning that the low curvature of this kind of nanotubes does not act anymore on the covalence. This is in accordance with the nondependence of the potential according to the fluorine content in fluorinated CNFs; nearly the same potential has been measured whatever the fluorinated wall, outer wall for a low F/C ratio or inner wall when the fluorination progresses toward the core [13]. The quasilinearity is verified in the overall series of fluorinated nanotubes and graphites.

As fluorinated fullerenes contain several kinds of C–F bonds, depending on the position in the pentagon or hexagon and also in their neighborhing, this case must be separately discussed. A linear dependence of the potentials with the chemical shift is also observed.

The discharge potentials in Li batteries are unambiguously correlated with the diameter of the outer tube, that is, the curvature of the carbon lattice. This has been

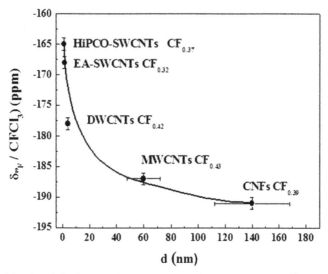

FIGURE 12.3 Correlation between the discharge potentials ($E_{1/2}$ or E_{C-F}) and ^{19}F NMR chemical shifts for a series of fluorinated C60 (●), nanocarbons and graphites (○).

done with precaution, that is, similar fluorine content, exclusive fluorination of the outer tubes, and same electrochemical operating conditions. The higher the curvature, the higher is the potential because the covalence is weakened. The potential of fluorinated nanotubes would be tuned in primary lithium batteries. The increase in the energy density, resulting from the increase of the potential because of curvature, is very promising. Nevertheless, another condition is required to achieve high-energy densities: the increase in the capacity by the adequate choice of the number of walls. A compromise must be found between curvature to obtain a high potential and number of walls to favor the capacity. Our study, with a low fluorine content ($0.3 > F/C < 0.4$) for the reasons discussed before, gives limited capacities ($500\,mAh\,g^{-1}$). A recent work has explored such a method with MWCNTs of different diameters [37]. The fluorine content was increased to compositions close to CF_1. Fluorinated MWCNTs with an average diameter >50 nm exhibit a high energy density, that is, $1923\,Wh\,kg^{-1}$. Moreover, the power density is also improved with a value of $7114.1\,W\,kg^{-1}$, with the batteries operating at a high discharge rate up to 5C. As demonstrated by our work on nanofibers, both the conductive networks of intimately contacting MWCNTs at the nanoscale level and the intrinsic fast rate capability of one-dimensional nanostructures explain those high power densities. As a general rule, the dispersion of fluorinated and nonfluorinated parts is of primary importance for electrochemical performance. The fluorination conditions must be optimized in order to reach a more efficient dispersion.

12.3 NEW FLUORINATION WAYS TO INCREASE THE POWER DENSITY OF A PRIMARY LITHIUM BATTERY

The low power density of carbon fluorides is due to the combination of kinetic limitations and the poor electrical conductivity of a strongly covalent $(CF)_n$ material. To increase this electrochemical performance, the enhancement of the intrinsic electrical conductivity of CF_x cathode materials thanks to the control of the fluorination can be an objective. Development of new fluorination methods is then necessary.

So, a process, called subfluorination, has been developed in the Blaise Pascal University and patented. The subfluorination process consists of combining a minute control of the fluorination conditions, that is, temperature T_T, fluorine gas flow rate (g min^{-1}), and time [38] in order to get fluorinated materials with nanodomains of unfluorinated carbon atoms in their core. These unfluorinated domains facilitate the electron flow within particles. The subfluorination process has been applied on various carbons, and the highest electrochemical densities have been first obtained with CNFs as a precursor of this new fluorination method.

High-purity (>90%) CNFs that are 8–20 mm in diameter with a 2- to 20-μm length have been supplied by MER Corporation; they have been obtained by the CVD method and then posttreated under an argon atmosphere at 1800 °C. Fluorinated CNFs (denoted as D-T$_F$, D for direct fluorination) were prepared with 200 mg of CNFs at temperatures T_F ranging between 380 °C and 480 °C in F_2 atmosphere for a reaction time of 16 h. The targeted composition "x" $0.2 < x < \sim 1.0$ has been achieved. The fluorination level "x" (i.e., F:C molar ratio) of fluorinated CNFs was determined by gravimetry (weight uptake) and by quantitative ^{19}F NMR measurements. Because of the nanostructuration of the fibers, which can be described as MWCNTs with a large diameter, and of its posttreatment graphitization, the fluorination temperature T_F is higher than the one needed for graphite in 1 atm. of pure F_2 gas. Such a nanostructured carbon lattice needs a progressive fluorination which processes from the outer tubes toward its core (supported by scanning electron microscopy (SEM) and transmission electron microscopy (TEM) images). In fact, four fluorination temperature zones can be pointed out for subfluorinated CNFs [39]. Only one zone is of interest for use in batteries; the one for fluorination temperature T_F ranged in between 435 and 450 °C because the fluorination level of CNFs is nearly constant x ~0.7–0.8. The fluorine atoms have progressed from the outer walls toward the core forming a $(C_2F)_n$ type of a carbon fluoride structure ("stage-2" compound with an FCCF slab-stacking sequence) of the fluorinated part (Figure 12.4).

Unfluorinated domains are still present as clearly evidenced by ^{13}C NMR analysis (Figure 12.5). This is underlined by the line at 120 ppm related to sp^2 C atoms. The other lines are assigned to covalent C–F bonds (at ~88 ppm), sp^2 C in a weak interaction with neighboring C–F bonds (140 ppm), and nonfluorinated

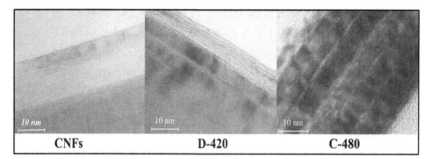

FIGURE 12.4 TEM images of unfluorinated nanofibers CNFs, fluorinated nanofibers by direct process D-420 and controlled process C-480. Micrographs were recorded on a FEI CM200 operating at 200 kV.

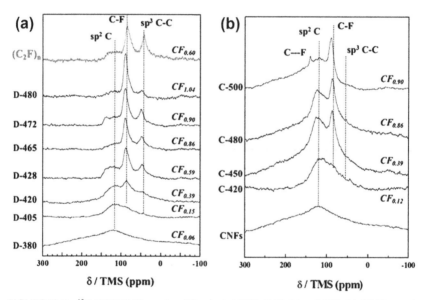

FIGURE 12.5 ^{13}C MAS NMR spectra of fluorinated CNFs D-T_F (a) and C-T_F (b). NMR experiments were carried out at room temperature on a Bruker AVANCE DSX 300 spectrometer operating at 75.47 MHz. For ^{13}C NMR, the external reference was TMS.

sp^3 C (42 ppm). The latter exist only in the $(C_2F)_n$ structural type and is then a good indicator of this phase, which is formed in a wide fluorination temperature range (420–480 °C).

For higher fluorination temperatures, the $(C_2F)_n$ phase is irreversibly converted into the $(CF)_n$-type one ("stage-1" compound with an FCF slab-stacking sequence), and this conversion leads unfortunately to some partial exfoliation. For $435 < T_F < 450$ °C, the fluorinated domains, which are the electrochemically active parts of the cathode material, are intermixed at the nanoscale level with unfluorinated domains, which ensure the electron flow in the electrochemical processes. This last point plays a key role in enhancing the intrinsic electrical conductivity of the D-T_F materials and is the origin of high energy density, as discussed in the following part.

Whatever the samples, at a low discharge current density, the average discharge potential is set at a very narrow range (2.45–2.6 V) [35,13]. This quasistability is in agreement with the covalent C–F bonding, which does not change upon fluorination. As discussed before, curvature effect is negligible with a high diameter. Although the discharge voltage and the fluorine content of CNFFs are close to those reported in a $(C_2F)_n$-type graphite fluoride, the achieved discharge capacity of the D-T_F materials synthesized between 435 and 450 °C is about 30% higher than those of $(C_2F)_n$. Moreover, an impressive 8057 W kg^{-1} power density associated with a high 1749 Wh kg^{-1} energy density was achieved in D-T_F with F/C equal to 0.76 (obtained both by quantitative NMR and gravimetry); its galvanostatic discharge is presented in Figure 12.6. Here, the power density is six times higher than the one of conventional graphite fluorides. Then, for F/C > 0.80, the maximum power density decreases but is still higher than the one of conventional graphite fluorides. Such enhanced performances are due to the presence of nonfluorinated carbons in the D-T_F core with a percentage > 10%, which favors a fast electron flow and accordingly enhances electrode reaction kinetics [13]. Actually, the effect of enhanced conductivity is more pronounced at higher discharge rates at which the power density difference of the new materials with conventional CF_1 becomes obvious. In fact, whatever the D-T_F cathode material, including

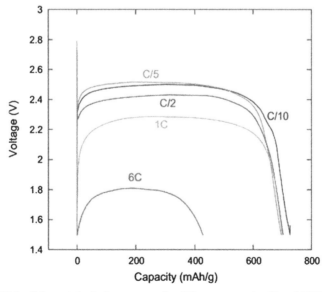

FIGURE 12.6 Galvanostatic discharge curves for different current densities of CNFs subfluorinated at 450 °C (electrolyte: PC:DME LiBF$_4$ 1M).

a lightly fluorinated $CF_{0.21}$, both the achieved C rate and power density are higher than those for CF_1. Going further on this strategy of subfluorination, Yue et al. [40] have found that fluorinated CNTs with an F/C ratio of 0.75 exhibited the best energy and power densities not yet reported, that is, 1147 Wh kg^{-1} and 8998 W kg^{-1}, respectively, at a current density of 4 A g^{-1}. For this case also, the explanation for the good performance is related to fluorine atom dispersion: fluorine atoms were located in the outer part of the CNTs initially where graphene layers were coaxial within a distance of 0.60 nm. In contrast, the inner part of the CNTs remains unchanged.

With the final aim to increase the power density, another solution for the enhancement of the intrinsic electrical conductivity of CF_x cathode materials consists in favor of a better distribution of the nonfluorinated part in between the fluorinated one. This can be achieved using pure molecular fluorine by a static process (a defined amount of F_2 in a closed reactor for a long duration in order to achieve a homogenous distribution of F atoms) or by using more diffusive and reactive atomic fluorine $F^{\bullet}$ species formed during the decomposition of a solid fluorinating agent such as TbF_4. This fluorination process was called "controlled fluorination." For such a purpose, TbF_4 powder was obtained from TbF_3 (Aldrich, 99.9%) fluorination in pure F_2 gas at 500 °C. Its purity (i.e., the absence of residual TbF_3) must be systematically checked by X-ray diffraction (XRD). The thermogravimetric analysis of freshly prepared TbF_4 indicated that exactly 1 mol of $F^{\bullet}$ was released per mole of TbF_4 between 300 and 500 °C. For the CNF fluorination by TbF_4, a closed nickel reactor was used in order to preserve the defined fluorine amount released by the thermal decomposition of TbF_4. A two-temperature oven was used: the part containing TbF_4 was heated at 450 °C whatever the experiment, whereas CNFs were heated at temperatures T_F that ranged between 420 and 500 °C (the samples are denoted as C-T_F; C is for controlled). A reaction time of 16 h was used. Prior to the heating, a primary vacuum ($\sim 10^{-2}$ atm) was applied into the reactor. The reactions involved during the fluorination are the following:

$$TbF_4 \xrightarrow{\Delta} TbF_3 + F^{\bullet} \text{ or } \frac{1}{2}F_2$$

$$C + xF^{\bullet} \text{ or } \frac{x}{2}F_2 \rightarrow CF_x$$

The total conversion of TbF_4 to TbF_3 was systematically checked by both weight loss and XRD analysis.

TbF_4 decomposition at temperatures around 450 °C releases exactly 1 mol of pure atomic fluorine per 1 mol of initial TbF_4, which recombines with carbon and more particularly CNFs. As the reactive species are different for subfluorination and fluorination, the fluorination mechanisms differ. In the reaction involving TbF_4, because the fluorine amount is related to the equilibria $TbF_4(s) \Leftrightarrow TbF_3(s) + F^{\bullet}$ and $2F^{\bullet}(g) \Leftrightarrow F_2(g)$, the fluorination is more progressive and homogeneous leading to the formation of $(CF)_n$-type phase whatever the fluorination conditions, that is, the highly fluorinated phase (Figure 12.5(b)). No $(C_2F)_n$-type phase is formed as in the case of subfluorination or more generally for a direct process with graphitized starting materials. During this process, with increasing T_F, a progressive densification of the $(CF)_n$-type compound takes place, and very few CF_2 and CF_3 groups are formed as shown in Figure 12.5(b).

Room temperature ^{19}F MAS NMR (14 kHz) is well adapted to underline the relative amounts of CF_2 and CF_3 groups, which exhibit chemical shifts of −60/-90 and approximately −120 ppm versus. $CFCl_3$, respectively. The spectra of two representative samples with a similar fluorine content (F/C~0.6) obtained either from direct process or fluorination by TbF_4 decomposition were compared. The lines related to CF_2 and CF_3 groups are less intense for the sample obtained with TbF_4. Such an observation can be made whatever the fluorine content. The CF_3 groups, can be localized on the fluorocarbon sheet edges and probably possess a spinning motion around the C–C bonds, explaining the narrowness of the resonance line. Moreover, the ^{19}F MAS NMR spectra of CNFs fluorinated by TbF_4 decomposition and direct processes exhibit a similar isotropic line located at −190 ppm/$CFCl_3$. The latter is attributed to fluorine atoms covalently bonded to carbon atoms (C–F) [27,41–43]. The similarities of the main isotropic line reveal the same C–F bonding whatever the fluorination strategy and the temperature. Nevertheless, another difference between the fluorinated carbon nanomaterials obtained using the two routes exists: first, the presence of a line at −178 ppm for the intermediate fluorination temperature by the direct process and another at −120 ppm also for the direct fluorination. The line at −178 ppm appears only for fluorinated carbons that contain fluorinated parts with a $(C_2F)_n$ structure and can be attributed to fluorine atoms bonded with the nonfluorinated sp^3 carbon atom itself bonded with another sp^3 carbon atom in accordance with the $(C_2F)_n$ structure. Another defect, dangling bonds, has been found using electron paramagnetic resonance to be present in less amounts after fluorination by TbF_4 decomposition whatever the fluorine content. Finally, according to TEM images, the fluorinated parts are homogeneously dispersed in to the carbon matrix using TbF_4 as the fluorinating agent contrary to the sample obtained using F_2, where only the outer tubes are fluorinated for a low and medium x ratio [44–47]. The last difference between direct (F_2) and controlled fluorination ($F^{\bullet}$), the fluorinated parts of C-T_F series were homogeneously dispersed in the whole volume of the fiber (Figure 12.5 for C-480) contrary to the D-T_F series (exemplified by D-420 in Figure 12.5).

In order to compare the electrochemical performances of fluorinated CNFs obtained by TbF_4 and direct fluorination, materials with a similar fluorine level have been used as the cathode in primary lithium batteries. We will focus on an F/C of about 0.7–0.8 because of its good performance. Different current densities have been applied to the two materials and Table 12.2 summarizes the

TABLE 12.2 Comparative Performances of D-T$_F$ and C-T$_F$ Series

	Current	F/C	Power density (W kg^{-1})
D-T$_F$	C	0.68	1296
	C/100	0.68	22.7
C-480	C	0.70	1360
	C/100	0.70	23.2

electrochemical performances obtained for the extreme current densities, that is, 10 mA g^{-1} and 720 mA g^{-1} corresponding to C/100 and 1C, respectively.

Whatever the electrochemical conditions, both the potential and the capacity are higher for CF$_x$ prepared by the TbF$_4$ method by comparison with the direct process, and it results in a faradic yield close to 100% for materials prepared by fluorination by TbF$_4$ decomposition. The low increase of potential cannot be explained by the variation of the covalence of the C–F bonding as we already mentioned that solid-state NMR has revealed exactly the same C–F bonding whatever the fluorination strategy. It can be explained by a lowering of internal resistive effect such as lithium or fluorine barrier diffusion due to CF$_2$, CF$_3$ type groups located at the edges of the fluorinated compounds or dangling bonds. The higher the amount of such groups, the higher the internal resistance is. The higher capacity of the compound prepared by fluorination by TbF$_4$ decomposition can also be explained by an indirect contribution of CF$_2$ or CF$_3$ type groups. Indeed, they contribute to the fluorine level value that is used to the prediction of the theoretical capacity. But as they cannot be reduced, the true active mass is actually overestimated like the capacities.

We can conclude from this section that fluorination conditions can be chosen in order to enhance one particular property: (1) catalytic fluorination combined with a postfluorination under pure fluorine gas for higher energy density, subfluorination process is a valuable route for very high power density, and fluorination using fluoride decomposition (fluorinating agents) for high faradic yield. This latter property is probably the most difficult to improve because it depends on many parameters, structure (presence of defects), the specific surface area (especially for the case of nanomaterials), the nature of the electrolyte (salt, solvents, additives). The following section will give a contribution to this important challenge. We will first focus on the discharge mechanism. This latter property is probably the most difficult to improve because it depends on many parameters, structure (presence of defects), the specific surface area (especially for the case of nanomaterials), the nature of the electrolyte (salt, solvents, additives). The following section will give a contribution to this important challenge. We will first focus on the discharge mechanism.

12.4 INCREASING THE FARADIC YIELD

Although fluorinated carbons are used since a long time as electrode material in primary Li batteries, the mechanism is still discussed. This discussion does not concern the nature of the products after the electrochemical defluorination, namely, amorphous carbon and LiF, but the process to form those products is still under discussion. A GIC intermediate with solvated lithium is formed on the graphite sheet edges and acts as a diffusion layer [48]. Moreover, the concentration of lithium ions decreases rapidly with the distance from the electrode surface meaning that lithium ions stay out of the fluorocarbon matrix. On the sheet edges, the GIC subsequently decomposes into the final discharge products, carbon and LiF. The discharge is accompanied by significant electrode swelling due to the formation of volumetric LiF crystals as reported by Abraham et al. [49]. Based on the discharge and open circuit voltage (OCV) recovery characteristics, Zhang et al. [50] proposed a discharge through a "shrinking core" model consisting of a CF_x core and a product shell. The product shell is composed of a GIC intermediate, carbon, and lithium fluoride. The product shell grows with the discharge process, and its composition varies with the decomposition of GIC intermediate. The location of LiF particles formed either on the CF_x grain surface (related to the location of the GIC intermediate on sheet edges) or in the whole volume (as expected by a core/shell process) was never observed, to the best of our knowledge. In order to go further on those mechanisms, partial galvanostatic discharges were acquired with graphite fluoride of $CF_{1.05}$ composition. The latter denoted as HTGF was synthesized in F_2 gas at 600°C for 3h. The average electrochemical data for the full discharge OCV, experimental capacity, potential at half discharge, energy density, and faradic yield (the ratio $100 \times Q_{exp}/Q_{theoretical}$) were equal to 3.17V, 800mAh g^{-1}, 2.5V, 2000Wh kg^{-1}, and 91%, respectively (Figure 12.7(a)). The low faradic yield is to be noted.

The electrodes were discharged at different depths of discharge (DoDs; 0, 10, 25, 50, 75% and 100%) in LiPF$_6$ 1M EC/PC/3DMC electrolyte (Figure 12.7(a)). A DoD of zero means that the electrode composite HTGF/polyvinylidene difluoride (PVDF)/Carbon black (80/10/10 w%) deposited on the current collector was put into the electrolyte during the relaxation time, which is usually imposed before the current is applied, that is, 5h.

Figure 12.7b shows the ^{19}F MAS NMR spectra (spinning rate of 34kHz) of HTGF discharged at different DoDs (0, 25, 50, 75, and 100% DoD i.e., 0, 80, 200, 400, 600, and 800mAh g^{-1}). The electrochemical defluorination is underlined by the decrease of the ^{19}F NMR line at −190ppm assigned to covalent C–F bonds. Its relative intensity decreases with increasing DoDs. Moreover, its line width significantly decreases for high DoDs because of the weakening of the ^{19}F–^{19}F homonuclear dipolar coupling, which occurs with the dilution of the C–F bonds into the carbon matrix. The intensities were normalized using the line of PVDF (used as a binder and containing CF_2 groups), which is electrochemically inactive. The amount of CF_2 groups underlined with a chemical

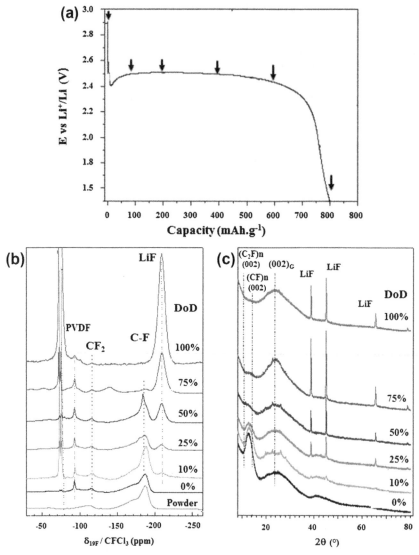

FIGURE 12.7 (a) Galvanostatic discharge curve of the HTGF sample showing the studied DoD, (b) ^{19}F NMR MAS spectra (34 kHz) of discharged HTGF in EC/PC/3DMC-LiPF$_6$ 1M at different DoDs, (c) XRD diagrams of discharged S-420 in EC/PC/3DMC LiPF$_6$ 1M at different DoDs. X-ray diffractograms were recorded using a Philips diffractometer with a Cu(K$_\alpha$) radiation ($\lambda = 1.5406$ Å). ^{19}F NMR experiments were performed at room temperature using a Bruker Avance spectrometer with working frequencies of 282.2 MHz. A MAS probe (Bruker) operating with 2.5-mm rotors was used. For the MAS spectra, a simple sequence was performed with a single $\pi/2$ pulse length of 4.0 µs. ^{19}F chemical shifts were referenced with respect to CFCl$_3$.

shift of −114 ppm/CFCl$_3$ on the ^{19}F NMR spectra does not change upon discharge. These groups are electrochemically inactive. LiF particles are progressively formed during the discharge by the association of F$^−$ ions (coming from C–F reduction) and solvated Li$^+$ ions from the electrolyte. As a matter of fact, the line at −205 ppm on the ^{19}F NMR spectra is assigned to LiF; its integrated surface increases with the DoD. A similar evolution was reported on fluorinated fiber, petroleum coke and graphite [51] and CF$_x$/Ag$_2$V$_4$O$_{11}$ hybrid [16]. This is in accordance with XRD data (LiF peaks at 39, 45, and 65° in 2θ values in Figure 12.7(c)). The formation of crystalline LiF particles is underlined by the sharpness of the XRD peaks. The diffraction peaks of the fluorinated phase (CF$_n$ and (C$_2$F)$_n$) disappear progressively during the discharge mechanism (Figure 12.7(c)). During the electrochemical defluorination, a carbonaceous phase seems to be formed, as evidenced by the appearance of a broad peak centered at 26° assigned to the (002) phase of disordered carbons. The formation of LiF and destructuration of the fluorocarbon matrix were underlined by both XRD and solid-state NMR, as expected. To answer the question of the location of LiF particles, SEM images were recorded (Figure 12.8). The raw material consists of stacked layers. On the contrary, the stacking is significantly decreased in the fully discharged sample. Thin multilayers appear as isolated and covered with LiF particles (Figure 12.8(b)). The latter are located between the multilayers (Figure 12.8(c)). According to XRD and ^{19}F NMR data, the destacking may be explained as follows: the electrochemical defluorination forms by the breaking of C–F bonds F$^−$ ions that diffuse within the interlayer distance. F$^−$ ions then combine with lithium ions from the electrolyte outside the carbon lattice. Such a combination occurs on the sheet edges, which are the release gate of F$^−$. NMR data (both ^{7}Li and ^{19}F spin–lattice relaxation time T$_1$) suggest that LiF formed upon discharge is excluded from the structure of the material [48]. Since LiF is an insulator, the kinetics of its exclusion from the carbon lattice acts on the electrochemical performances. Because the width of the sheet is much higher than its thickness, F$^−$ cumulates on the sheet edges and consequently LiF particles grow at this location, resulting in a progressive exfoliation of the carbon lattice. The splitting of the layers opened larger release gates for F$^−$ ions, and both the growth of LiF and exfoliation go further. During this partial exfoliation, the electrical contact of some partially defluorinated regions may be lost. In other words, those parts cannot be further defluorinated, and the fluorine atoms are trapped. The theoretical capacity is then not reached, and the faradic yield decreases. This explains the recorded value of 91%.

One solution to avoid the partial exfoliation during the discharge consists of dispersing LiF particles, not only on the sheet edges but also on the whole surface of the carbon lattice. The use of fluorinated carbons with nanometric size is one possible way. Such a concept has been tested for validation with fluorinated nanofibers. The SEM images were recorded for different DoDs with two different electrolytes LiClO$_4$ 1M PC and LiPF$_6$ 1M EC/PC/3DMC.

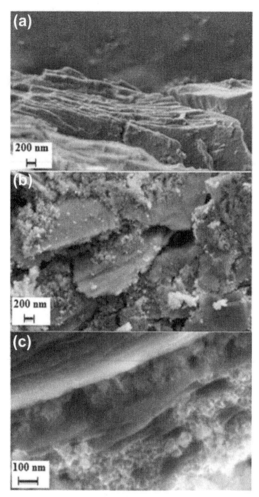

FIGURE 12.8 SEM images of raw HTGF (a) and fully discharged in the EC/PC/3DMC-LiPF$_6$ 1M electrolyte (b, c). Micrographs were recorded using a Cambridge Scan 360 SEM operating at 1 kV.

The starting material was nanofibers fluorinated at 420 °C in static conditions (a closed reactor filled with F_2 at 420 °C), as described before. The F/C ratio was 0.74. Figure 12.9(a) shows its galvanostatic curve in two electrolytes, and the DoDs are marked by arrows. Only one plateau of the potential is observed at 2.71 V with LiPF$_6$ 1M EC/PC/3DMC contrary to LiClO$_4$ 1M PC where two plateaus are observed at 2.49 and 2.27 V. Experimental capacity, faradic yield, and energy density values are higher when the EC/PC/3DMC mixture was used underling an enhancement of the performances with LiPF$_6$ 1M EC/PC/3DMC (and LiClO$_4$ 1M EC/PC/3DMC, not shown here) electrolyte in comparison with LiClO$_4$ 1M PC. This suggests that the nature of the lithium salt used is not critical, but it is rather the effect of the solvent that carries the electrochemical performance toward higher values.

First, it is to be noted that the discharge mechanism is similar to that for graphite fluoride HTGF as seen by NMR and XRD regardless of the electrolyte [52–54]. Only SEM images on the products discharged at different DoDs show some differences according to the electrolyte used (Figure 12.9). As for graphite fluoride (HTGF), SEM allows one to observe LiF particles and some differences of both LiF texture and nanostructuration. The latter were formed as a contiguous layer covering the whole surface of the fibers in PC-LiClO$_4$ 1M electrolyte (Figure 12.9(g) and (h)), contrary to well-defined cubic LiF particles in addition to the covering layer in EC/PC/3DMC-LiPF$_6$ 1M electrolyte seen at different DoDs (Figure 12.9(c)–(f)).

Both NMR and XRD underlined the progressive defluorination of the fluorocarbon matrix with a continuous formation of crystalline LiF particles. These particles seem to be formed at the edges and on the surface of the fibers as revealed by SEM images (Figure 12.9(d)). The accommodation of LiF particles outside the matrix leads to the maintaining of the tubular shape of the nanofibers whatever the DoD (even for 100% DoD) (Figure 12.9(f)). Cubic particles of LiF are clearly observed at 25% DoD (see the arrows in Figure 12.9(c)). Their dimensions and amounts increase gradually as the discharge progresses. Up to 75% DoD, defluorinated fibers are still observed (Figure 12.9(e)). The LiF is formed outside the fibers at any site of removal for F$^-$ ions. For instance, at 50% DoD (380 mAh g^{-1}) (Figure 12.9(d)), LiF particles are located along fractures on the fiber surface, which may be formed during the fluorination as seen at 0% DoD where fibers are in simple contact with the electrolyte. In order to compare the two electrolytes, an SEM image of CNF with 43% DoD in LiClO$_4$ 1M PC electrolyte is shown. Although the DoD is low, the fibers are uniformly covered by a thick layer of LiF (Figure 12.9(g)). This thick layer may induce an internal resistive phenomenon for F$^-$ and Li$^+$ ionic diffusion, resulting in an additional overpotential as seen on the galvanostatic curve (two plateaus of potential). At high DoDs (80% DoD), fibers are covered by a continuous and dense layer of LiF in the case of LiClO$_4$ 1M PC electrolyte (Figure 12.9(h)), whereas for the EC/PC/3DMC-LiPF$_6$ 1M electrolyte, cubic LiF nanoparticles are intermixed with a less dense covering layer (Figure 12.9(f)).

It is to be noted that the EC/PC/3DMC-LiPF$_6$ 1M electrolyte spontaneously reacts with the surface of the electrode (0% DoD) as revealed by NMR and the presence of the line at −205 ppm. The initial LiF particles may act as germs for the further growth of particles. This may explain the presence of well-defined cubic particles only for this electrolyte. Moreover, at a given amount of LiF, that is, according to the DoD, the thickness of the LiF layer is lower with LiPF$_6$ 1M EC/PC/3DMC electrolyte than for LiClO$_4$ 1M PC for two reasons: (1) compact cubic particles present only for LiPF$_6$ 1M EC/PC/3DMC are denser than a continuous layer and (2) LiF is more dissolved in the latter electrolyte.

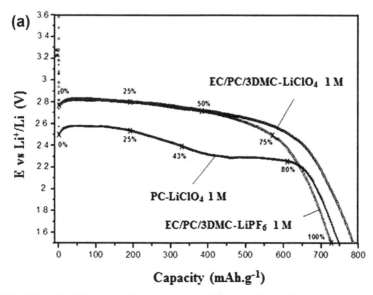

FIGURE 12.9 (a) Galvanostatic discharge curves of fluorinated nanofibers showing the DoD studied in different electrolytes (b–h) SEM images as a function of the DoD. Images were collected using a Cambridge Scan 360 SEM operating at 1 kV. (See Ref. 51 for details of acquisition.)

Solubility of LiF in PC is lower than in the mixture EC/PC/3DMC; solubility values are 0.13 and 9.08 g L^{-1}, respectively [55,56]. In addition to deriving a benefit for the use of nanostructured fluorinated carbon thanks to the better dispersion of LiF particles (without exfoliation), a potential improvement was achieved with the LiPF$_6$-based electrolyte. LiF particles grow massively on the fiber surface without any dissolution in PC-LiClO$_4$ 1M, contrary to a nanometric distribution of these particles in EC/PC/3DMC thanks to the good solubility of LiF in this mixture. As a result, the useful potential delivered by the battery is higher with the EC/PC/3DMC-based electrolyte compared with the PC electrolyte because of the overvoltage caused by LiF formation is smaller therewith.

Another solution to avoid the damage caused by the LiF growth is the use of porous fluorinated carbons. Up to now, the benefit of the porosity as container for LiF is not reported in the literature. Nevertheless, significant improvements with discharge potentials higher than covalent graphite fluorides and with both energy and power densities have been found in a recent work on porous CF$_x$ [57]. Soft-templated mesoporous carbons and activated mesoporous carbons were fluorinated using elemental fluorine between room temperature and 235 °C. Mesoporous carbon fluorides with narrow distributions of mesopores from 6 to 11 nm in width and specific surface areas as large as 852 m^2 g^{-1} were then prepared. Such a sample, for example, mesoporous activated carbon, exhibits three main advantages: high concentrations of sp^3 carbon atoms linked

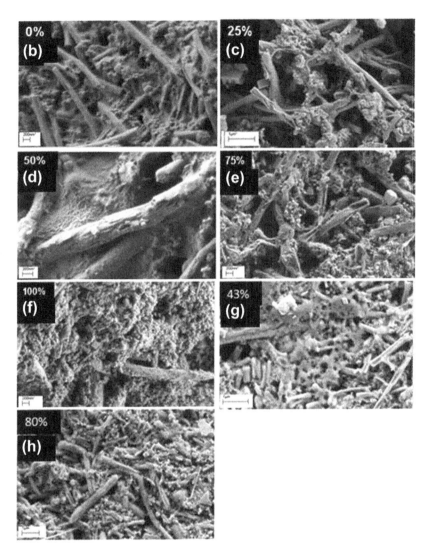

FIGURE 12.9 Continued

to fluorine (F site for Li storage) and C–C bonding for electrical conduction and high surface areas to maximize the number of F sites (related to capacity). When compared with $(C_2F)_n$ conventional graphite fluoride, the enhancement in power and energy densities with fluorinated mesoporous activated carbon (of $CF_{0.55}$ composition) are 28.6% and 29.0% at a C/20 discharge rate and increased to 36.4% and 50.2% at a 1C discharge rate and 52.8% and 63.9% at a 5C discharge rate, respectively.

Taking into account these promising results, the pore size distribution may be adjusted in order to serve as an LiF container without damaging the carbon lattice during its growth. Nevertheless, increase in the specific surface area results also in possible side reactions with the electrolyte. The effect of both solvent and salt must be carefully investigated. Our study on fluorinated nanofibers underlined that this effect could be very efficient on electrochemical properties. The strategy that takes advantage of the surface involves also research to prepare new fluorinated carbons starting from graphene and porous carbons or postfluorination process such as ball milling. Some examples underline the potential improvements. The work of Fulvio et al. on fluorinated mesocarbon is already discussed in the text [55]. Fluorinated graphene appears to be very promising. Fluorinated graphene with an F/C equal to 0.47, 0.66, and 0.89 exhibited a faradic yield of 75–81% at moderate rates when used as a cathode material for primary lithium batteries [58]. Among the samples, the one with a $CF_{0.47}$ composition maintained a capacity of 356 mAh g^{-1} at a 5C rate, higher than that of the traditional fluorinated graphite. A significant improvement was observed using ball milling in the electrochemical performance of graphite fluoride in both gravimetric energy density (800 Wh kg^{-1}) and power density (9860 W kg^{-1}) [59].

12.5 NEXT-GENERATION CARBON FLUORIDES FOR PRIMARY LITHIUM BATTERIES: SOME KEY POINTS

Fluorinated carbons are fascinating materials because of, among other advantages, the versatility of the C–F bonding. The covalence can be weakened by the curvature of the carbon lattice or by the presence of nonfluorinated carbon atoms (hyperconjugation). The prefect control of the fluorination conditions with either atomic or molecular fluorine according to the gas flux (dynamic or static), the temperature, and the duration allows one to exclude some carbon atoms from the fluorination (subfluorination) and the dispersion of the fluorinated and nonfluorinated parts to be optimized in order to reach a high power or energy densities. These advantages may be further enhanced using nanostructured carbons by taking advantage of either the high surface to homogeneously disperse LiF (avoiding partial exfoliation) or the nanostructure, that is, curvature of the nanotubes to increase the discharge potential or the packing of tubes in DWCNTs and MWCNTs. In the latter case, adequate dispersion of conductive nonfluorinated parts into the fluorinated regions, electrochemically active but insulating, ensures the efficiency of the electron flux at high current densities, resulting in very high power densities. An increase of the interface with the electrolyte because of the nanometric size may result in side reactions and a careful investigation according to the nature of the salt, solvent, or additive is necessary to reach the optimum performance. Those investigations must be carried out together with works on the fluorination of porous carbons and graphene (e.g., from exfoliation of graphite fluoride) in order to optimize the next generation of fluorinated carbons as electrode materials in primary lithium batteries.

REFERENCES

[1] [a] N. Watanabe, T. Nakajima, H. Touhara (Eds.), Graphite Fluorides, Elsevier, Amsterdam, 1988.
[b] T. Nakajima, N. Watanabe (Eds.), Graphite Fluorides and Carbon-fluorine Compounds, CRC Press, Boca Raton, 1991.
[2] A. Hamwi, K. Guérin, M. Dubois (Eds.), Fluorinated Materials for Energy Conversion, Elsevier, Oxford, U.K, 2005.
[3] N. Watanabe, R. Hagiwara, T. Nakajima, J. Electrochem. Soc. 131 (1984) 1980–1984.
[4] A. Morita, N. Eda, T. Ijima, H. Ogawa, in: J. Thompson (Ed.), Power Sources, 9, Academic Press, New York, 1983, p. 435.
[5] N. Watanabe, M. Fukuda, US Patent 3, 536, 532, 1970 and 3, 700, 502, 1972.
[6] M. Fukuda, T. Iijima, in: D.H. Collins (Ed.), Power Sources, vol. 5, Academic Press, New York, 1975, p. 713.
[7] I. Palchan, D. Davidov, H. Selig, J. Chem. Soc. Chem. Commun. 12 (1983) 657.
[8] T. Nakajima, N. Watanabe, I. Kameda, M. Endo, Carbon 24 (1986) 343–351.
[9] A. Hamwi, M. Daoud, J.C. Cousseins, Synth. Met. 26 (1988) 89–98.
[10] K. Guérin, J.P. Pinheiro, M. Dubois, Z. Fawal, F. Masin, R. Yazami, A. Hamwi, Chem. Mater. 16 (2004) 1786–1792.
[11] M. Dubois, K. Guérin, J.P. Pinheiro, F. Masin, Z. Fawal, A. Hamwi, Carbon 42 (2004) 1931–1940.
[12] Y.G. Guo, J.S. Hu, L.J. Wan, Adv. Mater. 20 (2008) 2878–2887.
[13] R. Yazami, A. Hamwi, K. Guérin, Y. Ozawa, M. Dubois, J. Giraudet, et al., Electrochem. Comm. 9 (2007) 1850–1855.
[14] T. Nakajima, J. Fluorine Chem. 149 (2013) 104–111.
[15] H. Groult, C.M. Julien, A. Bahloul, S. Leclerc, E. Briot, A. Mauger, Electrochem. Comm. 13 (2011) 1074–1076.
[16] P.J. Sideris, R. Yew, I. Nieves, K. Chen, G. Jain, C.L. Schmidt, S.G. Greenbaum, J. Power Sources 254 (2014) 293–297.
[17] T. Nakajima, N. Watanabe, Graphite Fluorides and Carbon-fluorite Compounds, CRC Press, Boca Raton, 1991.
[18] Y. Sato, K. Itoh, R. Hagiwara, T. Fukunaga, Y. Ito, Carbon 42 (2004) 3243.
[19] [a] O. Ruff, O. Breitschneider, E. Ebert, Z. Anorg, Allg. Chem. 217 (1934) 1.
[b] W. Rudorff, G. Rudorff, Z. Anorg. Allg. Chem. 293 (1947) 281.
[20] A. Hamwi, M. Daoud, J.C. Cousseins, Synth. Met. 26 (1988) 89.
[21] C. Delabarre, M. Dubois, J. Giraudet, K. Guérin, A. Hamwi, Carbon 44 (2006) 2543–2548.
[22] C. Delabarre, K. Guérin, M. Dubois, J. Giraudet, Z. Fawal, A. Hamwi, J. Fluorine Chem. 126 (7) (2005) 1078–1087.
[23] J. Giraudet, C. Delabarre, K. Guérin, M. Dubois, F. Masin, A. Hamwi, J. Power Sources 158 (2006) 1365–1372.
[24] C. Delabarre, J. Giraudet, K. Guérin, M. Dubois, A. Hamwi, J. Phys. Chem. Solids 67 (2006) 1157–1161.
[25] S. Niyogi, M.A. Hamon, H. Hu, B. Zhao, P. Bhowmik, R. Sen, M.E. Itkis, R.C. Haddon, Acc. Chem. Res. 35 (2002) 1105–1113.
[26] S. Banerjee, T. Hemraj-Benny, S.S. Wong, Adv. Mater. 17 (2005) 17–29.
[27] J. Kürti, V. Zolyomi, M. Kertesz, G. Sun, New J. Phys. 5 (2003) 125.1–125.21.
[28] W. Zhang, M. Dubois, K. Guérin, P. Bonnet, H. Kharbache, F. Masin, A.P. Kharitonov, A. Hamwi, Phys. Chem. Chem. Phys. 12 (2010) 1388.

[29] A.V. Krestinin, A.P. Kharitonov, Y.M. Shul'ga, O.M. Zhigalina, E.I. Knerel'man, M. Dubois, M.M. Brzhezinskaya, A.S. Vinogradov, A.B. Preobrazhenskii, G.I. Zvereva, M.B. Kislov, B.M. Martynenko, I.I. Korobov, G.I. Davydova, V.G. Zhigalina, N.A. Kiselev, Nanotech.Russ. 4 (2009) 60–78.
[30] A.M. Panich, Synth. Met. 100 (1999) 169–185.
[31] A. Hamwi, C. Latouche, V. Marchand, J. Dupuis, R. Benoit, J. Phys. Chem. Solids 57 (1996) 991.
[32] N. Liu, H. Touhara, F. Okino, S. Kawasaki, Y. Nakacho, J. Electrochem. Soc. 143 (1996) 2267.
[33] Y. Matsuo, T. Nakajima, Electrochim. Acta 41 (1996) 15.
[34] F. Okino, S. Yajima, S. Suganuma, R. Mitsumoto, K. Seki, H. Touhara, Synth. Met. 70 (1995) 1447.
[35] R.G. Compton, R.A. Spackman, R.G. Wellington, L.H. Malclm, J. Turner, J. Electroanal. Chem. 327 (1992) 337.
[36] Y. Chabre, D. Djurado, M. Armand, W.R. Romanov, N. Coustel, J.P. Mc Cauley, J.E. Fischer, A.B. Smith III, J. Am. Chem. Soc. 114 (1992) 764.
[37] Y. Li, Y. Feng, W. Feng, Electrochim. Acta 107 (2013) 343–349.
[38] R. Yazami, A. Hamwi, WO2007098478 (2007), WO2007126436(2007) and EP1999812 (2008), EP1976792 (2008).
[39] F. Chamssedine, M. Dubois, K. Guérin, J. Giraudet, F. Masin, D.A. Ivanov, L. Vidal, R. Yazami, A. Hamwi, Chem. Mater. 19 (2007) 161–172.
[40] H. Yue, W. Zhang, H. Liu, et al., Nanotechnology 24 (2013) 424003.
[41] J. Giraudet, M. Dubois, K. Guérin, J.P. Pinheiro, A. Hamwi, W.E.E. Stone, P. Pirotte, F. Masin, J. Solid State Chem. 178 (2005) 1262–1268.
[42] J. Giraudet, M. Dubois, K. Guérin, C. Delabarre, A. Hamwi, F. Masin, J. Phys. Chem. B 111 (2007) 14143–14151.
[43] H. Touhara, F. Okino, Carbon 38 (2000) 241–267.
[44] W. Zhang, L. Moch, M. Dubois, K. Guerin, A. Hamwi, J. Nanosci. Nanotech. 9 (2009) 4496–4501.
[45] W. Zhang, K. Guerin, M. Dubois, Z. Fawal, D. Ivanov, L. Vidal, A. Hamwi, Carbon 46 (2008) 1010–1016.
[46] W. Zhang, K. Guérin, M. Dubois, A. Houdayer, F. Masin, A. Hamwi, Carbon 46 (2008) 1017–1024.
[47] W. Zhang, M. Dubois, K. Guérin, P. Bonnet, H. Kharbache, F. Masin, P. Thomas, J.-L. Mansot, A. Hamwi, Eur. Phys. J. B 75 (2010) 133–139.
[48] M.S. Whittingham, J. Electrochem. Soc. 122 (1975) 526.
[49] K.M. Abraham, D.M. Pasquariello, Primary and secondary lithium batteries, in: K.M. Abraham, M. Salomon (Eds.), The Electrochemical Society Proceedings, 1991, pp. PV91–PV93.
[50] S.S. Zhang, D. Foster, J. Wolfenstine, J. Read, J. Power Sources 187 (2009) 233–237.
[51] J.H.S.R. DeSilva, R. Vazquez, P.E. Stallworth, T.B. Reddy, J.M. Lehnes, R. Guo, H. Gan, B.C. Muffoletto, S.G. Greenbaum, J. Power Sources 196 (2011) 5659–5666.
[52] Y. Ahmad, K. Guérin, M. Dubois, W. Zhang, A. Hamwi, Electrochim. Acta 114 (2013) 142–151.
[53] S.S. Zhang, D. Foster, J. Wolfenstine, J. Read, J. Power Sources 187 (2009) 233–237.
[54] K. Guérin, M. Dubois, A. Hamwi, J. Phys. Chem. Solids 67 (2006) 1173–1177.
[55] J. Jones, M. Anouti, M. Caillon-Caravanier, P. Willmann, D. Lemordant, Fluid Phase Equilib. 285 (2009) 62.

[56] J. Jones, M. Anouti, M. Caillon-Caravanier, P. Willmann, D. Lemordant, J. Mol. Liq. 153 (2010) 146.
[57] P.F. Fulvio, S.S. Brown, J. Adcock, R.T. Mayes, B. Guo, X.-G. Sun, S.M. Mahurin, G.M. Veith, S. Dai, Chem. Mater. 23 (2011) 4420–4427.
[58] P. Meduri, H. Chen, J. Xiao, J.J. Martinez, T. Carlson, J.-G. Zhang, Z.D. Deng, J. Mater. Chem. A 1 (2013) 7866–7869.
[59] A.M. Reddy, B. Breitung, M. Fichtner, ACS Appl. Mater. Interfaces 5 (2013) 11207–11211.

Chapter 13

Recent Advances on Quasianhydrous Fuel Cell Membranes

Benjamin Campagne, Ghislain David and Bruno Ameduri
Institut Charles Gerhardt, Ingénierie et Architectures Macromoléculaires, UMR CNRS 5253, Ecole Nationale Supérieure de Chimie de Montpellier, Montpellier, France

Chapter Outline
13.1 Introduction 290
13.2 Fluorinated Copolymers Based on Nitrogen Heterocycles 295
    13.2.1 Introduction and Challenges 295
    13.2.2 Membranes Based on Fluorinated Copolymers Containing Triazole Groups and Sulfonated Poly(Ether Ether Ketone) 297
13.3 Proton Mobility in Membranes Based on Nitrogenous Heterocycles and s-PEEK 302
13.4 Crosslinking of Membranes Based on Nitrogenous Heterocycles 305
    13.4.1 Thermal Stabilities of Crosslinked Membranes Composed of s-PEEK-Na (B) and Poly(CTFE-*alt*-IEVE)$_{94\%}$-*g*-1H-1,2,4-triazole-3-thiol$_{90\%}$-*co*-poly(CTFE-*alt*-GCVE)$_{6\%}$ Terpolymer Crosslinked by DiA or TEPA 309
    13.4.2 Mechanical Properties of Crosslinked Membranes 309
    13.4.3 Protonic Conductivities of Membranes Composed of s-PEEK and Poly(CTFE-*alt*-GCVE)$_{6\%}$-poly(CTFE-*alt*-IEVE)$_{94\%}$-*g*-1H-1,2,4-triazole$_{90\%}$ Terpolymer Crosslinked by Diamines DiA or TEPA 315
    13.4.4 Comparison of Conductivities of A/B-*ret*-DiA and A/B-*ret*-TEPA Membranes with Uncured Membranes Composed of Poly(CTFE-*alt*-IEVE)-*g*-1H-1,2,4-triazole$_{95\%}$ Copolymer/s-PEEK Blends 316
13.5 Conclusion 317
Acknowledgments 320
References 320

13.1 INTRODUCTION

A fuel cell is an energy convertor (exchanger) that converts, in an electrochemical process, the energy of an oxidoreduction reaction to electrical energy, heat, and water as the only waste products. A fuel cell is composed of a stack and various elementary cells (>40) that consist of a membrane located between an anode and a cathode. Fuel cell technology offers an attractive combination of high-energy conversion efficiency and a potential for large reductions in power source emissions, including CO_2 [1–6]. When the fuel used is hydrogen, the device is called a Hydrogen Fuel Cell. Polymer electrolyte membranes (PEMs) for fuel cells (PEMFCs) are ideally suited for transportation, combined heat and power, and mobile auxiliary power applications. Among the many attractive features, the high power density, rapid start-up, and high efficiency make the PEMFC the system of choice for transport manufacturers.

Different types of fuel cells have demonstrated their efficiency according to the choice of conditions (temperature), nature of electrolyte, and other main characteristics (listed in Table 13.1): solid oxide (SOFC), molten carbonate (MCFC), phosphoric acid AFC, PEMFC, and alkaline fuel cell (AFC). Thanks to their high flexibility and easy handling, fuel cells are one of the most attractive approaches to energy conversion. Actually, fuel cells display a number of key advantages over conventional energy conversion devices, including low emissions and noise, flexibility in fuel selection, cogeneration capability, and economy of scale [7].

In particular, the successful development of PEMFCs has generated widespread interest in these types of potentially viable fuel cells for automotive applications [8,9]. However, the commercialization of PEMFCs is still limited by high cost, unsatisfactory lifetime [1–6], and a lack of hydrogen supply network development. Proton exchange membranes remain one of the most crucial components of such devices [9]. Actually, the most efficient membranes were developed >50 years ago. In the early 1960s, the General Electric (GE) company developed the first membranes based on poly(styrene sulfonic acid) that were used in National Aeronautics and Space Administration(NASA) space programs. However, these membranes do not display any suitable thermal stability (e.g., the C_{III}–H bond of styrene), and their durability was thus limited (500 h at 60 °C). On the other hand, the Dupont Company produces a Nafion® membrane (Scheme 13.1) [3], which was also tested for space missions by GE in 1966 and 1969. The development of research in the topic of fuel cell membranes (PEMFCs) [1–6] is, probably more than others, linked to exceptional chemical, thermal, and oxidative stabilities of fluoropolymers.

To afford similar structures than that of Nafion®, various syntheses are possible: (1) either by direct copolymerization, (2) or by chemical modification, by irradiation of various fluorinated polymers (polytetrafluoroethylene (PTFE), polyvinylidene difluoride (PVDF), ethylene tetrafluoroethylene (ETFE), fluorinated ethylene propylene (FEP), and poly-co-perfluoroalkylvinyl ether (PFA))

TABLE 13.1 Different Types of Fuel Cells and Their Main Characteristics (M, T, and S Stand for Mobile, Transportation, and Stationary, Respectively)

Type of Fuel Cell	Polymer Electrolyte Membranes for Fuel Cell	Alkaline Fuel Cell	Phosphoric Acid Fuel Cell	Molten Carbonate Fuel Cell	Solid Oxide Fuel Cell
Electrolyte	Ion-exchange membrane	Mobile or immobilized KOH	Immobilized liquid	Immobilized molten carbonates	Ceramic
Temperature (°C)	60–120	65–220	190–205	650	700–1000
Charge carriers	H^+	OH^-	H^+	CO_3^{2-}	O_2^-, H^+
External reforming	Possible/yes	Possible/yes	Possible/yes	No	No
Catalyst	Pt	Ni, Ag, Pt	Pt	Ni	Perovskites, Ni
Application	M, T	M, T	M, T	T, S	T, S

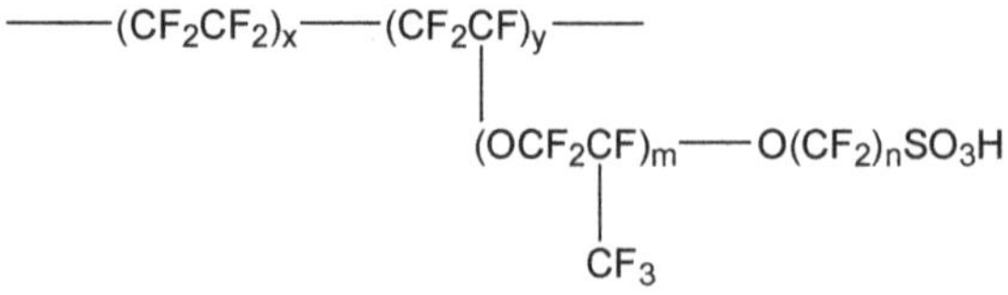

$$—(CF_2CF_2)_x—(CF_2CF)_y—$$
$$|$$
$$(OCF_2CF)_m—O(CF_2)_nSO_3H$$
$$|$$
$$CF_3$$

SCHEME 13.1 Chemical structure of Nafion® for commercially available perfluorosulfonic acid (PFSA) membranes (more details are given in Table 13.2).

TABLE 13.2 Names, Structures, and Characteristics of the Main Commercially Available PFSA Polymers

Structural Parameter	Tradename and Type	Equivalent Weight	Thickness (μm)
$m=1; x=5-13, 5;$ $n=2; y=1$	DuPont Nafion® 120 Nafion® 117 Nafion® 115 Nafion® 112	1200 1100 1100 1100	250 175 125 50
$m=0.1; n=1-5$	Asahi glass Flemion®-T Flemion®-S Flemion®-R	1000 1000 1000	120 80 50
$m=0; n=2-5;$ $x=1, 5-14$	Asahi chemicals Aciplex®-S	1000–1200	25–100
$m=0; n=2; x=3, 6-10$	Solvay-specialty polymers Aquivion®	800	125
$m=0; n=4; x=4-9$	3M® membrane	1000	80–100

followed by the grafting of polystyrene or other monomers that are potentially functionalizable. These strategies have been reviewed in the first edition of this book [10] and thus will not be mentioned in this present chapter, while more recent (nonexhaustive) reviews or book chapters can also be suggested [11]. Aromatic and heteroaromatic hydrocarbon polymer membranes with high ionic exchange capacities (IECs) do exhibit excellent conductivities in the fully hydrated state, especially for applications >80 °C, but they are often subject to excessive swelling and are very brittle in the dry state. As for the hydrocarbon polymer membranes with aliphatic main chains, they are only profitable supplements to PEMs for applications <80 °C. Phosphoric acid-doped polybenzimidazole (PBI) membrane-based fuel cells are not good for a pure hydrogen feed, and they provide lower performances than those of Nafion®-based fuel cells. However, they provide a performance superior to that of Nafion®-based cells above a certain CO level in the gas feed. Anionic exchange membranes are still not competitive with acidic PEMs in the present stage or in the near future, even at low temperatures (i.e., <60 °C), but the promising binding material casts a new light on the development of anion exchange membrane fuel cell. However, several other recent routes enable an increase of the conductivities or improvement in the electrochemical, mechanical, and thermal properties: addition of doping agents, nanofillers [12], or crosslinking agents that will be mentioned in Section 13.4. Currently, the most commonly applied polymer membranes are based on PFSA polymers such as Nafion®, Flemion®, Aquivion®, Fumion®, 3M®, or Aciplex® [13], and used at low temperatures (LT-PEMFC). They have been processed from copolymers, the overall chemical structure of which is displayed in Scheme 13.1.

In the fully hydrated form, these membranes provide a high proton conductivity of about $0.1\,S\,cm^{-1}$ for a relative humidity (RH)>75 to 80% but at a temperature <100 °C for Nafion® [3,14], which allows high power densities and efficiencies in fuel cell applications for these low temperature PEMFCs (LT-PEMFCs).

However, proton conduction drops for Nafion®- and PFSA-related membranes when the temperature is increased as liquid water evaporates from the membranes above approximately 85–100 °C at atmospheric pressure. High-temperature operation of fuel cells offers several advantages [15] such as an increase of catalyst CO tolerance (at the anode), which simplifies the hydrogen purification process [16] (the LT-PEMFC has a very low tolerance to impurities in the fuel, thus requiring 99.99999% pure hydrogen that is costly to produce), improved kinetics, and reduced heat elimination issues [17]. In addition, the heat produced from the LT-PEMFC is also of a low temperature and is thus difficult to transfer away for use in other processes. Due to the nature of the membrane, a water management system is needed to prevent flooding/drying out of the membrane electrode assembly (MEA), both of which lead to a loss in performance. Increasing the temperature of the PEMFC will allow for existing cooling infrastructures present in transport vehicles to be used, thus increasing the weight and mass specific energy densities and the overall energy efficiency. The efficiency can be further increased when cogeneration [18] and on-board reforming [9,19,20] are considered. These issues can be overcome through the use of a high temperature PEMFC (HT-PEMFC). By switching to higher temperatures, the oxygen reduction reaction rate is significantly increased [20], thus improving the performance of the PEMFC as a whole. Other effects of higher temperatures are (1) to allow for cogeneration of heat and power, (2) high tolerance to fuel impurities, (3) simpler system design, and (4) the reactant and product gases are expected to have increased diffusion rates [21].

As a matter of fact, a 2005 cost analysis [22] of an 80 kW HT-PEMFC system projected a cost of 56 US $/kW for the MEA, assuming a production of 500,000 units, which represents 83% of the cost of the stack. In 2009, the actual cost of a fuel cell stack and balance of the plant was 61 US $/kW [23], still short of the 30 US $/kW US Department of Energy target, but it can be seen from Figure 13.1 that the initial estimated costs in 2002 are continually dropping (Figure 13.1). In 2012, the DOE reported the cost of an 80-kW$_{net}$ automotive PEM fuel cell system based on a 2012 technology and operating on direct hydrogen was projected to be $47/kW when manufactured at a volume of 500,000 units/year [24] (Figure 13.2).

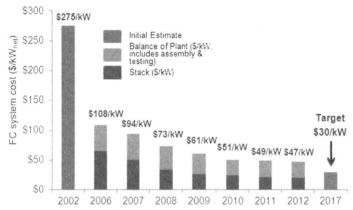

FIGURE 13.1 Projected transportation fuel cell system cost, based on an 80 kW system and 500,000 production units per year [24]. *Permission from Strategic Analysis (Brian James) and funded by the US Department of Energy, Washington, 2012.*

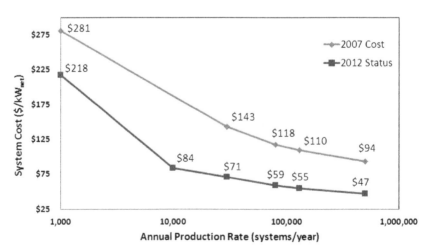

FIGURE 13.2 Projected cost of 80-kWnet transportation fuel cell systems at 1000; 10,000; 30,000; 80,000; 130,000; and 500,000 units/year [24]. *Permission from Strategic Analysis (Brian James) and funded by the US Department of Energy, Washington, 2012.*

A PEM is situated at the heart of the fuel cell in the MEA. Research into HT-PEMFCs has increased in the last few years, with approximately 140 papers published on the topic in 2011, mostly on the development of novel membranes [21].

13.2 FLUORINATED COPOLYMERS BASED ON NITROGEN HETEROCYCLES

13.2.1 Introduction and Challenges

Fluoropolymers [13,25] are endowed with remarkable properties because of their high electronegativity and small radius of the fluorine atom that induce a high stability of the C–F bond. The PTFE backbone ensures a very good thermostability and chemical inertness, even in oxidative media, which is the case for fuel cell applications. The membrane conductivities reach up to $0.1\,S\,cm^{-1}$ at 80 °C and 100% RH. In the case of Nafion®117, the conductivity corresponds to a specific resistance of $0.2\,Ohm\,cm^{-2}$, that is, a tension loss of 150 mV at a current density of $0.75\,A\,cm^{-2}$. By their use in fuel cells, these ionomers can form an interface between the electrodes and the electrolyte. Although many technological issues must be overcome, these membranes do not stick to the catalytic sites and do not disturb the oxidoreduction reaction. The excellent review of Mauritz and Moore [3] summarizes the exponential amount of studies.

The use of this type of PFSA copolymers displays several drawbacks. Besides its high price and the absence of a recycling technique, the functionalization of fluorocopolymers is difficult after these syntheses, a high methanol crossover though it has been proved that these membranes display high stabilities in constant, may degrade fast if operation cycles change. The use of fluoropolymers at high temperatures is limited by the arrangement of molecules, favoring a flow at approximately 100 °C (Figure 13.3). Fuel cells should operate

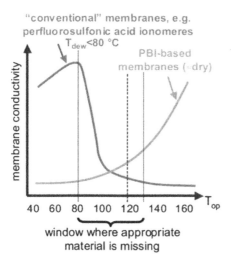

FIGURE 13.3 Type of membranes used versus temperature range. *Reproduced with permission from Wiley VCH [26].*

at least 500 h at T>100 °C for 5000 h imposed by the requirement of car manufacturers. These observations highlight the strong need to work with protogenic groups at temperatures >100 °C that allow overcoming flowing/creep limitations as shown for Nafion®.

So far, on either side of the temperature scale, Nafion®- and PBI-doped polymeric membranes with phosphoric acid (PBI/H_3PO_4) are the most efficient proton-conducting membranes as shown in Figure 13.3 [26].

PBIs are heterocyclic aromatic polymers. These are basic polymers ($pK_a = 5.5$) and can be complexed by strong acids or by very strong bases. PBIs exhibit excellent thermal and mechanical stabilities [27] and are inexpensive. In addition, the hydrogen crossover is lower than that of Nafion®. Pure PBI is an electronic and ionic insulator, but it becomes a good ionic conductor when it is doped with an acid.

Kreuer's group [28,29] reported that imidazole, benzimidazole, and tetrazole groups enabled to behave as water in order to carry out protons and the compounds derived from them could be processed as fuel cell membranes, reaching satisfactory conductivities for temperatures ranging between 100 °C and 150 °C. Hence, one of the attractive research trends consists of inserting imidazole or benzimidazole functions within (co)polymers or membranes. PBI is an excellent example, but it requires to be used at T>130 °C to display satisfactory conductivities.

Nonexhaustive strategies of copolymer synthesis containing nitrogenous heterocycles and their electrochemical properties have been listed and discussed in a comprehensive and quasiexhaustive review by Bozkurt's group [30], including major actors such as Kreuer [31], Jannasch [32], Coughlin [33], Nunes [34], and those from Toyota company [35]. This review also highlights the main results from the literature according to the architecture of the polymeric precursors (graft, telechelic, block, or statistic copolymers), the polymeric backbone (polydimethylsiloxane, poly(acrylate), poly(styrene), poly(ethylene), and poly(ethylene oxide). Tables 13.3 and 13.4 list out the most interesting results.

Model molecules synthesized by Kreuer's team [36] display the structure of oligo(ethylene oxide), EO, of various chain lengths, and these EO units enable a significant chain softness to favor structural rearrangements. These telechelic oligomers exhibit negative Tg values that surprisingly decreased with the number of EO units. In this case, the temperature used is higher than the Tg value and thus favors more mobility, which should induce a greater protonic transport. In addition, the decrease of hydrogen bonding resulting in an increase in the size of the chains is compensated by the decrease in the Tg of polymers since the volumic fraction of imidazole functions decreases. Results of conductivity show that oligomers conduct the proton better at higher temperatures than those of longer chains. The increase in the temperature of analysis nullifies the advantage of the Tg value and favors the heterocycle concentration and thus the amount of hydrogen bond present [29,31].

TABLE 13.3 Structures and Chemical Properties of Imidazole, Benzimidazole, and 1H-1,2,4-Triazole

Azole	Chemical Structure	Melting Point (°C)	Boiling Point (°C)	pK_a
Imidazole (Im)		90	257	7.2
Benzimidazole (BnIm)		170	–	12.9
1H-1,2,4-Triazole (Tri)		120	256	2.4

13.2.2 Membranes Based on Fluorinated Copolymers Containing Triazole Groups and Sulfonated Poly(Ether Ether Ketone)

More recently, our group reported a family of novel proton-conducting membranes based on a blend comprising sulfonated poly(ether ether ketone) (s-PEEK; IEC = 1.3 meq g^{-1}) and fluorinated copolymers bearing different azole groups [37,38]. s-PEEK enabled the supply of proton conductivity at temperatures <80 °C while the azole group enhanced it at higher temperatures. Fluorinated copolymers, based on chlorotrifluoroethylene (CTFE) and 2-chloroethyl vinyl ether (CEVE), were achieved from the Acceptor–Donor copolymerization that led to alternated copolymers (Scheme 13.2). Thus, their microstructures indicate 50% of fluoromonomer and 50% of hydrogenated comonomer. The latter comonomer brings complementary properties after chemical change: indeed, the proton-conducting function arose from the grafting by azole functions (imidazole, benzimidazole, 1H-1,2,4-triazole) through grafting poly(iodoethyl vinyl ether–*alt*–CTFE) copolymers, poly(IEVE-*alt*-CTFE) by commercially available 4(5)-hydroxymethylimidazole (III), 2-mercaptobenzimidazole (IV), or 1H-1,2,4-triazole-3-thiol (V) (Scheme 13.2) in 53–83% yields.

Thermal degradation of poly(IEVE-*alt*-CTFE)-*g*-2-mercaptobenzimidazole$_{y\%}$ (IV) copolymers started around 240 °C for y = 53–83 mol%. Differential scanning calorimetry (DSC) measurements displayed an increase in the Tg value when the grafting degree increased (e.g., Tg = 88 °C when y = 53 mol% while it was 112 °C for y = 83 mol%, probably due to the increasing benzimidazole group content that limits the mobility of the copolymer.

TABLE 13.4 Temperatures of Degradation (2 wt% Sample in Air) of Membranes Composed of Sulfonated Poly(Ether Ether Ketone) (s-PEEK) and Poly(CTFE-alt-IEVE)$_{94\%}$-g-1H-1,2,4-triazole-3-thiol$_{90\%}$-co-poly(CTFE-alt-GCVE)$_{6\%}$ Terpolymer A Crosslinked by 1,3-Propanediamine (DiA) or Tetraethylenepentamine (TEPA)

DiA			TEPA		
A/B in wt%	$n = n_{-NH}/n_{-SO_3H}$	$T_{d\text{-}2\% wt}$ (°C)	A/B in wt%	$n = n_{-NH}/n_{-SO_3H}$	$T°_{d\text{-}2\% wt}$ (°C)
100% B (s-PEEK)	0	287	100% B (s-PEEK)	0	287
30% A/70% B-ret-DiA	1.0	261	30% A/70% B-ret-TEPA	1.2	238
40% A/60% B-ret-DiA	1.5	240	40% A/60% B-ret-TEPA	1.7	230
50% A/50% B-ret-DiA	2.3	235	50% A/50% B-ret-TEPA	2.7	215
60% A/40% B-ret-DiA	3.4	199	60% A/40% B-ret-TEPA	4.0	217
70% A/30% B-ret-DiA	5.3	ND	70% A/30% B-ret-TEPA	6.3	212
100% A-ret-DiA	∞	216	100% A-ret-TEPA	∞	216

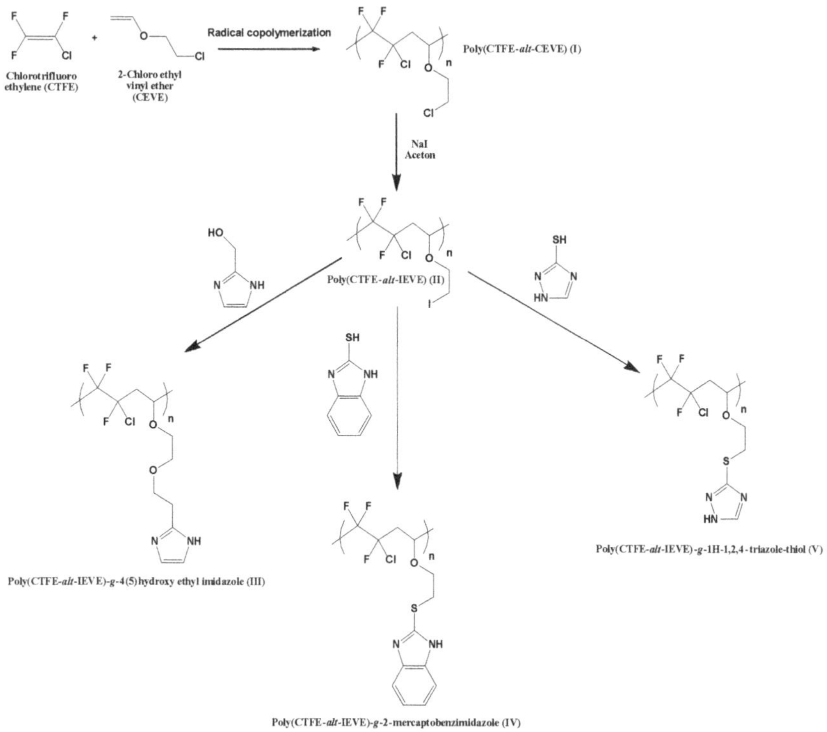

SCHEME 13.2 Various routes of syntheses of poly(CTFE-*alt*-CEVE) (I), poly(CTFE-*alt*-IEVE) (II), poly(CTFE-*alt*-IEVE)-*g*-4(5)-hydroxyethylimidazole (III), poly(CTFE-*alt*-IEVE)-*g*-2-mercaptobenzimidazole (IV), and poly(CTFE-*alt*-IEVE)-*g*-1H-1,2,4-triazole-3-thiol (V) copolymers [39].

The series of poly(IEVE-*alt*-CTFE)-*g*-1H-1,2,4-triazole-3-thiol$_{z\%}$ (V) copolymers was also prepared by the nucleophilic substitution of -I$^-$ by -S$^-$ (achieved from deprotonation of 1H-1,2,4-triazole-3-thiol) [39]. The degree of grafting, assessed by ^{1}H-nuclear magnetic resonance (NMR), ranged from 49% to 95%. These copolymers were stable under oxidative conditions from approximately 260 °C by thermogravimetric analysis (TGA) measurements. The increase in the grafting degree of 1H-1,2,4-triazole-3-thiol onto poly(IEVE-*alt*-CTFE) (II) copolymer induced a decrease in the glass transition temperature of the copolymer: Tg = 32 °C when z = 49 mol%, while it is 25 °C for y = 95 mol%. Poly(IEVE-*alt*-CTFE)-*g*-1H-1,2,4-triazole-3-thiol$_{z\%}$ (V-z) copolymers displayed a higher mobility than III and IV copolymers. An increase of the chain mobility [33] led to a lower glass transition temperature, which is expected to be in favor of proton conduction through a Grotthuss-type mechanism.

The properties of the membranes achieved from blends composed of s-PEEK and fluorinated copolymers (III), (IV), and (V) were comprehensively investigated to understand the effect of the nature of the azole group (imidazole,

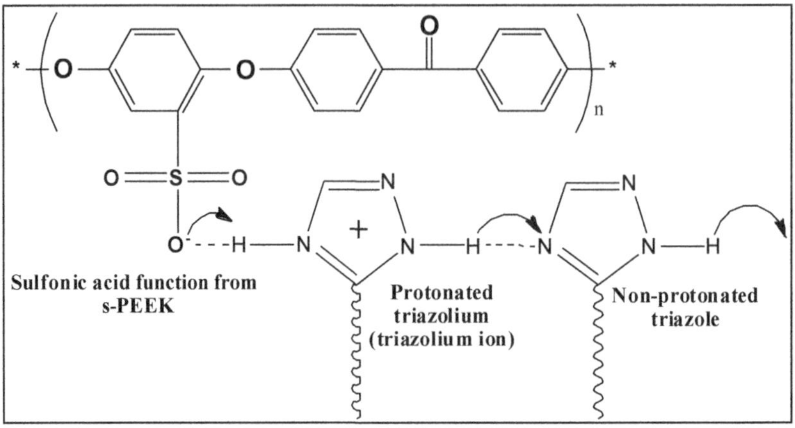

FIGURE 13.4 Protonation of 1,2,4-triazole functions grafted onto the fluorocopolymer by sulfonic acid functions of s-PEEK leading to triazolium formation and proton exchange between triazolium and nonprotonated triazole moieties.

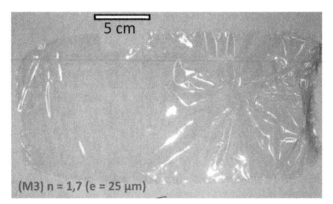

FIGURE 13.5 Picture of membrane M4: V-95 (40 wt%)/s-PEEK (60 wt%) (n = 1.7) obtained by casting from a glass substrate (thickness: ~20 μm).

benzimidazole, triazole) on membrane microstructure, water uptake, and thermal properties. Proton conductivities of the blend membranes based on triazole were significantly higher than those containing imidazole or benzimidazole moieties. Detailed characterizations allowed the identification of an optimal composition in terms of the basic (–NH):acidic (–SO$_3$H) groups ratio of 1.7 in this system, where the proton conductivity at 140 °C and at low RH (<25%) is 7 mS cm^{-1} for a blend membrane containing 60% wt of s-PEEK and 40% wt of poly(IEVE-*alt*-CTFE)-*g*-1H-1,2,4-triazole-3-thiol$_{95\%}$ copolymer (Figures 13.4–13.6) (Scheme 13.3).

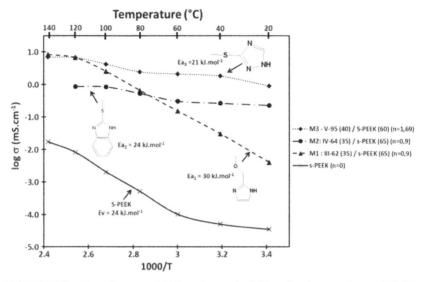

FIGURE 13.6 Alternating current (AC) proton conductivities of various membranes: M1: III-62 (35)/s-PEEK (65) (n=0.9), M2: IV-64 (35)/s-PEEK (65) (n=0.9), M3: V-95 (40)/s-PEEK (60) (n=1.7), and s-PEEK membranes versus reciprocal temperature at RH<25%. *Reproduced with permission of ACS [39].*

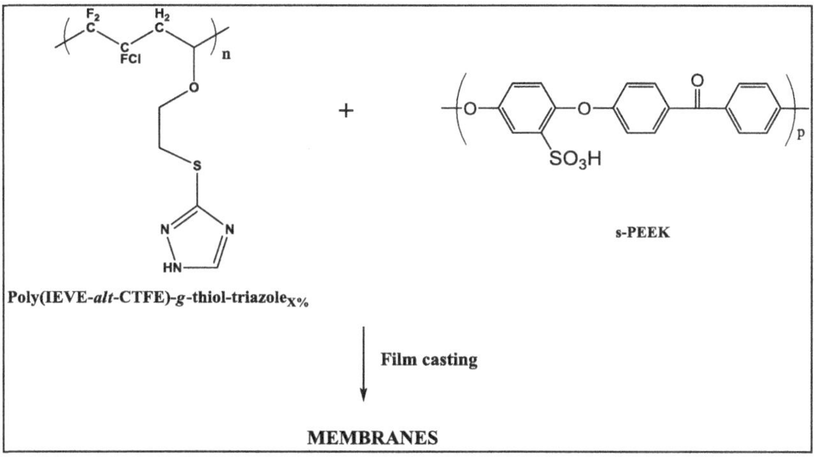

SCHEME 13.3 Processing of membranes composed of fluorinated poly(IEVE-*alt*-CTFE)-*g*-1H-1,2,4-triazole-3-thiol$_{X\%}$ copolymer (V) and sulfonated poly(ether ether ketone) (s-PEEK).

The nature of the nitrogen heterocycle has an influence on the proton conductivity. This recent study showed that grafted azole groups behave as immobilized solvents, which enables the reduction of the humidity dependence of the membranes. The proton conductivity depends upon the amount of azole group/amount of sulfonic acid function ratio (Figure 13.5).

The highest proton conductivities (7 mS cm^{-1} and 8 mS cm^{-1}, respectively, at 140 °C under a low RH < 25%) were reached for membranes based on 1,2,4-triazole and imidazole functions while those containing benzimidazole led to 2 mS cm^{-1} in similar conditions. Actually, the membranes containing imidazole displayed a higher conductivity at 140 °C but required more activation energy to enhance their proton transport mechanism. Such immobilized systems where both acid functions and azole moieties are grafted onto polymer chains allow us to overcome elution issues when membranes are operating under PEMFC conditions.

13.3 PROTON MOBILITY IN MEMBRANES BASED ON NITROGENOUS HETEROCYCLES AND s-PEEK

Regarding basic research, it is challenging to investigate the proton mobility in membranes and, among various methods, ^{1}H magic angle spinning (MAS) NMR spectroscopy has successfully been convincing. First, Traer and Goward [40] reported the 2D ^{1}H MAS NMR EXchange SpectroscopY (EXSY) spectra of telechelic oligomers that bear imidazole end groups synthesized by Schuster et al. [31a] and doped with phosphoric acid. Then, the same group [41] characterized a model system comprising imidazole with three methyl phosphonate groups by ^{1}H MAS spectroscopy to highlight an example of cooperative ionic conductivity. These studies were performed on model compounds that cannot be used as PEMFC membranes because of their hydrosolubility and also since these copolymers were not film-forming materials. However, Celik et al. [42] used double-quantum 2D H MAS solid-state NMR (SSNMR) spectroscopy to evidence the proton exchange of a copolymer based on 1-vinyl-1,2,4-triazole and vinylphosphonic acid. But these above-mentioned authors did not characterize their copolymers or compounds by ^{1}H MAS NMR EXSY, and to the best of our knowledge, only one study of 2D ^{1}H MAS NMR EXSY was recently carried out on polymer blend PEMFC membranes to investigate proton motion [43]. Actually, the proton mobility in PEMFC blend membranes containing 1,2,4-triazole groups (devoted to fuel cell operation at a high temperature and low RH, as mentioned above) was studied first by infrared spectroscopy and mainly by ^{1}H MAS SSNMR spectroscopy [43]. The membrane contained 40% wt of the fluorinated alternating poly(2-iodoethyl vinyl ether-*alt*-CTFE)-*g*-1H-1,2,4-triazole-3-thiol$_{95\%}$ copolymer (V) and 60% wt of s-PEEK (IEC = 1.3 meq g^{-1}) (n_{-NH}/n_{-SO_3H} = 1.7) [43]. This work confirmed that protonation acts in favor of proton mobility in the material at room temperature. Further, it provides experimental evidence for the increase in proton mobility due to triazole protonation from the sulfonic acid groups of s-PEEK. The 1D ^{1}H MAS spectra of the copolymer, achieved for the first time on such a membrane, were fully attributed. Following acidification of a suspension of this copolymer leading to acidified copolymer, the 1D ^{1}H MAS spectrum showed two populations corresponding to triazole and triazolium groups, respectively (Figure 13.7).

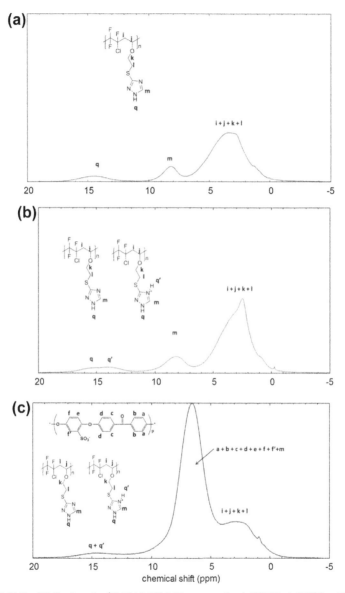

FIGURE 13.7 1D Single pulse ^{1}H MAS SSNMR spectra of poly(CTFE-*alt*-IEVE)-*g*-1H-1,2,4-triazole-3-thiol$_{95\%}$ (V) (a), poly(CTFE-*alt*-IEVE)-*g*-1H-1,2,4-triazole-3-thiol$_{95\%}$ partially protonated by hydrochloric acid (b), and of M4-A membrane based on 40% wt copolymer (V)/60% s-PEEK) (c), at 20 °C. These three spectra were normalized with respect to the signal at 13.8 ppm [43]. Reproduced with permission from Royal Society of Chemistry.

In addition, the 2D ^{1}H EXSY spectrum of copolymer (II') (Figure 13.8) displayed faster proton dynamics of the triazolium protonated form than in the copolymer containing only nonprotonated triazole groups. The off-diagonal

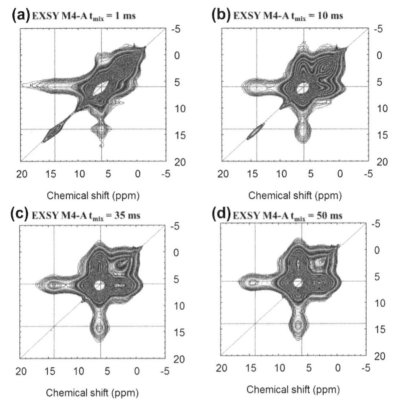

FIGURE 13.8 ¹H MAS SSNMR EXSY spectra of a membrane (based on 40% wt poly(IEVE-*alt*-CTFE)-*g*-1H-1,2,4-triazole-3-thiol copolymer/60% s-PEEK) (M3: V-95 (40)/s-PEEK (60) (n=1.7)) with t_{mix} = 1 ms (a), 10 ms (b), 35 ms (c), 50 ms (d) at 20 °C. *Reproduced with permission from RSC [43].*

correlation peaks (that evidence ¹H spins that remain in the same position or jump in a similar site associated to the same chemical shift) with the large signal centered at around 7 ppm means that they can be transferred onto the m sites or onto the aromatic ring protons of s-PEEK. The exchange observed between the (i, j, k, and l) protons (Figure 13.7) and the intense peak centered at 7 ppm confirm a suitable blend obtained between poly(CTFE-*alt*-IEVE)-*g*-1H-1,2,4-triazole-3-thiol$_{95\%}$ copolymer and s-PEEK.

Inhomogeneity of the (V′) sample was confirmed by the different spin dynamics of triazole and triazolium ring protons. In the homogeneous membrane, spin dynamics were further increased (i.e., very short mixing time to observe proton diffusion).

Signals in the 1D ¹H MAS spectrum of (V) copolymer were fully assigned and enabled us to observe the proton site of the secondary amine function of the triazole ring site at 13.8 ppm (not observed in the liquid state NMR). In its partially protonated triazolium form, a specific signal was noted at approximately 15.5 ppm in the 1D ^{1}H MAS spectrum of (V′) copolymer together with that centered at 13.8 ppm indicating that protonation is not homogeneous in the sample due to some acidification process (i.e., suspension of (II) copolymer in HCl). In membrane M3-A, the protons of the SO$_3$H groups were no longer observed, and this subsequently indicates a complete transfer of the protons from s-PEEK to the copolymer. Only one line corresponding to the triazolium form was observed indicating a homogeneous blend of copolymer V and s-PEEK together with fast proton exchange between the triazolium and triazole rings.

The 2D ^{1}H EXSY spectra of copolymer V showed that an intrachain slow regime exchange of protons is observed for $t_{mix} \geq 20$ ms. Under its protonated form, proton mobility was increased, and exchange was observed since $t_{mix} \geq 10$ ms; inhomogeneity of the sample protonation was confirmed by different spin dynamics of triazolium and triazole rings protons. In M3, V-95 (40)/s-PEEK (60) (n = 1.7) membrane, proton mobility was further increased and exchange was observed since $t_{mix} = 1$ ms. Even more for $t_{mix} \geq 35$ ms, all triazolium protons have switched to another position (Figure 13.8).

Hence, this work provides confirmation that protonation acts in favor of proton mobility in the material at room temperature, and provides experimental evidence for the increase of proton mobility due to triazole protonation from the sulfonic acid groups of s-PEEK. A deeper understanding of the mechanism of proton transport in the membranes will be obtained by varying sample temperatures to influence fast exchange dynamics.

13.4 CROSSLINKING OF MEMBRANES BASED ON NITROGENOUS HETEROCYCLES

A few studies on the crosslinking of fuel cell membranes have been reported and can be achieved physically or chemically. The former may occur under irradiation [44]. The radiations used can be electromagnetic, such as X and γ rays, charged particles, electron beam, or from ions of heavy metals (e.g., Kr, Xe). The most used source of γ rays is ^{60}Co, which emits radiations ranging from 1.17 to 1.33 MeV. The second methodology can be from various strategies for solid alkaline fuel cell (SAFC) [45] and PEMFCs [46–48] involving cure site monomers while other routes are based on an interpenetrated polymer network; IPN [49], one example [50] uses divinylbenzene or triallyl cyanurate (blended with PVDF) that can be added as crosslinking agents. Finally, these steps lie in functionalizing the dangling chain such as the sulfonation of the aromatic cycle by chlorosulfonic acid.

Yet, these techniques enable (1) increasing chemical inertness and mechanical properties in the course of various start/stop cycles when the fuel cell is operating and (2) significantly decreasing the water swelling of membranes that can also allow a better stability in oxidative media (by decreasing the radical penetration in solution through the membranes).

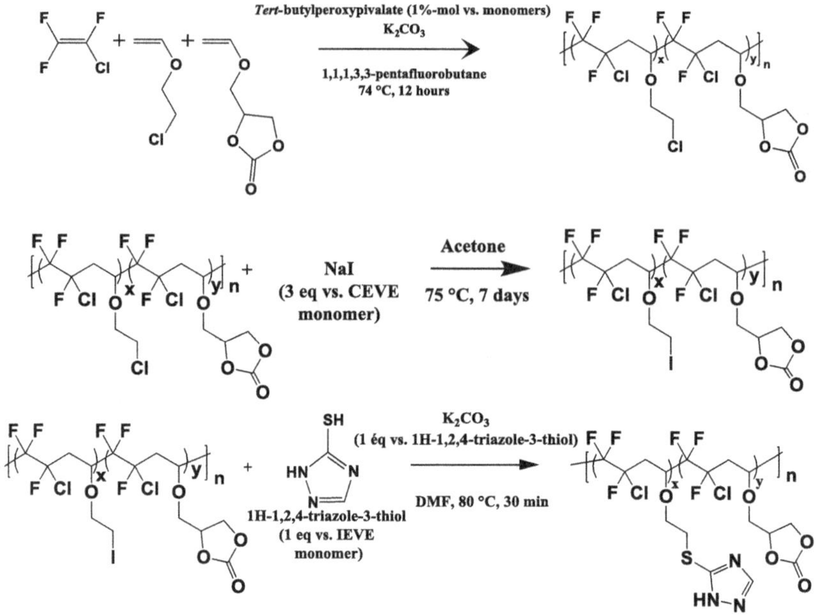

SCHEME 13.4 Radical terpolymerization of CGVE, chlorotrifluoroethylene (CTFE), and 2-chloroethyl vinyl ether (CEVE) lead to poly(CTFE-*alt*-IEVE)$_{x\%}$-*g*-1H-1,2,4-triazole-3-thiol$_{z\%}$-*co*-poly(CTFE-*alt*-GCVE)$_{y\%}$ by chemical modification of poly(CTFE-*alt*-IEVE)$_{x\%}$-*co*-poly(CTFE-*alt*-GCVE)$_{y\%}$ terpolymer [51].

To attempt reaching better conductivity and mechanical properties and a certain structuration of membranes composed of a fluoropolymer that bears 1H-1,2,4-triazole groups and s-PEEK, an original strategy [51] consisted of the elaboration of usual architectures, for example, semiinterpenetrated networks (pseudosemi-IPN). This decrease in the swelling rate can also allow a better stability in oxidative media.

Actually, the strategy involved the crosslinking of terpolymers based on (1) azole functions that allow proton conductivity and (2) cyclocarbonate groups that induce the crosslinking by polyaddition with telechelic amines [51]. These terpolymers were synthesized by radical terpolymerization of CTFE monomer with CEVE (that further enabled the grafting of 1H-1,2,4-triazole-3-thiol) and glycerine carbonate vinyl ether (GCVE) (Scheme 13.4). The synthesis of the last monomer was achieved by transetherification between glycerol carbonate and ethyl vinyl ether with a 45% yield (Scheme 13.5) and was reported to be successfully involved in radical copolymerization with CTFE, hexafluoropropylene, and perfluoromethyl vinyl ether leading to alternated structures [52].

A wide range of different poly(CTFE-*alt*-IEVE)$_{x\%}$-*g*-1H-1,2,4-triazole-3-thiol$_{z\%}$-*co*-poly(CTFE-*alt*-GCVE)$_{y\%}$ terpolymers was achieved from various initial molar ratios of the three monomers. The microstructures of the resulting

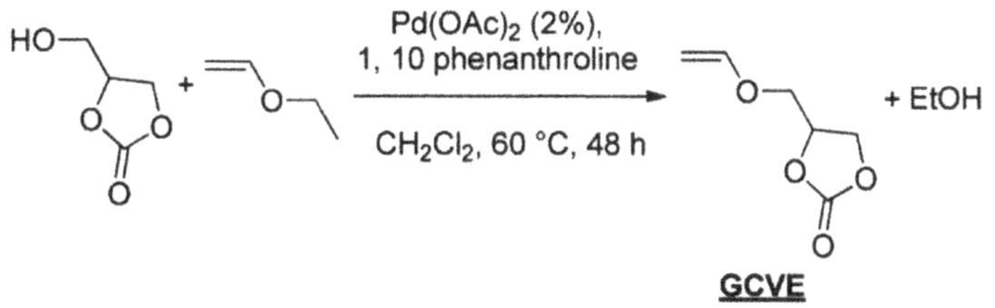

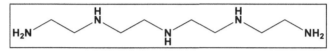

SCHEME 13.5 Synthesis of gylcerine carbonate vinyl ether (GCVE) by transetherification [52].

FIGURE 13.9 Chemical structure of TEPA as a crosslinking agent to obtain original crosslinked fuel cell membranes.

terpolymers, considering that alternating poly(CTFE-*alt*-CEVE) and poly(CTFE-*alt*-GCVE) copolymer units were produced, were confirmed by ^{1}H, ^{13}C, and ^{19}F NMR spectroscopy [51]. Our study focused on the crosslinking of two poly(CTFE-*alt*-CEVE)$_{x\%}$-*co*-poly(CTFE-*alt*-GCVE)$_{y\%}$ terpolymers in which targeted values of y and x were 5 and 20% in order to study the influence of the crosslinking rate [51].

Both poly(CTFE-*alt*-IEVE)$_{94}$-*g*-1H-1,2,4-triazole$_{90\%}$-*co*-poly(CTFE-*alt*-GCVE)$_6$ and poly(CTFE-*alt*-IEVE)$_{82}$-*g*-1H-1,2,4-triazole-3-thiol$_{86\%}$-*co*-poly(CTFE-*alt*-GCVE)$_{18}$ terpolymers were generated in 78% and 82% yields, with number average molecular weights of 15,200 and 11,400 g mol^{-1}, and polydispersity indices of 2.5 and 1.7, respectively. Their thermal properties were assessed by DSC (Tg ~ 74–81 °C) and TGA (T$_{dec}$ > ~210 °C under air).

The chemical modification of GCVE units by isopropylamine as a model reaction [53] enabled one to find out the best conditions for further successful crosslinking of these terpolymers via the reaction of cyclocarbonate function with two telechelic aliphatic diamines: 1,3-propanediamine (DiA) regarded as a rigid diamine that bears a C3 chain length and tetraethylenepentamine (TEPA, Figure 13.9) as a more longer and softer diamine that contains four ethylene bridges separated by secondary amino functions (Figure 13.9). Actually, two amino end groups are primary and are more reactive than the three secondary ones (in TEPA) about cyclocarbonate groups. The central secondary amine is more reactive than both secondary others (Figure 13.9).

The crosslinking reaction was first carried out with stoichiometric amounts of amine function/cyclocarbonate groups in poly(CTFE-*alt*-IEVE)$_{94}$-*g*-1H-1,2,4-triazole-3-thiol-*co*-poly(CTFE-*alt*-GCVE)$_6$, catalyzed by triethylamine [51]. The mixture was then cast and heated at 80 °C for 24 h followed by a final drying at 100 °C under vacuum. In both cases, brittle films were produced after drying (Figure 13.10).

Evidence of the crosslinking arises from both the self-standing films and their insolubility in polar solvents (which did solubilize the uncured terpolymer, e.g., acetone, dimethylformamide (DMF), dimethyl sulfoxide (DMSO), N-methyl-2-pyrrolidone (NMP)) even at 120 °C, and by Attenuated total reflection (ATR) Fourier transform infrared spectroscopic (FTIR) analyses [51]: increased absorbance in the 2400- to 3500-cm^{-1} range (assigned to –NH groups) after grafting of 1H-1,2,4-triazole-3-thiol onto poly(CTFE-*alt*-IEVE)$_{94\%}$-*co*-poly(CTFE-*alt*-GCVE)$_{6\%}$ terpolymer. The crosslinking reaction favors the presence of the hydroxyurethane bond (–NH–(=O)–O–), evidenced by absorption bands of carbonyl groups at approximately 1650 cm^{-1}.

Only membranes containing poly(CTFE-*alt*-IEVE)$_{94\%}$-*g*-1H-1,2,4-triazole-3-thiol$_{90\%}$-*co*-poly(CTFE-*alt*-GCVE)$_{6\%}$ terpolymer and s-PEEK-Na were processed. In the sections below, poly(CTFE-*alt*-IEVE)$_{94\%}$-*g*-1H-1,2,4-triazole-3-thiol$_{90\%}$-*co*-poly(CTFE-*alt*-GCVE)$_{6\%}$ terpolymer and s-PEEK-Na will be named A and B, respectively.

Five membranes composed of various molar ratios n (n = n_{-NH}/n_{-SO_3H}) were produced according to the scheme displayed in Figure 13.11. To study the influence of the length of the crosslinking agent of poly(CTFE-*alt*-GCVE)$_{6\%}$-(CTFE-*alt*-IEVE)$_{94\%}$-*g*-1H-1,2,4-triazole$_{90\%}$ terpolymer, two sets of membranes were processed: the first sets were based on s-PEEK-Na (B) and terpolymer A crosslinked with DiA while a second set was composed of s-PEEK-Na (B) and terpolymer A crosslinked with TEPA.

Membranes crosslinked with DiA or TEPA having a weight percentage of terpolymer As >50% led to brittle membranes. This fragility is enhanced when membranes are crosslinked by DiA.

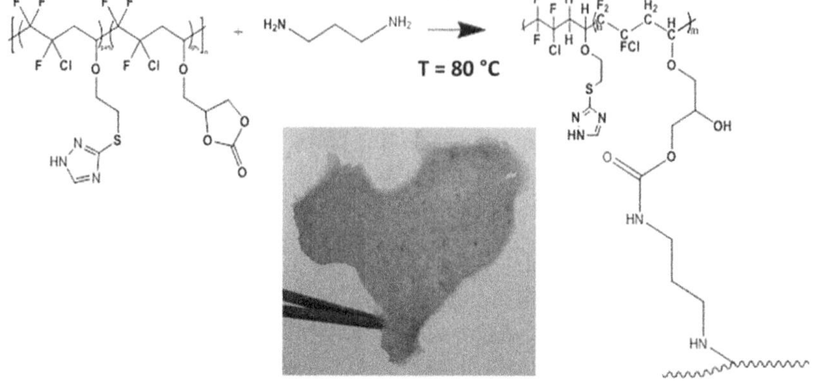

FIGURE 13.10 Crosslinking reaction of poly(CTFE-*alt*-IEVE)$_{94\%}$-*g*-1H-1,2,4-triazole-3-thiol-*co*-(CTFE-*alt*-GCVE)$_{6\%}$ terpolymer with DiA catalyzed by triethylamine (Et$_3$N) in dimethylformamide (DMF) at 80 °C and the photograph of the resulting film [51].

13.4.1 Thermal Stabilities of Crosslinked Membranes Composed of s-PEEK-Na (B) and Poly(CTFE-*alt*-IEVE)$_{94\%}$-*g*-1H-1,2,4-triazole-3-thiol$_{90\%}$-*co*-poly(CTFE-*alt*-GCVE)$_{6\%}$ Terpolymer Crosslinked by DiA or TEPA

The thermal stabilities of these membranes composed of poly(CTFE-*alt*-IEVE)$_{94\%}$-*g*-1H-1,2,4-triazole-3-thiol$_{90\%}$-*co*-poly(CTFE-*alt*-GCVE)$_{6\%}$ (A) terpolymer with s-PEEK-Na (B) crosslinked by DiA or TEPA have also been studied in air, and the results are listed out in Table 13.4.

13.4.2 Mechanical Properties of Crosslinked Membranes

Mechanical characterizations of these membranes crosslinked by primary diamines, assessed by dynamical mechanical analyses (DMA), display a pseudo-semi-IPN structure and have been compared to those of noncrosslinked ones [51]. Unfortunately, a majority of these membranes are quite brittle (% A>60 wt% crosslinked by DiA or TEPA) and thus could not be analyzed by DMA. Membranes (n=1.0, 2.3, 3.4, and n=0 crosslinked by DiA (Figure 13.11) and n=1.2, 1.7, 2.71, and n=0 for membranes crosslinked by TEPA (Figure 13.11)) were characterized by DMA.

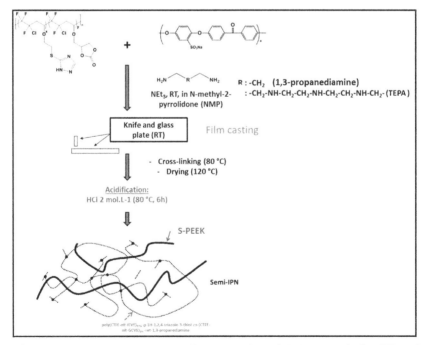

FIGURE 13.11 Process to prepare membranes (*pseudo*-semi-IPN) composed of s-PEEK and poly(CTFE-*alt*-IEVE)$_x$-*g*-1H-1,2,4-triazole-3-thiol-*co*-(CTFE-*alt*-GCVE)$_y$ terpolymer crosslinked by a primary diamine (DiA or TEPA) [51].

The measurement conditions were optimized to characterize them: linear deformation rate of 1% with a 1 Hz-frequency. Figure 13.12 (membranes crosslinked by DiA) and Figure 13.13 (those crosslinked by TEPA) represent the evolution of storage modulus, E′, and tan δ (damping factor or loss factor

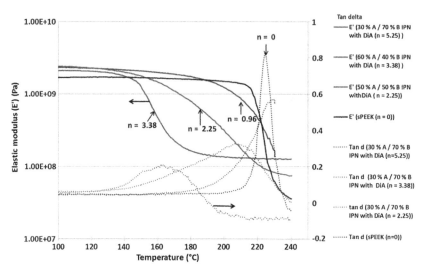

FIGURE 13.12 Mechanic–dynamic analyses of membranes composed of s-PEEK and poly(CTFE-*alt*-GCVE)$_{6\%}$-poly(CTFE-*alt*-IEVE)$_{94\%}$-g-1H-1,2,4-triazole$_{90\%}$ terpolymers crosslinked by DiA, deformation rate = 1%, frequency = 1 Hz.

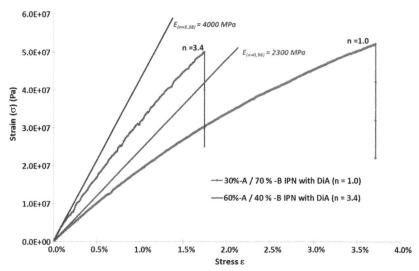

FIGURE 13.13 Strain–stress curves of membranes composed of s-PEEK and poly(CTFE-*alt*-GCVE)$_{6\%}$-poly(CTFE-*alt*-IEVE)$_{94\%}$-g-1H-1,2,4-triazole$_{90\%}$ crosslinked by DiA (Young modulus can be accessed from the tangent of the curves at the origin).

tan δ = E″/E′) versus the temperature. DMA analyses enabled one to show a glassy zone (before the transition zone of E′ modulus), a zone of transition of storage modulus E′, which corresponds to the maximum of tan δ, and a rubber zone (after the zone of transition). The maximum of tan δ is assigned to the glass transition temperature [51].

Figure 13.12 and Table 13.5 show that incorporation of terpolymer A crosslinked by DiA in membranes, the more the Tg determined by DMA of the material is lower about that of s-PEEK. Figure 13.12 shows the opposite tendency. Actually, in the case of membranes crosslinked by TEPA, the more the poly(CTFE-*alt*-GCVE)$_{6\%}$-poly(CTFE-*alt*-IEVE)$_{94\%}$-*g*-1H-1,2,4-triazole$_{90\%}$ terpolymer crosslinked by TEPA is incorporated, the more the glass transition temperature determined by DMA of the membranes is closer to that of s-PEEK. Such a phenomenon can be explained by the presence of secondary amino links of TEPA that are basic groups and able to be associated by ionic bonds (acid/basic type) with sulfonic acid function of s-PEEK. Hence, –SO$_3$H groups associated with TEPA's secondary amines are not associated with water molecules (i.e., ambient humidity) and as less water is trapped in membranes, the plasticizing effect of water onto the membrane is decreased. The Young modulus was assessed from the stress–strain curves (Figure 13.12) (membranes crosslinked by DiA) and Figure 13.13 (crosslinked membranes by TEPA). The main results of DMA analyses are summarized in Table 13.5.

As expected, E′ the storage modulus of the crosslinked membranes (30% A/70% B-*ret*-DiA (n = 1.0), 50% A/50% B-*ret*-DiA (n = 2.3) and 60% A/40% B-*ret*-DiA (n = 3.4) and 30% A/70% B-*ret*-TEPA (n = 1.2), 40% A/60% B-*ret*-TEPA (n = 1.7) and 50% A/50% B-*ret*-TEPA (n = 2.7)) is higher than those of s-PEEK and noncrosslinked M3-A membranes (n = 1.7) before and after the glassy zone.

Figure 13.13 shows that the more the membrane contains terpolymer A crosslinked by DiA, the less ductile and brittle it is, as evidenced by the increase in stiffness. Similar results were observed for membranes crosslinked by TEPA that possess a very low elastic ductility and are more rigid and brittle than s-PEEK. These membranes are less ductile than the s-PEEK membrane (elongation at break of 32% [51]). Membranes crosslinked by TEPA displayed elongation at break slightly higher than those of membranes crosslinked by DiA, arising from the softness of TEPA versus DiA.

The relative humidity (RH) (0–72% RH) [51] had a low influence on the mechanical properties of membrane 40%-A/60%-s-PEEK crosslinked by DiA and TEPA (Figures 13.13 and 13.14).

In addition, it was noted that whatever the RH (ranging from 0% to 72%), the storage modulus of the membrane is quasiconstant at 2.8×10^9 Pa (cf. 1.0×10^8 Pa for Nafion® 1100 at 70 °C and 50% RH reported by Zhao and Benziger [54]).

Crosslinking by DiA induced crosslinking knots between partially fluorinated chains of poly(CTFE-*alt*-GCVE)$_{6\%}$-poly(CTFE-*alt*-IEVE)$_{94\%}$-*g*-1H-1,2,4-

TABLE 13.5 Dynamical Mechanical Analyses (DMA) of Membranes Composed of Sulfonated Poly(Ether Ether Ketone) (s-PEEK) and Poly(CTFE-alt-GCVE)$_{6\%}$-poly(CTFE-alt-IEVE)$_{94\%}$-g-1H-1,2,4-triazole$_{90\%}$ Crosslinked by DiA or TEPA, and Membrane M3-A (E' = Storage Modulus, ND = Nondetermined)

A/B in wt%	Thickness (μm)	n = n$_{-NH}$/n$_{-SO_3H}$	E' (50°C) (Pa)	Tg DMA (°C)	E' (T > Tg) (°C)	Young Modulus (E) (Mpa)	Break Elongation Rate (%)
B (s-PEEK)	60 ± 1	0	1.7×10^9	234	3.6×10^7	1220 [53]	32
30% A/70% B-ret-DiA	60 ± 2	1.0	2.4×10^9	223	1.4×10^8	23	4
40% A/60% B-ret-DiA	40 ± 2	1.5	ND	ND	ND	ND	ND
50% A/50% B-ret-DiA	30 ± 3	2.3	2.4×10^9	206	7.5×10^7	ND	ND
60% A/40% B-ret-DiA	45 ± 1	3.4	2.4×10^9	162	1.3×10^8	40	1.7
70% A/30% B-ret-DiA	35 ± 2	5.3	ND	ND	ND	ND	ND

30% A/70% B-ret-TEPA	25 ± 2	1.2	2.8 × 10^9	123	9.3 × 10^8	18	3.4
40% A/60% B-ret-TEPA	35 ± 1	1.7	2.5 × 10^9	166	7.3 × 10^8	24	2.8
50% A/50% B-ret-TEPA	30 ± 1	2.7	2.4 × 10^9	182	7.5 × 10^8	23	2.8
60% A/40% B-ret-TEPA	35 ± 2	4.0	ND	ND	ND	ND	ND
70% A/30% B-ret-TEPA	30 ± 3	6.3	ND	ND	ND	ND	ND
M3-A	20 ± 1	1.7	1.6 × 10^9	224	1.3 × 10^7	32	5.8

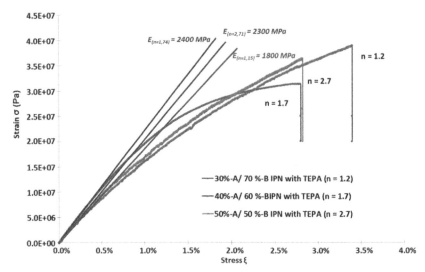

FIGURE 13.14 Strain–stress curves of membranes composed of s-PEEK and poly(CTFE-*alt*-GCVE)$_{6\%}$-(CTFE-*alt*-IEVE)$_{94\%}$-*g*-1H-1,2,4-triazole$_{90\%}$ crosslinked by DiA (Young modulus can be determined from the tangents of the curves at the origin).

triazole$_{90\%}$ terpolymer A. The denser the network, the more difficult it is for the water to solvate the sulfonic acid groups of s-PEEK and thus to swell the membranes (since water cannot penetrate within the membranes and thus cannot plasticize them). This induces a quasiconstant value of the storage modulus (E′) of the membrane 40%-A/60-B-*ret*-DiA (n = 1.5) whatever the RH (at 20 °C).

Figure 13.14 shows a higher dependence of the storage modulus (at 20 °C) with respect to the membrane crosslinked by DiA for the same composition (40%-A/60%-B-*ret*-TEPA, (n = 1.7)). However, this dependence is much lower than that of the uncrossed membrane that contains 40 wt% poly(CTFE-*alt*-IEVE)-*g*-1H-1,2,4-triazole$_{95\%}$ copolymer and 60 wt% of s-PEEK (n = 1.7).

Interestingly, this low influence of RH on the elastic modulus (E′) is a challenge. Actually, this is linked to the steady state of mechanical properties whatever the RH when operated in fuel cell conditions.

As expected, the crosslinked membranes composed of s-PEEK and poly(CTFE-*alt*-GCVE)$_{6\%}$-*co*-poly(CTFE-*alt*-IEVE)$_{94\%}$-*g*-1H-1,2,4-triazole$_{90\%}$ (A) display a twofold lower swelling rate (in water at 90 °C) than the uncrosslinked ones (membranes M-3). Crosslinking achieved via DiA gave greatly reduced swelling rates (up to 10 wt% for n = 5.3) compared to membranes crosslinked by TEPA (up to 20 wt% for n = 6.3) and far less than for noncrosslinked membranes (membranes M-3) (the minimum swelling rate was 28 wt% for n = 5.9) (Figure 13.15). This observation can be explained by the shorter DiA amine that gave a crosslinked network denser than that obtained from longer and more flexible TEPA, and thereby reducing the insertion of water within A/B-*ret*-DiA membranes.

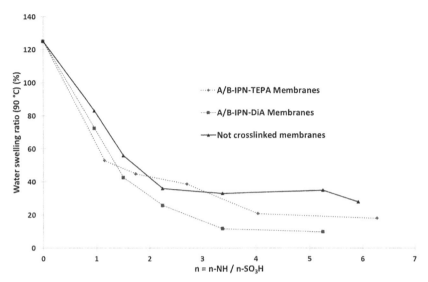

FIGURE 13.15 Comparison of swelling rates in water at 90 °C of various membranes: A/B-*ret*-DiA, A/B-*ret*-TEPA, and M3 membranes (noncrosslinked) composed of s-PEEK and poly(CTFE-*alt*-IEVE)-*g*-1H-1,2,4-triazole$_{95\%}$ copolymer versus molar ratio n.

Crosslinked membranes in the presence of DiA or TEPA display swelling rates up to twice lower than those of the uncured ones. This result is interesting since a low swelling rate in water enables the reduction of dimensional changes of membranes when it is operated in fuel cell conditions (start and stop cycles), which guarantees a better mechanical behavior than that of operating membranes. These results were obtained for a maximal RH (i.e., when the membranes were completely immersed in water) at 90 °C. They allow one to hypothesize that these swelling rates are even lower in the required conditions by the requirements (T = 120–200 °C for RH < 25%).

13.4.3 Protonic Conductivities of Membranes Composed of s-PEEK and Poly(CTFE-*alt*-GCVE)$_{6\%}$-poly(CTFE -*alt*-IEVE)$_{94\%}$-*g*-1H-1,2,4-triazole$_{90\%}$ Terpolymer Crosslinked by Diamines DiA or TEPA

Protonic conductivities of membranes composed of s-PEEK and poly(CTFE-*alt*-GCVE)$_{6\%}$-poly(CTFE-*alt*-IEVE)$_{94\%}$-*g*-1H-1,2,4-triazole$_{90\%}$ terpolymer crosslinked by DiA or TEPA were assessed by means of a conductivity cell equipped with two electrodes, first heated at 140 °C (P_{atm}) for 3 h, and the measurements were achieved by cooling (Figure 13.16).

Best protonic conductivities were achieved for membrane 40%-A/60-B-*ret*-DiA (n = 1.5) (σ = 4.3 mS cm^{-1} at 140 °C, HR < 25%) (membranes crosslinked by DiA) and for membrane 40%-A/60-B-*ret*-TEPA (n = 1.7) (σ = 2.1 mS cm^{-1} at

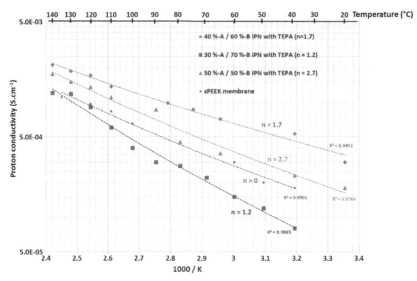

FIGURE 13.16 Protonic conductivity values through membranes composed of s-PEEK and poly(CTFE-*alt*-GCVE)$_{6\%}$-poly(CTFE-*alt*-IEVE)$_{94\%}$-*g*-1H-1,2,4-triazole$_{90\%}$ terpolymer crosslinked DiA and s-PEEK alone versus 1000/K (HR<25%) (via two point-cell measurements).

140 °C, HR<25%) (Figure 13.16 crosslinked by tetraethylenediamine (TEPA)). As in the case of membranes based on noncrosslinked poly(CTFE-*alt*-IEVE)-*g*-1H-1,2,4-triazole$_{95\%}$ copolymer/s-PEEK blend, an optimal value of the molar ratio n exists for which the highest protonic conductivity can be reached. In both cases of crosslinking (from DiA or TEPA), membrane conductivities follow an Arrhenius law (quasilinear evolution of log(conductivity) vs 1000/T), as for noncrosslinked membranes.

13.4.4 Comparison of Conductivities of A/B-*ret*-DiA and A/B-*ret*-TEPA Membranes with Uncured Membranes Composed of Poly(CTFE-*alt*-IEVE)-*g*-1H-1,2,4-triazole$_{95\%}$ Copolymer/s-PEEK Blends

Figure 13.17 displays the proton conductivities of 60%-A/40%-B-*ret*-TEPA (n=1.7), 60%-A/40%-B-*ret*-DiA (n=1.5), M3-V-95 (40%)/60%-s-PEEK (n=1.7) membranes and of s-PEEK. These values correspond to the best proton conductivities.

Figure 13.17 and Table 13.6 indicate that for crosslinked membranes of 60%-A/40%-B-*ret*-TEPA (n=1.7) and 60%-A/40%-B-*ret*-DiA (n=1.5), the protonic conductivities are lower than that of the noncrosslinked membrane, that is, M3-V-95 (40%)/60%-s-PEEK (n=1.7), at various temperatures. Interestingly, the crosslinking of partially fluorinated chains (that bear triazole groups) reduced the mobility of polymeric chains that decreased the kinetics of

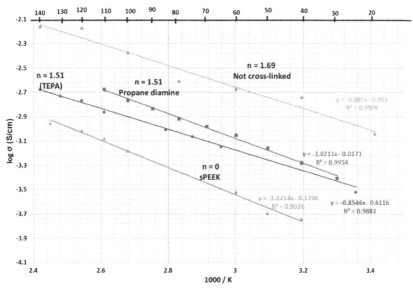

FIGURE 13.17 Comparison of proton conductivities of A/B-*ret*-DiA, A/B-*ret*-TEPA membranes and M3 membranes composed of s-PEEK of copolymer poly(CTFE-*alt*-IEVE)-*g*-1H-1,2,4-triazole$_{95\%}$ (non crosslinked), RH<25%.

orientation of nitrogenous heterocycles that ensure the protonic transport by a Grotthuss type mechanism (a proton hopping between triazole groups) through the membrane. Crosslinking via the DiA diamine enhances that phenomenon since that amine is shorter and thus brings more rigidity than TEPA (that has a longer and softer chain).

On the other hand, the pK_a values of TEPA (pK_a=9.7–9.5–8.3–5.0–3.3 [55]) indicate that acid–base interactions occur between s-PEEK's sulfonic acid functions (pK_a~−1) and TEPA, and hence, the number of triazole groups protonated by s-PEEK decreases. Thus, the network based on hydrogen bonds formed between protonated and nonprotonated azoles is modified by interactions between acidic functions borne by s-PEEK and amino groups of TEPA. For n equivalent values, these observations can be explained by conductivities of the membranes crosslinked by TEPA that are lower than both those crosslinked by DiA and those of noncrosslinked membranes.

13.5 CONCLUSION

PEMs represent an important part of fuel cell systems and a wide variety of copolymers have already been used as membranes in the process of manufacturing PEMFCs (hydrogenated, aromatic, fluorinated, and aliphatic). PFSA membranes have been reported in many surveys because they have been employed in mature PEMFC technologies. As a challenging characteristic, life times >50,000h for

TABLE 13.6 Comparison of the Main Proton Conductivities of A/B-ret-DiA, A/B-ret-TEPA Membranes and M3 Membrane Composed of s-PEEK and poly(CTFE-alt-IEVE)-g-1H-1,2,4-triazole$_{95\%}$ Copolymer (Noncrosslinked), RH < 25%

Membrane	Thickness (μm)	Proton Conductivity at 50°C (S cm^{-1})	Proton Conductivity at 100°C (S cm^{-1})	Proton Conductivity at 140°C (S cm^{-1})
s-PEEK	45	2.0×10^{-4}	8.1×10^{-4}	9.0×10^{-4}
M3-A (n=1.7) (Noncrosslinked)	55	1.8×10^{-3}	4.2×10^{-3}	6.6×10^{-3}
60%-A/40%-B-IPN-DiA (n=1.51)	40	5.1×10^{-4}	1.7×10^{-3}	4.3×10^{-3}
60%-A/40%-B-IPN TEPA (n=1.7)	35	5.0×10^{-4}	0.2×10^{-3}	2.1×10^{-3}

Nafion® and 10,000 h for Dow® membranes have been obtained. However, PFSA membranes are still expensive, permeable to methanol, and lose their properties from 80 °C at a low RH while radiation-grafted membranes, which can be manufactured on a large scale, require improvement of their durability.

Although a new generation of membranes can overcome these limitations, they have their own respective unique characteristics, but they are still far from perfect. Finding acceptable materials in terms of price and performance is still challenging. A more recent challenge deals with membranes that can maintain a high performance at high temperatures (>150 °C) and a low RH (<25–30%), in which phosphonic acid or nitrogenous heterocycles (e.g., imidazole, benzimidazole, or pyrrazole) are tethered to the polymeric backbone of a different nature. These functions enable the migration of protons through a diffusion mechanism that involves a proton hopping from one azole site to one another via an "proton-acceptor/proton-donor" interaction. This could be possible by various syntheses of copolymers that display a suitable flexibility/softness at these temperatures.

Various designs of copolymers (block, graft, random, or alternating) that bear nitrogenous cycles demonstrated a rich potential for preparing original fuel cell membranes that offer a satisfactory conductivity at "high" temperatures (100–140 °C) and low RHs (<30%).

More recent studies have shown that original partially fluorinated poly(IEVE-*alt*-CTFE)$_{94\%}$-*co*-poly(CTFE-*alt*-GCVE)$_{6\%}$ and poly(IEVE-*alt*-CTFE)$_{82\%}$-*co*-poly(CTFE-*alt*-GCVE)$_{18\%}$ terpolymers are crosslinkable by the addition reaction between a cyclocarbonate and an amine groups. Original crosslinked membranes were processed from an s-PEEK-Na (under –SO$_3$Na form)/poly(CTFE-*alt*-IEVE)$_{94\%}$-*g*-1H-1,2,4-triazole-3-thiol$_{90\%}$-*co*-poly(CTFE-*alt*-GCVE)$_{6\%}$ terpolymer blend in the presence of primary aliphatic diamines DiA or TEPA). Indeed, crosslinking improves the membranes' mechanical properties and reduces their swelling rates (at 90 °C) by total immersion in water. However, their protonic conductivities were unexpectedly low (for the 40%-A/60-B-*ret*-DiA (n=1.5) membrane, σ=4.3 mS cm^{-1} at 140 °C, HR<25%). Crosslinking reduces the mobility of polymeric chains of the resulting membranes, and this limits the reorientation of nitrogenous heterocycles that allow proton transport at high temperatures and low RH thereby inducing a loss of conductivity values.

To reduce the swelling and to improve the membrane durability, it is necessary to crosslink future PEMs and/or to reinforce them at the same time. It is impractical to expect that a single type of membrane can meet the requirements of all the possible automotive, stationary, and portable fuel cell applications. Mechanical properties, oxidative stabilities, and single cell performances of those materials will be investigated in further studies.

Another elegant strategy deals with doping of the polymers or membranes with various acidic components such as triflic acid, phosphoric acid, poly(phosphoric acid), and any other acidic derivatives that must not be leached out in fuel cell conditions.

To evaluate the intrinsic performances of these membranes, some optimization of the membrane/electrode interfaces and processing of MAE must be carried out to ensure that fuel cell performances are not limited by such an interface. One of the possible approaches for the "future" PEMs is to enhance well-defined phase separation by properly increasing the length of blocks or side chains (e.g., multiblock copolymers or comb-shaped copolymers). In such a highly interdisciplinary field, collaboration between experts from various fields including organic, inorganic, and polymer chemists; physicists; electrochemists; and process and chemical engineers is required. Thus, the educational and industrial communities will continue attempting to overcome challenges to develop more efficient and economical PEMs based on well-suited materials. Further studies are required, especially to assess the behavior in fuel cell conditions. It is expected that these targets will attract academic and industrial researchers' interest.

ACKNOWLEDGMENTS

The authors thank Peugeot Citroen PSA Company, which supported this work (especially Dr I. Roche and X. Glipa) and Dr D. Jones and J. Rozière from the University of Montpellier (AIME team), and Pr. G. Silly from the University of Montpellier.

REFERENCES

[1] [a] G. Alberti, M. Casciola, L. Massinelli, B. Bauer, J. Membr. Sci. 185 (2001) 73–81.
   [b] Kreuer, K. D, J. Membr. Sci. 185 (2001) 29–39.
   [c] D.E. Curtin, R.D. Lousenberg, T.J. Henry, P.C. Tangeman, M.E. Tisack, J. Power Sources 131 (2004) 41–48.
[2] M.A. Hickner, H. Ghassemi, Y.S. Kim, B.R. Einsla, J.E. McGrath, Chem. Rev. 104 (2004) 4587–4611.
[3] K.A. Mauritz, R.B. Moore, Chem. Rev. 104 (2004) 4535–4585.
[4] [a] F.A. De Bruijn, Green Chem. 7 (2005) 132–150.
   [b] F.A. de Bruijn, V.A.T. Dam, G.J.M. Janssen, Fuel Cells 8 (2008) 3–22.
[5] J.L. Zhang, Z. Xie, J.J. Zhang, Y.H. Tanga, C.J. Song, T. Navessin, Z.Q. Shi, D.T. Song, H.J. Wang, D.P. Wilkinson, Z.S. Liu, S. Holdcroft, J. Power Sources 160 (2006) 872–891.
[6] [a] K. Xu, C. Chanthad, M.A. Hickner, Q. Wang, Chem. Mater. 20 (2010) 6291–6298.
   [b] A. Collier, H. Wang, X. ZiYuan, J. Zhang, D.P. Wilkinson, Int. J. Hydrogen Energy 31 (2006) 1838–1854.
[7] [a] K. Prater, J. Power Sources 29 (1990) 239–248.
   [b] J.A. Kerres, J. Membr. Sci. 185 (2001) 3–27.
   [c] in: W. Vielstich, A. Hubert, M. Gasteiger, A. Lamm (Eds.), Handbook of Fuel Cells—Fundamentals, Technology and Applications; Fuel Cell Technology and Applications, Wiley, New York, 2003.
   [d] O. Savadogo, J. New Mater. Electrochem. Syst. 1 (1998) 47–66.
   [e] T. Xu, J. Membr. Sci. 263 (2005) 1–29.
   [f] C. Iojoiu, F. Chabert, M. Marechal, N.E. Kissi, J. Guindet, J.Y. Sanchez, J. Power Sources 153 (2006) 198–209.

[8] [a] R.C.T. Slade, J.R. Varcoe, Solid State Ionics 176 (2005) 585–597.
[b] J.R. Varcoe, R.C.T. Slade, Fuel Cells 5 (2005) 187–200 Weinheim, Germany.
[c] J.R. Varcoe, R.C.T. Slade, E.L.H. Yee, Chem. Commun. (2006) 1428–1429.
[9] Q. Li, R. He, J.O. Jensen, N.J. Bjerrum, Chem. Mater. 15 (2003) 4896–4915.
[10] R. Souzy, B. Ameduri, Functional fluoropolymers for fuel cell membranes, in: T. Nakajima, H. Groult (Eds.), Fluorinated Materials for Energy Conversion, Elsevier, Oxford, 2005, pp. 469–485 (Chapter 21).
[11] [a] K. Miyatake, M. Watanabe, Electrochemistry 73 (2005) 12–19.
[b] B. Smith, S. Sridhar, A.A. Khan, J. Membr. Sci. 259 (2005) 10–26.
[c] K. Miyatake, M. Watanabe, J. Mater. Chem. 16 (2006) 4465–4467.
[d] S.J. Hamrock, M.A. Yandrasits, Polym. Rev. 46 (2006) 219–244.
[e] Y. Shao, G. Yin, Z. Wang, Y. Gao, J. Power Sources 167 (2007) 235–242.
[f] R. Devanathan, Energy Environ. Sci. 1 (2008) 101–119.
[g] Q. Li, J.O. Jensen, R.F. Savinell, N.J. Bjerrum, Prog. Polym. Sci. 34 (2009) 449–477.
[h] R.H. Puffer, S.J. Rock, J. Fuel Cell Sci. Technol. 6 (2009).
[i] S.M.J. Zaidi, Research Trends in Polymer Electrolyte Membranes for PEMFC, Springer Science, Weinhem, 2009.
[j] N. Gourdoupi, J.K. Kallitsis, S. Neophytides, J. Power Sources 195 (2010) 170–174.
[k] K.E. Martin, J.P. Kopasz, Fuel Cells 9 (2009) 356–362.
[l] S. Bose, T. Kuila, T.X.H. Nguyen, N.H. Kim, K.-T. Lau, J.H. Lee, Prog. Polym. Sci. 36 (2011) 813–843.
[m] Y. Wang, K.S. Chen, J. Mishler, S.C. Cho, X.C. Adroher, Appl. Energy 88 (2011).
[n] A.-C. Dupuis, Prog. Mater. Sci. 56 (2011) 289–327.
[o] H. Bai, W.S.W. Ho, Polymer Int. 60 (2011) 26–41.
[p] C.H. Park, C.H. Lee, M.D. Guiver, Y.M. Lee, Prog. Polym. Sci. 36 (2011) 1443–1498.
[q] A. El-Kharouf, A. Chandan, M. Hattenberger, B.G. Pollet, J. Energy Inst. 85 (2012) 188–200.
[12] C. Laberty-Robert, K. Valle, F. Pereira, C. Sanchez, Chem. Soc. Rev. 40 (2011) 961–1005.
[13] [a] W.G. Grot, Makromol. Chem. 82 (1994) 161–172.
[b] M. Doyle, G. Rajendran, Perfluorinated membranes, in: W. Vielstich, A. Lamm, H. Gasteiger (Eds.), Handbook of Fuel Cells: Fundamentals, Technology and Applications, Fuel Cell Technology and Applications, vol. 3, Wiley, NewYork, 2003, pp. 351–395. Chapter 30.
[c] V. Arcella, A. Ghielmi, G. Tommmasi, Ann. N. Y. Acad. Sci. 984 (2003) 226–244.
[d] Y.S. Yang, A. Siu, T.J. Peckham, S. Holdcroft, Adv. Polym. Sci. 215 (2008) 55–126.
[e] M. Yoshitake, A. Watakabe, Adv. Polym. Sci. 215 (2008) 127–155.
[14] R. Dillon, S. Srinivasan, A.S. Arico, V. Antonucci, J. Power Sources 127 (2004) 112–126.
[15] H.P. Brack, M. Wyler, G. Peter, G.G. Scherer, J. Membr. Sci. 214 (2003) 1–19.
[16] S.J. Paddison, K.-D. Kreuer, J. Maier, J. Phys. Chem. Chem. Phys. 8 (2006) 4530–4538.
[17] H.L. Yeager, A.A. Gronowski, Membrane applications, in: M.R. Tant, K.A. Mauritz, G.L. Wilkes (Eds.), Ionomers, Synthesis, Structure, Properties, and Applications, 1995, pp. 333–364.
[18] M. Martinez, Y. Molmeret, L. Cointeaux, C. Iojoiu, J.-C. Leprêtre, N.E. Kissi, P. Judeinstein, J.-Y. Sanchez, J. Power Sources 195 (2010) 5829–5839.
[19] H. Ghassemi, G. Ndip, J.E. McGrath, Polymer 45 (2004) 5855–5862.
[20] J.A. Asensio, E.M. Sanchez, P. Gomez-Romero, Chem. Soc. Rev. 39 (2010) 3210–3239.
[21] A. Chandana, M. Hattenberger, A. El-kharouf, S. Du, A. Dhir, V. Self, B.G. Pollet, A. Ingrama, W. Bujalski, J. Power Sources 231 (2013) 264–278.
[22] E.J. Carlson, P. Kopf, J. Sinha, S. Sriramulu, Y. Yang, in: National Renewable Energy Laboratory, 2005.

[23] J. Spendelow, J. Marcinkoski, Report Record 9012 from the Department of Energy, USA. http://www.hydrogen.energy.gov/pdfs/review09/46600-02_intro.pdf, 2009 (accessed on 03.03.2014).
[24] J. Spendelow, J., Marcinkoski report from the Department of Energy, USA. Record 11012, http://www.hydrogen.energy.gov/pdfs/12020_fuel_cell_system_cost_2012.pdf, 2012 (accessed on 29.04.2014).
[25] [a] B. Ameduri, B. Boutevin, Well Architectured Fluoropolymers, Elsevier, Amsterdam, 2004.
[b] F. Boschet, B. Ameduri, Chem. Rev. 114 (2014) 927–980.
[26] C. Wieser, Fuel Cells 4 (2004) 245–250.
[27] D.J. Jones, J. Rozière, in: W. Vielstich, A. Lamm, H. Gasteiger (Eds.), Handbook of Fuel Cell Technology, vol. 3, John Wiley & Sons, 2003, pp. 447–455.
[28] K.D. Kreuer, New proton conducting polymers for fuel cell applications, in: B.V.R. Chowdari (Ed.), Solid State Ionics: Science and Technology, World Science Publishing, Singapore, 1998, pp. 263–274.
[29] K.D. Kreuer, A. Fuchs, M. Ise, M. Spaeth, J. Maier, Electrochim. Acta 43 (1998) 1281.
[30] [a] S.U. Celik, A. Bozkurt, Electrochim. Acta 56 (2011) 5961–5965.
[b] M.S. Boroglu, S.U. Celik, A. Bozkurt, I. Boz, J. Membr. Sci. 375 (2011) 157–164.
[c] S.U. Çelik, A. Bozkurt, S.S. Hosseini, Prog. Polym. Sci. 37 (2012) 1265–1291.
[31] [a] M. Schuster, W.H. Meyer, G. Wegner, H.G. Herz, M. Ise, K.D. Kreuer, J. Maier, Solid State Ionics 145 (2001) 85–92.
[b] M. Schuster, T. Rager, A. Noda, K.D. Kreuer, J. Maier, Fuel Cells 5 (2005) 355–365.
[c] G. Scharfenberger, W.H. Meyer, G. Wegner, M. Schuster, K.D. Kreuer, J. Maier, Fuel Cells 6 (2006) 237–250.
[32] [a] E.A. Munch, P. Jannasch, Polymer 46 (2005) 7896–7908.
[b] J.C. Persson, P. Jannasch, Macromolecules 38 (2005) 3283–3289.
[c] E.A. Munch, P. Jannasch, Solid State Ionics 177 (2006) 573–579.
[33] [a] S. Martwiset, R.C. Woudenberg, S. Granados-Focil, O. Yavuzcetin, M.T. Tuominen, E.B. Coughlin, Solid State Ionics 178 (2007) 1398–1403.
[b] R.C. Woudenberg, O. Yavuzcetin, M.T. Tuominen, E.B. Coughlin, Solid State Ionics 178 (2007) 1135–1141.
[34] [a] C.S. Karthikeyan, S.P. Nunes, K. Schulte, Macromol. Chem. Phys. 207 (2006) 336–341.
[b] D. Gomes, J. Roeder, M.L. Ponce, S.P. Nunes, J. Membr. Sci. 295 (2007) 121–129.
[c] M.L. Ponce, M. Boaventura, D. Gomes, A. Mendes, L.M. Madeira, S.P. Nunes, Fuel Cells 8 (2008) 209–216.
[35] S. Li, Z. Zhou, M. Liu, W. Li, WO2005/072413 A2, (assigned to Toyota Technical Center, USA Inc. and Georgia Tech Research Corporation).
[36] K.D. Kreuer, Solid State Ionics 145 (2001) 85–92.
[37] G. Frutsaert, L. Delon, G. David, B. Ameduri, D.J. Jones, X. Glipa, J. Roziere, J. Polym. Sci. Part A: Polym. Chem. 48 (2010) 223–231.
[38] G. Frutsaert, G. David, B. Ameduri, D.J. Jones, J. Roziere, X.J. Glipa, J. Membr. Sci. 387 (2011) 127–133.
[39] [a] I. Roche, J. Roziere, D.J. Jones, B. Campagne, B. Ameduri, G. David, (assigned to PSA and CNRS) FR Patent 2014/2992650. [b] B. Campagne, G. David, B. Ameduri, D.J. Jones, J. Roziere, I. Roche, Macromolecules 46 (2013) 3046–3057
[40] J.W. Traer, G.R. Goward, Phys. Chem. Chem. Phys. 12 (2010) 263–272.
[41] G.R. Goward, M.F.H. Schuster, D. Sebastiani, I. Schnell, H.W. Spiess, J. Phys. Chem. B 106 (2002) 9322–9334.

[42] S.U. Celik, U. Akbey, R. Graf, A. Bozkurt, H.W. Spiess, Phys. Chem. Chem. Phys. 10 (2008) 6058–6066.
[43] B. Campagne, G. Silly, G. David, D.J. Jones, J. Rozière, I. Roche, B. Ameduri, RSC advances, 4 (2014) 28769–28779.
[44] [a] J.R. Varcoe, R.C.T. Slade, Fuel Cells 5 (2005) 187–200.
[b] M.M. Nasef, Fuel cell membranes by radiation-induced graft copolymerization: current status, challenges, and future directions, in: S.M.J. Zaidi, T. Matsuura (Eds.), Polymer Membranes for Fuel Cells, Springer, Weinheim, 2009, ISBN: 978-0-387-73531-3, pp. 87–114 (Print) 978-0-387-73532-0 (Online).
[45] G. Couture, A. Alleaddine, F. Boschet, B. Ameduri, Prog. Polym. Sci. 36 (2011) 1521–1557.
[46] L. Sauguet, B.Ameduri, B. Boutevin, J. Polym. Sci. Part A: Polym. Chem. 44 (2006) 4566–4578.
[47] [a]M. Gao, M. Taillades-Jacquin; D.J. Jones, J. Rozière, M. DuPont, Y. Nedelec, Y.M. Zhang, in: Proceedings of Fluoropolymer 2010 Conference, Meze (France) June 13–16, 2010.
[b] S. Subianto, M. Pica, M. Casciola, P. Cojocaru, L. Merlo, G. Hards, D.J. Jones, J. Power Sources 233 (2013) 216–230
[48] Y.-M. Zhang, L. Li, J. Tang, B. Bauer, W. Zhang, H.-R. Gao, M. Taillades-Jacquin, D.J. Jones, J. Rozière, N. Lebedeva, R. Mallant, ECS Trans. 25 (2009) 1469–1472.
[49] L. Chikh, V. Delhorbe, O. Fichet, J. Membr. Sci. 368 (2011) 1–17.
[50] G.K.S. Prakash, M. Smart, Q.J. Wang, A. Atti, V. Pleynet, B. Yang, K. McGrath, G.A. Olah, S.R. Narayanan, W. Chun, M. Valdez, J. Fluorine Chem. 125 (2004) 1217.
[51] I. Roche, B. Campagne, G. David, B. Ameduri, D.J. Jones, J. Roziere, FR Demand (INPI n° 1450590) deposited on January 23, 2014. (assigned to Peugeot Citroën- PSA and CNRS).
[52] A. Alaaeddine, F. Boschet, B. Ameduri, B. Boutevin, J. Polym. Sci. Part A: Polym. Chem. 50 (2012) 3303–3312.
[53] R. Hamieh, A. Alaaddine, B. Campagne, S.M. GuillaumeB. Ameduri, S. Caillol, J-F. Carpentier, Polym. Chem., 5 (2014) 5089–5600. http://dx.doi.org/10.1039/C4PY00547C.
[54] Q. Zhao, J. Benziger, J. Polym. Sci. Part B: Polym. Phys. 51 (2013) 915–925.
[55] N. Sun, A. Avdeef, J. Pharm. Biomed. Anal. 56 (2011) 173–182.

Chapter 14

The Use of Per-Fluorinated Sulfonic Acid (PFSA) Membrane as Electrolyte in Fuel Cells

Madeleine Odgaard
IRD Fuel Cells A/S, Kullinggade, Svendborg, Denmark

Chapter Outline

14.1 Introduction 326	14.3.7 Degradation and Durability Aspects 341
14.2 Polymer Electrolyte Fuel Cells 328	**14.4 Application and Performance of PFSA Membranes in FCs** 345
14.2.1 Principle of Operation 328	14.4.1 Introduction 345
14.2.2 The Role of the Polymer Electrolyte 331	14.4.2 The PEMFC Stack 346
14.3 Properties of the PFSA Membrane 331	14.4.3 Application of PEMFC 348
14.3.1 Introduction 331	14.4.3.1 Automotive Applications 349
14.3.2 Perfluorinated Membranes in General 332	14.4.3.2 Stationary Application 351
14.3.3 The Nafion® Membrane 333	14.4.3.3 Portable Application 356
14.3.4 Morphology and Proton Conductivity 336	14.4.4 DMFC and Their Applications 356
14.3.5 Water and Methanol Transport a Technological Aspect 339	14.4.4.1 The Direct Methanol Fuel Cell 357
14.3.6 Other Properties Facing the FC Requirements 340	14.4.4.2 Applications of DMFC 360
	References 364

14.1 INTRODUCTION

The proton exchange membrane fuel cells (PEMFCs) are electrochemical devices that efficiently convert chemical energy of the fuel directly into electrical energy. They operate like batteries and are similar in characteristics and components; the principle of operation is described in more detail in Section 14.2.

Several types of fuel cells (FCs) exist classified after the type of electrolyte used, e.g., solid oxide fuel cell (SOFC), molten carbonate fuel cell (MCFC), phosphoric acid fuel cells (PAFCs), and the PEMFC. The operating temperature of the FCs is determined by the electrolyte used. This chapter covers the PEMFC only; the explanation and application of the other FC types is covered in several books and reviews published in the past decade [1–7], as well as in historical reviews [8–14].

The PEMFC operates at low temperatures (60–80 °C), allowing quick start-ups and immediate response to changes in the power demand. The use of FCs gives several advantages compared to conventional power generator systems. They offer a source of electrical energy that is continuous and environmentally safe and additional benefits include low maintenance, excellent load performance, etc. In FCs, the chemical energy is directly converted into electricity, without preliminary conversion to heat. Consequently, this conversion is not limited by the Carnot cycle and 100% efficiency can be achieved.

The direct methanol fuel cells (DMFCs) technology described in Section 14.4.4, using methanol derived from biomass or other renewable energy sources, gives the same advantages as PEMFC technology, e.g., high energy efficiency and low or zero emissions. Although the DMFC has a lower performance than hydrogen FCs, it has a number of additional advantages: it does not have hydrogen storage problems, its fuel supply infrastructure is cheaper than for hydrogen, and emissions are significantly lower than for petrol- or diesel-fueled generators.

The development of PEMFC, also called the solid polymer fuel cell (SPFC), has as the name indicates, been strongly related to improvements in performance of the polymer electrolyte membrane. The use of an ion exchange membrane as electrolyte was first suggested by W.T. Grubb in 1957 [15,16], and the first FC system based on a sulfonated polystyrene electrolyte was developed by General Electric in the 1960s for National Aeronautics and Space Administration for applications such as an on-board power source in the Gemini space program [17]. The fact that the polystyrene sulfonate membrane was not electrochemically stable and only limited power density (less than $50\,mW\,cm^{-2}$) was achieved with an excessive noble metal loading per square centimeters of electrode ($10–40\,mg\,Pt\,cm^{-2}$) resulted in a focus on alkaline FCs and only academic interest in the PEMFC.

A major breakthrough for PEMFC started with the invention of a perfluorinated sulfonic acid (PFSA) membrane, introduced by E. I. du Pont de Nemours and Company in the 1960s under the name Nafion®, for application in the chlor alkali industry. The advantage of the Nafion membranes is the chemical stability. The membrane specifically developed for requirements of chemical stability toward strong sodium hydroxide and wet chlorine, which is a strong oxidizing agent, also gave increased lifetime in the PEMFC application needed. Dupont's Nafion® is one of the most advanced studied commercially available proton conducting polymer material and has been the preferred electrolyte material for both hydrogen PEMFCs (H_2-PEMFCs) and DMFCs.

Development over the past 30 years, starting with work performed from the mid-1980s (under contracts by the Canadian Department of National Defense), resulted in significant improved PEMFC performance [3]. Development of a similar perfluorinated membrane from Dow Chemical, tested in PEMFC at Ballard Power Systems gave improved power density. Groups at Los Alamos National Laboratory, improving the utilization of catalyst, achieved a further step in the development, in the sense that the noble metal loading needed to obtain high power densities could be reduced.

Today power densities up to $1\,W\,cm^{-2}$ using low metal loading per square centimeter electrode ($\sim 0.3\,mg\,Pt\,cm^{-2}$) have been demonstrated and PEMFCs are becoming well established in a number of markets where they now are recognized as an improved technology option compared to conventional internal combustion engine (ICE) generators or batteries.

Owing to their high energy efficiency, convenient operation, and environment-friendly characteristics, the PEMFC is seen as a system of choice for automotive systems, for stationary application such as prime power for buildings, for small-scale grid-connected micro-combined heat and power (μCHP) generators for residential use, for off-grid backup power systems providing uninterruptable power supplies to critical infrastructures such as telecommunication backup power, and for portable applications, e.g., consumer electronics or industrial power backup systems.

Nevertheless, several barriers still exist for a widespread FC commercialization, in particular, cost and durability of FC system [18–26]. Due to the current rather expensive perflourinated proton exchange membranes there is considerable application-driven interest in lowering the membrane cost.

The lifetime required by a commercial FC is over 5000 operating hours for light-weight vehicles and up to 60,000 h for stationary power generation with less than a 10% performance decay [27,28].

Through intensive research and development efforts during the past decade highly focusing on understanding the degradation mechanism of the FC components including the role of the PFSA membrane, significant achievements have been demonstrated in the field of FC durability and mitigation. Lifetimes of >20.000 h of operation using FCs based on PFSA polymer membrane have been shown [29–36]. Current understanding of commonly reported FC degradation phenomena are published in several papers and reviews [37–47].

14.2 POLYMER ELECTROLYTE FUEL CELLS

The FC is related to some of the earliest electrochemical discoveries. The first description of the FC principle goes back to 1839, published by Friedrich Schoenbein, and with the first demonstration of an FC in 1843 by William Grove [48]. In spite of the fact that all essential FC components and their various functions have been known for this long time, we still have to solve many of the basic materials and material-processing problems, essential if we are to achieve the goal of a commercial PEMFC in large scale without using rare and expensive materials and with long lifetime.

Strong industrial and scientific efforts are being spent in United States, Japan, Asia, and Europe to commercialize the PEMFC technology. Owing to their high energy efficiency, convenient operation, and environment-friendly characteristics, PEMFCs are considered one of the most promising FC technologies for both stationary and mobile applications.

14.2.1 Principle of Operation

The polymer electrolyte membrane (PEM) FC converts hydrogen and oxygen electrochemically into electrical power, heat, and water. The heart of the FC is the membrane electrode assembly (MEA), where the electrochemical reaction takes place. The MEA consists of the ion-conducting polymer electrolyte membrane sandwiched between the anode and the cathode: each containing a macroporous diffusion backing and active catalyst layer.

Hydrogen is split into protons and electrons at the anode (the negative electrode). The polymer electrolyte membrane placed in the center allows protons to pass from the anode to the cathode (positive electrode), while the electrons pass through the external circuit to the cathode. The electrons combine at the cathode with the protons that have crossed the membrane and with oxygen from the air, forming water (Figure 14.1).

The electrodes are usually made of a porous mixture of carbon-supported platinum and/or platinum alloys for cathode and anode respectively and ionomer (i.e., electrolyte), Figure 14.2. The catalyst particles must have contact to both protonic and electronic conductors. Furthermore, these must have passage for reactants to reach the catalyst sites. The contacting point of the reactants, catalyst, and electrolyte is conventionally referred to as the three-phase interface.

The catalysts are usually deposited as nanoparticles on the high-surface-area carbon. The MEAs can be fabricated in various sizes depending on the aimed power and/or voltage output and the electrodes may differ in catalyst–ionomer composition tailored to specific operating conditions.

Figure 14.3 shows images of MEAs and an scanning electron microscopic (SEM) image taken from a cross-section of a PEM MEA.

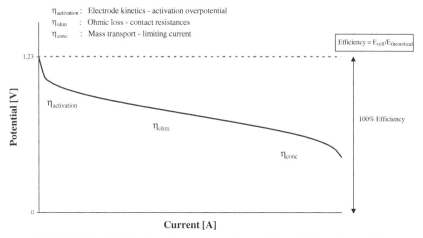

FIGURE 14.1 PEMFC electrode reactions and general fuel cell (FC) characteristic.

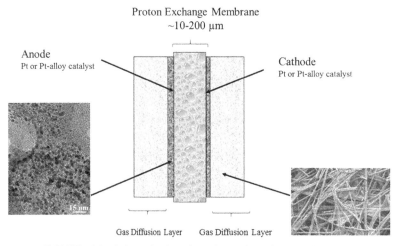

FIGURE 14.2 Schematic view of membrane electrode assembly (MEA).

During FC operation, a complex flow of reactants and reaction products exists in the pores of the electrodes. The cathode pores must allow gaseous oxygen to reach the catalyst surface and support efficient removal of product water to prevent flooding of the backing or catalyst layer. The anode pores must provide efficient transport of fuel from the flow field to the catalyst surface.

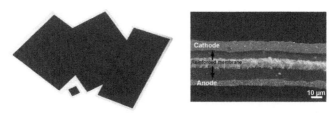

FIGURE 14.3 Membrane electrode assemblies (MAEs) in various sizes and cross-sectional SEM image of a PEM MEA using a 25-μm reinforced PFSA membrane with anode and cathode electrodes.

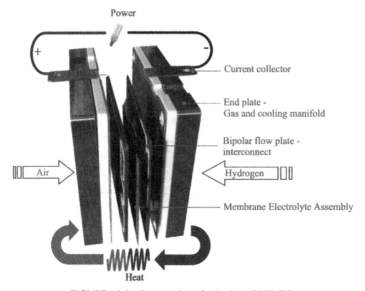

FIGURE 14.4 Presentation of a single-cell PEMFC.

The current is proportional to rate at which the reaction happens. The performance is thus often quoted in terms of current density (current per square centimeters) at a fixed voltage.

As long as fuel and oxygen (typically air) are supplied, the FC will continue to produce electricity and heat. The only byproduct when fueled with hydrogen is water. A single PEMFC (Figure 14.4) in operation generates a voltage <1 V.

Therefore, to obtain sufficiently high voltage levels for a specific application the FCs are stacked in series. They are electrically connected by bipolar flow plates, which also distribute the fuel to the MEA within each cell. The electrical efficiency, η_{el}, is defined as the FC voltage relative to the standard potential ($E^0 = 1.23\,V$) of the cell reaction.

The power density of a PEMFC depends on the operating conditions. The typical performance of a PEM single FC using a 25 μm PFSA (Nafion®) membrane operating at 70°C and 2.5 bar absolute pressure is shown in Figure 14.5.

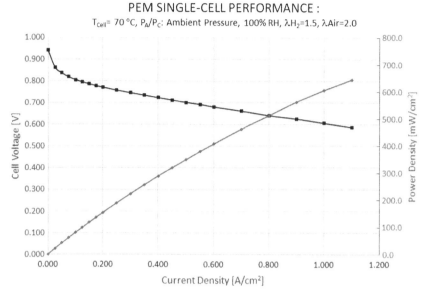

FIGURE 14.5 Single-cell performance obtained using a 25-μm PFSA membrane (Nafion®). The cell operating at 70 °C, at ambient pressure using hydrogen and air supplied at a stoichiometry of 1.5 and 2.0, respectively, and fully humidified (100% RH).

14.2.2 The Role of the Polymer Electrolyte

The major role of a polymer electrolyte in an FC is to provide ionic conductivity. The electrolyte must also prevent the passage of electrons, act as a barrier to the reactants, and maintain chemical, thermal, and mechanical stability. Chemical stability of the membrane in the oxidative and reductive electrochemical environments within the FC is crucial. It must be chemically inert to withstand possible degradations such as peroxide radical attachment on the polymer end groups. In the operating FC any gas or fuel permeating through the membrane, often quoted as fuel cross-over, is equivalent to an internal current (short) reducing the cell voltage. A reasonable mechanical strength and moderate dimensional changes are needed when the electrolyte membrane is used for making the MEA, incorporating the MEA into a stack, and during FC operation.

14.3 PROPERTIES OF THE PFSA MEMBRANE

14.3.1 Introduction

This chapter aims at describing the structure and properties of the PFSA membranes from an FC application point of view. A clear understanding of the structure of the fluorinated membrane and its relation to the electrochemical properties is necessary for full utilization of the membrane, and for future developments of the membrane. Transport of protons and associated water molecules is known to strongly depend on the membrane microstructure, although this is not necessarily well understood. The PFSA membrane reviewed and described in this chapter is by no means fully comprehensive. For further information attention is directed toward several excellent published reviews of the field of PFSA membranes [49–56].

The Nafion® brand, developed and introduced by DuPont in the 1960s [57,58], is one of the most extensively studied and most widespread of the commercially available proton conducting materials. Nafion® was originally developed for use in the chlor alkali industry, and was the largest application for Nafion membranes in the early 1980s [59]. Nafion membranes have many applications due to their high chemical and electrochemical stability, low permeability to reactant species, selective and high ionic conductivity, and ability to provide electronic insulation. Apart from the chlor alkaline cells applications cover water electrolysis, gas separation, sensors, dehydration/hydration of gas streams, recovery of precious metals, salt-splitting, and FCs [60–62]. Due to the fact that Nafion® is an expensive membrane, many attempts have been made to develop alternative materials. To date Nafion® or other similar PFSA membranes from other branches have been the preferred polymer electrolyte materials for FC application.

14.3.2 Perfluorinated Membranes in General

In general the perfluorinated membranes consist of a polytetrafluoroethylene (PTFE) backbone and side chains with acidic functionality. The success of Nafion has led to development of several variants of PFSA membranes since the 1960s by DuPont and other companies.

The synthesis and preparation of PFSA membranes with varied side chain types and/or length developed by the different companies vary from place to place and has continuously been changed and improved. In the mid-1980s significant improvements in FC performance was achieved with a membrane developed by Dow Chemical [63] and tested at Ballard in FCs [64]. The Dow membrane is similar to Nafion but prepared with shorter side chain, i.e., reduced equivalent weight (EW) 800 [65] corresponding to a higher SO_3H concentration. The process and manufacturing of the Dow membrane were complicated and expensive [66]. Other ion exchange perfluoropolymer membranes, with so-called long side chains (LSCs) were developed by companies such as Fumatech producing Fumion® membranes, Asahi Chemical that produces the Aciplex® membranes [67], and Asahi Glass Company producing the Flemion® membrane [68]. Membranes with short side chain (SSC) ionomers were later introduced by other companies, e.g., Aquivion™ (Solvay Solexis) and Fumapem® FS (Fumatech).

Figure 14.6 illustrates the chemical structure of different PFSA polymers, showing Nafion® and Fumion® as examples of the ones possessing the longest side chain and Aquivion™ as example of the shortest side chain.

A very significant improvement in FC performance was achieved by reducing the membrane thickness. Thinner membranes increase performance

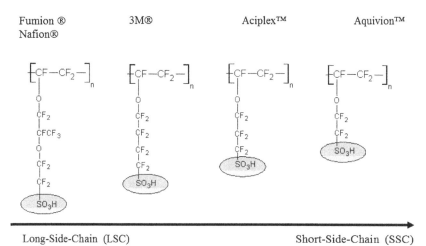

FIGURE 14.6 Structure of various PFSA ionomers with different length of the side chain.

efficiency partly due to lower cell resistance; however, they also have lower physical strength. Enhancing the durability of FCs is one of the key research and development goals being considered for widespread use of FCs. Membranes must be chemical stable for thousands of hours, over thousands of start–stop and humidity cycles, in a corrosive environment.

The advantage gained with the thinner membranes combined with a fundamental understanding of proton exchange membrane degradation has led to new approaches for making thin perfluorinated membranes using mechanical and chemical reinforcement [69–71]. Impregnating a Nafion solution into micro- or macroporous PTFE fabric was introduced by W.L. Gore and Associates under the trade name Gore-Select™ [72,73]. Dupont introduced a reinforced membrane Nafion® XL that combines the advantage of mechanical reinforcement with chemical stability (Section 18.3.5). Figure 14.7 shows the SEM images of a 40-μm-thick PFSA-reinforced membrane (Fumapem®940).

Many innovative attempts to develop alternative proton-conducting membranes for FC applications have taken place. The main motivations are (1) reduced membrane cost, (2) improved performance at higher temperatures, (3) lower requirement for humidification, and (4) reduced methanol permeability. The membranes vary from perfluorinated, partially fluorinated, or nonfluorinated aliphatic polymers to polymers with an aromatic backbone [52–54,74–79].

14.3.3 The Nafion® Membrane

The starting point for the DuPont technology is the perfluorination of the different monomers. The synthesis of Nafion is based on the copolymerization of tetrafluoroethylene (TFE) and a functional fluorinated monomer (vinyl ether) [80]. The sulfonic acid group functionality is introduced through the functional

FIGURE 14.7 SEM image of 40-μm reinforced Fumapem® F-940rf membrane.

$$-(CF_2-CF_2)_x-(CF_2-CF)_y$$
$$|$$
$$(O-CF_2-CF)_n-O-(CF_2)_p-SO_3H$$
$$|$$
$$CF_3$$

FIGURE 14.8 The general structure of the Nafion® membrane, with x = 5–13, p = 2, and y, n = 1.

sulfonyl fluoride groups (SO_2F). The general chemical structure of the Nafion perfluorosulfonic acid ionomer polymer is shown in Figure 14.8.

The length of the side chain, the composition of the polymer backbone, and the processing into a film determine the final polymer electrolyte membrane properties [60]. The perfluorinated backbone provides chemical and mechanical stability, the ether groups provides flexibility while the sulfonic acid groups yields high ionic conductivity [75]. The acid groups are fixed to the polymer and cannot leach out, while the counter ions (H^+) are free to migrate and can readily be exchanged with other ions, according to the general reaction scheme:

$$-SO_3-M_1^+ + M_2^+ \leftrightarrow -SO_3-M_2^+ + M_1^+$$

The ion exchange capacity of the polymer is directly related to the EW. The EW is defined as the molar mass of the polymer per sulfonic acid group:

$$EW = \frac{1}{IEC}$$

The "concentration" of the fixed ionic groups in the membrane determines the hydration of the polymer and hence the ionic conductivity and selectivity. As the length of the side chain decreases, i.e., lower value of EW, the number of sulfonic acid groups per mass increases resulting in higher conductivity. The desired EW is achieved by varying the ratio of vinyl ether monomer to PTFE. The Nafion® is commercially available in different forms such as extrusion- or dispersion-cast nonreinforced membranes, powders, tubes, and solutions.

TABLE 14.1 Nafion Properties, All Values Taken with Membrane Conditioned at 23 °C, 50% RH.

Membrane Type	Typical Thickness (µm)
N-111	25
N-112	51
NE-1135	89
N-115	127
N-117	183
NE-1110	254
Other Properties	
Conductivity (S cm^{-1})	0.083
Acid capacity (meq g^{-1})	0.89
Specific gravity	1.98
Tensile strength, max. (MPa)	43 in MD, 32 in TD
Tear resistance: initial (gm m^{-1})	6000 in MD, TD

MD, Machine direction; TD, Transverse direction [30].

The first commercially available Nafion membrane was Nafion® 120 (1200 EW, 250 µm thick), prepared by an extrusion-cast membrane manufacturing process [81]. Since then DuPont has been active in developing the membranes' performance by varying the EW and thickness. To date the Nafion membranes are manufactured and available in different EWs in the range between 1500 and 800, corresponding to ion exchange capacities in the range of 0.6–1.25 meq g^{-1}. The thickness ranges from 25 to 250 µm. The Nafion® 120 was followed by Nafion® 117 (1100 EW, 180 µm). Nafion® 115, 112, 111, and 105 are the latest developed membranes for FC application. The most common membranes used today in H$_2$ PEMFC are the thinnest Nafion® 112 and 111, and Nafion® 117 and Nafion® 115 are preferred for DMFC (Section 14.4). Table 14.1 shows the various Nafion® membranes and their physical properties [82].

Early work by Raistick [83] and Gottesfeld [84,85] showed that significant improvement of FC performance could be achieved by incorporating the electrolyte in the catalyst layer as a solution. This increased the utilization of the catalyst and thereby to a considerable reduction of the noble metal loading needed [86]. Today, dissolved PFSA in terms of dispersion is commonly used in the catalyst layers to prepare thin-film electrode layers [87–89]. The dispersions typically have solid contents ranging from 5 to 20% by weight.

14.3.4 Morphology and Proton Conductivity

PFSA materials and Nafion membranes have been extensively studied to clarify the correlation between morphology and proton transport [90–95]. A summary of various proposed models was described in the first edition of this book [82].

Using a number of techniques, including small- and wide-angle X-ray scattering and neutron scattering [65,96–103], electron microscopy [104–106], atomic force microscopy (AFM) [107], nuclear magnetic resonance (NMR) [108–113], and electron spin resonance (ESR), infrared (IR), and Raman spectroscopy [114–116], several models have been proposed to explain the structure of the Nafion membranes [117–119].

Different models have been proposed to interpret results from these experiments. Eisenberg [120] presented the existence of ion cluster and the most widely referenced model for PFSA was proposed by Gierke [117,121,122]. Followed by this work other models were proposed, e.g., the "channel network model" by Kreyer [101] and the "parallel cylindrical channel" model by Schmidt-Rohr and Chen [123]. Although these models differ in the geometry and spatial distribution of the ionic clusters, their common feature is that there exists a phase separation between the hydrophobic backbones and hydrophilic sulfonic-acid-terminated side chains which are located in water clusters. The size and shape of the hydrophilic clusters strongly depend on the water content in the material.

The most important electrolyte membrane property for the PEMFC application is proton conductivity. The membranes must provide an ionic pathway for the protons produced at the anode to the cathode. Low proton conductivity of the membrane results in high ohmic resistance. The role of using thinner membranes in FC applications is to achieve higher FC performance; reduced ohmic resistance of the PFSA membrane results in a significantly reduced slope in the pseudolinear region of the FC performance curve, i.e., cell potential vs current density.

Nafion is an excellent proton conductor with conductivity similar to 1.0 M sulfonic acid. The Nafion polymer behaves similar to an aqueous acid in the sense that it is limited to operate at temperatures below the boiling point of water. The conductivity of Nafion is $0.08\,S\,cm^{-1}$ at 25 °C and 100% relative humidity (RH) [124].

The proton conductivity of the polymer membrane has been studied extensively in the past 30 years, using AC impedance spectroscopy and DC techniques [67,100,109,125–142]. Several groups have examined the effect of temperature on the conductivity of Nafion in detail. In general the proton conductivity increases with temperature but it is also depends on the variation in water content, i.e., proton mobility, and hence the specific resistance is related to the hydration level. At elevated temperatures the membranes tend to dehydrate and conductivity tends to decrease [100,128,130,131,136,143–149]. Proton conduction can only occur at desired rates when the PFSA membrane is hydrated with water. During FC operation, the hydration level might change due to a transient state of operation. Maintaining optimal hydration of the membrane requires additional water management in an FC system such as air and fuel humidifiers, water recirculators, etc., adding cost and complexity to the entire

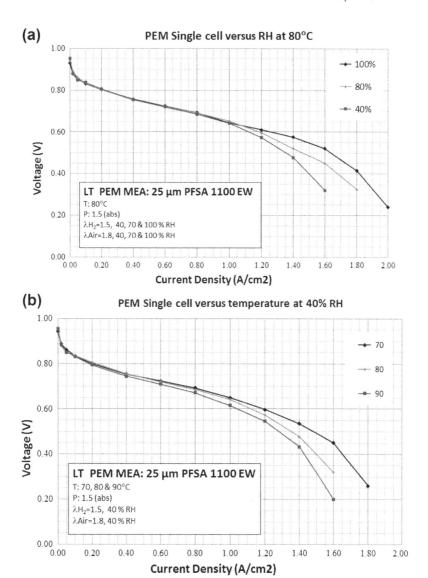

FIGURE 14.9 Single-cell polarization curves for an membrane electrode assembly (MEA) based on 25-μm Nafion® 1100-EW membrane measured at 80 °C with (a) different levels of humidification ranging from 40% RH to 100% RH and (b) at different temperatures at 40% RH.

FC system. This trend will be reflected in the performance of the operating FC. Figure 14.9 shows the single-cell FC polarization curves for an MEA based on 25-μm Nafion® 1100-EW PTFE-reinforced membrane measured at 80 °C with (1) different levels of humidification ranging from 40% RH to 100% RH and (2) at different temperatures at 40% RH.

Comprehensive reviews and papers covering the proton transport properties and more fundamental understanding of the proton transport mechanism of PFSA membranes and Nafion combining experimental data with nonequilibrium statistical mechanical transport models has been published [90,101,150–160].

The hydrated hydrophilic domains provide the high conductivity and depend very much on the presence of water. Due to the hygroscopic nature of the sulfonic acid group the Nafion® membrane adsorbs water. The water in the membrane exists at least in three different stages (ref til mig selv). The first two to three water molecules (per sulfonic acid group) adsorbed by the polymer interact almost entirely with the acid groups forming its primary hydration shell. These first water molecules are tightly bound to the ions. The next water molecules entering the matrix are more weakly bound and able to solvate the protons, and the final stage is bulk water filled in the open pore structure. Proton transfer in solid polymer electrolytes follow two principle mechanisms: a Grotthus-like hopping and a "vehicle" mechanism [161–165].

The mechanism of proton conduction in full-hydrated perfluorinated membranes has been suggested by Kreuer et al. [151] to be comparable to liquid water. The proton conductivity diffusion coefficient D_σ roughly follows the water diffusion coefficient D_{H_2O} for Nafion as function of temperature and water content. The protons are mainly located in the central part of the hydrated hydrophilic nanochannels. In this region the water is bulk-like with proton transport similar to protons in liquid water. Proton transfer is thought to occur along the hydrogen bonds of water network and Grotthus hopping becomes an important transport mechanism. For decreasing water contents, where the water is more bound through hydrophilic/electrostatic interactions with the $-SO_3^-$ group, the proton mobility decreases, and becomes dependent and cooperational with the water movement. This mode of proton movement is sometimes described as the "vehicle mechanism," since movement of protons occurs as complexes, e.g., H_3O^+.

For the past 40 years, LSC PFSA membranes such as Nafion® have been the benchmark of PFSA ionomer membrane due to their superior chemical and mechanical stability. Membranes possessing high proton conductivity at lower RH at elevated temperatures are currently the focus of a lot of research. SSC-PFSA-based membranes show promising potential for PEMFC application. SSC ionomers can theoretically be prepared with lower EW, and hence exhibit higher proton conductivity without causing excessive swelling or dissolution. For a given EW more TFE units are present in the backbone of the sort side chain compared to its LSC analogs, providing a higher degree of crystallinity, thereby improving the resistance to swelling in water [65,146,166–179].

Figure 14.10 shows the conductivity measured at various humidification levels for an SSC-based membrane (750 EW) compared to LSC-based membrane (1100 EW). At lower RH the membrane with low EW shows higher conductivity than the high-EW membrane. At 40% RH the 1100-EW membrane had a conductivity of $0.019\,S\,cm^{-1}$, while the conductivity of the 750-EW membrane under the same condition was $0.057\,S\,cm^{-1}$.

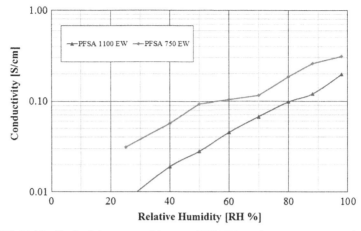

FIGURE 14.10 Conductivity compared between 1100-EW membrane over a range of relative humidity (RH) versus 750-EW-based membrane measured at 80 °C.

The challenge in membrane development is to achieve excellent proton conductivity without sacrificing mechanical and chemical properties. Although SSC PFSA membranes show improved FC performance with relatively good mechanical properties the stability of SSC-based membranes could be a possible concern for their widespread application in PEMFC applications [71,180–182].

14.3.5 Water and Methanol Transport a Technological Aspect

Water transport inside perfluorinated membranes is a complex phenomenon. The water balance of the membrane has been extensively studied [183–197] as the protonic diffusivity and conductivity of the membrane depend on the hydration level. In practical FCs the membrane is typically operated in a partially dehydrated form.

Factors affecting the water balance in a PEMFC are: (1) water adsorption from the vapor phase, (2) electro-osmotic drag, and (3) back diffusion of water. The transfer of protons from anode to cathode is associated with the electro-osmotic drag of water in the same direction. Each proton will drag at least one water molecule in water-vapor-saturated membranes and more (~2–2.5) in liquid-water-saturated membranes [130,184]. Figure 14.11 illustrates the transport mechanism in the membrane in an operating FC.

Without addition of water to the anode this would lead to dehydration of the membrane and very reduced membrane conductivity. The dehydration is partially balanced by back diffusion of water produced in the cathode. The water balance is highly dependent on the operating conditions of the PEMFC, such as temperature, pressure, humidity of the gasses, and cell current. Effective operation of the PEMFC thus requires delicate control over the water balance in the entire stack.

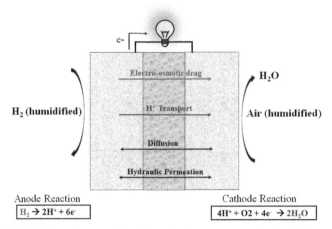

FIGURE 14.11 Schematic presentation illustrating the transport mechanism in the membrane of an operating fuel cell.

In contrast to H_2-fueled PEMFC, the DMFC anode is supplied with a liquid methanol/water mixture maintaining full hydration of the electrolyte membrane.

A shortcoming of the PFSA (Nafion®) membrane related to the DMFC is high methanol permeability, which drastically reduces the performance [14,198–205]. Water is needed for anode reaction in the DMFC (Section 14.4). This implies that pure methanol cannot be used, but a mixture of methanol/water is necessary. Methanol mixes very easily with water and methanol readily diffuses through the Nafion® membrane as well as other commercially available proton exchange membranes. This results in low fuel utilization, low cell voltage, and excess catalyst loading (Section 14.4.4).

14.3.6 Other Properties Facing the FC Requirements

Besides being a good proton conducting membrane, it must also serve as an electronic insulator. Any gas or fuel permeation from the anode to the cathode side through the polymer causes an internal chemical short circuit to the FC, and decreases the power and energy density. If fuel hydrogen/methanol are simultaneously present at the cathode, reactions run in parallel and the net current is the sum of the anodic oxidation and the cathodic oxygen and the "fuel cross-over" is often expressed as mass flow, and converted into the corresponding parasitic electronic current, using Faraday's law [206].

Investigations of mass transport parameters through Nafion membranes, like permeation, diffusion, and solubility of different gasses and methanol, has therefore been carried out by several authors [128,134,207–209]. Gas permeation is found to be a function of the RH, or the water content in the membrane and permeation can be separated into a solubility and a diffusive component in the aqueous and the hydrophobic perfluoro phase of the membrane [208]. It is also found that the permeability through the water-saturated sample is higher than that through the sample equilibrated with water vapor. From the literature it is known that when Nafion® takes up water from the liquid phase; the number of moles of water per mole of sulfonic acid is significantly larger than the number obtained by equilibration with saturated water vapor. This so-called Schroeder's paradox was explained by Zawodzinski et al. [128], as the difficulty in condensing vapor within the pores of the membrane. Due to the different water contents different permeabilities are to be expected. The pronounced dependency of the permeability also indicates that transport of the gases primarily takes place through the hydrophilic regions in Nafion® [207].

14.3.7 Degradation and Durability Aspects

To commercialize PEMFC technology, one of the most critical criteria is to achieve long lifetime of the overall system, especially for the residential stationary power market, requiring durable performance for many years of operation. Several reviews have been published in the field of FC durability issues of the various FC components and mitigation [37–42,44–47,210–212]. Numerous publications cover ex situ materials testing as well as in situ and postmortem single-cell and stack analysis.

Membranes must be durable for thousands of hours, over thousands of start–stop and humidity cycles, in a corrosive environment. As mentioned in Sections 18.3.5 and 18.3.6 performance of the FC is strongly related to water movement. In a number of studies the hydration level, pressure, and temperature have shown to have influence on the membrane durability with respect to long-term operation [181,213–220]. Variations in the hydration level within the MEA can lead to membrane failure which is believed to be caused by chemical and mechanical effects acting together.

PEMFC durability is often evaluated by monitoring the cell voltage over time. The voltage is commonly seen to, at first, decrease gradually but nondramatically, which has mainly been associated with the degradation of electrodes through the mechanisms of platinum catalyst dissolution, particle growth, and corrosion of the carbon support [37]. After a certain time, at which the gradual performance decay may or may not have become intolerable for practical applications, the PEMFC suddenly fails. This "sudden death" is associated with significant reactant gas cross-over through pinholes or tears formed in the membrane. Membrane degradation can occur through viscoelastic creep (mechanical degradation) and through chemical degradation, with either mechanisms able to cause thinning of the membrane as well as crack formation and propagation. The concomitant deterioration of the barrier properties of the membrane leads to an increase in reactant gas cross-over, which in turn decreases the open circuit voltage. Eventually, hydrogen and oxygen can permeate the membrane to an extent where the exothermic combustion reaction between these two gases can bring about temperatures high enough to melt or combust the membrane locally, which leads to pinhole formation and failure of the membrane and the FC as a whole.

For PFSA-based membranes, the major factors contributing to chemical failure of the membrane involves an unzipping mechanism via hydroxyl (HO·) or hydroperoxyl (HOO·) radical attack of the carboxylic acid end group and cleavage of the C–O or C–S bonds in the side chain of the membrane [59,212,221]. The formation of radicals responsible for the chemical degradation is compounded by degradation of barrier properties, since the radicals are formed through processes at the electrodes, especially when oxygen permeates to the low potential of the anode [222]. More specifically, hydrogen peroxide may form in connection with oxygen diffusion through the electrolyte toward the anode followed by incomplete reduction. Subsequent peroxide reaction with Fenton-active transition metal ion impurities, e.g., Fe^{2+}, leads to formation of the detrimental peroxide radicals upon homolytic cleavage of the oxygen–oxygen bond in the hydrogen peroxide molecule. In practice, transition metal trace amounts may be present as impurities in the MEA, originating from components such as end plates [42]. In summary the chemical failure of the membrane will cause membrane thinning, release of fluoride ions, increased gas cross-over, and voltage degradation.

Mechanical degradation primarily arise from variations in RH, e.g., cyclic stresses and strains during start–stop of the FC. Such variations cause swelling and dehydration of the membrane, which translates into compressive and tensile stresses when the membrane is constrained in an assembled FC stack. Furthermore, anisotropic current density, hot spots, and poking/piercing action of catalyst particles or carbon fibers can give rise to localized stresses either directly or through anisotropies in water content. These stresses can lead to viscoelastic creep of the material, i.e., stress relief by stretching of the polymer chains, and slip of chains out of entanglements [47]. Mechanical degradation is often the cause of early FC life failures, especially for very thin membranes. Mechanical and chemical reinforcement of the membranes have proven to be superior with respect to durability.

The need for being able to conduct experiments and materials testing in accelerated tests can easily be understood considering the target lifetimes of FC systems as presented in the United States Department of Energy (DoE) Fuel Cell Technologies Office Multi–Year Research, Development and Demonstration Plan and in the Annual Implementation Plans published from the Fuel Cells and Hydrogen Joint Undertaking in Europe [27,28].

Various accelerated stress test (AST) protocols are designed to rapidly induce FC degradation, while some of them are intended to emphasize certain degradation modes or mechanisms over others [223–225]. Beyond component testing, ASTs can also be designed to provide information about FC durability on a cell, stack, or system level under various operating conditions and environments, e.g., repeated start–stop cycles, load cycles replicating a certain application, etc. Such tests are valuable supplements to the vastly expensive and time-demanding nonaccelerated tests (e.g., steady–state operation, field tests).

Membrane ASTs are often distinguished by whether they intend to do damage by a chemical or mechanical route. Mechanical degradation is generally studied by exposing the membrane, in situ, to rapid variations in RH from zero to unity at temperatures close to the operating temperature of a typical low temperture (LT)–PEMFC (Figure 14.12). Chemical degradation, on the other hand, is studied in situ by subjecting the cell to open cell voltage (OCV) conditions at reduced humidity (e.g., 30%) for an extended period of time (Figure 14.12). This is intended to provide optimal conditions for radical formation via oxygen permeation to the low potential of the anode, which is accomplished by (1) maximizing the concentration gradients for chemical diffusion of the reactant gases by having no net reactant consumption at the electrodes as well as less dilution of the reactants by water vapor, and (2) a possibly higher gas permeability of the not fully hydrated membrane. In addition to the dedicated mechanical or chemical degradation tests, combined tests have been proposed [226–228].

Hydrogen cross-over and fluoride ion release is a good indicator of overall membrane degradation and often used as diagnostic for membrane durability test. The membrane degradation mechanism is believed to be loss of material as the membrane's polymeric structure deteriorates. When a PFSA membrane degrades, it releases peroxides and hydrogen fluoride (HF).

The published accelerated stress component test protocols have generally not been correlated with actual life under "normal" operating conditions. Furthermore, it is recognized that some of the ASTs overestimate the problem of membrane thinning when compared to field tests that reproduced the degradation phenomena observed in the field. The protocols mainly describe the course of the actual test, but for many test types, preconditioning procedures are equally important for obtaining reproducible results [229,230].

Figure 14.12 shows steady-state OCV hold test at 90 °C at 30% RH in hydrogen/air for a standard 1100-EW Nafion® membrane compared to a PTFE-reinforced Nafion® membrane. OCV was interrupted every c. 50 h for diagnostics of the evolution of the electrochemical hydrogen cross-over. Figure 14.13 presents the H_2 cross-over rate measured as a function of lifetime during the OCV hold test.

The mechanical reinforcement had a large effect on membrane life. Nafion® membranes with reinforcement offered much longer sufficient OCV during the test compared with nonreinforced membranes. It can be seen clearly that the trend of H_2 cross-over rate with time is different between PTFE-reinforced Nafion® and nonreinforced membrane. For the case of the nonreinforced membrane the H_2 cross-over rate gradually increased until failure, a drastic jump appeared after approximately 200 hours life test which is characteristic of a catastrophic failure. During the OCV hold test, effluent water from anode and cathode sides was collected during FC operation used to measure the fluoride ion concentration arising from degradation of the membrane structure. By

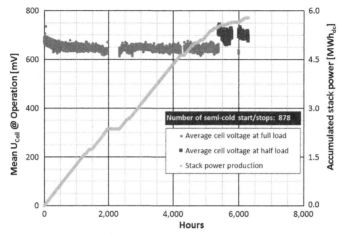

FIGURE 14.12 Steady-state open cell voltage (OCV) hold test at 90 °C at 30% RH in hydrogen/air for a standard 1100-EW Nafion® membrane compared to a PTFE-reinforced Nafion® membrane.

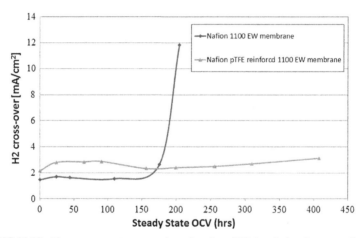

FIGURE 14.13 H_2 cross-over rate measured as a function of lifetime during the open cell voltage (OCV) hold test (Figure 14.12).

measuring the total amount of water produced, the fluoride ion release rate and the cumulative amount of released fluoride ions could be determined. The fluoride ion release for the reinforced Nafion® membrane during the OCV hold test was measured at a level of 0.012 µmol h^{-1} cm^{-2}.

Example of the hydrogen cross-over measured during RH cycling between 30% and 90% at 80 °C is shown in Figure 14.14. The total cycle time was 5 min. The criteria for the electrochemical cross-over ≥2 mA cm^{-2} is often used for these tests [223]. The test in Figure 14.14 was conducted using a 1100-EW Nafion® membrane and shows that this threshold was passed after 14,000 cycles.

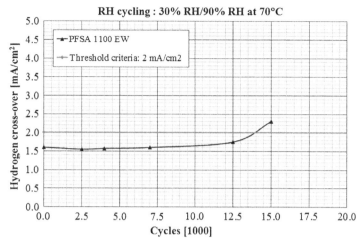

FIGURE 14.14 Hydrogen cross-over measured during relative humidity (RH) cycling between 30% RH and 90% RH at operating temperature of 80 °C.

14.4 APPLICATION AND PERFORMANCE OF PFSA MEMBRANES IN FCs

14.4.1 Introduction

The most unique feature of the FC technology is its flexibility; the FC is modular and the power therefore scalable, with the potential to find application in products over a wide power range from few milliwatts (mW) to megawatts (MW). In contrast to liquid acid electrolytes, the PFSA membrane has the advantage of being a solid polymer, which makes it easy to handle. The technology has reached a stage in its development where commercial exploitation now seems a reality, although some basic material and material-processing problems yet have to be solved, the significant focus over the past 30 years on adapting the PEMFC for use in consumer and industrial applications. Papers and reports including surveys and summaries of the current technological and commercial status have recently been published [76,231–240].

The PEMFC is seen as a system of choice for: automotive systems, for stationary application such as micro-CHP generators, as well as for portable applications i.e., power backup systems. The PEMFC operates at low temperatures (60–80 °C), allowing quick start-ups and immediate response to changes in the power demand. The use of FCs gives several advantages compared to conventional power generator systems. They offer a source of electrical energy that is continuous and environmentally safe and additional benefits include low maintenance, excellent load performance, etc.

An analysis of the current energy consumption and associated emissions in the industrialized countries shows significant contributions from heating and power plants in the form of CO_2 and SO_2. This provokes the development of not only more advanced heat and power technologies but also the use of alternative fuel supplies. Advanced technologies will help solve this problem through, for example, the use of excess electricity from renewable energy sources for the production of hydrogen via the electrolysis of water. A life cycle analysis of hydrogen so produced leads to global warming emissions, which are a whole order of magnitude less than for fossil-fuel-produced hydrogen, even considering the effort needed to produce these systems and to transport the energy produced. However, such an environment-friendly technology is possible but not yet available. PEMFCs today are therefore fueled by hydrogen from reformed fossil-fuel, which are the present most realistic possibility. FCs have the potential to reduce the fuel consumption. They have a higher efficiency than even most efficient combustion alternatives, resulting in fuel savings between 15% and 50%, depending on the application, and a significant reduction in CO_2 emission.

Another attractive solution is to use liquid methanol as fuel. DMFC technology, using methanol derived from biomass or other renewable energy sources, gives the same advantages as PEMFC technology, e.g., high energy efficiency and low or zero emissions. Although the DMFC technology has a lower performance than the hydrogen FCs, it has a number of other advantages: it does not have hydrogen storage problems and is easy to store, its fuel supply infrastructure is simpler than for hydrogen, and emissions are significantly lower than for petrol- or diesel-fueled generators. Eliminating the need for a reformer also facilitates technical simplifications, and thereby reduces the size and weight of portable power generator systems. This leads the DMFC well to applications where the energy density must be high. FCs can substitute batteries entirely or used in an FC/battery hybrid, where the FC acts as the battery charger.

14.4.2 The PEMFC Stack

A single PEMFC in operation (Figure 14.2, Section 14.2.1) delivers a voltage <1 V. To obtain practical useable voltages FCs are therefore stacked in series. High current is realized by enlarging the active area of the MEA. The cell typically consists of bipolar plates, which are pressed against the MEA. The bipolar plates providing the electrically conductive path for the generated current to the adjacent cell also serve as distributors of the reactant fuel and oxidant to the entire surface of the electrodes, having a manifold of grooves.

The choice of material for commercial PEMFC stacks is dictated by several factors, to some extent depending on the specific market. These factors concern not only performance, like current conduction, heat conduction, and mechanical strength, but also lifetime and cost issues. The bipolar plate material differs from sheet metal, graphite foil, and graphite polymer composites which all are potentially low-cost materials. By volume the bipolar plates make up the largest part of the PEMFC, and for some applications low-density materials are crucial for maximizing the power to weight ratio ($kW\,kg^{-1}$). For applications

FIGURE 14.15 A 47-cell PEMFC stack by IRD Fuel Cell A/S designed for stationary Combined Heat and Power (CHP) application. The bipolar plate is graphite based for long lifetime. The stack end plate manifolds have a polymer-based sandwich construction which, together with polymer frames around the single cells, ensures thermal insulation and thermal management of the stack.

aiming at long lifetime, graphite materials provide good chemical resistance, whereas metals have shown some limitations due to low corrosion in the harsh electrochemical environment inside the FC stacks [76,241,242]. Other stack components include seals, cooling elements, current collectors, and end plates. End plates give the FC stack mechanically stability and enable sealing of the components by compression. Figure 14.15 shows an example of a FC stack made up 47 cells connected in series.

The power output of a PEMFC based on MEAs using Nafion® membrane is highly dependant on the experimental conditions. Gas pressure and utilization are the most important factors assuming a perfect gas humidification and an optimal stack temperature. Elevated pressure is preferential from a performance point of view because of an easier water management especially at high currents.

Figure 14.16 shows the performance of a 47-cell PEMFC stack run operated at a cell temperature of 70 °C, 85% fuel utilization, and 50% oxygen utilization at ambient pressure operation.

Effective water and heat management has a major impact on FC performance [243–247]

Thermal management is required to remove the heat in order to prevent excessive operating temperatures, i.e., prevent dehydration of the membrane, and due to chemical degradation Nafion is limited to a maximum operating temperature of 130°–140°. Nafion and other similar PFSA membranes exhibit the highest proton conductivity when fully hydrated; this implies humidification of the gasses. Increasing the temperature above 100 °C requires a pressurized system to maintain a water-saturated environment. The water management is not only to ensure that the membrane remains fully hydrated but also to prevent excessive water accumulation within the cells and stack manifolds. Water flooding can block the oxygen transport to the electrode.

In the past decade there has been focus in developing FC systems for backup power application with decreased complexity and cost by allowing the water

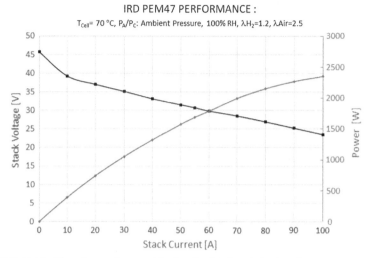

FIGURE 14.16 The voltage and power of the 47-cell PEMFC stack developed by IRD Fuel Cells A/S.

and thermal management systems to be simplified or eliminated altogether. This implies challenge for the PFSA membranes with respect to performance and durability (Section 14.3.3).

14.4.3 Application of PEMFC

Worldwide numerous prototype FCs and FC systems have been built and demonstrated. The common driving force for applying FC technology is the overall environmental benefit to conserve energy with high efficiency and reduce chemical and noise pollution on both a local and a global scale. Some applications will benefit in the use of the waste heat, and others consider the main advantage as their potential use to backup or replace batteries allowing extended runtime for portable electronic devices, due to high energy density.

Today more than 150,000 FC systems have been operated worldwide; in 2011–2012 alone a growth of 85% of overall FC systems shipped has been reported [237].

The biggest challenge for widespread FC commercialization to the developers is a substantial reduction of the cost of the FC systems and durability. Significant cost reduction has been achieved in the past decade [236]. The long-term target price of European Union (FP6) for stationary PEMFC systems is 100€ kW_e^{-1} and for automotive PEMFC systems the cost target by 2020 currently proposed by US DoE is \$40 kW_e^{-1} in order to compete with conventional technology [27,28,223].

The present high cost is mainly ascribed to the use of materials such as the PFSA membrane and the catalyst materials but components like graphite-based bipolar plates are also expensive due to the present piece-by-piece production and the limited number of component manufactures. The stringent requirements in terms of compactness, high energy density, performance stability, and low cost will change the research direction toward optimizing the different aspects of PEMFC system including the membrane.

14.4.3.1 Automotive Applications

Governmental legislations such as emission regulations especially in regions with air quality problems, e.g., urban areas, has created considerable interest in using FCs for providing power for cars and buses. In terms of volume, the automobile market represents the largest opportunity for high volume production of FCs and systems.

Electric vehicles in general can be fueled by a wide variety of primary energy sources—reducing oil dependency and enhancing security of energy supply. Well-to-wheel efficiency analysis also shows that electric vehicles are more energy efficient than ICEs over a broader range of primary energy sources. Fuel cell electric vehicles (FCEVs) have a driving performance (similar acceleration and range ~ around 600 km) and refueling time comparable to ICE vehicles. They are therefore a feasible low-emission substitute potentially achieving 80% carbon dioxide (CO_2) reduction by 2030 compared to today's ICEs, calculated for medium cars and longer trips. The benefits of lower CO_2 emissions, lower local emissions, diversification of primary energy sources, and the transition to renewable energy all come at an initial cost. These will ultimately marginalize with the reduction in battery and FC costs, economies of scale, and potentially increasing costs for fossil fuels and ICE specifications [248,249].

A challenge for the PFSA development is to meet the requirement of low price in the automotive market. To meet the proposed DoE 2020 price target of $40 kW$_e^{-1}$ for the total PEMFC system, it is anticipated that the MEA including the catalyst must be available for less than $20 m^{-2} [27,223].

Several other barriers need to be to overcome for widespread FC commercialization in the automotive sector. Firstly, the distribution and refueling infrastructure for hydrogen is not yet in place. Secondly, the amount of rare metals in electrocatalysts (i.e., Pt) must be reduced before mass production. Overall, the capital costs are high, and cost reduction targets still require improved materials and high-volume production methods to be established. A hydrogen supply infrastructure for around 1 million FCEVs by 2020 requires an investment of $3 billion for production, distribution, and retail infrastructure and building on existing infrastructure [248].

Nevertheless, the development within FC systems for automotive applications has within the past 5 years reached a premarket stage with several low-series production realized. All the major automotive manufactures have an FC vehicle either in development or in testing and as mentioned several have begun low-series production. The FC industry is progressing from an investment-led technology industry to a market-led commercial industry [238]. In the past years notable collaborations have been entered into the industry. Automakers look to pool their resources to accelerate cost reduction and ease mass manufacture of FC systems.

For automobile application a shortcoming of PEMFC based on the PFSA membranes is their limited temperature range of operation, conventionally around 60–80 °C. The incentive to develop membranes for higher temperature operation is not only enhanced reaction activity but also easier thermal management with potential to lower the additional volume and cost for the total FC system. The thinnest PFSA membranes (usually <20 μm) are preferred in FC stack for automotive application due to high power density demand. The thin membranes provide the lowest resistance. A consequence of the high power density is a possible shortening of the FC lifetime. The system target level for durability of an integrated transportation FC power system is 5000 h. Membranes must therefore be durable for thousands of hours and over thousands of start–stop and humidity cycles, in a corrosive environment [250].

The need to simultaneously achieve chemical stability, mechanical strength, durability, and cost targets poses a significant challenge. Improvements in one characteristic can often negatively impact another characteristic.

Most work on membrane development for automotive applications has focused on the use of low-EW ionomers, which are known to have higher conductivity under dry conditions than ionomers typically used in FCs [169,182,250,251].

Figure 14.17 shows the single-cell PEMFC polarization curves of two MEAs based on SSC membrane with low EW (750) and LSC membrane with higher EW (1100). Both membranes have a thickness of 20 μm with mechanical reinforcement built into their structure and they are composed of the exact same anode and cathode electrodes.

At 80 °C and 100% RH the 1100-EW-based MEA shows comparable performance with the 750-EW-based MEA cells (Figure 14.17(a)). The performance for the same MEAs shown at elevated temperature and decreased RH (Figure 14.17(b)) the 750-EW-based MEA outperformed the FC assembled with high EW membrane specially in the intermediate and high current density region. In the intermediate and high current density region, the FC performance is largely determined by the proton conductivity and water diffusion properties in the membrane. This property dependence is more pronounced with the increase of the current density.

Stability during start–stop operation and operation at low humidification levels is extensively studied. The development of reinforced membranes has improved the overall technology in the past decade. Figure 14.18 shows 600-h steady-state OCV hold test for an MEA based on the 40-μm reinforced Fumion®940rf pictured in Figure 14.7 in Section 14.3.2 at even very dry operating conditions with only 10% RH. The hydrogen cross-over measured after 600 h was only 0.2 mA cm^{-2} (refer to Section 14.3.7).

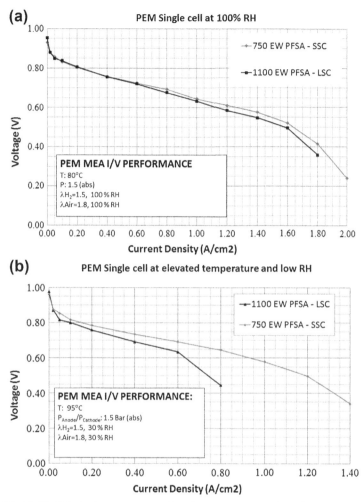

FIGURE 14.17 Polarization curve comparison between 750-EW short side chain (SSC) and 1100-EW long side chain (LSC) membrane at (a) 80 °C and 100/100% relative humidity (RH) anode and cathode inlet condition and (b) 95 °C and 25/25% RH for anode and cathode.

Field tests look promising; Honda introduced the FCX Clarity in 2008, a PEMFC car fueled with compressed hydrogen (700 bar) and a range of about 500 km. General Motors reports having reached 100,000 miles with a vehicle in their 119-vehicle fleet and Honda has demonstrated operation of FCs down to −20 °C [257,258].

14.4.3.2 Stationary Application

Stationary FC systems have been under development for several decades and have been demonstrated for a number of years across Asia, Europe, and North

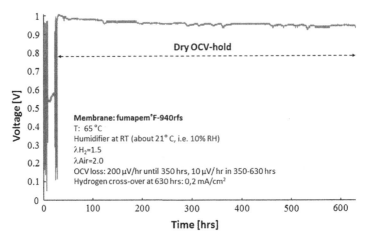

FIGURE 14.18 Steady-state open cell voltage (OCV) hold test for an membrane electrode assembly (MEA) based on Fumion®940rf. *Data provided from Bernd Bauer fumatech.*

America. Commercialization of these systems is now accelerating with small- and large-scale systems being installed worldwide [237].

Stationary FC systems incorporating PEM technology, is expected to evolve into three different markets; large-scale prime power generation, residential μCHP, and small-scale stationary power generation for grid or off-grid support. The telecom market has gained prominence using FCs as backup power systems. The advantage using PEMFC technology beside environmental benefit is their low maintenance.

PEMFCs for large power generation (>100 kW) have been demonstrated; however, the efficiency of PEMFC seems not to be sufficient for large-scale baseload applications, which probably will be the domain of other FC types like SOFC, PAFC, and MCFC technology.

PEMFCs are generally believed to cover the lower end of the power output scale, where FCs are expected to be economically viable. There has been substantial progress in using PEMFC in small-scale stationary power units between 1 and 10 kW. The exact size of needed FC system differs within geographical areas and their respective needs [259–268].

Currently FC-based μCHP plants are installed and demonstrated in several countries like Japan (ENE-FARM), Germany (Callux), Europe (ene.field), and Denmark (Danish Micro Combined Heat & Power Based on Fuel Cells). The Japanese government has supported the residential-based ENE-FARM combined heat and power (CHP) FCs since the technology debuted in 2009. By the end of 2012 more than 34,000 of the natural-gas (NG)-powered μCHP plants based on FC systems had already been installed [237,238,269,270].

In Germany the "Callux" program, set up by the German Ministry for Transport, Construction and Urban Development, and nine partners from industry, which will deploy more than 800 residential FC units. In January 2013 the "ene.

field" project was launched as the largest European demonstration of FC-based μCHP. The demonstration, which is cofunded by the European Commission's Fuel Cells and Hydrogen Joint Undertaking [271] aims to deploy up to 1000 residential FC installations across 12 European countries in the period from 2012 to 2017.

Although mass commercialization has not yet been achieved by any of the FC developers, some already have quite many systems that are nearing deployment to the market from a wide variety of international manufacturers and system integrators. Most of the systems are built for NG fuel.

A Danish development and demonstration project with μCHP systems based on different types of FCs and fuels was launched back in 2006 and more than 60 μ-CHP were deployed between 2010 and 2014 [252–256].

Calculations have shown that approximately 100% of the annual electricity demand in a standard Danish single-family home can be produced by a microcogeneration unit in a thermal-load-following operation mode. 50% of this electricity is exported to the grid.

As part of the demonstration project 30 units have been successfully demonstrated in the hydrogen-based community in the Vestenskov village in Lolland, Denmark. The concept of the Vestenskov hydrogen society is illustrated in Figure 14.19. This demonstration is the first of its kind worldwide.

An example of a 1-kW μCHP unit is shown in Figure 14.19. The unit is designed to operate in parallel with the existing power grid. The power is sized for an average heat need in a north European single-family house. The key component in the CHP system is the PEMFC stack, using MEAs based on reinforced Nafion® membranes as electrolyte. The durability of PEMFC stack in the μCHP system was greatly improved partly by implementing reinforced PFSA membranes in the design. The system was operated through several start/stops without any failures of the MEAs.

The complete CHP system comprises the following components:

- PEMFC stacks
- Cell voltage monitor system
- Hydrogen fuel supply system
- Air supply system (compressor)
- Humidifier
- Cooling system
- Grid connected DC→AC inverter
- FC control unit system

The electric efficiency is 50% (lower calorific value (LCV)) at full load ($H_2 \rightarrow AC$), and the overall nominal efficiency is 94%. The hydrogen was supplied from a distribution grid directly to the houses by underground pipes, and is produced by electrolyzers to demonstrate the potential use (and storage) of local surplus wind-power-based electricity (Figure 14.20). The hydrogen is stored in 6-bar pressure vessels before entering the 4-bar distribution grid. The units

(a)

Single family H_2 based renewable energy system

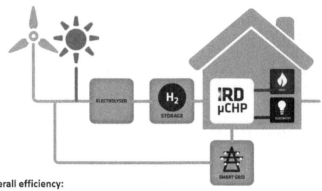

Overall efficiency:
$P_{AC} \rightarrow H_2 \rightarrow P_{AC}$: 37%
$P_{AC} \rightarrow H_2 + P_{TH} \rightarrow P_{AC} + P_{TH}$: 85%

(b)

Nominal Power	1.2 kW$_{AC}$
Power range	0.9 – 2.0 kW$_{AC}$
Nominal Heat	1.2 kW$_{TH}$
Heat range	0.8 – 2.0 kW$_{TH}$
Electrical efficiency ($H_2 \rightarrow P_{AC}$)	50%
Heat efficiency ($H_2 \rightarrow P_{TH}$)	44%
Combined efficiency	94%
Ready-mode Power	15 W$_{AC}$

FIGURE 14.19 (a) The concept of the hydrogen society that has been demonstrated in Vestenskov, Denmark. (b) Specification of the micro-combined heat and power (μCHP) unit developed by IRD Fuel Cell A/S developed by the partial support and acknowledgment to several research contracts [252–256].

were installed as the primary heat source, with a 200-L heat storage tank that includes an electrical coil for peak load heating. Furthermore, since the units are operated directly on pure hydrogen, there is no need for fuel reforming. This is an advantage in relation to start-up time (<1 min) and load response. It is also an advantage regarding the μCHP unit's electricity production efficiency, which is close to 50% with LCV reference. For the hydrogen test series the CO_2 savings were approximately 4.5 tons for each house, with an annual full-load operating time of 6000 h [272].

Figure 14.21 shows example of the field test data of the μCHP over a period of 1 year in total. The MEAs in the FC stack were based on 20-μm reinforced

Use of Per-Fluorinated Sulfonic Acid Membrane Chapter | 14 355

FIGURE 14.20 (a) One of the host of micro-combined heat and power (μCHP) installations in Vestenskov. The only visual sign of the installation in the houses is the μ-CHP located between the two persons which have the same size as an oil burner. (b) The hydrogen was supplied from a distribution grid, and is produced by electrolyzers to demonstrate hydrogen production by local surplus wind power-based electricity. The hydrogen is stored in 6-bar pressure vessels before entering the 4-bar distribution grid. (c) An installation at one of the hydrogen-fueled μCHP test sites. The floor-standing heat storage tank with integrated supplementary heating coil can be seen in the forefront of the photograph.

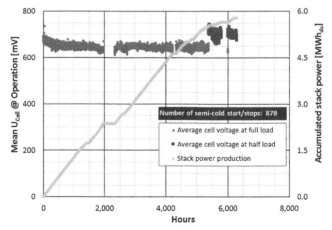

FIGURE 14.21 Field test of one of the micro-combined heat and power (μCHP) installations in Vestenskov. The test data is collected over a period of 1 year in total. The average cell voltage is at operation at full load. During the field test the system had approximately 878 start–stops and data is cleaned for data in standby mode as well as during start and stops.

Nafion® membrane. The operating strategies are heat controlled and electricity controlled. The cogeneration output is controlled by the thermal energy demand in the heat-controlled operation. Surplus electricity is exported to the grid. The operation is securing a high overall efficiency. The house has an annual electricity consumption of 5000 kWh, a space heating demand of 12,000 kWh, and an annual hot water demand of 5000 kWh. The heat-to-power ratio of the cogeneration unit is assumed to be 2. It corresponds roughly to a unit with 30–35% electrical efficiency. The average degradation calculated from the voltage decay over the entire period was lower than 3 µV per hour.

The future challenges include FC degradation and lifetime, unit size, and cost reduction. Installation and heating system integration aspects are also important, especially when a microcogeneration unit is added to an existing heating system rather than a full appliance replacement.

Another challenge strongly related to the technical challenges mentioned above is the competition of other emerging energy technologies such as photovoltaics and electric and gas heat pumps and a wider use of district heating using renewable energy sources.

14.4.3.3 Portable Application

Application of small PEMFC systems (ranging from a few Watts up to ~1 kW) for portable devices, provide a potential alternative to batteries and diesel or gasoline generators. The use of FCs is motivated by several factors including high energy to weight ratio and long run time for portable electronic equipment [273,274]. The fuel could be hydrogen from a metal hydride storage system, sized to meet the desired runtime between fueling, or a liquid methanol/water mixture [275]. The hydrogen FCs give high power density and would be excellent for portable power but complicated because of hydrogen storage and the fact that hydrogen distribution to the public today remains unsolved. The DMFC is a simpler and more near future possibility. The focus in this field is no longer concentrating on the consumer electronics sector only but has changed to development of FCs for emergency backup power and auxiliary power units (APUs) in the industrial sector. Military organizations around the world also show interest in FC technology evaluating it as a means to significantly reduce the weight carried by soldiers in the field [276,277]

14.4.4 DMFC and Their Applications

DFMCs convert liquid methanol directly into electrical energy and heat.

In order to be competitive within the portable and distributed energy markets, the DMFCs must be reasonably cheap, they should be characterized by high durability and capable of delivering high power densities. New generation high EW PFSA membranes and stabilized versions as well as membranes with mixed functionality comprising PFSA (sulfonic acid component) and a long chain phosphonic acid, and other composite PFSA-based polymers is being developed and explored be used for DMFC applications [55,278–280].

In the past 5 years significant progress has been achieved in the field of DMFC and the technology has reached a stage in its development stage where commercial exploitation is now seen as a reality. However, one major challenging problems to solve for the practical application of DMFCs is poor kinetics of the methanol reaction and the fact that it suffers from severe fuel crossover through the PFSA polymer electrolyte, such as Nafion®. The cross-over of methanol from anode to cathode causing a partial chemical short circuit of the FC, and decreases the power and energy density. Even at low methanol concentrations at the anode, sufficient methanol and water diffuse through the electrolyte to significantly impact the performance of the cathode. Any methanol reaching the cathode chemisorb on the electrode surface with subsequent oxidation to CO_2 at the high potentials reducing the cell voltage.

The overall increased demand for electrical power and equipment with convenient rechargeability for high-technology electronic applications leads to applications of the DMFC where the energy density must be high. FCs can substitute batteries entirely or partly as in a FC/battery hybrid, where the FC acts as the battery charger [281,282]. Despite energy-conversion efficiencies in the range of 30–40%, FCs still has an overall advantage over batteries in energy density terms. The specific energy of methanol is $4.28\,\text{Wh}\,\text{L}^{-1}$. Making it viable to have DMFC system >$1000\,\text{Wh}\,\text{L}^{-1}$ or $\text{Wh}\,\text{Kg}^{-1}$ giving considerable savings in weight and volume compared to 350–$400\,\text{Wh}\,\text{L}^{-1}$ or 150–$200\,\text{Wh}\,\text{Kg}^{-1}$ for the best battery systems.

Like in the case of LT-PEMFCs, understanding the performance degradation mechanism and development of methods that would either prevent or reverse (fully in part) performance losses in an operating DMFC system are crucial to both the demonstration of practical viability and the commercialization potential of DMFC power systems. A minimum durability of 10,000 operating hours for the DMFC stack is a prerequisite for commercialization.

14.4.4.1 The Direct Methanol Fuel Cell

The structures and materials of the MEAs used in PEMFC and DMFC are very similar (Section 18.2.1). In the DMFC the electrode reactions can be represented by the electrode reactions and the following overall cell reaction as shown in Figure 14.22.

The practical cell voltages obtained using liquid methanol are considerable lower, and the losses higher than for the hydrogen-fed PEMFC.

Reviewing the literature [233,283–291], the relatively low performance compared to the hydrogen PEMFC is thus caused by:

- Poor kinetics of the methanol reaction
- Methanol permeation through the Nafion® membrane

The anode reaction shows that water is needed with methanol to enable the oxidation reaction. This implies that pure methanol cannot be used, but a mixture of methanol/water is necessary. The methanol oxidation reaction is quite

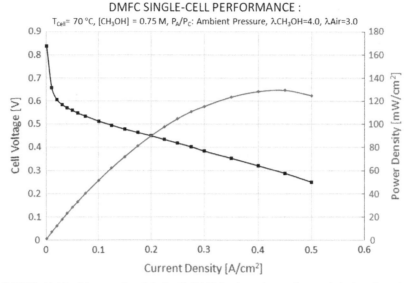

FIGURE 14.22 Direct methanol fuel cell (DMFC) electrode reactions and single-cell performance.

complicated in which surface intermediates play a key role [288,292,293]. The cathode reaction in the DMFC is the same as that for the hydrogen feed FC the protons and electrons recombine with oxygen.

Studies examining the extent and impact of methanol transport, in DMFCs using Nafion® membrane, as function of various operating conditions have frequently been published [108,204,206,279,284,294–297].

The methanol cross-over can be driven by three transport mechanisms: (1) electro-osmotic drag by proton transport, (2) diffusion by methanol concentration gradient, and (3) convection by the hydraulic pressure gradient between the anode and cathode. During typical DMFC operation diffusion is the predominant transport mechanism. It has been widely demonstrated that the methanol cross-over is close related to membrane structure, morphology, and thickness and operating conditions such as temperature, pressure, and fuel concentration. Methanol arriving at the cathode catalyst will react with the oxygen present to form CO_2. The methanol reaching the cathode therefore causes depolarization due to the competing electrochemical processes of oxygen reduction and methanol oxidation. The rate of methanol permeation decreases with increasing current density. At higher currents the methanol concentration at the anode/electrolyte interface is reduced due to consumption in the porous electrode structure. This reduces the driving force for diffusion through the membrane.

Figure 14.23 shows the methanol crossover measurements expressed in milliamperes per square centimeters obtained at 40–80 °C in (a) 1.0 and (b) 2.0 M CH_3OH for a complete MEA based on Nafion®115.

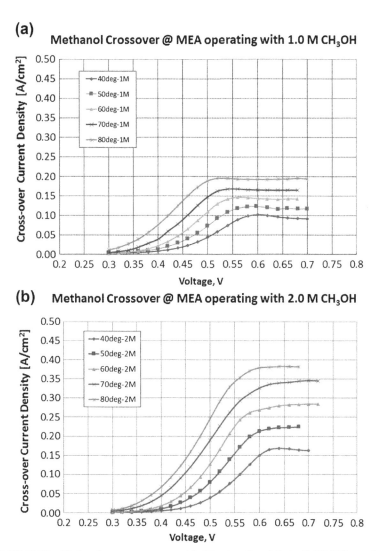

FIGURE 14.23 Methanol cross-over measured in Direct methanol fuel cell (DMFC) membrane electrode assembly (MEA) based on chemically stabilized Nafion® 115 membrane measured circulating (a) 1.0M and (b) 2.0M methanol.

The actual amount of fuel crossing over the membrane electrolyte from anode to cathode can be measured electrochemically. During the measurement the fuel passes through the anode and inert gas such as nitrogen passes through the cathode side. An external power supply applies a positive voltage at the cathode and once the fuel passes through the membrane to reach the cathode it is oxidized. The current is proportional to the amount of fuel crossing over according to Faraday's law [298]. The methanol cross-over has been

360 Advanced Fluoride-Based Materials for Energy Conversion

shown to increase with increased cell voltage corresponding to low current density.

The methanol cross-over measured on an MEA depends on not only membrane properties but also a combination of membrane, electrode structure, and gas diffusion layer and fuel delivery design [299–304]. In practice, the effects of methanol cross-over are counteracted by careful MEA and system design.

The state of art has been using the thicker Nafion® 117 (180 μm), in contrast to Nafion® 112 (50 μm) used in the hydrogen feed PEMFC. Although the thicker membrane increases the cell resistance, the gain from reduced methanol cross-over results in improved performance. The fuel must be supplied as a diluted methanol solution (typically 0.5–2.0 M) as a consequence of the high methanol cross-over, adding to the size and complexity of the system [288]. A combination improved chemical stabilization of the thinner PFSA membrane (50–100 μm), and by tailoring the operating strategies has enabled the use of PFSA membranes with thickness of 50–125 μm. The thinner membranes provide higher conductivity and lower cost.

14.4.4.2 Applications of DMFC

The use of portable devices worldwide has grown dramatically in recent years and continues to do so. Also, the overall increased demand for electrical power, estimated at >10% per year in the United States (similar increases have been foreseen in Europe), has resulted in a corresponding increased demand for uninterrupted power supply systems. This market is currently dominated by batteries combined with gasoline- or diesel-fueled generators, which pollute excessively, through their high emission levels and noise. Current trends toward higher fuel prices and greater environmental awareness strongly support a shift in consumer demands.

DMFCs are considered as suitable systems for power generation in the field of portable power sources, remote and microdistributed energy generation, as well as for APUs in stationary and mobile applications. The DMFC systems are promising candidates because of their high energy density, light weight, compactness, simplicity, as well as easy and fast recharging. Nevertheless, in order to be competitive in the portable and distributed energy markets, the DMFCs must be reasonably cheap, they should be characterized by high durability, and should be capable of delivering high power densities.

Figure 14.24 shows a 500-W DMFC stack developed and designed for battery charging applications. The stack consists of 35 cells connected in series. The MEA is based on a chemically stabilized Nafion® membrane, and gives a maximum power density of 130 mW cm^{-2} operating at a temperature of 70 °C. The flow distribution/interconnect plate is graphite based for long lifetime. The cells and stacks are designed with a low pressure loss reducing the power used for the auxiliary pumps, etc. The electrical efficiency is 35% at 45 A but can be varied according to the load (Figure 14.25).

Unlike, a hydrogen feed PEMFC the DFMC does not require components such as fuel processor or humidifier system.

FIGURE 14.24 A 35-cell DMFC stack developed by IRD Fuel Cells A/S. The DMFC stack provides 600 W_{el} nominal power output and is integrated with balance-of-plant components into the DMFC generator module aimed for battery charging applications designed as 500-W DMFC system.

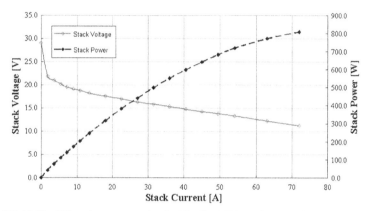

FIGURE 14.25 I/V performance graph of a 35-cell DMFC stack operated at 70 °C with 1.0 M CH_3OH at ambient pressure developed by IRD Fuel Cells A/S.

The advantage compared to conventional batteries is the prospect of longer runtime and the potential for instantaneous refueling. This makes them attractive and well suited for use in electronic equipment such as portable computers, mobile phones, and other handheld electronic equipment. Other potential applications are: powering remote telecommunication transmission equipment and remote scientific investigation equipment, emergency AC power for hospitals, and other consumer leisure applications such as camping, sailing, etc. Figure 14.26 provides the key features of a 500-W DMFC generator system for charging batteries and powering remote telecommunication equipment and other remote equipment.

Understanding the performance degradation mechanism and development of methods that would either prevent or reverse (fully in part) performance losses in an operating DMFC system are crucial to both the demonstration of practical viability and the commercialization potential of DMFC power systems.

DMFC GENERATOR

Nominal Power output:	500 or 800 W
Nominal DC Voltage:	24/48V
Fuel consumption:	0,5 – 0,9 l/hour
Fuel:	50- 100% MeOH
Ambient temperature:	-10 to +40 C
Weight:	45 kg
Size (LxWxH):	600x381x482 mm
Lifetime (min.):	3000 h

FIGURE 14.26 A 500-W Direct methanol fuel cell (DMFC) generator developed by IRD Fuel Cells A/S.

Although the routes of performance degradation in DMFC have similarity to those found in hydrogen–air systems, the major causes of DMFC performance loss are often different due to the use of different materials and cell operating conditions (liquid rather than gaseous fuel, generally lower FC power, significant fuel cross-over, etc.).

In the past 5 years significant progress has been achieved in the field of DMFC durability by investigating and understanding the degradation mechanisms.

DMFC stacks have today proven operation times of more than 10,000 which show good prospects for successful commercialization. Recently results of a 20,000-h life test with a realistic dynamic load profile of a 7-kW DMFC hybrid system have been published and represent a milestone for the commercialization of DMFC systems [305].

The experimental investigations found in the literature show that degradation has both permanent and temporary contributions; the latter can be recovered with application of suitable operation strategies [306–312].

The DMFC lifetime performance achieved using MEAs based on 125-µm-thick Nafion® membrane is shown in Figure 14.27.

The durability test of DMFC MEAs (single cells) were carried out using constant current mode. Data are acquired at constant $200\,mA\,cm^{-2}$; load cycles were completed every 1 min and consisted of a 15-s hold at OCV and break in air supply. During the life test, in addition to measuring the cell voltage, the high-frequency resistance (HFR) and OCV are monitored. In order to analyze anode and cathode performance losses half-cell polarization curves were measured every 250 h during the life test.

Certain performance losses incurred by the cell during the steady-state operation were recovered, fully or in part, after the OCV hold at 15 s and particular after intermittent characterization test. Such "recoverable" performance losses are usually associated with reversible phenomena occurring in the FC, for example, cathode catalyst surface oxidation, cell dehydration, and incomplete water removal from the catalyst layer and/or gas diffusion layer. The performance losses that were not reversed to a significant degree in an operating cell

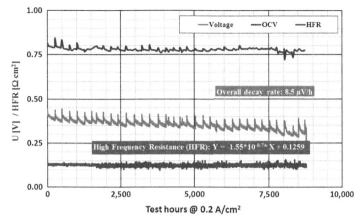

FIGURE 14.27 DMFC MEA durability test at cell operating at 75 °C, with 1.0 M CH$_3$OH and stoichiometry of air and methanol of $\lambda_{Air} = 3.0$, $\lambda_{CH_3OH} = 4.0$. The catalyst of the anode based on 1.8 mg Pt Ru cm^{-2} and for the cathode 1.2 mg Pt cm^{-2}.

are referred to as "permanent" performance losses. During the life test the OCV and the HFR were measured. Both show very stable levels during the life test, no indication of membrane degradation or interfacial contact between catalyst layer and membrane. The permanent degradation calculated during the 8500-h test equals 8.5 µV per hour.

Progress in the field of improving the stability and lifetime for DMFC systems has also been realized by several studies on the influence of contamination with both organic and inorganic impurities. Contaminants, involving different inorganic (metal ions and anions) or organic substances that may poison the catalyst or reduce the protonic conductivity of the membrane, are described in the literature [49,313–320].

The DMFC systems are typically designed with a closed water balance loop, i.e., regenerating the water from the cathode into the anode methanol/water circuit. Metal ions in the fuel will accumulate in the electrolyte membrane and degrade cell performances. Such degradation has shown to be proportional to the amount of metal ions trapped in the electrolyte membrane [314,315].

Improvement over the past years has therefore also been achieved by a careful choice of materials for the entire construction of the DMFC systems in order to avoid performance losses.

At present, there are very few commercial DMFC systems available. Commercial DMFC systems currently available in Europe guarantee at least 3000–5000 operating hours. SFC Energy AG produces portable and stationary systems that have an electrical output of 40–105 W for off-grid power [321] and 500- to 800-W system from IRD Fuel Cells A/S, Denmark [322]. Oorja Protonics in the United States markets the OorjaPac Models with power output in the range from 750 to 9000 W for a wide variety of applications with 3500–5000 total operating hours [323]. These devices are used for stationary power supply in remote areas. In Asia, commercial DMFC systems are mainly used for low-power portable applications supplying power to mobile digital consumer products. An example of an external power source is the Dynario that was launched in 2009 in a limited edition of 3000 units from Toshiba [324], Japan, which can deliver a continuous output of 2.5 W.

REFERENCES

[1] A.J. Appelby, R.L. Foulkes (Eds.), Fuel Cell Handbook, Van Nostrand, New York, 1989.
[2] K. Kordesch, G. Simader (Eds.), Fuel Cells and Their Applications, VCH Verlaggsgesellschaft mbH, 1996.
[3] J. Larminie, A. Dicks (Eds.), Fuel Cell Systems Explained, second ed., John Wiley & Sons Ltd, 2003.
[4] S. Srinivasan (Ed.), Fuel Cells – from Fundamentals to Applications, Springer, 2006.
[5] T. Zhao, K.D. Kreyer, T. van Nguyen (Eds.), Advances in Fuel Cells, Elsevier, 2007.
[6] P. Hoffmann (Ed.), Tomorrow's Energy: Hydrogen, Fuel Cells and the Prospects for a Cleaner Planet, second ed., MIT Press, 2012.
[7] F. Barbir (Ed.), PEM Fuel Cells, second ed., Academic Press, 2013.
[8] K. Kordesch, J. Electrochem. Soc. 125 (2004) 77C–91C.
[9] A.J. Appelby, J. Power Sources 29 (1990) 3–11.
[10] M.L. Perry, T.F. Fuller, J. Electrochem. Soc. 149 (2002) S59–S67.
[11] K. Scott, C. Xu, X. Wu, Energy Environ. 3 (2014) 24–41.
[12] Z. Sharaf, M.F. Orhan, Renewable Sustainable Energy Rev. 32 (2014) 810–853.
[13] M. Ptacek, T. Pavelka, Tomas, J. Novotny, Electric power engineering (EPE), in: Proccedings of the 2014 15th International Scientific Conference, 2014, pp. 363–366.
[14] P.S. Kauranen, E. Skou, J. Appl. Electrochem. 26 (1996) 909.
[15] W.T. Grubb, L.W. Niedrach, J. Electrochem. Soc. 107 (1960) 131–135.
[16] W.T. Grubb, (General Electric Company) US Patent 2,913,511 (November 17, 1959).
[17] W. Vielstich, Fuel Cells: Modern Processes for the Electrochemical Production of Energy, Wiley & Sons Ltd, 1970.
[18] D. Wheeler, G. Scerdrup, 2007 Status of Manufacturing: Polymer Electrolyte Membrane (PEM) Fuel Cells, Technical Report NREL/TP-560-41655, National Renewable Energy Laboratory, Colorado, USA, 2008.
[19] J. Pander, Hamburg Speeds up Preparation for Fuel-Cell Cars, Spiegel, SPIEGELnet GmbH Hamburg, 2009.
[20] J. DeMatio, Kia's Big Fuel Cell Plans, Automobile Magazine, 2009. http://www.automobilemag.com/features/news/kias-big-fuel-cell-plans-135290/.
[21] Toyota, Toyota Advanced Fuel Cell Hybrid Vehicle Completes Government Field Evaluation, PR Newswire, 2009. http://multivu.prnewswire.com/mnr/toyota/39419/.
[22] C.D.M. Gittleman, S. Jorgensen, J. Waldecker, S. Hirano, M. Mehall, Automotive fuel cell R&D needs, in: DOE Fuel Cell Pre-Solicitation Workshop, Department of Energy, Lakewood, Colorado, 2010.
[23] J. Marcinkoski, et al., J. Power Sources 196 (2011) 5282–5292.
[24] I. Staffell, R. Green, The Cost of Domestic Fuel Cell Micro-CHP Systems, Discussion paper Imperial College Business School, 2012. https://spiral.imperial.ac.uk/bitstream/10044/1/9844/6/Green%202012-08.pdf.

[25] U.S. Department of Energy (Hydrogen Program), Record 12020: Fuel Cell System Cost –2012. http://www.hydrogen.energy.gov/pdfs/12020_fuel_cell_system_cost_2012.pdf.

[26] B.D. James, A.B. Spisak, Mass Production Cost Estimation of Direct H_2 PEM Fuel Cell Systems for Transportation Applications: 2012 Update. http://www1.eere.energy.gov/hydrogenandfuelcells/pdfs/sa_fc_system_cost_analysis_2012.pdf (2012).

[27] U.S. Department of Energy, Fuel Cell Technologies Office Multi-Year Research, Development and Demonstration Plan – 3.4 Technical Plan – Fuel Cells, 2012. http://www1.eere.energy.gov/hydrogenandfuelcells/mypp/pdfs/fuel_cells.pdf.

[28] Fuel Cells and Hydrogen Joint Undertaking (FCH JU), Multi - Annual Implementation Plan 2008–2013, Document FCH JU 2009.001, 2009. http://ec.europa.eu/research/fch/pdf/fch_ju_multi_annual_implement_plan.pdf.

[29] O. Savadogo, J. New, Mat. Electrochem. Syst. 1 (1998) 47–66.

[30] A.J.L. Verhage, J.F. Coolegem, M.J.J. Mulder, M.H. Yildirim, F.A. de Bruijn, Int. J. Hydrogen Energy 38 (2013). http://www.sciencedirect.com/science/journal/03603199/38/114714-4724.

[31] R. Beith (Ed.), Small and Micro Combined Heat and Power (CHP) Systems: Advanced Design, Performance, Materials and Applications, Woodhead Publishing Limited, 2011.

[32] F.N. Büchi, M. Inaba, T.J. Schmidt (Eds.), Polymer Electrolyte Fuel Cell Durability, Springer, 2009.

[33] D. Stolten (Ed.), Hydrogen and Fuel Cells - Fundamentals, Technologies and Applications, Wiley-VCH Verlag GmbH & Co. KGaA, 2010.

[34] S. Yoshioka, A. Yoshimura, H. Fukumoto, O. Hiroi, H. Yoshiyasu, J. Power Sources 144 (2005) 146–151.

[35] A.S. Feitelberg, J. Stathopoulos, Z. Qi, C. Smith, J.F. Elter, J. Power Sources 47 (2005) 203–207.

[36] S.D. Knights, K.M. Colbow, J. St-Pierre, D.P. Wilkinson, J. Power Sources 127 (2004) 127–134.

[37] J. Wu, X.Z. Yuan, J.J. Martin, H. Wang, J. Zhang, J. Shen, S. Wu, W. Merida, J. Power Sources 184 (2008) 104–119.

[38] S. Zhang, X. Yuan, H. Wang, W. Merida, H. Zhu, J. Shen, S. Wu, J. Zhang, Int. J. Hydrogen Energy 34 (2009) 388–404.

[39] W. Schmittinger, A. Vahidi, J. Power Sources 180 (2008) 1–14.

[40] Y. Shao, G. Yin, Z. Wang, Y. Gao, J. Power Sources 167 (2007) 235–242.

[41] Q. Li, J.O. Jensen, R.F. Savinell, N.J. Bjerrum, Prog. Polym. Sci. 34 (2009) 449–477.

[42] R.L. Borup, J. Meyers, B. Pivovar, Y.S. Kim, R. Mukundan, N. Garland, D. Myers, M. Wilson, F. Garzon, D. Wood, P. Zelenay, K. More, K. Stroh, T.A. Zawodzinski, J. Boncella, J.E. McGrath, M. Inaba, K. Miyatake, M. Hori, K. Ota, Z. Ogumi, S. Miyata, A. Nishikata, Z. Siroma, Y. Uchimoto, K. Yasuda, K.-I. Kimijima, N. Iwashita, Chem. Rev. 107 (2007) 3904–3951.

[43] M.P. Rodgers, L.J. Bonville, H.R. Kunz, D.K. Slattery, J.M. Fenton, Chem. Rev. 112 (2012) 6075–6103.

[44] X.-Z. Yuan, H. Li, S. Zhang, J. Martin, H. Wang, J. Power Sources 196 (2011) 9107–9116.

[45] J.C. Meier, C. Galeano, I. Katsounaros, A.A. Topalov, A. Kostka, F. Schu, K.J.J. Mayrhofer, ACS Catal. 2 (2012) 832–843.

[46] P.J. Ferreira, G.J. la O', Y. Shao–Horn, D. Morgan, R. Makharia, S. Kocha, H.A. Gasteiger, J. Electrochem. Soc. 152 (2005) A2256–A2271.

[47] V. Stanic, M. Hoberecht, in: Murthy, Ota, V. Zee, Narayanan, Takeuchi (Eds.), Prot. Conduct. Membr. Fuel Cells IV, The Electrochemical Society, Honolulu, Hawaii, 2006, pp. 391–401.

[48] U. Bossel, The Birth of the Fuel Cell, European Fuel Cell Forum, Göttingen, Germany, 2000.

[49] W. Vielstich, A. Lamm, H.A. Gasteiger, Handbook of Fuel Cells: Fundamentals, Technology, and Applications, vol. 5, Jon Wiley & Sons, 2010.
[50] S.J. Peighambardoust, S. Rowshanzamir, M. Amjadi, Int. J. Hydrogen Energy 35 (2010) 9349–9384.
[51] W. Grot (Ed.), Fluorinated Ionomers, second ed., Elsevier Inc., 2010.
[52] A.A. Franco, Polymer Electrolyte Fuel Cells – Science Applications and Challenges, CRC Press, 2013.
[53] H. Ghassemi, T. Zawodzinskib, D. Schiraldia, S. Hamrockc, in: K.A. Page, C.L. Soles, J. Runt (Eds.), Polymers for Energy Storage and Delivery: Polyelectrolytes for Batteries and Fuel Cells, vol. 1096, American Chemical Society, 2012, pp. 201–220.
[54] Y. Wang, K.S. Chen, J. Mishler, S.C. Cho, X.C. Adrohe, Appl. Energy 88 (2011) 981–1007.
[55] R. Wycisk, P.N. Pintauro, J.W. Park, Curr. Opin. Chem. Eng. 4 (2014) 71–78.
[56] K.D. Kreyer (Ed.), Fuel Cells -Selected Entries from the Encyclopedia of Sustainability Science and Technology, Springer, 2013.
[57] H.H. Gibbs, V.W. Va, R.N. Giffin, (E.I.du Pont de Nemours and Company) US Patent 3,041,317, (June 26, 1962).
[58] D.J. Connolly, W.F. Gresham, (E.I.du Pont de Nemours and Company) US Patent 3,282,875, (November 1, 1966).
[59] D.E. Curtin, R.D. Lousenberg, T.J. Henry, P.C. Tangeman, M.E. Tisack, J. Power Sources 131 (2004) 41–48.
[60] T.A. Davis, J.D. Genders, D. Pletcher (Eds.), A First Course in Ion Permeable Membranes, The Electrochemical Consultancy, 1997.
[61] C. Heitner-Wirguin, J. Membr. Sci. 120 (1996) 1–33.
[62] M.D. Le Van, X. Ye, J. Membr. Sci. 221 (2003) 147–161.
[63] B.R. Ezzell, L. Jackson, W.P. Carl, (Dow Chemical Company) US Patent 4,940,525, (July 10, 1990).
[64] K. Prater, J. Power Sources 29 (1990) 239–250.
[65] R.B. Moore, C.R. Martin, Macromolecules 22 (1989) 3594–3599.
[66] G. Hinds, Performance and Durability of PEM Fuel Cells : A Review, NRL Report DEPC-MPE 002, National Physical Laboratory, Teddington, UK, 2004.
[67] M. Wakizoe, O.A. Velev, S. Srinivasan, Elecrochim. Acta 40 (1995) 335–344.
[68] M. Yamabe, H. Miyake, K. Arai, (Asahi Glass Co.), Japanese Patent 52-28 588, (March 3, 1977).
[69] H.L. Tang, M. Pan, F. Wang, P.K. Shen, S.P. Jiang, J. Phys. Chem. B 111 (2007) 8684–8690.
[70] Y. Liu, T. Nguyen, N. Kristian, Y. Yu, X. Wang, J. Membr. Sci. 330 (2009) 357–362.
[71] P. Xiao, J. Li, H. Tang, Z. Wang, M. Pan, J. Membr. Sci. 442 (2013) 65–71.
[72] B. Bahar, A.R. Hobson, J.A. Kolde, D. Zuckerbrod, (W.L. Gore & Associates, Inc.), US Patent 5,547,551, (August 20, 1996).
[73] W. Liu, K. Ruth, G. Rusch, J. New Mater. Mater. Electrochem. Syst. 4 (2001) 227–231.
[74] S. Renaud, B. Ameduri, Functional fluoropolymers for fuel cell membranes, in: T. Nakajima, H. Groult (Eds.), Fluorinated Materials for Energy Conversion, Elsevier Science, 2005, pp. 469–512.
[75] A.E. Steck, Membrane materials in fuel cells, in: Proceedings of the First International Symposium on New Materials for Fuel Cell Systems I, 1995, pp. 74–94.
[76] N.P. Brandon, S. Skinner, B.C. Steele, Annu. Rev. Mater. Res. 33 (2003) 183–213.
[77] G. Inzelt, M. Pineri, J.W. Schultze, M.A. Vorotyntsev, Elecrochim. Acta 45 (2000) 2403–2421.
[78] M. Hogarth, X. Glipa, High Temperature Membranes for Solid Polymer Fuel Cells, ETSU F/02/00189/REP, U.K. Department of Trade and Industry, 2001.

[79] J. Roziére, D. Jones, Annu. Rev. Mater. Res. 33 (2003) 503–555.
[80] W. Grot, Chemie Ing. Tech. 47 (1975) 617.
[81] R.A. Smith, (E.I. du Pont de Nemours and Company), US Patent 4,437,952, (March 20, 1984).
[82] M. Odgaard, The use of Nafion in fuel cells, in: T. Nakajima, H. Groult (Eds.), Fluorinated Materials for Energy Conversion, Elsevier Science, 2005, pp. 439–468.
[83] D. Raistrick, in: Proceedings of the Symposium on Diaghrams, Separators, and Ion Exchange Membranes, The Elctrochemical Society, Princeton, NJ, 1986, p. 172.
[84] M.S. Wilson, S. Gottesfeld, J. Appl. Electrochem. 22 (1992) 1–7.
[85] M.S. Wilson, S. Gottesfeld, J. Electrochem. Soc. 139 (1992) L28.
[86] P.G. Pickup, G. Li, J. Electrochem. Soc. 150 (2003) C745–C752.
[87] M.S. Wilson, J.A. Valerio, S. Gottesfeld, Elecrochim. Acta 40 (1995) 355–363.
[88] S. Srinivasan, E.A. Ticianelli, C.R. Derouin, A. Redondo, J. Power Sources 22 (1988) 359–375.
[89] S. Litster, G. McLean, J. Power Sources 130 (2004) 61–76.
[90] S.J. Paddison, Annu. Rev. Mater. Res. 33 (2003) 289–319.
[91] K.D. Kreuer, S.J. Paddison, E. Spohr, M. Schuster, Chem. Rev. 104 (2004) 4637–4678.
[92] J.A. Elliott, S.J. Paddison, Phys. Chem. Chem. Phys. 9 (2007) 2602–2618.
[93] D. Wu, S.J. Paddison, J.A. Elliott, Energy Environ. Sci. 1 (2008) 284–293.
[94] D. Brandell, J. Karo, A. Liivat, J.O. Thomas, J. Mol. Model. 13 (2007) 1039–1046.
[95] J. Karo, A. Aabloo, J.O. Thomas, D. Brandell, J. Phys. Chem. B. 114 (2010) 6056–6064.
[96] R.B. Moore, C.R. Martin, Macromolecules 21 (1988) 1334–1339.
[97] J.J. Fontanelli, M.C. Wintersgill, R.S. Chen, Y. Wu, S.G. Greenbaum, Elecrochim. Acta 40 (1995) 2321–2326.
[98] S.W. Yeo, A. Eisenberg, J. Appl. Polym. Sci. 21 (1977) 875–898.
[99] E.J. Roche, M. Pineri, R. Duplessix, A.M. Levelut, J. Polym. Sci. Polym. Phys. 20 (1982) 107–116.
[100] J. Halim, F.N. Buchi, O. Haas, M. Stamm, G.G. Sherer, Elecrochim. Acta 39 (1994) 1303–1307.
[101] K.D. Kreuer, J. Membr. Sci. 185 (2001) 29–39.
[102] E.M. Lee, R.K. Thomas, D.Y.B.A.N. Bugess, A.K. Soper, A.R. Rennil, Macromolecules 25 (1992) 3106–3109.
[103] B. Loppinet, G. Gebel, C.E. Wiliams, J. Phys. Chem. B 101 (1997) 1884–1892.
[104] J. Ceynowa, Polymer 19 (1978) 73–76.
[105] T. Xue, Y.S. Trent, K. Osseo-Asare, J. Membr. Sci. 45 (1989) 261–271.
[106] R.A. Zoppi, I.V.P. Yoshida, S.P. Nunes, Polymer 39 (1998) 1309–1315.
[107] P.J. James, T.J. McMaster, J.M. Newton, M.J. Miles, Polymer 41 (2000) 4223–4231.
[108] T.A. Zawodzinski, M. Neeman, L.O. Sillerud, S. Gottesfeld, J. Phys. Chem. 95 (1991) 6040–6044.
[109] J.J. Fontanelli, C.A. Edmonson, M. Wintersgill, Y. Wu, S.G. Greenbaum, Macromolecules 29 (1996) 4944–4951.
[110] G. Xu, Y.S. Pak, Solid State Ionics 50 (1992) 339–343.
[111] N.G. Boyle, V.J. McBrierty, D.C. Douglas, A. Eisenberg, Macromolecules 16 (1983) 80–84.
[112] S. Schlick, G. Gebel, M. Pineri, F. Volino, Macromolecules 24 (1991) 3517–3521.
[113] N.G. Boyle, V.J. McBrierty, D.C. Douglas, Macromolecules 16 (1983) 75–80.
[114] S. Schlick, M.G. Alonso-Amigo, Phys. Chem. 90 (1986) 6353–6358.
[115] M. Laporta, M. Pegoraro, L. Zanderighi, Phys. Chem. Chem. Phys. 1 (1999) 4619–4628.
[116] D.I. Ostrovski, A.M. Brodin, L.M. Torell, Solid State Ionics 85 (1996) 323–327.
[117] T.D. Gierke, G.E. Munn, F.C. Wilson, J. Polym. Sci. Polym. Phys. 19 (1981) 1687–1704.
[118] M. Eikerling, A.A. Kornyshev, U. Stimming, J. Phys. Chem. B 101 (1997) 10807–10820.

[119] H.L. Yeager (Ed.), Perfluorinated Ionomer Membrane, American Chemical Society, Washington, DC, 1982.
[120] A. Eisenberg, Macromolecules 3 (1970) 147–154.
[121] W.Y. Hsu, T.D. Gierke, J. Membr. Sci. 13 (1983) 307–326.
[122] W.Y. Hsu, T.D. Gierke, Macromolecules 15 (1982) 101–105.
[123] K. Schmidt-Rohr, Q. Chen, Nat. Mater. 7 (2008) 75–83.
[124] Product Information Sheet by E.I.du Pont de Nemours and Company.
[125] S. Oshi, O. Kamishima, J. Mizusaki, J. Kawamura, Solid State Ionics 180 (2009) 580–584.
[126] S. Hink, N. Wagner, W.G. Bessler, E. Roduner, Membranes 2 (2012) 237–252.
[127] F.C. Walsh, S. Slade, S.A. Cambell, T.R. Ralph, J. Electrochem. Soc. 149 (2002) A1556–A2002.
[128] T.A. Zawodzinski, C. Derouin, S. Radzinski, R.J. Sherman, V.T. Smith, T.E. Springer, S. Gottesfeld, J. Electrochem. Soc. 140 (1993) 1041–1047.
[129] T.A. Zawodzinski, T.E. Springer, J. Davey, R. Jestel, C. Lopez, J. Valerio, S. Gottesfeld, J. Electrochem. Soc. 140 (1993) 1981–1985.
[130] T.A. Zawodzinski, T.E. Springer, F. Uribe, S. Gottesfeld, Solid State Ionics 60 (1993) 199–211.
[131] A.V. Anantaraman, C.L. Gardner, J. Electroanal. Chem. 414 (1996) 115–120.
[132] M. Cappadonia, J.W. Erning, U. Stimming, J. Electroanal. Chem. 376 (1994) 189–193.
[133] A. Pozio, R.F. Silva, M. De Francesco, J. Power Sources 134 (2004) 18–26.
[134] N. Yoshida, T. Ishisaki, A. Watanabe, M. Yoshitake, Elecrochim. Acta 43 (1998) 3749–3754.
[135] C.A. Edmonson, P.E. Stallworth, M.C. Wintersgill, J.J. Fontanelli, Y. Dai, S.G. Greenbaum, Elecrochim. Acta 43 (1998) 1295–1299.
[136] Y. Sone, P. Ekdunge, D. Simonsson, J. Electrochem. Soc. 143 (1996) 1254–1259.
[137] M.W. Verbrugge, R.F. Hill, J. Electrochem. Soc. 137 (2004) 1131–1139.
[138] C.L. Gardner, A.V. Anantaraman, J. Electroanal. Chem. 449 (1998) 209–214.
[139] F.N. Buchi, G.G. Scherer, J. Chem. Soc. Faraday Trans. 404 (1996) 37–43.
[140] V. Tricoli, N. Caretta, M. Bartolozzi, J. Electrochem. Soc. 147 (2000) 1286–1290.
[141] M. Verbrugge, R.F. Hill, J. Electrochem. Soc. 137 (1990) 3770–3777.
[142] J.A. Kolde, B. Bahar, M.S. Wilson, T.A. Zawodzinski, S. Gottesfeld, PV 95-23, in: The Electrochemical Society Proceedings Series, Pennington, NY, 1995, p. 193.
[143] F. Opekar, D. Svozil, J. Electroanal. Chem. 385 (1995) 269–271.
[144] A. Parthasarathy, S. Srinivasan, A.J. Appelby, C.R. Martin, J. Electrochem. Soc. 139 (1992) 2530–2537.
[145] M. Cappadonia, J.W. Erning, S.M.S. Niaki, U. Stimming, Solid State Ionics 77 (1995) 65–69.
[146] J.J. Sumner, S.E. Creager, J.J. Ma, D.D. DesMarteau, J. Electrochem. Soc. 145 (1998) 107–109.
[147] S. Slade, S.A. Campbell, T.R. Ralph, F.C. Walsh, J. Electrochem. Soc. 149 (2002) A1556–A1564.
[148] P. Costamagna, C. Yang, A.B. Bocarsly, S. Srinivasan, Electrochim. Acta 47 (2002) 1023.
[149] R. He, Q. Li, G. Xiao, N.J. Bjerrum, J. Membr. Sci. 226 (2003) 169–184.
[150] K.D. Kreuer, Chem. Mater. 8 (1996) 610–641.
[151] K.D. Kreuer, Solid State Ionics 97 (1997) 1–15.
[152] S.J. Paddison, R. Paul, Phys. Chem. Chem. Phys. 4 (2002) 1158–1163.
[153] R. Jinnouchi, K. Okazaki, J. Electrochem. Soc. 150 (2003) E66–E73.
[154] P. Choi, N.H. Jalani, R. Datta, J. Electrochem. Soc. 152 (2005) E123–E130.
[155] Y.K. Choe, E. Tsuchida, T. Ikeshoji, S. Yamakawa, S. Hyodo, J. Phys. Chem. B 112 (2008) 11586–11594.
[156] Y.K. Choe, E. Tsuchida, T. Ikeshoji, S. Yamakawa, S. Hyodo, Phys. Chem. Chem. Phys. 11 (2009) 3892–3899.
[157] S. Pitchumani, P. Sridhar, A.K. Shukla, Bull. Mater. Sci. 32 (2009) 285–294.

[158] S. Ahadian, A. Ranjbar, H. Mizuseki, Y. Kawazoe, Int. J. Hydrogen Energy 35 (2010) 3648–3655.
[159] J.X. Wu, X. Wang, G. He, J. Benziger, J. Polym. Sci. B Polym. Phys. 49 (2011) 1437–1445.
[160] M. Phonyiem, S. Chaiwongwattana, C. Leo-ngam, K. Sagarik, Phys. Chem. Chem. Phys. 13 (2011) 10923–10939.
[161] K.D. Kreuer, A. Rabenau, W. Weppner, Angew. Chem. Int. Ed. Engl. 21 (1982) 208–209.
[162] T.E. Springer, T.A. Zawodzinski, S. Gottesfeld, J. Electrochem. Soc. 138 (1991) 2334–2342.
[163] N. Agmon, Chem. Phys. Lett. 224 (1995) 456–462.
[164] D. Marx, M.E. Tuckerman, J. Hütter, M. Parrinello, Nature 397 (1999) 601–604.
[165] A.A. Kornyshev, A.M. Kuznetsov, E. Spohr, J. Ulstrup, J. Phys. Chem. B 107 (2003) 3351–3366.
[166] G.A. Eisman, J. Power Sources 29 (1990) 389–398.
[167] C. Ma, L. Zhang, S. Mukerjee, D. Ofer, B. Nair, J. Membr. Sci. 219 (2003) 123–136.
[168] L. Merlo, A. Ghielmi, L. Cirillo, M. Gebert, V. Arcella, Sci. Technol. 42 (2007) 2891–2908.
[169] K.D. Kreuer, M. Schuster, B. Obliers, O. Diat, U. Traub, A. Fuchs, U. Klock, S.J. Paddison, J. Maier, J. Power Sources 178 (2008) 499–509.
[170] I.H. Hristov, S.J. Paddison, R. Paul, J. Phys. Chem. B 112 (2008) 2937–2949.
[171] J. Peron, D. Edwards, M. Haldane, X.Y. Luo, Y.M. Zhang, S. Holdcroft, Z.Q. Shi, J. Power Sources 196 (2011) 179–181.
[172] S. Subianto, S. Cavalière, D. Jones, J. Rozière, ECS Trans. 41 (2011) 1517–1520.
[173] A. Stassi, I. Gatto, E. Passalacqua, V. Antonucci, A.S. Arico, L. Merlo, C. Oldani, E. Pagano, J. Power Sources 196 (2011) 8925–8930.
[174] E. Moukheiber, G. DeMoor, L. Flandin, C. Bas, J. Membr. Sci. 389 (2011) 294–304.
[175] X. Luo, S. Holdcroft, A. Mani, Y. Zhang, Z. Shi, Phys. Chem. Chem. Phys. 13 (2011) 18055–18062.
[176] Z. Tu, H. Zhang, Z. Luo, J. Liu, Z. Wan, M. Pan, J. Power Sources 222 (2012) 277–281.
[177] E. Moukheiber, G. De Moor, L. Flandin, C. Bas, J. Membr. Sci. 389 (2012) 294–304.
[178] S. Subianto, S. Cavaliere, D.J. Jones, J. Rozière, J. Polym. Sci. Part A Polym. Chem. 51 (2012) 118–128.
[179] M.C. Ferrari, J. Catalano, M.G. Baschetti, M.G. DeAngelis, G.C. Sarti, Macromolecules 45 (2012) 1901–1912.
[180] S.R. Samms, S. Wasmus, R.F. Savinell, J. Electrochem. Soc. 143 (1996) 1498–1504.
[181] H.L. Tang, P.K. Shen, S.P. Jiang, W. Fang, P. Mu, J. Power Sources 170 (2007) 85–92.
[182] M.S. Schaberg, J.E. Abulu, G.M. Haugen, M.A. Emery, S.J. O'Conner, P.N. Xiong, S. Hamrock, ECS Trans. 33 (2010) 627–633.
[183] T. Okada, G. Xie, M. Meeg, Elecrochim. Acta 43 (1998) 2141–2155.
[184] T.A. Zawodzinski, J. Davey, J. Valerio, S. Gottesfeld, Elecrochim. Acta 3 (1995) 297–302.
[185] F.N. Buchi, G.G. Scherer, J. Electrochem. Soc. 148 (2001) A183–A188.
[186] X. Ren, S. Gottesfeld, J. Electrochem. Soc. 148 (2001) A87–A93.
[187] S. Motupally, A.J. Becker, J.W. Weidner, J. Electrochem. Soc. 147 (2000) 3171–3177.
[188] R.J. Bellows, M.Y. Lin, M. Arif, A.K. Thompson, D. Jacobsen, J. Electrochem. Soc. 146 (1999) 1099–1103.
[189] E.R. Gonzales, T.J.P. Freire, J. Electroanal. Chem. 503 (2001) 57–68.
[190] A.Z. Weber, J. Newman, J. Electrochem. Soc. 150 (2003) A1008–A1015.
[191] M. Ji, Z. Wei, Energies 2 (2009) 1057–1106.
[192] W. Dai, H. Wang, X.Z. Yuan, J.J. Martin, D. Yang, J. Qiao, J. Ma, Int. J. Hydrogen Energy 34 (2009) 9461–9478.
[193] T. Berning, M. Odgaard, S. Kær, J. Electrochem. Soc. 156 (2009) B1301–B1311.
[194] T. Berning, M. Odgaard, S. Kær, J. Power Sources 195 (2010) 4842–4852.
[195] T. Berning, M. Odgaard, S. Kær, ECS Trans., vol. 33, Las Vegas, NV, 2010.

[196] T. Berning, M. Odgaard, S. Kær, J. Power Sources 196 (2011) 6305–6317.
[197] S. Bhatta, B. Guptaa, V.K. Sethib, M. Pandey, Int. J.Curr. Eng. Technol. 2 (2012) 219–226.
[198] V.M. Barragán, A. Heinzel, J. Power Sources 84 (1999) 70–74.
[199] X. Ren, T.E. Springer, T.A. Zawodzinski, S. Gottesfeld, J. Electrochem. Soc. 147 (2000) 466–474.
[200] N. Jia, M.C. Lefebvre, J. Halfard, Z. Qi, P.G. Pickup, Electrochem. Solid State Lett. 3 (2000) 529–531.
[201] A. Casalegno, R. Marchesi, J. Power Sources 185 (2008) 318–330.
[202] T.S. Zhao, C. Xu, R. Chen, W.W. Yang, Prog. Energy Combust. Sci. 35 (2009) 275–292.
[203] F. Lufrano, V. Baglio, P. Staiti, V. Antonucci, A.S. Arico, J. Power Sources 243 (2013) 519–534.
[204] M. Ahmed, I. Diner, Int. J. Energy Res. 35 (2011) 1213–1228.
[205] Y.S. Kim, B.S. Pivovar, J. Electrochem. Soc. 157 (2010) B1608–B1615.
[206] H. Dohle, J. Divisek, J. Mergel, H.F. Oetjen, C. Zingler, D. Stolten, J. Power Sources 105 (2002) 274–282.
[207] K. Broka, P. Ekdunge, J. Appl. Electrchem. 27 (1997) 117–123.
[208] F.N. Buchi, M. Wakizoem, S. Srinivasan, J. Electrochem. Soc. 143 (1996) 927–932.
[209] A.G. Guzman-Garcia, P.N. Pintauro, M.W. Verbrugge, J. Appl. Electrochem. 22 (1992) 204.
[210] J. Healy, C. Hayden, T. Xie, K. Olson, R. Waldo, M. Brundge, H. Gasteiger, J. Abbott, Fuel Cells 5 (2005) 302–308.
[211] A. Collier, H. Wang, X.Z. Yuan, J. Zhang, D.P. Wilkinson, Int. J. Hydrogen Energy 31 (2006) 1838–1854.
[212] T. Fuller, C. Hartnig, V. Ramani, H. Uchida, H.A. Gasteiger, S. Cleghorn, P. Strasser, T. Zawodzinski, D. Jones, P. Shirvanian, T. Jarvi, P. Zelenay, C. Lamy, P. Bele (Eds.), Proton Exchange Membrane Fuel Cells 9, Issue 1, The Electrochemical Society, Pennington, NJ, USA, 2009.
[213] G. Pourcelly, A. Oikonomou, C. Gavach, J. Electroanal. Chem. 287 (1998) 43–59.
[214] R. Baldwin, M. Pham, A. Leonida, J. Mcelroy, T. Nalette, J. Power Sources 29 (1990) 399–412.
[215] Z. Qi, A. Kaufman, J. Power Sources 109 (2002) 38–46.
[216] Z. Qi, A. Kaufman, J. Power Sources 109 (2002) 469–476.
[217] J. Yu, T. Matsuura, Y. Yoshikawa, Md N. Islam, M. Hori, Phys. Chem. Chem. Phys. 7 (2004) 373–378.
[218] S. Kundu, L.C. Simon, M. Fowler, S. Grot, Polymer 46 (2005) 11707–11715.
[219] J. Zhang, Y. Tang, C. Song, Z. Xia, H. Li, H. Wang, Electrochim. Acta 53 (2008) 5315–5321.
[220] E. Endoh (Ed.), Highly Durable PFSA Membranes. Handbook of Fuel Cells, John Wiley and Sons Inc., 2010.
[221] M. Pianca, E. Barchiesi, G. Esposto, S. Raice, J. Fluorine. Chem. (1999) 71–84.
[222] M.J. Larsen, E.M. Skou, J. Power Sources 202 (2012) 35–46.
[223] The United States Council for Automotive Research, U. S. DRIVE Partnership Fuel Cell Technical Team - Fuel Cell Technical Roadmap – Revised, June, 2013.
[224] United States Department of Energy, DOE Cell Component Accelerated Stress Test Protocols for PEM Fuel Cells (Electrocatalysts, Supports, Membranes, and Membrane Electrode Assemblies), 2007.
[225] R.L. Perry, FY 2012 Annual Progress Report – DOE Hydrogen and Fuel Cells Program – V.E.7 Analysis of Durability of MEAs in Automotive PEMFC Applications, Wilmington, DE, USA, 2012.
[226] T.G. Benjamin, in: High Temp. Membr. Work. Gr. Meet, 2007.

[227] T. Patterson, V. Srinivasamurthi, T. Skiba, FY 2012 Annual Progress Report – DOE Hydrogen and Fuel Cells Program – V.E.4 Improved Accelerated Stress Tests Based on Fuel Cell Vehicle Data, South Windsor, CT, 2012.
[228] T.C. Jao, G.B. Jung, S.C. Kuo, W.J. Tzeng, A. Su, Int. J. Hydrogen Energy 37 (2012) 13623–13630.
[229] M.P. Rodgers, R.P. Brooker, N. Mohajeri, L.J. Bonville, H.R. Kunz, D.K. Slattery, J.M. Fenton, J. Electrochem. Soc. 159 (2012) F338–F352.
[230] R. Mukundan, R.L. Borup, J. Davey, R. Lujan, D. Torraco, D. Langlois, F. Garzon, D. Spernjak, J. Fairweather, S. Balasubramanian, A. Weber, M. Brady, K. More, G. James, D. Ayotte, S. Grot, FY 2012 Annual Progress Report – DOE Hydrogen and Fuel Cells Program – V.E.5 Accelerated Testing Validation, Los Alamos, NM, USA, 2012.
[231] D.M. Jollie, M.A.J. Cropper, S. Geiger, J. Power Sources 131 (2004) 57–61.
[232] E. Kahn, J. Power Sources 135 (2004) 212–214.
[233] R. Dillon, S. Srinivasan, A.S. Arico, V. Antonucci, J. Power Sources 127 (2004) 112–126.
[234] A. Kirubakaran, S. Jain, R.K. Nema, Renewable Sustainable Energy Rev. 13 (2009) 2430–2440.
[235] H. Maru, S.C. Singhal, C. Stone, D. Wheeler, 1–10 KW Stationary Combined Heat and Power Systems Status and Technical Potential, NREL, 2010. http://www.hydrogen.energy.gov/pdfs/48265.pdf.
[236] Y.W. Ken, S. Chen, J. Mishler, S.C. Cho, X.C. Adroher, Appl. Energy 88 (2011) 981–1007.
[237] Fuel Cells Today, The Fuel Cell Industry Review, 2013. http://fuelcelltoday.com/media/1889744/fct_review_2013.pdf.
[238] U.S. Department of Energy, 2012 Fuel Cell Technologies Market Report, October 2013. http://www1.eere.energy.gov/hydrogenandfuelcells/pdfs/2012_market_report.pdf.
[239] A. Brouzgou, A. Podias, P. Tsiakaras, J. Appl. Electrochem. 43 (2013) 119–136.
[240] D. Kingshuk, K. Piyush, D. Suparna, P.P. Kundu, Polym. Rev. 54 (2014) 1–32.
[241] E. Middelman, W. Kout, B. Vogelaar, J. Lenssen, E. de Waal, J. Power Sources 118 (2003) 44–46.
[242] D.P. Davies, P.L. Adcock, M. Turpin, S.J. Rowen, J. Power Sources 86 (2000) 237–242.
[243] H.C. Lim, J.-H. Koh, H.-K. Seo, C.G. Lee, Y.-S. Yoo, J. Power Sources 115 (2003) 54–65.
[244] R.G. Reddy, A. Kumar, J. Power Sources 113 (2003) 11–18.
[245] F.A. de Bruijn, V.A.T. Dam, G.J.M. Janssen, Fuel Cells 8 (2008) 3–22.
[246] N. Yousfi-Steinera, P. Mocoteguya, D. Candusso, J. Power Sources 183 (2008) 260–274.
[247] A. Casalegno, L. Colombo, S. Galbiati, R. Marchesi, J. Power Sources 195 (2010) 4143–4148.
[248] The Role of Battery Electric Vehicles, Plug-in Hybrids and Fuel Cell Electric Vehicles a Portfolio of Power-Trains for Europe: A Fact-based Analysis, December (2011). http://www.fch-ju.eu/sites/default/files/documents/Power_trains_for_Europe.pdf.
[249] D. Garraín, Y. Lechón, C. de la Rúa, Smart Grid Renewable Energy 2 (2011) 68–74.
[250] C. Houchins, G.J. Kleen, J.S. Spendelow, J. Kopasz, D. Peterson, N.L. Garland, D.L. Ho, J. Marcinkoski, K.E. Martin, R. Tyler, D.C. Papageorgopoulos, Membranes 2 (2012) 855–878.
[251] J. Fenton, Lead Research and Development Activity for DOE's High Temperature, Low Relative Humidity Membrane Program, U.S. DOE Hydrogen Program Annual Merit Review and Peer Evaluation; U.S. Department of Energy, Arlington, VA, USA, 2011.
[252] Danish Public Service Obligations, Contract no. 2006-1-6295.
[253] The Danish Energy Authority, EFP-Akt.167 J.no. 033001/33033–0151.
[254] The Danish Energy Authority, EFP-Akt.167 J.no. 033001/33033-0333.
[255] KeePEMAlive, F.C.H-J.U.No. 245113.

[256] Competitive μCHP for H2omes, E.U.D.P-11-I.J.No. 64011-70051.
[257] General Motors, Fuel Cell Equinox Tops 100,000 Miles in Real-World Driving, 2013. http://media.gm.com/media/us/en/gm/news.detail.html/content/Pages/news/us/en/2013/Oct/1022-fc-equinox.html (Last accessed 14.01.14).
[258] K. Goto, I. Rozhanskii, Y. Yamakawa, Polym. J. 41 (2009) 95–104.
[259] P. Dondi, D. Bayoumi, C. Haederli, D. J ulian, M. Suter, J. Power Sources 106 (2002) 1–9.
[260] O. Yamazaki, M. Echigo, N. Shinke, T. Tabata, Lucerne, Switzerland, in: Proceedings of the Fuel Cell Home, 2001, pp. 93–102.
[261] H. Ren, W. Guo, Energy Build. 42 (2010) 853–861.
[262] L. Barrelli, G. Bidini, F. Gallorini, A. Ottaviano, Appl. Energy 88 (2011) 4334–4342.
[263] N. Briguglio, M. Ferraro, G. Brunaccini, V. Antounucci, Int. J. Hydrogen Energy 36 (2011) 8023–8029.
[264] J.J. Hwang, J. Power Sources 223 (2013) 325–335.
[265] S. Sevencan, T. Guan, G. Lindberg, P. Alvfors, B. Ridell, Int. J. Hydrogen Energy 38 (2013) 3858–3864.
[266] A. Arsalis, M.P. Nielsen, S.K. Kær, Appl. Therm. Eng. 50 (2013) 704–713.
[267] M. Hosseini, I. Dincer, M.A. Rosen, Progress in Exergy, Energy, and the Environment, Springer, 2014. 181–191.
[268] M. Gandiglio, A. Lanzini, M. Santarelli, P. Leone, Energy Build. 69 (2014) 381–393.
[269] The Alternative eMagazine, Residential-Scale Power Generation Issue October/November, 2013. http://www.altenergymag.com/emagazine/2013/10/residential-scale-power-generation-/2162.
[270] S. Campanari, L. Rosés, Fuel Cell-Based Micro-CHP System, November 2009. http://www.powergenworldwide.com/index/display/articledisplay/6333305363/articles/cogeneration-and-on-site-power-production/volume-10/issue-6/features/fuel-cell-based_micro-chp.html.
[271] Enefiel, F.C.H-J.U.N. 303462.
[272] J. de Wit, M.M. Melchiors, L.G. Madsen, Cogeneration & On–Site Power Production, May–June 2014. 14–19.
[273] C.K. Dyer, J. Power Sources 106 (2004) 31–34.
[274] M. Cropper, Fuel Cells 4 (2004) 236–240.
[275] A. Heinzel, C. Hebling, M. Müller, M. Zedda, C. Müller, J. Power Sources 105 (2004) 250–255.
[276] F.N. Büchi, M. Inaba, T.J. Schmidt, Y. Seung, P. Zelany (Eds.), Polymer Electrolyte Fuel Cell Durability, Springer Science + Business Media, LLC, 2009.
[277] A.R. Jha (Ed.), Next-Generation Batteries and Fuel Cells for Commercial, Military, and Space Applications, CRC Press, 2012.
[278] DURAMET, F.C.H-J.U. No. 278054.
[279] V. Neburchilov, J. Martin, H. Wang, J. Zhang, J. Power Sources 169 (2007) 221–238.
[280] M.H.D. Othman, A.F. Ismail, A. Mustafa, Malays. Polym. J. 5 (2010) 1–36.
[281] B.D. Lee, D.H. Jung, Y.H. Ko, J. Power Sources 131 (2004) 207–212.
[282] J. Han, E.S. Park, J. Power Sources 112 (2002) 477–483.
[283] H.R. Corti, E.R. Gonzales (Eds.), Direct Alcohol Fuel Cells – Materials, Performance, Durability and Applications, Springer, 2014.
[284] X. Li, A. Faghri, J. Power Sources 226 (2013) 223–240.
[285] N.K. Shrivastava, S.B. Thombre, Int. J. Eng. Sci. Technol. 3 (2011) 6000–6007.
[286] A. Casalegno, Direct Methanol Fuel Cells: Experimental Analysis and Model Development (Ph.D. thesis), 2007.

[287] A. Casalegno, R. Marchesi, F. Rinaldi, J. Fuel Cell Sci. Technol. 4 (2007) 418–424.
[288] M.P. Hogarth, T.R. Ralph, Platinum Met. Rev. 46 (2002) 146–164.
[289] T. Schultz, S. Zhou, K. Sundmacher, Chem. Eng. Technol. 24 (2001) 1223–1233.
[290] S. Wasmus, A. Kuver, J. Electroanal. Chem. 461 (1999) 14–31.
[291] M.P. Hogarth, P.A. Christensen, A. Hamnet, A. Shukla, J. Power Sources 69 (1997) 113–124.
[292] C. Lamy, J.-M. Leger, S. Srinivasan, V. Antonucci, in: J.O.´M. Bockris, B.E. Conway, R.E. White (Eds.), Modern Aspects of Electrochemistry, vol. 34, 2001.
[293] A.S. Arico, S. Srinivasan, V. Antonucci, Fuel Cells 1 (2001) 133–161.
[294] X. Ren, T.A. Zawodzinski, F. Uribe, H. Dai, S. Gottesfeld, PV 95-23, in: The Electrochemical Society Proceedings Series, Pennington, New Jersey, 1995, pp. 284–298.
[295] J. Kallo, W. Lehnert, R. von Helmolt, J. Electrochem. Soc. 150 (2004) A765–A769.
[296] K. Scott, W.M. Taama, P. Argyropoulos, K. Sundmacher, J. Power Sources 83 (1999) 204–216.
[297] E.S. Smotkin, B. Gurau, J. Power Sources 112 (2002) 339–352.
[298] S. Ma, M. Odgaard, E. Skou, Solid State Ionics 176 (2005) 2923–2927.
[299] A.S. Hollinger, R.J. Maloney, R.S. Jayashree, D. Natarajan, L.J. Markoski, P.J.A. Kenis, J. Power Sources 195 (2010) 3523–3528.
[300] X.Y. Li, W.W. Yang, Y.L. He, T.S. Zhao, Z.G. Qu, Appl. Therm. Eng. 48 (2012) 392–401.
[301] Y.L. He, Z. Miao, T.S. Zhao, W.W. Yang, Int. J. Hydrogen Energy 37 (2012) 4422–4438.
[302] M.Z.F. Kamaruddin, S.K. Kamarudin, W.R.W. Daud, M.S. Masdar, Renewable Sustainable Energy Rev. 24 (2013) 557–565.
[303] Q.X. Wu, L. An, X.H. Yan, T.S. Zhao, Electrochim. Acta 133 (2014) 8–15.
[304] C.H. Wan, M.T. Lin, C.H. Lin, B.J. Su, Int. J. Hydrogen Energy 39 (2014) 2516–2525.
[305] N. Kimiaie, K. Wedlich, M. Hehemann, R. Lambertz, M. Müller, C. Korte, D. Stolten, Energy Environ. Sci. (2014) (accepted 16.06.14, first published online 20.06.14).
[306] P. Zelenay, ECS Trans. 1 (2006) 483–495.
[307] M. Schulze, N. Wagner, T. Kaz, K.A. Friedrich, Electrochim. Acta 52 (2007) 2328–2336.
[308] S. Kundu, M. Fowler, L.C. Simon, R. Abouatallah, J. Power Sources 182 (2008) 254–258.
[309] C. Eickes, P. Piela, J. Davey, P. Zelenay, J. Electrochem. Soc. 153 (2006) A171–A178.
[310] A.A. Kulikovsky, H. Schmitz, K. Wippermann, J. Mergel, B. Fricke, T. Sanders, D.U. Sauer, J. Power Sources 173 (2007) 420–423.
[311] J.Y. Park, M.A. Scibioh, S.K. Kim, H.J. Kim, I.H. Oh, T.G. Lee, H.Y. Ha, Int. J. Hydrogen Energy 34 (2009) 2043–2051.
[312] F. Bresciani, A. Casalegno, J.L. Bonde, M. Odgaard, R. Marchesi, Int. J. Energy Res. 38 (2014) 117–124.
[313] X. Zhao, G. Sun, L. Jiang, W. Chen, S. Tang, B. Zhou, Q. Xin, Electrochem. Solid State Lett. 8 (2005) A149–A151.
[314] K. Yasuda, Y. Nakano, Y. Goto, ECS Trans. 5 (2007) 291.
[315] Y. Nishimura, M. Matsuyama, I. Nagai, M. Yamane, M. Yanagida, Y. Miyazaki, ECS Trans. 17 (2009) 511–515.
[316] X. Jie, Z.-G. Shao, J. Hou, G. Sun, B. Yi, Electrochim. Acta 55 (2010) 4783–4788.
[317] X. Jie, Z.-G. Shao, B. Yi, Electrochem. Commun. 12 (2010) 700–702.
[318] J.Y. Park, K.Y. Park, K.B. Kim, Y. Na, H. Cho, J.-H. Kim, J. Power Sources 196 (2011) 5446–5452.
[319] T. Tsujiguchi, T. Furukawa, N. Nakagawa, J. Power Sources 196 (2011) 9339–9345.
[320] N. Kimiaiel, C. Trappmann1, H. Janßen1, M. Hehemann1, H. Echsler, M. Müller, Fuel Cells 1 (2014) 64–75.

[321] http://www.efoy-omfort.com/sites/default/files/140414_Data_Sheet_EFO_Comfort_US.pdf.
[322] http://www.ird.dk/getattachment/60d6608d-fc05-455c-bc36-41dfad5b0345/IRD_DMFC.aspx.
[323] http://oorjafuelcells.com/services/.
[324] http://www.toshiba.co.jp/about/press/2009_10/pr2201.htm.

Chapter 15

Surface-Fluorinated Carbon Materials for Supercapacitor

Young-Seak Lee
Department of Applied Chemistry and Biological Engineering, Chungnam National University, Daejeon, Republic of Korea

Chapter Outline
15.1 Introduction 375
15.2 Fluorinated Activated Carbons for Supercapacitor 376
15.3 F-AC Fibers for Supercapacitor 379
15.4 Fluorinated Carbon Nanotubes for Supercapacitor 382
References 385

15.1 INTRODUCTION

Supercapacitors, which are often called electrical double-layer capacitors (EDLCs), pseudocapacitors, ultracapacitors, power capacitors, gold capacitors, or power caches, have attracted research interest worldwide because of their potential applications as energy storage devices in many fields [1,2]. Supercapacitors have considerably higher specific powers and longer cycle lifetimes compared to most rechargeable batteries, such as lead-acid, nickel-metal hydride, and Li-ion batteries. Hence, supercapacitors have attracted considerable interest because of the ever-increasing demands of electric vehicles, portable electronic devices, and power sources for memory backup [3–5].

On the basis of their charge storage mechanisms, supercapacitors can be divided into two general classes: EDLCs and pseudocapacitors. Each class is characterized by its own unique mechanism for storing charge, which are non-faradaic and faradaic mechanisms [6,7]. Faradaic processes, such as oxidation–reduction reactions, involve the transfer of charge between the electrode and electrolyte. By contrast, a non-faradaic process does not use a chemical mechanism and the charges are quite distributed on surfaces through physical processes that do not involve any forming or breaking of chemical bonds.

Carbon materials such as activated carbon (AC) or carbon nanotubes (CNTs) are commonly used for supercapacitor electrodes [8]. Theoretically, as the specific surface area of the carbon material increases, it becomes more capable of accumulating charge; however, specific capacitance does not necessarily increase proportionally with the specific surface area of carbon materials [9]. As previously mentioned, the performance of a supercapacitor is based on the accumulation of charge at the electrode/electrolyte interface; thus, the surface characteristics of the carbon material are very important with respect to its performance [10,11]. Therefore, surface modification of the carbon material is critical for improving its capacity.

Surface modification can affect the interaction between an electrolyte and an electrode at their interface. Various surface modification methods have been investigated, such as chemical treatments, plasma, flame, corona discharge, and direct fluorination [12–17]. Of these methods, fluorination is one of the most effective for modifying the carbon surface. Fluorination can control textural properties and introduce fluorine-containing functional groups, which increase the polarity of the carbon surface because of the high electronegativity of fluorine atoms [18,19].

Many previous studies have investigated fluorinated carbon materials for use as electrode materials for lithium-ion batteries, primary lithium batteries, etc. [20–23]. However, fluorinated carbon materials have been investigated less frequently as supercapacitor electrodes than as electrodes for other batteries. According to some studies, fluorinated carbon materials can be used as electrode materials for supercapacitors. The present chapter reviews the recent results on the physical, chemical, and electrochemical behaviors of fluorinated carbon materials for supercapacitors.

15.2 FLUORINATED ACTIVATED CARBONS FOR SUPERCAPACITOR

AC is the most widely used material for supercapacitor electrodes [24]. So, several studies have been performed on fluorinated AC (F-AC) for supercapacitor electrodes than have been performed for other carbon materials. Lee et al. investigated the electrochemical properties of F-AC at room temperature under different fluorine partial pressures [25]. According to other previous studies, fluorination is concomitant with decreases in the specific surface areas and pore volumes of porous carbon materials, which results from the formation of functional groups [26]; however, in Lee's study, F-AC (i.e., phenol-based AC) had a higher specific surface area and pore volume than raw AC (R-AC), as shown in Table 15.1. After fluorination, 0.04–1.13 at% fluorine was introduced onto the AC surface, depending on the fluorine partial pressure [25]. Cyclic voltammograms (CVs) of F-AC electrodes are shown in Figure 15.1. The specific capacitance of the F-AC electrode increased from 375 and 145 F g^{-1} to 491 and 212 F g^{-1} at scan rates of 2 and 50 mV s^{-1}, respectively, in comparison to that of a R-AC electrode when the fluorination process was optimized with a F_2 gas partial pressure of 0.2 bar (fluorine content: 0.77 at%). This enhancement in

TABLE 15.1 Textural Properties of Raw Activated Carbons (R-ACs) and Fluorinated Activated Carbons (F-ACs) (Gas-Phase Reaction) [25]

Sample[a]	Fluorine Gas Partial Pressure (bar)	SSA[b] ($m^2\,g^{-1}$)	T-PV[c] ($cc\,g^{-1}$)	Mic-PV[d] ($cc\,g^{-1}$)	Mes-PV[e] ($cc\,g^{-1}$)	Mes-PV/T-PV[f] (%)
R-AC	—	1875	0.80	0.72	0.08	10
F1-AC	0.1	2036	0.94	0.74	0.20	21
F2-AC	0.2	2338	1.27	0.97	0.30	24
F3-AC	0.3	2286	0.97	0.75	0.22	23

[a]R-AC: raw activated carbon. FNo.-AC: fluorinated activated carbon at given partial pressure of fluorine gas.
[b]BET-specific surface area.
[c]Total pore volume.
[d]t-Plot micropore volume.
[e]Mesopore volume.
[f]Mesopore volume/total pore volume (%).

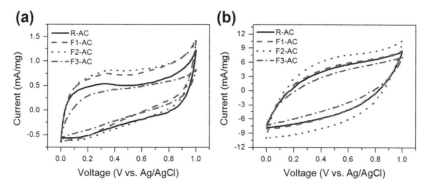

FIGURE 15.1 Cyclic voltammograms of untreated and fluorinated activated carbon (F-AC) based electrodes obtained at (a) 2 mV s^{-1} and (b) 50 mV s^{-1} [25].

capacitance can be attributed to the synergistic effect of increased polarization on the AC surface, increased specific surface area, and increased micro- and mesopore volumes, all of which were induced by the fluorination process. The increased surface area and pore volume of the AC resulted from the physical function of the fluorine functional group.

In the CV curves shown in Figure 15.1, no other peaks are observed from the F-AC electrodes, suggesting that no pseudo-faradaic reaction of the fluorine functional groups occurred. This result demonstrates that the fluorine functional groups are electrochemically active and increase the specific capacitance of the F-AC electrode, whereas other electrochemically active functional groups, such as oxygen or nitrogen, induce a pseudo-faradaic reaction [25,27–29].

Kim et al. reported that F-AC that was treated with a hydrofluoric acid (HF) solution exhibited significantly improved electrochemical capacitive performance [30]. According to their X-ray photoelectron spectroscopy (XPS) results (Figure 15.2), the F-AC that was treated in a 1.0 M HF solution contained 0.7 at% fluorine. Figure 15.3 presents CVs obtained under various scan rates for supercapacitors with R-AC and F-AC electrodes. The F-AC electrode exhibited a specific capacitance approaching 19.8 F cm^{-3}, whereas the capacitance of the R-AC electrode was 18.4 F cm^{-3} in a nonaqueous electrolyte (1.8 M triethylmethylammonium tetrafluoroborate in acetonitrile). Moreover, the kinetic enhancement of the F-AC electrode was also notable, which indicated that the F-AC tended to rapidly form electric double layers on its surface. This formation was the result of the increased electrical conductivity attributed to the semi-ionic C–F bond on the AC surface [30].

Although Kim et al. did not mention changes in the textural properties of F-AC, treating F-AC with a HF solution might change its surface area and pore volume. Lee et al. investigated the changes in the properties of F-AC treated with HF solution [31]. The specific surface area and total pore volume of F-AC treated with a 0.1 M HF solution increased compared to those of R-AC (in Table 15.2), and the AC surface was observed to contain 0.4 at% fluorine. The specific

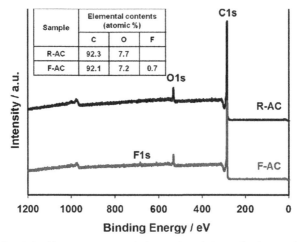

FIGURE 15.2 XPS wide-scan spectra and elemental analysis results (inset) of raw activated carbon (R-AC) and fluorinated activated carbon (F-AC) [30].

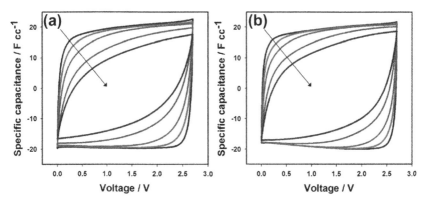

FIGURE 15.3 Cyclic voltammograms obtained at various scan rates for supercapacitors with (a) raw activated carbon (R-AC) and (b) fluorinated activated carbon (F-AC) between 0.0 and 2.7 V with a nonaqueous electrolyte at various scan rates (arrow indicates the increasing direction of scan rate, from 5 to 100 mV s^{-1}) [30].

surface area and pore volume of F-AC also suggested that an etching effect might be induced by the reaction between the HF solution and the carbon surface. Figure 15.4 shows the CVs of F-AC. At a scan rate of 10 mV s^{-1}, the specific capacitance of the F-AC was 381 F g^{-1}, which is 58% greater than that of R-AC.

15.3 F-AC FIBERS FOR SUPERCAPACITOR

Activated carbon fibers (ACFs) commonly exhibit a narrower pore size distribution than the usual granulated ACs [32]. Therefore, ACFs have primarily been

TABLE 15.2 Textural Properties of Raw Activated Carbons (R-ACs) and Fluorinated Activated Carbons (F-ACs) (Liquid-Phase Reaction) [31]

Sample	BET-Specific Surface Area ($m^2\ g^{-1}$)	Total Pore Volume ($cc\ g^{-1}$)	t-Plot Micropore Volume ($cc\ g^{-1}$)	Average Pore Diameter (nm)
Raw AC	1924	0.87	0.51	1.78
Fluorinated AC	2331	1.01	0.94	1.74

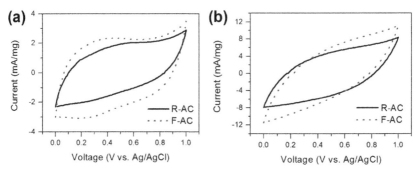

FIGURE 15.4 Cyclic voltammograms of the untreated and treated activated carbons (ACs) obtained at scan rates of (a) 10 mV s^{-1} and (b) 50 mV s^{-1} [31].

investigated for their adsorption properties toward various gases or harmful substances. Likewise, many researchers have investigated the adsorption properties of fluorinated ACFs, i.e., the changes in surface and textural properties that result from fluorination [33–35]. Even though the interest in ACFs for the use as porous electrodes in supercapacitors has emerged because of its good textural and electrical properties [36], fluorinated ACFs gave good evidences as a supercapacitor electrode in limited conditions compared with F-ACs.

Lee et al. reported fluorinated activated carbon nanofibers (ACNFs) as supercapacitor electrodes to investigate the role of fluorination in improving the electrochemical properties of carbon nanofibers [37]. ACNFs with various textural properties were prepared using polyacrylonitrile through electrospinning and chemical activation and were subsequently fluorinated with a 0.2 bar partial pressure of F_2 gas at room temperature. Table 15.3 shows the XPS elemental analysis of the ACNFs before and after fluorination. After fluorination, the specific surface area and the total pore volume of the ACNFs decreased because of the etching effect of fluorine. The fluorination process introduced 7.0–21.7 at% of fluorine into the ACNFs and also increased the specific surface area and the

TABLE 15.3 XPS Surface Elemental Analysis Parameters of Activated Carbon Nanofibers (ACNFs) and Fluorinated ACNFs [37]

Sample[a]	Elemental content (At%)			O/C (%)	F/C (%)
	C	O	F		
1KACNF	90.5	9.6	–	11.0	–
4KACNF	92.0	8.0	–	9.0	–
6KACNF	93.5	6.5	–	7.0	–
F-1KACNF	83.9	9.1	7.0	11.0	8.0
F-4KACNF	76.2	11.4	12.4	15.0	16.0
F-6KACNF	71.4	6.9	21.7	10.0	3.0

[a]1KACNF: activated carbon nanofiber with 1 M KOH, F-1KACNF: fluorinated activated carbon nanofiber with 1 M KOH.

TABLE 15.4 Textural Properties of Activated Carbon Nanofibers (ACNFs) and Fluorinated ACNFs [37]

Sample[a]	S_{BET}[b] ($m^2\,g^{-1}$)	V_T[c] ($cm^3\,g^{-1}$)	V_{micro}[d] ($cm^3\,g^{-1}$)	V_{meso}[e] ($cm^3\,g^{-1}$)	V_{meso}/V_T[f] (%)
1KACNF	777	0.38	0.29	0.09	24
4KACNF	1694	0.84	0.63	0.21	25
6KACNF	2369	1.29	0.75	0.54	42
F-1KACNF	590	0.30	0.25	0.05	17
F-4KACNF	1102	0.57	0.48	0.09	16
F-6KACNF	1209	0.66	0.57	0.09	14

[a]1KACNF: activated carbon nanofiber with 1 M KOH, F-1KACNF: fluorinated activated carbon nanofiber with 1 M KOH.
[b]BET-specific surface area.
[c]Total pore volume.
[d]Micropore volume.
[e]Mesopore volume.
[f]Mesopore volume/total pore volume (%).

total pore volume of the ACNFs. The carbon at the ACNF surface is hypothesized to primarily react with fluorine because of the formation of C-F bonds.

As shown in Table 15.4, both the specific surface area and the total pore volume of these ACNFs decreased after fluorination. These changes are attributed

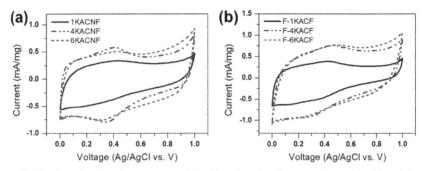

FIGURE 15.5 Cyclic voltammograms of the (a) activated carbon nanofibers (ACNFs) and (b) fluorinated ACNFs obtained at scan rates of 5 mV s^{-1} [37].

to the enhanced etching effect of the ACNF surface with more reaction sites, which are provided by the greater specific surface area, the greater total pore volume, and the greater number of fluorine gas diffusion pathways provided by the pores in the ACNFs. In addition, the greater decrease in the mesopore volumes compared with the decrease in the micropore volumes indicates that the mesopores reacted more with fluorine, which resulted in the collapse of more mesopores than micropores as a result of the fluorine-induced etching [37].

Figure 15.5 shows the CVs of the ACNFs obtained at a scan rate of 5 mV s^{-1} in H_2SO_4 electrolyte before and after the ACNFs were fluorinated. The CV profiles of the ACNFs and fluorinated ACNFs exhibited approximately rectangular shapes with small redox peaks caused by the pseudo-faradaic reaction of oxygen-containing quinone functional groups [38]. The specific capacitances were calculated to be 199 and 230 F g^{-1} for ACNFs and fluorinated ACNFs activated with 6 M KOH, respectively. Interestingly, the specific capacitances of the ACNF electrodes increased by 15.8–47.3% after fluorination, although the specific surface area and the total pore volume of the ACNF electrodes decreased significantly. These results demonstrate that the introduction of C-F functional groups (especially semicovalent C-F) is more important than the textural properties with respect to the electrochemical properties of the fluorinated ACNF electrodes [39,40].

15.4 FLUORINATED CARBON NANOTUBES FOR SUPERCAPACITOR

Recently, CNTs have attracted considerable attention as new materials because of their excellent mechanical strength, high electric conductivity, and reasonable specific surface area [41]. Notably, the CNT sheet electrode provides a discharge capacitance of 10–15 F g^{-1} even at an extremely high current density (200 A g^{-1}), whereas no discharge capacitance is obtained with a typical high-performance AC electrode at such a high current density [42]. For this reason, CNTs have been used as capacitor electrode materials. However, fluorinated

CNTs also give good evidences as supercapacitor electrodes as well as ACFs in limited conditions.

Lee et al. reported that fluorinated single-walled carbon nanotubes (SWCNTs) that were treated with 0.2 bar of F_2 gas at 200 °C exhibited lower capacitance than the raw SWCNTs [43]. Figure 15.6 shows transmission electron microscopic images of raw SWCNTs and fluorinated SWCNTs [44]. The untreated SWCNT bundles with a diameter of 1.5 nm are clearly observed, and the SWCNTs remain largely intact after fluorination. Figure 15.7 shows that the radial breathing mode located near 178 cm^{-1} disappears and that the intensity of the defect-related D-band peak near 1250 cm^{-1} increases in intensity in the spectrum of the fluorinated SWCNTs. The I_G/I_D peak intensity ratios deceased with increasing fluorination temperature, indicating a change in their bulk structural properties after fluorination. It was considered that the resistivity of fluorinated SWCNTs increased with structural changes [44].

Charge–discharge tests of SWCNT electrodes were performed in a 7.5 N KOH aqueous solution as the electrolyte. The specific capacitances obtained at 10 mA g^{-1} were 72 and 73 F g^{-1} for raw SWCNTs and fluorinated SWCNTs, respectively. Additionally, the specific capacitances obtained at 150 mA g^{-1} were 69 and 62 F g^{-1} for raw SWCNTs and fluorinated SWCNTs, respectively [43]. The specific capacitances with a low charge time and low discharge current density were similar for the raw SWCNTs and fluorinated SWCNTs. In contrast, the specific capacitances decreased more rapidly in the fluorinated SWCNTs than in the raw SWCNTs with a long charge time and high discharge current, although the trend should be the opposite because of the increase in the micropore area and the decrease in the average pore diameter in the fluorinated SWCNTs [43]. Abundant micropores with smaller pore diameters are covered by the fluoride ions and can be more easily blocked, particularly under large discharge currents.

In the case of multiwalled carbon nanotubes (MWCNTs), the MWCNTs fluorinated with 0.2 bar of F_2 gas at 100 °C exhibited a surface fluorine content of 13.0 at%; in addition, Fe, which was used as a catalyst for the preparation of the MWCNTs, was detected [45]. The specific surface area of the fluorinated MWCNTs increased from 238 to 366 m^2 g^{-1}. These results were attributed to the etching and digging of the MWCNT surface as a consequence of fluorination. As shown in Figure 15.8(a), the discharge characteristics exhibited almost linear behavior, which further confirmed the capacitive behavior. As shown in Figure 15.8(b), the fluorinated MWCNTs exhibited the highest specific capacitance of 94 F g^{-1} at a current density of 0.2 A g^{-1} in an aqueous electrolyte of 1 M H_2SO_4. Unlike fluorinated SWCNTs, the capacitances of MWCNTs fluorinated at 200 °C were higher than those of raw MWCNTs. The significant enhancement in the specific capacitance of the fluorinated MWCNTs was attributed to an increase in the specific surface area and surface polarity induced by the fluorine functional groups; these increases were caused by the increased affinity between the electrode surface and electrolyte ions [39,40].

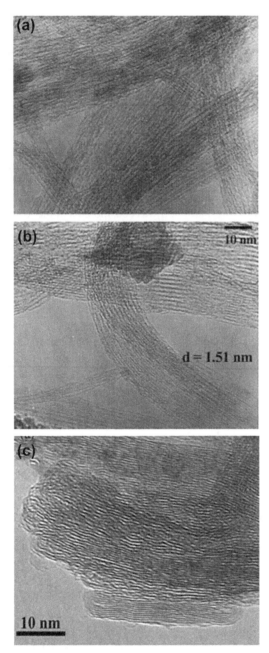

FIGURE 15.6 High-resolution transmission electron microscopic images of the SWCNTs: untreated (a) and fluorinated at 150 °C (b) and 300 °C (c) [44].

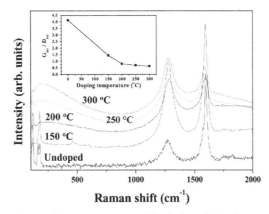

FIGURE 15.7 Fourier transform-Raman spectra of the single-walled carbon nanotubes (SWCNTs) fluorinated at various fluorination temperatures [44].

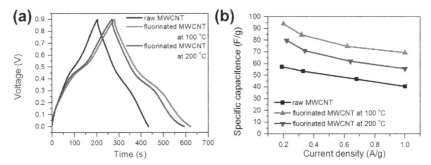

FIGURE 15.8 (a) Galvanostatic charge–discharge curves of 0.2 A g^{-1} and (b) specific capacitance at various current densities of MWCNTs and fluorinated multiwalled carbon nanotubes (MWCNTs).

REFERENCES

[1] A. Lewandowski, M. Galinski, J. Power Sources 173 (2007) 822–828.
[2] O. Bohlen, J. Kowal, D.U. Sauer, J. Power Sources 172 (2007) 468–475.
[3] J.R. Miller, Electrochim. Acta 52 (2006) 1703–1708.
[4] J.P. Zheng, J. Electrochem. Soc. 152 (2005) A1864–A1869.
[5] C.Z. Yuan, B. Gao, X.G. Zhang, J. Power Sources 173 (2007) 606–612.
[6] A.K. Shukla, A. Banerjee, M.K. Ravikumar, A. Jalajakshi, Electrochim. Acta 84 (2012) 165–173.
[7] S. Mitra, S. Sampath, Solid-State Lett. 7 (2004) A264–A268.
[8] G. Yu, X. Xie, L. Pand, Z. Baod, Y. Cui, Nano Energy 2 (2013) 213–234.
[9] D. Zhai, B. Li, F. Kang, H. Du, C. Xu, Microporous Mesoporous Mater. 130 (2010) 224–230.
[10] Y. Liu, Z. Hu, K. Xu, X. Zheng, Q. Gao, Acta Phys. Chim. Sin. 24 (2008) 1143–1148.
[11] T. Momma, X. Liu, T. Osaka, Y. Ushio, Y. Sawada, J. Power Sources 60 (1996) 249–253.
[12] H. Shamsijazeyi, T. Kaghazchi, J. Ind. Eng. Chem. 16 (2010) 852–858.
[13] T. Savage, S. Bhattacharya, B. Sadanadan, J. Gaillard, T.M. Tritt, Y.P. Sun, J. Phys. Condens. Matter 15 (2003) 5915–5921.

[14] A. Felten, C. Bittencourt, J.J. Pireaux, Nanotechnology 17 (2006) 1954–1959.
[15] T. Hoshida, D. Tsubone, K. Takada, H. Kodama, T. Hasebe, A. Kamijo, T. Suzuki, A. Hotta, Surf. Coat. Technol. 202 (2007) 1089–1093.
[16] Z. Hruska, X. Lepot, J. Fluorine Chem. 105 (2000) 87–93.
[17] J.M. Lee, S.J. Kim, J.W. Kim, P.H. Kang, Y.C. Nho, Y.S. Lee, J. Ind. Eng. Chem. 15 (2009) 66–71.
[18] Y.S. Lee, Y.H. Kim, J.S. Hong, J.K. Suh, G.J. Cho, Catal. Today 120 (2007) 420–425.
[19] A. Bismarck, R. Tahhan, J. Springer, A. Schulz, T.M. Klapötke, H. Zell, W. Michaeli, J. Fluorine Chem. 84 (1997) 127–134.
[20] R. Yazami, A. Hamwi, K. Guérin, Y. Ozawa, M. Dubois, J. Giraudec, F. Masin, Electrochem. Commun. 9 (2007) 1850–1855.
[21] T. Nakajima, V. Gupta, Y. Ohzawa, H. Iwata, A. Tressaud, E. Durand, J. Fluorine Chem. 114 (2002) 209–214.
[22] J. Li, K. Naga, Y. Ohzawa, T. Nakajima, A.I. Shames, A.M. Panich, J. Fluorine Chem. 126 (2005) 265–273.
[23] T. Nakajima, V. Gupta, Y. Ohzawa, M. Koh, R.N. Singh, A. Tressaud, E. Durand, J. Power Sources 104 (2002) 108–114.
[24] E. Frackowiak, Phys. Chem. Chem. Phys. 9 (2007) 1774–1785.
[25] M.J. Jung, E. Jeong, S. Kim, S.I. Lee, J.S. Yoo, Y.S. Lee, J. Fluorine Chem. 132 (2011) 1127–1133.
[26] M.J. Jung, J.W. Kim, J.S. Im, S.J. Park, Y.S. Lee, Ind. Eng. Chem. 15 (2009) 410–414.
[27] E. Frackowiak, F. Béguin, Carbon 39 (2001) 937–950.
[28] S.J. Gregg, K.S.W. Sing, Adsorption Surface Area and Porosity, second ed., Academy Press, London, 1982. 195.
[29] T. Nakajima, L. Li, K. Naga, K. Yoneshima, T. Nakai, Y. Ohzawa, J. Power Sources 133 (2004) 243–251.
[30] M.H. Kim, J.H. Yang, Y.M. Kang, S.M. Park, J.T. Han, K.B. Kim, K.C. Roh, Colloids Surf. A: Physicochem. Eng. Aspects 443 (2014) 535–539.
[31] E. Jeong, M.J. Jung, S.H. Cho, S.I. Lee, Y.S. Lee, Colloids Surf. A: Physicochem. Eng. Aspects 377 (2011) 243–250.
[32] A.B. Fuertes, G. Marbán, D.M. Nevskaia, Carbon 41 (2003) 87–96.
[33] J.S. Im, M.J. Jung, Y.S. Lee, J. Colloid Interf. Sci. 339 (2009) 31–35.
[34] S.C. Kang, J.S. Im, S.H. Lee, T.S. Bae, Y.S. Lee, Colloids Surf. A 384 (2011) 297–303.
[35] N. Setoyama, G. Li, K. Kaneko, F. Okino, R. Ishikawa, H. Touhara, Adsorption 2 (1996) 293–297.
[36] M.J. Jung, E. Jeong, Y. Kim, Y.S. Lee, J. Ind. Eng. Chem. 19 (2013) 1315–1319.
[37] E. Jeong, M.J. Jung, Y.S. Lee, J. Fluorine Chem. 150 (2013) 98–103.
[38] V. Gupta, N. Miura, Electrochim. Acta 52 (2006) 1721–1726.
[39] T. Nakajima, K. Hashimoto, T. Achiha, Y. Ohzawa, A. Yoshida, Z. Mazej, B. Žemva, Y.S. Lee, M. Endo, Collect. Czech. Chem. Commun. 73 (2008) 1693–1704.
[40] K. László, E. Tombácz, K. Josepovits, Carbon 39 (2001) 1217–1228.
[41] T. Katakabe, T. Kaneko, M. Watanabe, T. Fukushima, T. Aida, J. Electrochem. Soc. 152 (2005) A1913–A1916.
[42] Y. Honda, T. Haramoto, M. Takeshige, H. Shiozaki, T. Kitamura, M. Ishikawa, Electrochem. Solid-State Lett. 10 (2007) A106–A110.
[43] J.Y. Lee, K.H. An, J.K. Heo, Y.H. Lee, J. Phys. Chem. B 107 (2003) 8812–8815.
[44] Y.S. Lee, T.H. Cho, B.K. Lee, J.S. Rho, K.H. An, Y.H. Lee, J. Fluorine Chem. 120 (2003) 99–104.
[45] M.J. Jung, E. Jeong, J.S. Jang, Y.S. Lee, in: Proceedings of the 8th International Carbon Festival, Korea Carbon Society, Jeonju, 2013, p. 124.

Chapter 16

Fluorine Chemistry for Negative Electrode in Sodium and Lithium Ion Batteries

Mouad Dahbi[1,2] and Shinichi Komaba[1,2]
[1]*Department of Applied Chemistry, Tokyo University of Science, Tokyo, Japan;* [2]*Elements Strategy Initiative for Catalysts and Batteries, Kyoto University, Kyoto, Japan*

Chapter Outline

16.1	Introduction	387	16.7.1 Tin and Antimony as Alloy Materials	404
16.2	Why Na-Ion Battery?	388	16.7.2 Red and Black Phosphorus	405
16.3	Hard-Carbon as Potential Negative Electrode	390	16.8 Silicon for Lithium-Ion Battery	407
16.4	Fluorinated Electrolyte and Additive	392	16.9 Conclusive Remarks	409
16.5	Poly Vinylidene Fluoride and CMC-Based Binder	397	Acknowledgments	411
16.6	Aluminum Corrosion Inhibitor	401	References	411
16.7	Na Alloys and Compounds	403		

16.1 INTRODUCTION

Research interest in sodium-ion batteries (NIB) has increased rapidly because of eco-friendliness and environmental friendliness of sodium compared to lithium. Fluorine chemistry plays an important role in the development of materials for lithium-ion batteries. Throughout this chapter, we shed light on fluorine chemistry for NIB, especially carbonaceous materials and sodium alloy/compounds as negative electrode materials. These electrode materials have different reaction mechanisms for electrochemical sodiation/desodiation processes. Electrochemical formation of lithium silicide is also described. Moreover, not only sodiation-active materials but also binders, electrolytes, electrode/electrolyte interphase, and its stabilization are essential for long cycle life. Indeed, fluorinated ethylene carbonate is an effective electrolyte additive. This chapter also addresses the corrosion of the aluminum current collector at high potential, depending on the nature on the electrolyte and fluorinated anions salts.

16.2 WHY NA-ION BATTERY?

Rechargeable lithium (Li) batteries, often called as an Li-ion battery (LIB) as was first named by Sony, have been recognized to be the most successful and sophisticated energy storage devices since their first commercialization in 1991. LIBs have been originally developed as a high-energy and safety power source for portable electronic devices. Furthermore, they are used now as an alternative power source for electric motors instead of combustion engines equipped with a fuel tank. Electric vehicles equipped with large-scale lithium batteries as power sources have been introduced to the automotive market, promising to reduce the dependence of transportation on fossil fuels in the future. In addition, LIBs are now being used for electrical energy storage (EES) [1]. An essential element, lithium, is widely distributed in the Earth's crust, but is not regarded as an abundant element [2–5]. Indeed, the cost of the material increased after the commercialization of Li batteries. In contrast, sodium resources are in principle unlimited as they are abundant in sea water and as salt deposits [6]. Moreover, sodium is the second lightest and smallest alkali metal next to lithium, and in fact chemistries of Li and Na are in general similar as alkali metals.

There are a few fundamental differences between the two elements: sodium atom is three times heavier than lithium atom. Na^+ has a longer ionic radius than Li^+ as shown in Table 16.1. Electrochemical standard potential of Na^+/Na is 0.34 V higher than that of Li^+/Li. As a result, gravimetric and volumetric energy density based on the metallic Na is inevitably much lower than that of the metallic lithium when they are used as negative electrode. One can note

TABLE 16.1 Comparison of Chemical Nature of Sodium and Lithium

	$_3Li$	$_{11}Na$
Cation radius (Å)	0.68	0.97
Atomic weight (g mol^{-1})	6.9	23.0
E° (V vs. SHE)	−3.04	−2.70
Melting point (°C)	180.5	97.7
Capacity (mAh g^{-1}), metal	3829	1165
Cost, carbonates ($/ton)	4000	120
Distribution	70% in South America	Everywhere

that difference in gravimetric energy density based on the insertion materials becomes much smaller, which is due to the weight of host materials. Additionally, safety concerns could be raised due to the lower melting temperature of Na metal (97.7 °C) as compared to Li metal (180.5 °C) when Na/Li metal is employed as the negative electrode.

NIB, named as LIB counterpart, consists of two distinct electrodes composed of Na-insertion materials without metallic Na, as shown in Figure 16.1. NIB possesses two sodium insertion materials, positive and negative electrodes, which are electronically separated by electrolyte (in general, electrolyte salts dissolved in aprotic polar solvents) as a pure ionic conductor. NIBs are a promising candidate for the use of EES because the abundance and cost-effectiveness of Na is essential for large-scale applications when we consider limited availability of lithium resource.

Aluminum as current collector forms a binary alloy with Li. Copper is, therefore, used as a current collector for negative electrode materials for rechargeable Li batteries, while aluminum current collector is used for positive electrode. In contrast, Na does not form an alloy with aluminum at ambient temperature, which can be used as the current collector for rechargeable Na batteries. The use of cost-effective aluminum is an additional practical advantage of the NIB system as it promises to reduce the total cost of batteries, compared to LIBs.

The major obstacle to realize NIB was the absence of suitable negative electrodes. In the middle of 1980s, carbonaceous materials were found to be potential candidates for Li insertion (intercalation) hosts, which are now commercially utilized as negative electrode materials for practical LIBs, e.g., disordered carbons [7,8] and graphite [9–11]. Research interests of LIB further accelerated thanks to finding graphite, which theoretically delivers high-reversible capacity with low and flat operating voltage of 0.1–0.2 V vs. Li^+/Li. Unfortunately, graphite cannot be utilized as an insertion host of Na ions [12–15].

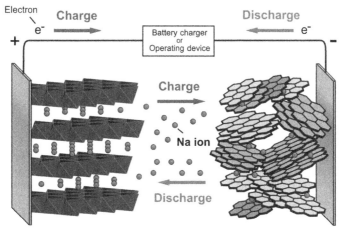

FIGURE 16.1 Schematic illustration of sodium ion batteries.

As the demand of large-scale batteries for EES is now increasing, the research interest for NIBs has been renewed in the 2010s. Indeed, the field of NIB battery materials, such as active materials, electrolytes, and their characterization, is rapidly growing. In particular, the numbers of publications on NIB have drastically increased in recent years. On the basis of the long history of Li batteries over 30 years, we can confidently say that the electrochemical performance of negative electrodes for NIBs has also improved rapidly, thanks to extensive research efforts throughout the world.

In this chapter, recent research progress on advanced negative electrodes and fluorine chemistry for NIBs is reviewed and provided. Electrode performance of different electrode materials in Na cells is given in terms of the reaction mechanisms for sodiation/desodiation processes. Since the electrode potential is generally low around 0–1 V vs. Na^+/Na (i.e., $-3 \sim -2$ V vs. NHE), the decomposition of electrolyte solution at the negative electrode is a serious issue facing the importance of passivation for long cycle life of battery. The selection of binders, additives, and electrolytes, therefore, has significant impacts on the life cycle of negative electrodes, associated with the formation of surface passivation layers, namely solid electrolyte interphase (SEI) [16].

16.3 HARD-CARBON AS POTENTIAL NEGATIVE ELECTRODE

Graphite is widely used as negative electrode materials for LIB, in comparison with other carbon materials because of its high gravimetric and volumetric capacity. Graphite electrodes deliver reversible capacity of more than 360 mAh g^{-1} comparable to the theoretical capacity of 372 mAh g^{-1} [17]. By electrochemical reduction, Li^+ ions are inserted in the van der Waals gap between graphene layers, and Li-graphite intercalation compounds are formed with stage transformations through 8th, 4th, 3rd, 2nd, and 1st stages [18]. At the first stage, all graphite layers are completely filled by Li, forming LiC_6 at the end of electrochemical reduction process [19].

However, the graphite is electrochemically less active in Na cells. Although a small amount of Na atoms seems to be inserted into the graphite by heating with a Na metal under helium or vacuum atmosphere and by electrochemical reduction, resulting in the formation of NaC_{64} [12,20], the Na insertion amount is quite smaller than that for Li and K insertions into the graphite. However, lower crystalline carbons, such as soft and hard-carbons, show higher electrochemical activity in nonaqueous sodium cells. The larger reversible capacity of carbonaceous electrode materials in Na cells was first reported by Doeff et al. [13] in 1993 using a soft carbon prepared by pyrolysis of petroleum cokes; however, cells were operated as heating at 86 °C because of the use of solid polymer electrolyte. Another disordered carbon material, hard-carbon (so-called nongraphitizable carbon) is also intensively studied as a negative electrode material for high-power LIBs. Hard-carbon has a disordered structure. The detailed structure of hard-carbon is, however, still a debatable subject even though many structural models have been proposed [21–31]. In all the proposed models, hard-carbon is composed of two domains, that is, carbon layers (graphene-like) and micropores (nano-sized pores) formed between disorderly stacked carbon layers. Detailed structures, domain size, fraction of carbon layers, and micropores in hard-carbon highly depend on the carbonization conditions.

It is commonly known that hard-carbon was utilized as electrodes in the first commercial Li-ion cells and that hard-carbon electrodes deliver comparable reversible capacity without a staging transition due to the disordering. In some cases, this reversible capacity exceeds the theoretical capacity of graphite [32]. Indeed, electrochemical reversibility of Na extraction/insertion from/into hard-carbon at room temperature was first reported by Stevens and Dahn in 2000 [33]. They prepared a hard-carbon sample by carbonization of glucose at 1000 °C. The hard-carbon electrodes delivered ca. 300 mAh g^{-1} of reversible capacity in Na cells. In contrast to graphite, much higher amount of Na$^+$ ions can reversibly insert into hard-carbon by electrochemical reduction in Na cells.

Initial galvanostatic charge/discharge tests for electrode preparation commercially available hard-carbon (Carbotron P[J], Kureha Co. [34]) in ethylene carbonate (EC), propylene carbonate (PC), and butylene carbonate (BC) solution containing 1 mol dm^{-3} NaClO$_4$ are shown in Figure 16.2. An irreversible reduction plateau at 0.9–1.2 V is observed at only an initial reduction (sodiation) process, indicating the decomposition of electrolyte solvent and salts associated with the formation of a passivation layer on the surface of hard-carbon. The potentiogram is sloping from 1.2 to 0.1 V during the initial reduction, followed by a long plateau region at about 0.1 V vs. Na$^+$/Na, approaching 220 mAh g^{-1} of reversible capacity. During the following oxidation, reversible capacity of 100–130 mAh g^{-1} is observed near 0.1 V vs. Na$^+$/Na, and then the potential gradually increases to 1.2 V,

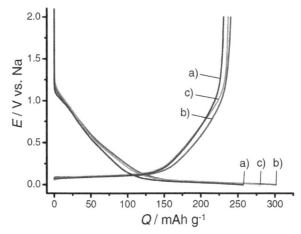

FIGURE 16.2 Initial galvanostatic reduction/oxidation curves for hard-carbon electrodes in (a) ethylene carbonate, (b) propylene carbonate, and (c) butylene carbonate solution containing 1 mol dm^{-3} NaClO$_4$ tested at 25 mA g^{-1} in beaker-type cells Ref. [35]. *(Reproduced from Shinichi Komaba et al. Electrochemical Na insertion and solid electrolyte interphase for hard-carbon electrodes and application to Na-ion batteries. In: Advanced Functional Materials, WILEY-VCH Verlag (August 11, 2011).)*

due to reversible Na insertion/extraction for hard-carbon [35]. Moreover, hard-carbon electrodes exhibit good capacity retention during the subsequent cycles with fluoroethylene carbonate (FEC) as fluorinated solvent additive. Details on electrolyte and additives are described in the next section in this chapter.

Therefore, electrode performance of hard-carbon will be possibly further improved through the optimization of structures. Indeed, recently, reversible capacity higher than 300 mAh g^{-1} for different hard-carbon derived from a sucrose precursor in Na cells has been reported [36–38].

16.4 FLUORINATED ELECTROLYTE AND ADDITIVE

Sodium batteries require electrolytes that contain mobile Na$^+$ ions, such as organic electrolyte solution. The liquid electrolytes are obtained by dissolving sodium salts in an aprotic solvent such as EC, PC, dimethyl carbonate (DMC), diethyl carbonate (DEC), ethyl methyl carbonate (EMC), tetrahydrofuran, 1,2-dimethoxyethane, or mixture of them. Many sodium salts can be dissolved in this solvents to make electrolyte such as sodium perchlorate (NaClO$_4$), sodium hexafluorophosphate (NaPF$_6$), sodium trifluoromethanesulfonate (NaCF$_3$SO$_3$), sodium bis(trifluoromethanesulfonyl)imide (NaN(SO$_3$CF$_3$)$_2$) and so on. Note that all of these salts have fluorinated anions except NaClO$_4$.

By designing better electrolytes, it is possible to enhance the energy/power density and cycle life of hard-carbon electrodes in Na cells. Bhide et al. [39] compared the physicochemical properties of nonaqueous electrolytes based on NaPF$_6$, NaCF$_3$SO$_3$, and NaN(SO$_3$CF$_3$)$_2$ and NaClO$_4$ salts in the EC/DMC mixture solution. They measured the ionic conductivity of the electrolytes as a function of salt concentration. The change in ionic conductivity with the salt concentration is considerable for NaCF$_3$SO$_3$ and NaClO$_4$ compared to NaPF$_6$-based electrolytes. The maximum ionic conductivity values of the electrolytes were 6.8 mS cm^{-1} for 0.6 mol dm^{-3} NaPF$_6$ and 5.0 mS cm^{-1} for 1.0 mol dm^{-3} NaClO$_4$; however 0.8 mol dm^{-3} NaCF$_3$SO$_3$ showed low conductivity.

The selection of suitable electrolytes is essential for realizing and maximizing the practical electrochemical performance of hard-carbon electrodes because one of the reasons is the fact that the surface passivation is rather difficult for the Na system compared with the Li system.

We think the difficulty originates from the difference of Lewis acidity of Li$^+$ and Na$^+$ ions. The Lewis acidity of Li$^+$ ion is stronger than Na$^+$, so that Coulombic interaction of Li$^+$ with negatively charged species or lone pair electrons is notable compared with that of Na$^+$ ion, resulting in a general tendency of weaker solvation and higher ion transport number for Na$^+$ ion. One can note that Li$_2$CO$_3$ is hardly soluble in water, whereas Na$_2$CO$_3$ is water-soluble [40]. According to previous literature [41,42] about SEI in Li cells, several lithium compounds, such as Li$_2$CO$_3$, lithium alkylcarbonate(s), lithium alkoxide(s), LiF, and so on, are found in SEI layer as insoluble components on carbon deposited by electroreductive reaction at ca. 0.8 V vs. Li$^+$/Li and on Li metal.

Considering the difference of Lewis acidity and solubility between lithium and sodium cells, it is reasonable to point out that the formation of SEI passivation layer is more difficult in Na cell because the higher solubility of the decomposed products in an Na^+-based electrolyte solution results in more difficult passivation with sodium compounds. Therefore, there is no doubt that the electrolyte is very important to exploit potential battery performance and SEI formation for not only hard-carbon but also any electrode materials for NIB.

In the case of lithium battery, various electrolytes, such as an organic electrolyte solution, solid polymer electrolyte, gel polymer electrolyte, inorganic solid electrolyte, molten salt including ionic liquid, aqueous solution, have been studied to date, and their development is now in progress. Concerning wide potential window, high ionic conductivity, good temperature performance, low toxicity, produciblity of practical cell, and so on, an organic electrolyte solution based on carbonate ester solvent is thought to be one of the most successful electrolyte media. In organic electrolyte solution, sufficient amount of lithium salt is dissolved with a small portion of functional additive(s) for the practical lithium-ion battery. Therefore, it is reasonably expected that a carbonate ester-based solution containing sodium salt is one of the most appropriate electrolyte media for battery applications.

Now, one can focus on the impact of carbonate ester electrolyte solutions including sodium salt and additives on the electrode performance of hard-carbon in Na cells.

Alcántara et al.[43] tested and compared the performance of hard-carbon electrodes in different nonfluorinated electrolyte solutions: $1\,mol\,dm^{-3}$ $NaClO_4$ dissolved in EC/DMC as a mixed cyclic and linear carbonate system, which is one of conventional electrolyte solvents in the lithium system. However, the capacity degradation of hard-carbon electrode within a few initial cycles was unavoidable regardless of the difference in the electrolyte solutions.

In 2011, systematic studies on electrolyte solvent were reported with $NaClO_4$ as the electrolyte salt [35]. The electrochemical cycling test of commercially available hard-carbon was carried out in PC and binary solvent electrolyte based on EC and linear carbonate esters containing $NaClO_4$ in beaker type cells as shown in Figure 16.3(a). When PC and EC/DEC solutions are used, the hard-carbon electrodes exhibit high reversible capacity of more than $200\,mAh\,g^{-1}$ with excellent capacity retention over 100 cycles. On the other hand, in the case of EC/DMC and EC/EMC, the cycle performance in beaker-type cells was insufficient. Figure 16.3(b) compares the cycle performance of hard-carbon electrodes in coin-type cells, using PC-based electrolyte solution containing different electrolyte salts. Despite the fact that the beaker-type cell with $NaClO_4$ in PC shows good capacity retention, decrease in capacity was observed when using a coin-type cell in which thin glass fiber separator soaked with the minimal amount of electrolyte solution was sandwiched between the Na metal foil and hard-carbon electrodes. On the other hand, the cells with fluorinated salts $NaPF_6$ and $NaN(SO_3CF_3)_2$, dissolved in PC solutions, exhibited better cycle performance.

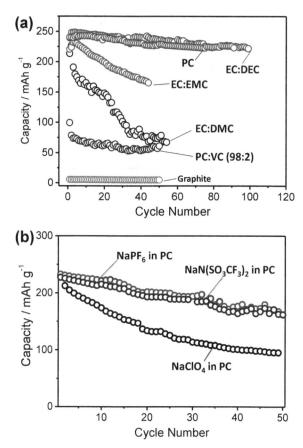

FIGURE 16.3 (a) Capacity retention for hard-carbon electrodes with 1 mol dm^{-3} NaClO$_4$ dissolved in different solvent mixtures, (b) capacity retention for hard-carbon electrodes with PC-based electrolyte solvent containing different Na salts; 1 mol dm^{-3} NaPF$_6$, NaN(SO$_3$CF$_3$)$_2$, and NaClO$_4$ at a rate of 25 mA g^{-1} Ref.[98]. *(Reproduced from Mouad Dahbi et al. Negative electrodes for Na-ion batteries. In: Physical chemistry chemical physics, Royal Society of Chemistry (June 4, 2014).)*

Another key material to improve the reversible capacity of hard-carbon electrodes and its retention is electrolyte additives. In this section, the electrochemical performance of Na cells was compared with several additives, such as FEC, transdifluoroethylene carbonate (DFEC) [44], ethylene sulfite (ES) [45], and vinylene carbonate (VC) (Figure 16.4 shows their molecular structures), which are well known to be the efficient electrolyte additives for LIBs. The film-forming additives into electrolyte solutions are widely known to improve the electrode properties for lithium insertion materials [46–49].

Indeed, VC is the most widely used film-forming additive in the practical LIBs with the graphite negative electrode [50,51]. In Na cell, however, the PC solution containing NaClO$_4$ clearly became colored in yellow after the first cycle of hard-carbon cells with VC addition compared to that of the VC-free

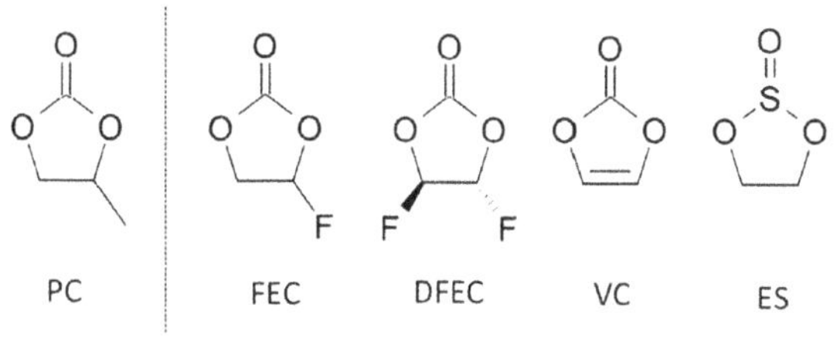

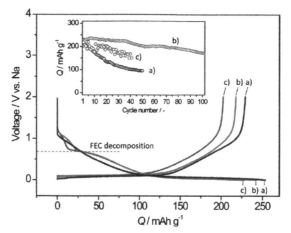

FIGURE 16.4 Molecular structures of electrolytes additives used in sodium cells.

FIGURE 16.5 Initial galvanostatic reduction/oxidation curves for hard-carbon electrodes in NaClO$_4$/PC solution (a) without and with (b) 2% and (c) 10% FEC. Inset shows oxidative capacity retention for the hard-carbon electrodes Ref. [98]. *(Reproduced from Shinichi Komaba et al. Fluorinated ethylene carbonate as electrolyte additive for rechargeable Na batteries. In: Applied materials, American Chemical Society (November 1, 2011).)*

electrolyte (Figure 16.3(a)). FEC is also known as an effective electrolyte additive for improving SEI films, especially when used for lithium metal and silicon-based electrode materials [47,48,52]. As shown in Figure 16.5, the addition of fluorinated additive FEC into electrolyte solution has an effect on the electrode performance of hard-carbon electrodes [53]. When a small amount of FEC is added into 1 mol dm^{-3} NaClO$_4$/PC solution, an additional voltage plateau at ca. 0.7 V appears during the first reduction due to the decomposition of FEC. Moreover, as shown in Figure 16.6, the FEC addition is noticeably effective to achieve the comparable reversibility of the Na insertion even in the PC/DMC solution, which is easily decomposable in the sodium cells. By adding this fluorinated carbonate into the PC/DMC solution, capacity degradation as observed for electrolyte without FEC is sufficiently suppressed [53]. Indeed, the FEC addition is noticeably effective to achieve good reversibility of the Na insertion, presumably associated with the formation of a stable passivation layer at the surface of hard-carbon.

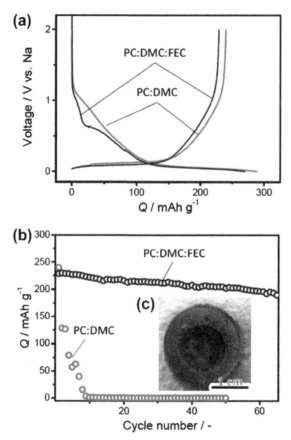

FIGURE 16.6 (a) The initial reduction/oxidation curves and (b) reversible capacity variation for hard-carbon in NaClO$_4$ PC:DMC solutions with and without 2% FEC additive tested at 25 mA g^{-1} in coin-type Na cells. (c) A photo of separator tested in the FEC-free PC:DMC after 50 cycles Ref. [53]. *(Reproduced from Shinichi Komaba et al. Fluorinated ethylene carbonate as electrolyte additive for rechargeable Na batteries. In: Applied materials, American Chemical Society (November 1, 2011).)*

The XPS measurement reveals that the peak assigned to sodium fluoride is evidently observed in F 1s XPS spectra as increase in the amount of added FEC while there is no remarkable difference in C 1s XPS spectra as shown in Figure 16.7. We believe that the improved cycle performance by FEC addition originates from the modification of the surface passivation layer with fluoride. In addition, we can infer that water contamination might affect the different influence of FEC. Indeed, our experience clearly confirms that both purity of electrolyte solution and contamination from atmosphere of Ar-filled glove box, in which Na cells are assembled, are very important to collect good and reproducible electrochemical data. Our data also show an increase in polarization by 5% FEC addition into 1 mol dm^{-3} NaPF$_6$ PC solution when Na carboxymethylcellulose (CMC) binder is used as described below. Further optimization of the additive amount with the

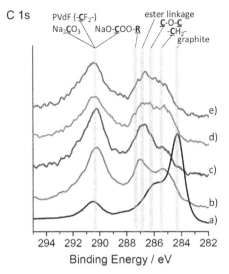
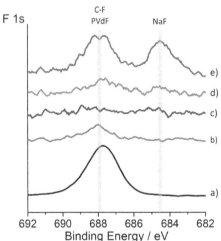

FIGURE 16.7 C 1s and F 1s XPS spectra for hard-carbons: (a) pristine and hard-carbon electrodes after the first galvanostatic cycle test in NaClO$_4$ PC (b) without and with (c) 0.5, (d) 2, and (e) 10 vol% FEC addition in Na cells. *(Reproduced from Shinichi Komaba et al. Fluorinated ethylene carbonate as electrolyte additive for rechargeable Na batteries. In: Applied materials, American Chemical Society (November 1, 2011).)*

suitable combination of fluorinated solvent and salts under investigation may be necessary for each electrode material to enhance the electrode performance for battery applications.

Contrary to what we thought, when a small quantity of DFEC is added in the PC, any beneficial effects are not found in the Na/hard-carbon cells as shown in Figure 16.8, and the ES addition causes the detrimental effect on the Na cells [53].

16.5 POLY VINYLIDENE FLUORIDE AND CMC-BASED BINDER

Since commercialization of LIBs, poly(vinylidene fluoride) (PVdF) as fluorinated binder has been widely used because of its good chemical and electrochemical

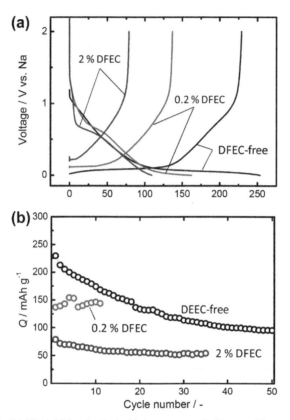

FIGURE 16.8 (a) The initial reduction/oxidation curves and (b) reversible capacity variation for hard-carbon electrodes in NaClO$_4$ PC solutions with and without 0.2% and 2% DFEC examined at a rate 25 mA g^{-1} in coin-type Na cells Ref. [53]. *(Reproduced from Shinichi Komaba et al. Fluorinated ethylene carbonate as electrolyte additive for rechargeable Na batteries. In: Applied materials, American Chemical Society (November 1, 2011).)*

stability for long time use. However, this polymer is relatively costly and requires the use of volatile organic solvents to make slurry paste such as N-methylpyrrolidone that are often toxic in the processing. Moreover, defluorination from PVdF is observed when the polymer is attacked by alkali hydroxides [54]. In contrast, water-soluble binders, such CMC, which is derived from cellulose as natural polymers, provide environmental friendliness and cost-effectiveness for battery manufacturing [55–57]. Additionally, adhesive strength as binders and dispersion of active materials in the slurry are efficiently improved by using water soluble binders having a carboxylic group such as CMC, polyacrylates (PAA) etc. [52,58–62]. The selection of suitable binders is critically important to maximize the electrode performance of active materials in Na cells, similar to Li cells.

The effects of PVdF and CMC binders on electrode performance of hard-carbon were examined in the NaPF$_6$ PC solution without or with 2% FEC additive as shown in Figure 16.9 [63]. The initial capacity of hard-carbon with PVdF

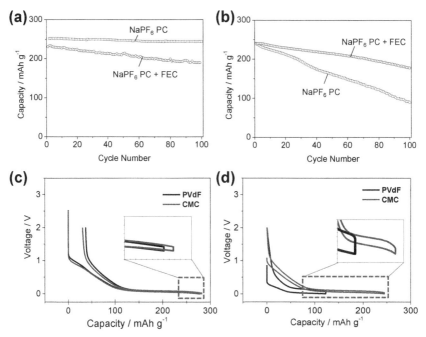

FIGURE 16.9 Cycle performance of hard-carbon electrodes in NaPF$_6$ PC solutions with or without 2% FEC addition: (a) carboxymethylcellulose and (b) poly vinylidene fluoride binders. Galvanostatic charge/discharge curves at (c) 1st and (d) 80th cycles are magnified to compare the difference of polarization Ref. [63]. *(Reproduced from Mouad Dahbi et al. Sodium carboxymethyl cellulose as a potential binder for hard-carbon negative electrodes in sodium-ion batteries. In: Electrochemistry communications, Elsevier (July 2014).)*

was 250 mAh g^{-1} tested in the FEC-free electrolyte, and the capacity was gradually decayed, which is consistent with our previous results tested in the NaClO$_4$ PC solution [53]. Whereas the initial capacity of approximately 250 mAh g^{-1} was similarly observed for all electrodes, the capacity retention depended on the binder and additives. The superior capacity retention above 97% over 100 cycles is achieved for the electrode with CMC in the additive-free electrolyte. The average Coulombic efficiency (in 11–100th cycles) and the initial Coulombic efficiency, 99.8% and 89.3%, respectively, for the CMC binder are higher than those (99.5% and 86.7%, respectively) for the PVdF binder [63].

The addition of FEC in electrolyte influences the capacity retention and the Coulomb efficiency. Figure 16.9(b) indicates that the FEC addition remarkably improved the cyclability in the case of PVdF binder as we reported [53]. However, it is noted that the FEC additive has an adverse effect on the CMC electrode as seen in Figure 16.9(a). Figures 16.9(c) and 16.9(d) compare the evolution of polarization at the potential close to 0 V tested in the FEC-free solution. Although the polarization and the reversible properties are almost the same during the first cycle for the CMC and PVdF electrodes, the PVdF one suffers from the increase in polarization after 80 cycles.

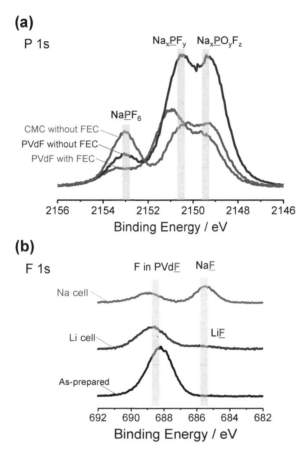

FIGURE 16.10 Hard X-ray photoemission spectroscopy (HAXPES) spectra of (a) P 1s for the hard-carbon tested in NaPF$_6$ PC after 1st cycle and (b) F 1s for the hard-carbon as prepared and after 1st cycle in NaClO$_4$ or LiClO$_4$ PC solution in lithium or sodium cell, respectively Ref. [63]. *(Reproduced from Mouad Dahbi et al. Sodium carboxymethyl cellulose as a potential binder for hard-carbon negative electrodes in sodium-ion batteries. In: Electrochemistry communications, Elsevier (July 2014).)*

Hard X-ray photoemission spectroscopy (HAXPES) of P 1s spectra as shown in Figure 16.10(a) clearly verifies the difference of phosphorus species. The spectra can be deconvoluted at least into three components: residual NaPF$_6$ and decomposed products (Na$_x$PF$_y$, and Na$_x$PO$_y$F$_z$). For the PVdF electrode, intensity of Na$_x$PF$_y$ and Na$_x$PO$_y$F$_z$ becomes lower by FEC additive and is further decreased for the CMC electrode. From these results, it is thought that the decomposition of electrolyte on hard-carbon is most successfully suppressed by CMC binder. The electrochemical performance of PVdF electrodes is enhanced by FEC due to its passivation ability [53]. Meanwhile, the decomposition is sufficiently suppressed only by CMC binder, suggesting the role of preformed

SEI for CMC, similar to the carboxyl polyanion binder in lithium insertion electrodes [64]. It is also thought that the combined use of both CMC and FEC causes the gradual increase of polarization because of the massive growth of resistive layer for the sodium insertion, resulting in capacity degradation.

The stability of PVdF binder is examined by HAXPES analysis for the hard-carbon tested in the fluorine-free electrolytes, $1\,mol\,dm^{-3}$ $LiClO_4$ and $NaClO_4$ PC solutions, showing comparable reversible capacity in lithium and sodium cells, respectively [35]. In Figure 16.10(b), the peak at ca. 688.8 eV, attributed to the $-CH_2-\underline{C}F_2-$ in PVdF, appears for as-prepared and tested electrodes. The significant photoemission peak around 685.7 eV, responsible for alkali fluoride, is also noted only for the sodium-inserted electrode while it is hardly observed for the lithium case. This result agrees with the fact that PVdF is stable binder for LIBs. Consideration of defluorination of PVdF and polytetrafluoroethylene by using sodium metal [65,66] NaF is formed by defluorination of PVdF binder by hard-carbon during sodium insertion. When the hard-carbon electrode is easily damaged by PVdF decomposition, hard-carbon particles should be partly loosened and electrically isolated, resulting in the capacity degradation and/or lower efficiency. Indeed, loosened composite electrode was visually observed only for sodium cells when the composite electrodes with PVdF were taken out from the tested cells. This damage is fairly avoided by the passivation enhancer of FEC; therefore, better capacity retention. Due to the preformed SEI and improved passivation by CMC binder, the damage is believed to be completely suppressed [63].

16.6 ALUMINUM CORROSION INHIBITOR

Aluminum is commonly used as current collector for positive electrode in LIBs and NIBs. The native passive Al_2O_3 layer existing on the metal surface provides protection against corrosion. It has been reported that aluminum collectors oxidize between 3.5 and 3.7V in electrolytes containing $Li/Na-N(SO_3CF_3)_2$ [67–69], which leads to the corrosion of the metal, limits the voltage windows, and hence the specific energy of the LIB/NIBs. As corrosion may occur in the presence of $Li/Na-N(SO_3CF_3)_2$, the anodic behavior of aluminum collectors in the presence of $Li/Na-N(SO_3CF_3)_2$-based electrolyte has been studied by cyclic voltammetry as shown in Figure 16.11. In order to decrease or even suppress the aluminum corrosion, adding $Li/Na-PF_6$ to the $Li/Na-N(SO_3CF_3)_2$-based electrolyte was proposed. Indeed, the impact of $Li/Na-PF_6$ as additive on aluminum passivation was investigated by addition of controlled amount of $Li/Na-PF_6$ into pure $Li/Na-N(SO_3CF_3)_2$-based electrolyte. Figure 16.11 compares the cyclic voltammograms of Al electrode in 1.0M $Li/Na-N(SO_3CF_3)_2$, 1.0M $Li/Na-PF_6$, and 1.0M $Li/Na-N(SO_3CF_3)_2$ containing 1% molar, 5% and 10% of $LiPF_6$ in a mixture of EC/DMC [68], or 5% $NaPF_6$ in PC electrolyte [69]. In the $NaN(SO_3CF_3)_2$-based solutions, the voltammograms in the first cycle show a hysteresis loop initiated at about 3.5–3.7V vs. Li^+/Li or Na^+/Na, with a large irreversible oxidative current peak. In the case of $LiPF_6$ (1.0M) in

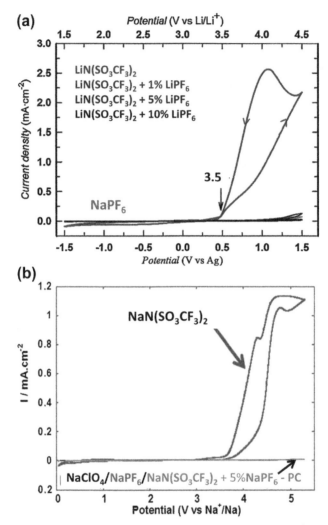

FIGURE 16.11 (a) Cyclic voltammograms of aluminum foil in 1.0 mol dm^{-3} LiN(SO$_3$CF$_3$)$_2$, 1.0 mol dm^{-3} LiPF$_6$, and 1.0 mol dm^{-3} LiN(SO$_3$CF$_3$)$_2$ containing 1%, 5%, and 10% of LiPF$_6$ in EC/DMC mixture electrolyte. Ref. [68] *(Reproduced from Mouad Dahbi et al. Comparative study of EC/DMC LiTFSI and LiPF$_6$ electrolytes for electrochemical storage. In: J. Power Sources, Elsevier (November, 2011).* (b) Cyclic voltammograms of aluminum in 1.0 mol dm^{-3} NaClO$_4$, 1.0 mol dm^{-3} NaN(SO$_3$CF$_3$)$_2$, 1.0 mol dm^{-3} NaPF$_6$, and 1.0 mol dm^{-3} LiN(SO$_3$CF$_3$)$_2$ containing 5% of NaPF$_6$ in PC electrolyte Ref. [69]. *(Reproduced from Alexander Ponrouch et al. In search of an optimized electrolyte for Na-ion batteries. In: Energy & Environmental Science, Royal Society of Chemistry (June, 2012).)*

EC/DMC as expected the initiation potentials of the hysteresis loop move significantly to more anodic side while the peak current sharply decreases. Ponrouch et al. demonstrated the same behavior is obtained for NaPF$_6$ added in 1.0 mol dm^{-3} NaN(SO$_3$CF$_3$)$_2$-based electrolyte. These results suggest that aluminum passivation would be achieved in the bis(trifluoromethanesulfonyl)imide anion-based electrolytes for both Na and Li system, with small amount of Li/Na-PF_6 as additive. Additionally, it is thought that fluoride ions (F^-) produced from the equilibrium equation of decomposition of PF_6^- ion ($PF_6^- \rightarrow PF_5 + F^-$) are responsible to the formation of a passivation layer (fluoroaluminates) on the Al current collector [68].

16.7 Na ALLOYS AND COMPOUNDS

In the past decade, Li alloys and binary compounds as electrode materials have been extensively studied, especially for the Li-Sn and Li-Si systems because of their higher reversible capacity. Indeed, this reversible capacity is much higher than that of the Li-graphite system, which has now almost reached its theoretical limit in practical LIB [70–73].

In the former section, Na insertion materials, which show relatively small volume expansion by sodiation, are described. In this section, we will discuss electrode materials that accompany an Na insertion reaction with the destruction of framework structure, such as Na-Me alloying electrodes. Group 14 and 15 elements of periodic table, including metals (Sn, Pb, Bi), metalloids (Si, Ge, As, Sb), and polyatomic nonmetal (P), are known to form binary compounds with Na. These electrode materials, alloying with Na or forming Na binary compounds, have been recently studied as potential negative electrodes for rechargeable NIBs [74–78].

These materials have the particularity to interact with a larger number of Na (compared to materials insertion) and to show much higher capacity than hard-carbon. The main problem of these materials lies in the volume expansion of the electrode correlated with the destruction of the structure of the starting material in the reaction with Na. Current research in the field of the Li-Si and Li-Sn systems is essentially devoted to controlling at least the effects of volume changes of the electrodes during cycling and to maintain the electronic conduction. Indeed, these two issues greatly penalize the life and the speed of charge/discharge batteries based on these electrode materials. Optimization of the alloy composition and passivation and a careful study of the structural change for the different alloy phases during the electrochemical reactions will lead to further improvement of the electrochemical performances [76–82].

Such huge volume expansion could restrict the long-term continuous cycling as electrode materials. The large volume expansion, depending on the amount of Na^+ ions incorporated into the structures, is unavoidable in this system while in many cases the volume expansion for Na insertion materials does not exceed 120%. Therefore, when successive cycles of charge/discharge, the materials, alloying with Na or forming Na binary compounds, suffer from high mechanical stress and repeated formation of passivation layer. The cracking in the composite electrodes leads to a loss of electrical contact and results in capacity loss.

As shown in the next section, several strategies, including binders, electrolyte additives, and so on, have been developed to overcome problems related to the large volume change. According to an early pioneering research published in

1987 by Jow et al. [83], an Na-Pb binary alloy, showing large volume expansion (>400%) on electrochemical cycling, shows surprisingly stable cycles in Na cells. A lead composite electrode with a conductive polymer, poly(*p*-phenylene), and a nonfluorinated polymer as the binder shows excellent cyclability in Na cells [83]. Although lead as a heavy element could limit its use in the practical application, the achievement of excellent cyclability using the Na alloy is the first important discovery in this topic. Another recent important discovery is that reversible three-electron redox of phosphorus has been demonstrated in Na cells [78,79,84].

16.7.1 Tin and Antimony as Alloy Materials

Tin and antimony form binary alloys with sodium. The electrochemical alloying/dealloying for the Na-Sn system occur in a series of steps, Sn, $NaSn_5$, NaSn, Na_9Sn_4, and $Na_{15}Sn_4$ [85], with a two-phase reaction mechanism. Recently, it has been demonstrated that the reversibility of the allowing/dealloying for Sn electrodes is effectively improved by the use of a PAA binder [86]. The Sn-PAA electrode delivers higher reduction and oxidation capacities of 880 and 760 mAh g^{-1}, respectively (Figure 16.12). When the Sn forms $Na_{15}Sn_4$ by sodiation, the theoretical capacity reaches 847 mAh g^{-1}. These results agree well with each other, and the formation of $Na_{15}Sn_4$ phase after full reduction of Sn in the Na cell has been experimentally observed [86]. The specific capacity and Coulombic efficiency are effectively improved by adopting the PAA binder instead of PVdF as seen in Figure 16.12. The averaged operating potential is remarkably low for the Sn electrode, which is advantageous to increase the energy density. Nevertheless, the Sn-PAA electrode shows insufficient capacity retention, most probably due to the inevitable volume change of the electrode material and electrolyte decomposition at the active surface of Na-Sn alloys. When a small amount of fluorinated carbonate FEC is added to electrolyte, the performance of the Sn electrode with PAA is further improved, similar to hard-carbon as discussed above. As shown in Figure 16.12, the Sn electrode delivers 700 mAh g^{-1} of reversible capacity for more than 20 cycles, which is more than two times larger reversible capacity compared with a hard-carbon electrode.

Qian et al. [87] prepared an Sb/C composite electrode by mechanical ball-milling of Sb powder with conductive carbon. They found that the Sb/C electrode delivers a reversible capacity of 610 mAh g^{-1}, which corresponds to the capacity based on three electron redox of Sb and alloying with Na. Good rate capability is also demonstrated; 50% of capacity is retained at a high current of 2000 mA g^{-1} with long-term cycle stability. Although the Sb/C electrode in the FEC-free electrolyte can only be cycled in the first 50 cycles, the cyclability is highly improved by adding 5% FEC into the electrolyte solution, similar to the Sn electrode. In the optimized electrolyte, the Sb/C electrode demonstrates a long-term cycling stability with 94% capacity retention over 100 cycles.

The reaction mechanism of Sb in Na cells has been studied by in situ XRD method [88]. Crytalline Sb is electrochemically reduced to form Na_xSb as an

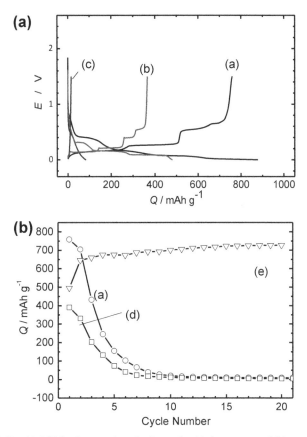

FIGURE 16.12 (a) Initial galvanostatic reduction and oxidation curves and (b) reversible capacity variation of (a) Sn-PAA, (b) Pb-PAA, (c) Si-PAA, (d) Sn-PVdF, and (e) Sn-SNO3-PANa. Reproduced from Ref. [86] with permission. *(Reproduced from Mouad Dahbi et al. Negative electrodes for Na-ion batteries. In: Physical chemistry chemical physics, Royal Society of Chemistry (June 4, 2014).)*

amorphous phase. Amorphous Na_xSb is further reduced to crystalline Na_3Sb with a hexagonal lattice through the formation of cubic Na_3Sb as a metastable intermediate phase. The crystalline Na_3Sb is oxidized in Na cells, and changes into amorphous Sb as the oxidation product.

16.7.2 Red and Black Phosphorus

Phosphorus, a nonmetallic element of the group 15 in the periodic table, has three main allotropes: white, red, and black phosphorus. Red phosphorus is a relatively stable allotrope of phosphorus, and both amorphous and crystalline phases are known to be present. As electrode materials for Na cells, in 2012, we reported the three-electron reversible redox process of red phosphorus electrodes in Na cells (theoretical capacity is calculated to be approximately 2600 mAh g^{-1}), and its potential application in rechargeable Na batteries [89]. Very recently, similar results using phosphorus–carbon composites have also been reported as electrode materials for Na cells [78,79]. One can note that Na$_3$P as the reduction product releases flammable and toxic phosphine, PH$_3$, upon hydrolysis. This will inevitably restrict and limit its practical use, similar to Na/S batteries. Nevertheless, phosphorus electrodes have interesting and important characteristics as electrode materials in Na cells. Briefly, it is the anomalously small volume expansion during the sodiation process that is different from other Na-based alloys [84,85]. Since Na atoms have a covalent character in Na$_3$P as the reduction product, the apparent molar volume of Na is reduced to 77% and 59%, based on the Na–Me alloys and metallic Na, respectively [38,84].

Although the expected volume expansion by sodiation is lower than other Na alloys, a severe problem of electrolyte decomposition has also been found at the highly reactive surface of Na$_3$P. Cyclability is much better for a Li counterpart, Li$_3$P [84]. Similar to the Sn and Sb systems as Na alloys, a reversibility of the formation of sodium phosphide in Na cells is successfully improved by utilizing nonfluorinated binders, such as sodium polyacrylate (PANa) [78,84] and CMC with FEC as fluoridated electrolyte additives [84,90] as shown in Figure 16.13(a) and (b). As

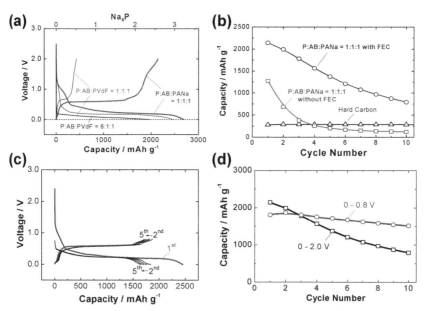

FIGURE 16.13 Electrochemical properties of red phosphorus//Na cells at a rate of 50 mA g^{-1}: (a) Galvanostatic charge/discharge curves with different phosphorus concentration and binders (PVdF or PANa); (b) comparison of capacity retention with or without FEC; (c) electrode performance of the phosphorus electrodes with the upper cut-off voltage of 0.8V; (d) capacity retention of the Na cells Ref. [84]. *(Reproduced from Naoaki Yabuuchi et al. Phosphorus electrodes in sodium cells: small volume expansion by sodiation and the surface-stabilization mechanism in aprotic solvent. In: ChemElectroChem, John Wiley and Sons (August, 2015).)*

we reported previously, HAXPES reveals that phosphorus species are incorporated in the surface film and that electrolyte decomposition is effectively suppressed by the use of FEC [84]. Cyclability is further improved by lowering the cut-off voltage to 0.8 V as shown in Figure 16.13(c) and (d). Our group found by HAXPES that surface layers containing phosphorus species are easily oxidized during anodic scan from 0.0 to 2.0 V, resulting in the loss of reversible capacity. Cyclability is also significantly improved by using $NaPF_6$ as salts of electrolyte instead of $NaClO_4$. Acceptable capacity retention more than 50 cycles and good rate-capability have been reported using the "fluoridated salt" $NaPF_6$-based electrolyte solution with well-optimized phosphorus/carbon composite electrodes [78,79].

Black phosphorus is the least reactive allotrope and has higher electrical conductivity compared with white and red phosphorus [91,92]. Black phosphorus can be prepared from red phosphorus by high-pressure synthesis under 4.5 GPa at high temperature (900 °C). It is also prepared by mechanical milling with mixer mill and planetary ball mill apparatuses [93]. Similar to red phosphorus, black phosphorus is reported to show high lithiation capacity as negative electrodes for Li cells [94,95]. For instance, Park and Sohn [95] reported that black phosphorus with an orthorhombic lattice shows a large initial capacity above 2000 mAh g^{-1}. Similar to this work, Marino and co-workers [96] developed a phosphorus/carbon composite electrode with ordered mesoporous carbon, which showed improved capacity retention as electrode materials in Li cells.

In 2013, electrode performance of black phosphorus in Na cells has been also reported [79]. However, electrode performance seems to be insufficient compared with the amorphous phase. Very recently, our group has demonstrated that a black phosphorus electrode [97] with the PANa binder shows high specific capacity and excellent cycling performance in Na cells. The black phosphorus electrode with PANa binder shows high initial charge and discharge capacities of 2050 and 1620 mAh g^{-1}, respectively, at rate of 125 mAh g^{-1} in 1 mol dm^{-3} $NaPF_6$ EC:DEC. Averaged Columbic efficiency reaches 97% from the second cycle [98]. The charge/discharge capacities of black phosphorus electrode slowly decrease during continuous cycle testing (Figure 16.14(a)). Cyclability in 1 mol dm^{-3} $NaPF_6$ EC:DEC solution is further improved by using the FEC additive as shown in Figure 16.14(b). Electrolyte decomposition is expected to be suppressed by the relatively stable surface film [84]. The black phosphorus electrode with PANa delivers approximately 60% of the theoretical capacity, presumably because of kinetic limitations for the sodiation process of phosphorus. Further optimization for the preparation of composite electrodes could increase the available reversible capacity of phosphorus-based electrode materials.

16.8 SILICON FOR LITHIUM-ION BATTERY

For the application of silicon electrode as negative electrode for LIB, electrochemical lithiation of silicon to form lithium silicide, $Li_{15}Si_4$ ($Li_{3.75}Si$), is known as the most Li-rich phase, which has been evidenced experimentally in

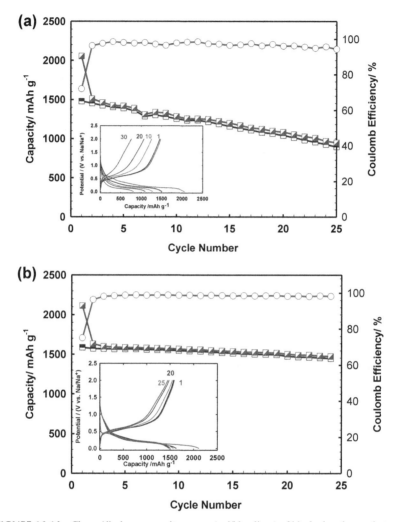

FIGURE 16.14 Charge/discharge capacity curves (red/blue lines) of black phosphorus electrode in 1 mol dm^{-3} NaPF$_6$/EC/DEC electrolyte (a) without additive and (b) with 5% FEC in the voltage range of 0–2 V. The insets shows charge/discharge profiles Ref. [98]. *(Reproduced from Mouad Dahbi et al. Negative electrodes for Na-ion batteries. In: Physical chemistry chemical physics, Royal Society of Chemistry (June 4, 2014).)*

numerous studies, whereas NaSi is known as the most Na-rich phase of Na–Si binary compounds [99]. The capacity estimated based on the formation of NaSi is inevitably lower than that of the Li system with Li$_{3.75}$Si. Furthermore, to the best of our knowledge, electrochemical formation of NaSi in an Na cell at room temperature has not been reported experimentally so far [86,100].

In case of powder silicon materials, polymeric binder is required to prepare composite electrode on copper foil. Selection of polymer binders plays a significant role to maximize its electrochemical performance because mechanical and adhesion properties depend on the selection. While the conventional binder, PVdF, is not sufficient for the silicon electrode, one of the efficient binders for silicon composite electrode is poly(acrylic acid), one of polyelectrolytes having carboxylic group. Furthermore, the impact of neutralization degrees of poly(acrylic acid) as a binder on the silicon-based electrode performance in an Li cell is remarkable [101]. The electrochemical activity of lithiation for the silicon–graphite composite electrode is significantly enhanced as shown in Figure 16.15(a) and (b) by using the 80% NaOH-neutralized polyacrylate ($PAH_{0.2}Na_{0.8}$) binder in comparison to the PAH and PANa binders. The silicon–graphite composite electrode with $PAH_{0.2}Na_{0.8}$ can deliver a rechargeable capacity of more than $1000\,mAh\,g^{-1}$ for a 100 cycle test (Figure 16.15(c)). The graphite–silicon–$PAH_{0.2}Na_{0.8}$ composite electrode has preferable porous structures when compared to the PAH and fully neutralized PANa binders. The self-formed porous structure of $PAH_{0.2}Na_{0.8}$ is effective to buffer the strain induced by the volume expansion of silicon [101]. As seen in Figure 16.15(c), FEC as electrolyte additive is an effective electrolyte additive to further enhance cycle life of the graphite–silicon $PAH_{0.2}Na_{0.8}$ composite electrode. The fluorinated additive is clearly efficient to improve the battery performance for LIB similar to NIB as described above. The FEC additive is decomposed by electrochemical reduction, leading to improvement of passivation properties of the negative electrodes. It is believed that one of important chemicals contained in passivation layer is LiF and NaF for LIB and NIB, respectively.

16.9 CONCLUSIVE REMARKS

Developments in NIB and LIB technology and their understanding of electrode process are strongly influencing and accelerating the development of NIBs. Further understanding the electrode materials, electrolyte, and interface/interphase is certainly required before commercializing practical NIBs.

Before 2010, the most difficult issue facing the demonstration of NIB was to achieve sodium-insertion negative electrodes showing high capacity good cycle life. In the 2010s, new highly performing negative electrode materials for NIBs have been discovered. Further studies are needed not only for active materials but also for electrolytes, additives, and binders, which can highly influence electrode performance. The approaches used for improving cycling performance of negative electrode materials for NIBs include fluorinated electrolyte additive. Indeed, the surface film and electrolyte decomposition is sufficiently suppressed by the use of FEC. Moreover, further understanding of the passivating interphase between electrode/electrolyte is needed to realize further breakthroughs. Such approach with comparative studies between Li and Na systems will facilitate the further development of the NIBs, even though we still have many tough challenges to overcome before accomplishment of high-energy storage system based on sodium as charge carriers in the future.

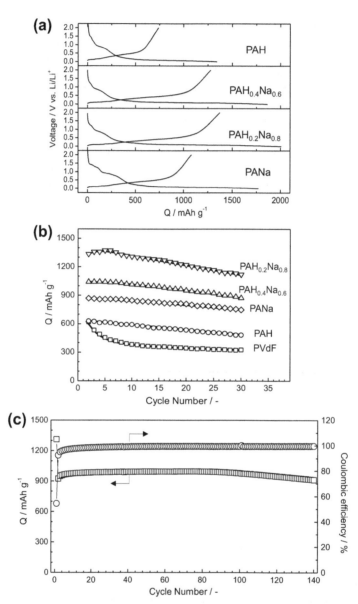

FIGURE 16.15 (a) Initial reduction–oxidation curves of silicon–graphite electrodes prepared with 10 wt% PAH, $PAH_{0.4}Na_{0.6}$, $PAH_{0.2}Na_{0.8}$, and PANa binders at a rate of 50 mA g^{-1}. (b) Capacity retention (2nd to 30th cycles) of the silicon–graphite electrodes with 10 wt% PVdF, PAH, $PAH_{0.4}Na_{0.6}$, $PAH_{0.2}Na_{0.8}$, and PANa binders in 1 mol dm^{-3} LiPF$_6$ EC:DMC (1:1 by volume) without additive. (c) Cyclability of the $PAH_{0.2}Na_{0.8}$ electrode with the electrolyte containing 2 vol% FEC. Reproduced from Ref. [101], with permission. *(Reproduced from Zhen-Ji Han et al. High-capacity Si–graphite composite electrodes with a selfformed porous structure by a partially neutralized polyacrylate for Li-ion batteries. In: Energy & Environmental Science, Royal Society of Chemistry (July 23, 2012).)*

ACKNOWLEDGMENTS

The authors would like to acknowledge all of students, collaborators, and industrial partners who have contributed to this series of research over the past several years. This study was in part granted by JSPS through KAKENHI (21750194) and the "Funding for NEXT Program," LiEAD project, NEDO and Elements Strategy Initiative for Catalyst and Batteries project by MEXT.

REFERENCES

[1] B. Dunn, H. Kamath, J.-M. Tarascon, Science 334 (2011) 928–935.
[2] N. Yabuuchi, M. Kajiyama, J. Iwatate, H. Nishikawa, S. Hitomi, R. Okuyama, R. Usui, Y. Yamada, S. Komaba, Nat. Mater. 11 (2012) 512–517.
[3] M.J. McDonald, J.W.H. Smith, J.R. Dahn, Carbon 68 (2014) 452–461.
[4] J.R. Dahn, T. Zheng, Y. Liu, J.S. Xue, Science 270 (1995) 590–593.
[5] C. Luo, Y. Zhu, Y. Xu, Y. Liu, T. Gao, J. Wang, C. Wang, J. Power Sources 250 (2014) 372–378.
[6] N. Yabuuchi, M. Yano, H. Yoshida, S. Kuze, S. Komaba, J. Electrochem. Soc. 160 (2013) A3131–A3137.
[7] R. Kanno, Y. Takeda, T. Ichikawa, K. Nakanishi, O. Yamamoto, J. Power Sources 26 (1989) 535–543.
[8] M. Mohri, N. Yanagisawa, Y. Tajima, H. Tanaka, T. Mitate, S. Nakajima, M. Yoshida, Y. Yoshimoto, T. Suzuki, H. Wada, J. Power Sources 26 (1989) 545–551.
[9] R. Fong, U. von Sacken, J.R. Dahn, J. Electrochem. Soc. 137 (1990) 2009–2013.
[10] K. Sawai, T. Ohzuku, T. Hirai, Chem. Express 5 (1990) 837–840.
[11] T. Ohzuku, Y. Iwakoshi, K. Sawai, J. Electrochem. Soc. 140 (1993) 2490–2498.
[12] P. Ge, M. Fouletier, Solid State Ionics 28-30 (1988) 1172–1175.
[13] M.M. Doeff, Y.P. Ma, S.J. Visco, L.C. Dejonghe, J. Electrochem. Soc. 140 (1993) L169–L170.
[14] S. Komaba, T. Itabashi, M. Watanabe, H. Groult, N. Kumagai, J. Electrochem. Soc. 154 (2007) A322–A330.
[15] S. Komaba, T. Itabashi, B. Kaplan, H. Groult, N. Kumagai, Electrochem. Commun. 5 (2003) 962–966.
[16] E. Peled, J. Electrochem. Soc. 126 (1979) 2047–2051.
[17] M. Wakihara, and O. Yamamoto, Wiley, 2008.
[18] T. Ohzuku, Y. Iwakoshi, K. Sawai, J. Electrochem. Soc. 140 (1993) 2490–2498.
[19] D. Aurbach, B. Markovsky, I. Weissman, E. Levi, Y. Ein-Eli, Electrochim. Acta 45 (1999) 67–86.
[20] R.C. Asher, S.A. Wilson, Nature 181 (1958) 409–410.
[21] R.E. Franklin, in: Proceedings of the Royal Society of London. Series a. Mathematical and Physical Sciences, The Royal Society 209, (1951) 196–218.
[22] G.M. Jenkins, K. Kawamura, Nature 231 (1971) 175.
[23] L.L. Ban, D. Crawford, H. Marsh, J. Appl. Crystallogr. 8 (1975) 415–420.
[24] M. Shiraishi, in: Introduction to Carbon Materials, The Carbon Society of Japan, Tokyo (1984) 29–40.
[25] J.R. Dahn, W. Xing, Y. Gao, Carbon 35 (1997) 825–830.
[26] H. Azuma, H. Imoto, S. Yamada, K. Sekai, J. Power Sources 81 (1999) 1–7.
[27] J. Conard, P. Lauginie, TANSO 2000 (2000) 62–70.
[28] P.J.F. Harris, Philos. Mag. 84 (2004) 3159–3167.
[29] M.A. Smith, H.C. Foley, R.F. Lobo, Carbon 42 (2004) 2041–2048.

[30] A. Kumar, R.F. Lobo, N.J. Wagner, Carbon 43 (2005) 3099–3111.
[31] M. Nagao, C. Pitteloud, T. Kamiyama, T. Otomo, K. Itoh, T. Fukunaga, K. Tatsumi, R. Kanno, J. Electrochem. Soc. 153 (2006) A914–A919.
[32] S. Yata, H. Kinoshita, M. Komori, N. Ando, T. Kashiwamura, T. Harada, K. Tanaka, T. Yamabe, Synth. Metals 62 (1994) 153–158.
[33] D.A. Stevens, J.R. Dahn, J. Electrochem. Soc. 147 (2000) 1271–1273.
[34] K. Gotoh, M. Maeda, A. Nagai, A. Goto, M. Tansho, K. Hashi, T. Shimizu, H. Ishida, J. Power Sources 162 (2006) 1322–1328.
[35] S. Komaba, W. Murata, T. Ishikawa, N. Yabuuchi, T. Ozeki, T. Nakayama, A. Ogata, K. Gotoh, K. Fujiwara, Adv. Funct. Mater. 21 (2011) 3859–3867.
[36] A. Ponrouch, A.R. Goni, M.R. Palacin, Electrochem. Commun. 27 (2013) 85–88.
[37] J. Zhao, L.W. Zhao, K. Chihara, S. Okada, J. Yamaki, S. Matsumoto, S. Kuze, K. Nakane, J. Power Sources 244 (2013) 752–757.
[38] K. Chihara, N. Chujo, A. Kitajou, S. Okada, Electrochim. Acta 110 (2013) 240–246.
[39] A. Bhide, J. Hofmann, A. Katharina Durr, J. Janek, P. Adelhelm, Phys. Chem. Chem. Phys. 16 (2014) 1987–1998.
[40] M. Moshkovich, Y. Gofer, D. Aurbach, J. Electrochem. Soc. 148 (2001) E155–E167.
[41] J. Jones, M. Anouti, M. Caillon-Caravanier, P. Willmann, P.-Y. Sizaret, D. Lemordant, Fluid Phase Equilibr 305 (2011) 121–126.
[42] K. Xu, Chem. Rev. 104 (2004) 4303–4418.
[43] R. Alcántara, P. Lavela, G.F. Ortiz, J.L. Tirado, Electrochem. Solid-State Lett. 8 (2005) A222–A225.
[44] M. Kobayashi, T. Inoguchi, T. Iida, T. Tanioka, H. Kumase, Y. Fukai, J. Fluorine Chem. 120 (2003) 105–110.
[45] G.H. Wrodnigg, J.O. Besenhard, M. Winter, J. Electrochem. Soc. 146 (1999) 470–472.
[46] S.-K. Jeong, M. Inaba, R. Mogi, Y. Iriyama, T. Abe, Z. Ogumi, Langmuir 17 (2001) 8281–8286.
[47] R. Mogi, M. Inaba, S.-K. Jeong, Y. Iriyama, T. Abe, Z. Ogumi, J. Electrochem. Soc. 149 (2002) A1578–A1583.
[48] H. Nakai, T. Kubota, A. Kita, A. Kawashima, J. Electrochem. Soc. 158 (2011) A798–A801.
[49] O. Borodin, W. Behl, T.R. Jow, J. Phys. Chem. C 117 (2013) 8661–8682.
[50] D. Aurbach, K. Gamolsky, B. Markovsky, Y. Gofer, M. Schmidt, U. Heider, Electrochim. Acta 47 (2002) 1423–1439.
[51] H. Ota, Y. Sakata, A. Inoue, S. Yamaguchi, J. Electrochem. Soc. 151 (2004) A1659–A1669.
[52] Z.-J. Han, N. Yabuuchi, K. Shimomura, M. Murase, H. Yui, S. Komaba, Energy Environ. Sci. 5 (2012) 9014–9020.
[53] S. Komaba, T. Ishikawa, N. Yabuuchi, W. Murata, A. Ito, Y. Ohsawa, ACS Appl. Mater. Interfaces 3 (2011) 4165–4168.
[54] R. Crowe, J.P.S. Badyal, J. Chem. Soc. Chem. Commun. (1991) 958–959.
[55] J. Drofenik, M. Gaberscek, R. Dominko, F.W. Poulsen, M. Mogensen, S. Pejovnik, J. Jamnik, Electrochim. Acta 48 (2003) 883–889.
[56] J. Li, R.B. Lewis, J.R. Dahn, Electrochem. Solid St 10 (2007) A17–A20.
[57] J.H. Lee, U. Paik, V.A. Hackley, Y.M. Choi, J. Electrochem. Soc. 152 (2005) A1763–A1769.
[58] N. Yabuuchi, K. Shimomura, Y. Shimbe, T. Ozeki, J.-Y. Son, H. Oji, Y. Katayama, T. Miura, S. Komaba, Adv. Energy Mater. 1 (2011) 759–765.
[59] H. Komaba, K. Moriwaki, S. Goto, S. Yamada, M. Taniguchi, T. Kakuta, I. Kamae, M. Fukagawa, Am. J. Kidney Dis. 60 (2012) 262–271.
[60] S. Komaba, T. Ozeki, N. Yabuuchi, K. Shimomura, Electrochemistry 79 (2011) 6–9.

[61] A. Magasinski, B. Zdyrko, I. Kovalenko, B. Hertzberg, R. Burtovyy, C.F. Huebner, T.F. Fuller, I. Luzinov, G. Yushin, ACS Appl. Mater. Interfaces 2 (2010) 3004–3010.
[62] I. Kovalenko, B. Zdyrko, A. Magasinski, B. Hertzberg, Z. Milicev, R. Burtovyy, I. Luzinov, G. Yushin, Science 334 (2011) 75–79.
[63] M. Dahbi, T. Nakano, N. Yabuuchi, T. Ishikawa, K. Kubota, M. Fukunishi, S. Shibahara, J.-Y. Son, Y.-T. Cui, H. Oji, S. Komaba, Electrochem. Commun. 44 (2014) 66–69.
[64] S. Komaba, N. Yabuuchi, T. Ozeki, K. Okushi, H. Yui, K. Konno, Y. Katayama, T. Miura, J. Power Sources 195 (2010) 6069–6074.
[65] C.-H. Huang, Y.-H. Chang, H.-W. Wang, S. Cheng, C.-Y. Lee, H.-T. Chiu, J. Phys. Chem. B 110 (2006) 11818–11822.
[66] O. Tanaike, H. Hatori, Y. Yamada, S. Shiraishi, A. Oya, Carbon 41 (2003) 1759–1764.
[67] M. Dahbi, F. Ghamouss, F. Tran-Van, D. Lemordant, M. Anouti, Electrochim. Acta 86 (2012) 287–293.
[68] M. Dahbi, F. Ghamouss, F. Tran-Van, D. Lemordant, M. Anouti, J. Power Sources 196 (2011) 9743–9750.
[69] A. Ponrouch, E. Marchante, M. Courty, J.-M. Tarascon, M.R. Palacin, Energy Environ. Sci. 5 (2012) 8572–8583.
[70] M.N. Obrovac, L. Christensen, D.B. Le, J.R. Dahn, J. Electrochem. Soc. 154 (2007) A849–A855.
[71] G. Jeong, Y.-U. Kim, H. Kim, Y.-J. Kim, H.-J. Sohn, Energy Environ. Sci. 4 (2011) 1986–2002.
[72] H. Li, X. Huang, L. Chen, Z. Wu, Y. Liang, Electrochem. Solid-State Lett. 2 (1999) 547–549.
[73] L.Y. Beaulieu, K.C. Hewitt, R.L. Turner, A. Bonakdarpour, A.A. Abdo, L. Christensen, K.W. Eberman, L.J. Krause, J.R. Dahn, J. Electrochem. Soc. 150 (2003) A149–A156.
[74] L. Wu, X. Hu, J. Qian, F. Pei, F. Wu, R. Mao, X. Ai, H. Yang, Y. Cao, J. Mater. Chem. A 1 (2013) 7181–7184.
[75] A. Darwiche, M.T. Sougrati, B. Fraisse, L. Stievano, L. Monconduit, Electrochemistry Commun. 32 (2013) 18–21.
[76] J.W. Wang, X.H. Liu, S.X. Mao, J.Y. Huang, Nano Lett. 12 (2012) 5897–5902.
[77] Schacklette, L. Wayne, Taiguang, R. Jow, Mcrae, Maxfield, Toth, J. Edward, and S. Gould, Patent EP0347952, (1989).
[78] Y. Kim, Y. Park, A. Choi, N.-S. Choi, J. Kim, J. Lee, J.H. Ryu, S.M. Oh, K.T. Lee, Adv. Mater. (Weinheim, Ger.) 25 (2013) 3045–3049.
[79] J. Qian, X. Wu, Y. Cao, X. Ai, H. Yang, Angew. Chem. 125 (2013) 4731–4734.
[80] L. Monconduit, J. Fullenwarth, A. Darwiche, A. Soares, B. Donnadieu, J. Mater. Chem. A 2 (2014) 2050–2059.
[81] M.K. Datta, R. Epur, P. Saha, K. Kadakia, S.K. Park, P.N. Kumta, J. Power Sources 225 (2013) 316–322.
[82] L.D. Ellis, T.D. Hatchard, M.N. Obrovac, J. Electrochem. Soc. 159 (2012) A1801–A1805.
[83] T.R. Jow, L.W. Shacklette, M. Maxfield, D. Vernick, J. Electrochem. Soc. 134 (1987) 1730–1733.
[84] N. Yabuuchi, Y. Matsuura, T. Ishikawa, S. Kuze, J.-Y. Son, Y.-T. Cui, H. Oji, S. Komaba, ChemElectroChem. 1 (2014) 580–589.
[85] V.L. Chevrier, G. Ceder, J. Electrochem. Soc. 158 (2011) A1011–A1014.
[86] S. Komaba, Y. Matsuura, T. Ishikawa, N. Yabuuchi, W. Murata, S. Kuze, Electrochemistry Commun. 21 (2012) 65–68.
[87] J. Qian, Y. Chen, L. Wu, Y. Cao, X. Ai, H. Yang, Chem. Commun. 48 (2012) 7070–7072.
[88] A. Darwiche, C. Marino, M.T. Sougrati, B. Fraisse, L. Stievano, L. Monconduit, J. Am. Chem. Soc. 134 (2012) 20805–20811.
[89] Y. Matsuura, T. Ishikawa, N. Yabuuchi, S. Kuze, S. Komaba, in: The 53rd Battery Symposium in Japan, Fukuoka, The Committee of Battery Technology, ECSJ 2012, 1E26.

[90] J.B. Goodenough, H.Y.P. Hong, J.A. Kafalas, Mater. Res. Bull. 11 (1976) 203–220.
[91] L. Pauling, M. Simonetta, J. Chem. Phys. 20 (1952) 29–34.
[92] R.W. Keyes, Phys. Rev. 92 (1953) 580–584.
[93] M. Nagao, A. Hayashi, M. Tatsumisago, J. Power Sources 196 (2011) 6902–6905.
[94] L.-Q. Sun, M.-J. Li, K. Sun, S.-H. Yu, R.-S. Wang, H.-M. Xie, J. Phys. Chem. C 116 (2012) 14772–14779.
[95] C.M. Park, H.J. Sohn, Adv. Mater. 19 (2007) 2465–2468.
[96] C. Marino, A. Debenedetti, B. Fraisse, F. Favier, L. Monconduit, Electrochemistry Commun. 13 (2011) 346–349.
[97] M. Dahbi, N. Yabuuchi, T. Nakano, M. Fukunishi, K. Tokiwa, S. Komaba, in: The 54rd Battery Symposium in Japan, The Committee of Battery Technology, ECSJ 2013, 2G25.
[98] M. Dahbi, N. Yabuuchi, K. Kubota, K. Tokiwa, S. Komaba, Phys. Chem. Chem. Phys 16 (2014) 15007–15028.
[99] H. Morito, T. Yamada, T. Ikeda, H. Yamane, J. Alloys Compounds 480 (2009) 723–726.
[100] L.D. Ellis, B.N. Wilkes, T.D. Hatchard, M.N. Obrovac, J. Electrochem. Soc. 161 (2014) A416–A421.
[101] Z.J. Han, N. Yabuuchi, K. Shimomura, M. Murase, H. Yui, S. Komaba, Energy Environ. Sci. 5 (2012) 9014–9020.

Chapter 17

Application of Carbon Materials Derived from Fluorocarbons in an Electrochemical Capacitor

Soshi Shiraishi[1] and Osamu Tanaike[2]

[1]*Graduate School of Science and Technology, Gunma University, Kiryu, Japan;* [2]*Research Center for Compact Chemical System, National Institute of Advanced Industrial Science and Technology, Sendai, Japan*

Chapter Outline

17.1 Introduction	415	of Porous Carbon from Fluorocarbon	422
17.2 Synthesis Method and Basic Characterization of Porous Carbon from Fluorocarbon	417	17.4 Conclusion	427
		Acknowledgment	429
		References	429
17.3 Performance of Electric Double Layer Capacitance			

17.1 INTRODUCTION

Carbon materials can be generally obtained by pyrolyzing organic materials. On the other hand, a special carbon can also be generated in some special cases through the phase separation of carbon and other elements within polymer materials using chemical reactions. For example, sugar is well known to be dehydrated in concentrated sulfuric acid to form a carbonaceous material. Another analogous example is found in the production of porous carbon by inducing defluorination reactions on fluoropolymer materials, of which polytetrafluoroethylene (PTFE) is a typical example. Figure 17.1 shows a conceptual diagram for the synthesis of porous carbon from a fluorocarbon material that contains a structure in which all the hydrogen in polymer has been replaced with fluorine. Because fluoropolymers are chemically stable, they must undergo reactions with a strong reducing agent, such as corrosive alkali metals, in order to allow their defluorination reactions to occur. In such cases, this reaction will end up producing a byproduct compound of an alkali fluoride within the carbonaceous matrix. Removing the fluoride through acid washing leaves pores in the

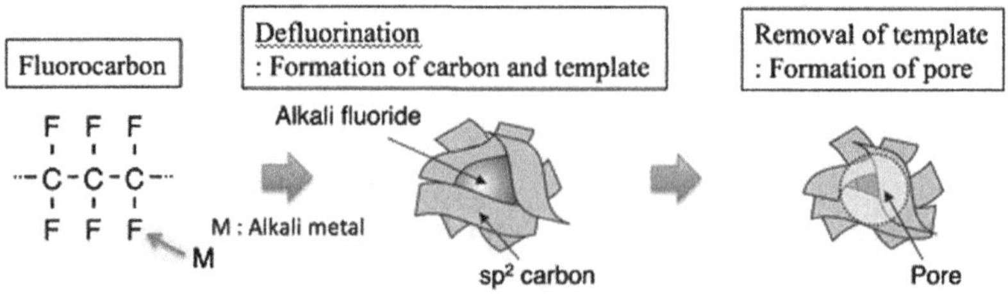

FIGURE 17.1 Schematic illustration of formation mechanism of porous carbon from PTFE by chemical defluorination.

domain of the fluoride, which will in effect produce a porous carbon. Jansta et al. reported that defluorination of PTFE using an alkali metallic amalgam followed by the removal of the alkali fluoride and mercury can produce porous carbon materials with high surface areas of more than 2000 m^2g^{-1} [1]. This mechanism of creating pores in the defluorination product is similar to the "template method" for porous carbon [2]. However, the method used here does not require any prior additional preparation of the template materials, and it can also facilitate the self-generation of the templates only through the chemical treatment of the fluoropolymer as the starting material. Additionally, some properties of the product carbon from this method will differ from those obtained by the conventional pyrolysis of a polymer precursor, because the defluorination reactions leave the carbon while retaining the molecular structure of the fluoropolymers without an overly gasification loss of the raw materials. For example, some reports showed that the so-called carbyne structures, like the one-dimensional sp-hybridized carbon chain, can be formed as an intermediary when using linear chain molecules like PTFE [3].

Recently, electrochemical capacitors, which have a higher capacitance than conventional dielectric capacitors, have been focused as an electric energy storage device like the rechargeable battery [4,5]. The electric double layer capacitor (EDLC), which is an electrochemical capacitor, has the advantages of a high-rate charge/discharge capability, long cycle life, etc. The capacitive performance of the EDLC is based on the dielectric property of the electric double layer at the interface between the electrolyte and a nanoporous carbon electrode such as activated carbon; therefore, carbon materials with a high surface area available for the adsorption of the electrolyte ions have been investigated to improve the capacitance of the EDLC. Especially, porous carbons with abundant mesopores (defined by the pore size of 2–50 nm), referred to as "mesoporous carbons," are well known to be effective in enhancing the high-rate charge/discharge performance [6].

The authors have examined the defluorination technique of fluorocarbon materials as a means for producing mesoporous carbon favorable for the use in the EDLC. In this chapter, the synthesis and structure of the carbon, pore structure control, and the capacitor electrode properties are highlighted.

17.2 SYNTHESIS METHOD AND BASIC CHARACTERIZATION OF POROUS CARBON FROM FLUOROCARBON

By applying various types of alkali metal species when performing the defluorination of PTFE, the authors obtained porous carbons that possess a porous structure dependent on the nature of the reducing agent. Although alkali metals in their solid state do not react very fast with PTFE, it promotes the defluorination reactions when PTFE is carefully allowed to come into contact with the alkali metals vapors. As shown in Figure 17.2, the authors sealed the PTFE and an alkali metal solid block separately in a single glass tube and then heated the tube very carefully to 473 K. Afterward, the authors could see the white color of the PTFE powder gradually change to a black color, wherein the carbonaceous product was produced. The obtained reactant was sufficiently exposed to the outside air by opening the tube, and the excess alkali metal was completely deactivated and then washed using diluted hydrogen fluoric or chloric acid. Porous carbons were then prepared by removing the alkali fluoride without any typical activation process of the carbon to evaluate the capacitance [7].

Figure 17.3 shows the nitrogen physisorption isotherms and pore size distribution of the porous carbons obtained from using commercially available PTFE

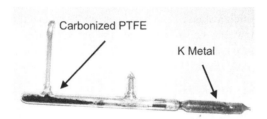

FIGURE 17.2 Products just after defluorination of PTFE by potassium metal. *Reprinted from Ref. [7] with permission from The Carbon Society of Japan.*

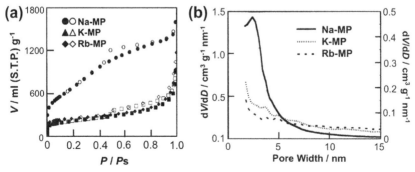

FIGURE 17.3 (a) Nitrogen adsorption (closed marks)/desorption (opened marks) isotherms (77 K) and (b) pore size distribution curves by Dollimore-Heal (DH) method of porous carbons from PTFE and alkali metals.

directly defluorinated by alkali metal vapors, and Figure 17.4 shows the transmission electron microscopy (TEM) pictures of a typical one [8,9]. All of the isotherms were type IV based on the IUPAC classification, which means both micropores and mesopores are developed given that the isotherms possessed hysteresis loops. When making a comparison of the pore size distributions among the alkali metal species, development of mesopores with 2–3 nm pore sizes was more significant for sodium than potassium and rubidium. The results also showed that sodium provided a BET-specific surface area up to three times greater (2000 m^2g^{-1}) than the other two. From the TEM pictures, because the authors can see images of relatively large pores with a circular morphology in the carbon particles, it can be assumed that the aggregation of the alkali fluoride at the time of defluorination is reflected in the pore size. When the temperature is held constant, the authors deduced that the aggregation of the fluoride particles will be further promoted due to the higher reactivity of the potassium. Thereby, it is possible to obtain a porous carbon with a high specific surface area with highly introduced mesopores.

In contrast, using alkali metals directly has disadvantages for the mass production operations and safety. It is possible to produce porous carbon in the same way from PTFE using a relatively easy-to-use alkali metal–naphthalene anion radical complex with extremely strong reducing properties as the defluorinating agent. In such instances, the basic concept would be the same as shown in Figure 17.1; however, because the naphthalene radical has the role of the reducing agent in this case, the equation for the actual defluorination reaction would be indicated by the following reaction.

$$-(CF_2\text{-}CF_2)_n- + 4n(C_{10}H_8)^{\cdot}Li^+ \rightarrow -(2C)_n- + 4nLiF + 4nC_{10}H_8$$

In the case using the naphthalene radical, the defluorination reactions progress when PTFE is immersed in an ether-type solvent (e.g., 1,2-dimethoxyethane)

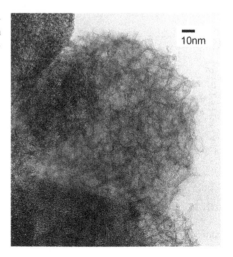

FIGURE 17.4 TEM images of porous carbons from PTFE defluorinated by sodium metal. *Reprinted from Ref. [8] with permission from Elsevier.*

solution containing a dissolved alkali metal and naphthalene. As shown in Figure 17.5, the authors used a variety of alkali metals–naphthalene anion radicals to defluorinate the PTFE and then compared the nitrogen physisorption isotherms of the final porous carbon products [7]. All of these are classified as type IV isotherms similar to those directly using metals, which mean that the porous carbons contain both micropores and mesopores. The BET-specific surface area was quite high at $1000 \sim 1800$ m^{2}g^{-1}, and there were no significantly large effects due to differences in the alkali metal species in contrast to the direct reactions with metals. In cases when direct reactions are induced with metals having different melting points, there may be significant differences in the reactivity between the metals and PTFE. However, when using naphthalene, its radical can contribute to the defluorination reactions independent of the alkali metal species. Hence, the authors concluded from the results that the dependence may originate when the produced template fluoride is in an aggregating or dispersing state. This technique using naphthalene is favorable for a large-scale synthesis and is capable of producing porous carbon from PTFE under defluorination conditions that provide a relatively easy handling without heating highly dangerous alkali metals.

In order to produce porous carbons with various porous structures, the authors are trying to modify the experimental conditions before or after the defluorination reactions take place. After the PTFE undergoes defluorination, the growth of the template fluoride particle is stimulated in the carbon matrix through annealing at high temperature before the fluoride is removed, which thereby results in control of the pore size distribution. Figure 17.6 shows X-ray diffraction (XRD) patterns of the carbon matrix products from the defluorinated PTFE using lithium-naphthalene anion complexes and then heat-treated at different temperatures, and Figure 17.7 shows a comparison of the nitrogen

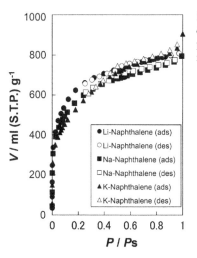

FIGURE 17.5 Nitrogen adsorption (closed marks)/desorption (opened marks) isotherms (77 K) of porous carbons from PTFE defluorinated by alkali metals–naphthalene anion radicals.

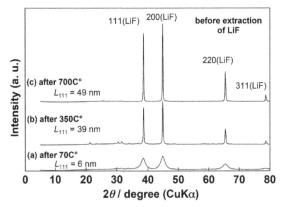

FIGURE 17.6 X-ray diffraction (XRD) patterns (CuKα) for the reaction product of PTFE with Li-naphthalenide (a) after heat treatment at 70 °C for 2 days in air, (b) after heat treatment at 350 °C for 2 h in N_2, and (c) after heat treatment at 700 °C for 1 h in N_2.

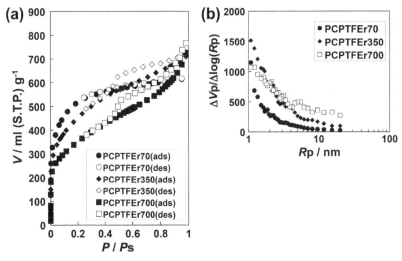

FIGURE 17.7 (a) Nitrogen adsorption/desorption isotherms (77 K) of PCPTFEr70 (heat treatment temperature: 70 °C), PCPTFEr350 (heat-treatment temperature: 350 °C), and PCPTFEr700 (heat-treatment temperature: 700 °C), obtained from the LiF extraction with 1M HCl aq. Filled marks: adsorption data, hollow marks: desorption data. (b) Pore size distribution (PSD) in the region of mesopores of PCPTFEr70, PCPTFEr350, and PCPTFEr700, calculated by DH method. R_p: pore radius.

physisorption isotherms and the pore size distributions of the porous carbons after removal of the lithium fluoride [11]. It is clear that the XRD peaks of LiF grow sharper when the heat treatment is applied at higher temperatures, which facilitates the growth of the LiF particles in the carbon matrix. The pore size distribution for the porous carbon obtained by the removal of LiF in the defluorinated product heated at a higher temperature clearly showed higher ratios in the mesoporous region of the curve. This means it is possible to stimulate the growth in size of the mesopores by applying a heat treatment after the defluorination.

By making adjustments to the structure of the raw fluorinated materials themselves, the structure of the porous carbon can also be modified. For example, when a PTFE oligomer with a low molecular weight evaporates and physically deposited on a quartz substrate in order to reconstruct the regular stacking of the molecular chain, the relatively ordered stacking of the PTFE molecules can be defluorinated to obtain porous carbons with a more ordered porous structure. Figure 17.8 shows scanning and transmission electron micrograph (TEM) pictures of the evaporated PTFE and the obtained carbon, respectively [7]. After evaporation, the PTFE gradually accumulated while the thin flakes vertically aligned on the substrate. The TEM observations revealed that the porous carbon obtained from these materials had a more uniform pore size of pore compared to those from the bulk PTFE polymers as in Figure 17.4. It can be postulated that the physical vapor deposition of the PTFE oligomer results in an ordered stacking arrangement of the PTFE molecules with a uniform density, thereby, the aggregation and growth of the fluoride template particles become uniform in the matrix during the defluorination reaction. Similarly, the authors also adjusted the molecular structure of the PTFE using γ-irradiation to obtain various types

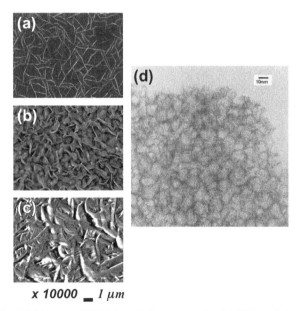

FIGURE 17.8 SEM images of PTFE deposited on quartz glass by PVD method with the rough thickness of (a) 0.1 μm, (b) 0.8 μm, and (c) 9 μm and (d) TEM image of porous carbon from the PTFE. *Reprinted from Ref. [7] with permission from The Carbon Society of Japan.*

of carbon porosities [12]. Various porous carbons can be also prepared by using fluorinated organic monomer materials such as octafluoronaphthalene, which is an aromatic monomer in contrast to the aliphatic PTFE [13]. Therefore, by controlling both the alkali fluoride as the "template" and the raw perfluorinated organic materials as the sources for the "carbon wall," it is possible to increase the degree of variation on the structure of the porous carbon.

17.3 PERFORMANCE OF ELECTRIC DOUBLE LAYER CAPACITANCE OF POROUS CARBON FROM FLUOROCARBON

The EDLC was commercialized as a memory back-up device in the 1970s and is currently expected to be an electric energy storage device because of its excellent power density and cycle life [4,5]. The energy density of the EDLC is lower than that of rechargeable batteries; therefore, the electric double capacitance as a key factor for the energy density should be further improved for energy saving applications.

The charge or discharge process of the EDLC is the adsorption/desorption of the electrolyte ions through the electric double layer on the porous carbon electrode as shown in Figure 17.9. The dependence of the electric double layer capacitance on the pore structure has been investigated since the capacitance is strongly influenced by the pore structure of the carbon electrode [14]. The electric double layer capacitance is almost in proportion to the accessible surface area of the electrolyte ions, so the activated carbons have been used as practical electrode materials [15]. The high capacitance of the activated carbon is due to the high specific surface area (>1000 m^2g^{-1}) provided by the many micropores (defined by the pore size of less than 2 nm). However, in the case of the activated carbons, the micropores with a narrow pore size often sieve the ions and prevent adsorption. The presence of mesopores can cancel the ion sieving effect and enhance the capacitance performance.

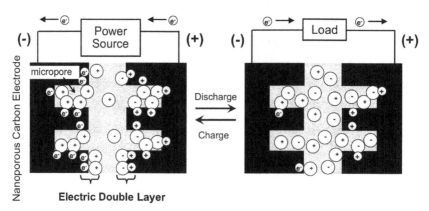

FIGURE 17.9 Schematic illustration of EDLC.

The porous carbons derived from defluorination of PTFE by an alkali metal have many mesopores, which can promote the permeation of electrolyte ions into the carbon matrix compared to the conventional activated carbons. Figure 17.10 shows the correlation between the specific surface area and the gravimetric capacitance (the double-layer capacitance normalized by the electrode carbon mass) for the porous carbons derived by the defluorination of PTFE (PTFE-based porous carbon) in the H_2SO_4 aqueous electrolyte [7,10,16]. All PTFE-based samples in Figure 17.10 were post-heated at 800–1000 °C in an inert gas, such as nitrogen, to improve the electric conductivity for the capacitance measurement. The PTFE-based porous carbons showed a higher capacitance than the conventional activated carbon sample (activated carbon fiber: ACF). The capacitance in the aqueous electrolyte is also sensitive to other factors (surface functionalities and carbon crystallinity of the carbon electrode) than the pore structure; therefore, it is difficult to quantitatively discuss the correlation between the pore structure and the capacitance in an aqueous electrolyte; however, the data of Figure 17.10 suggest that the mesopore contribution denoted as the mesopore ratio (S_{meso}/S_{BET}) is high. The authors also examined the effect of the post-heat treatment on the capacitance in the H_2SO_4 aqueous electrolyte for the PTFE-based porous carbons [9].

In the case of an organic electrolyte, the electrolyte ions and the solvent molecules are larger than those of the aqueous electrolyte, so the influence of the pore structure on the capacitance is more prominent. Additionally, a typical organic electrolyte for the EDLC, such as propylene carbonate solution, has a wider electrochemical window, which leads to a high voltage operation that is advantageous at energy density. The capacitive behavior of the PTFE-based porous carbon electrode is more characteristic in the organic electrolyte [17].

The pore structure and the pore size distribution (PSD) of the PTFE-based porous carbons and the activated carbon samples as reference materials are summarized in Table 17.1 and Figure 17.11, respectively. The pore structure was successfully controlled by the preheat treatment before the LiF removal process. The microporous type sample (micro-PTFE) prepared without the pre-heat

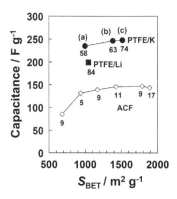

FIGURE 17.10 Correlation between surface area (S_{BET}) and the double layer capacitance (in 1M H_2SO_4 aq) of porous carbons obtained by reaction of (a) non-, (b) 1γ, or (c) 5γ-irradiated PTFE with K (PTFE/K). The reference samples are carbon obtained by reaction of PTFE with Li (PTFE/Li) and (ACFs) with various steam activations. Numbers in figure mean the ratio of mesopore surface area to BET surface area (S_{meso}/S_{BET}) for each sample.

TABLE 17.1 Pore Structure Parameters of PTFE-based Carbons and ACFs

Sample	S_{BET} [m²g⁻¹]	S_{meso} [m²g⁻¹]	V_{meso} [mlg⁻¹]	V_{micro} [mlg⁻¹]	w_{micro} [nm]	d [gcm⁻³]
ACF-1	650	70	0.06	0.28	0.64	1.00
ACF-2	930	30	0.03	0.37	0.73	0.87
ACF-3	1150	100	0.08	0.47	0.85	0.85
ACF-4	1480	140	0.09	0.60	0.94	0.74
ACF-5	1780	250	0.16	0.73	1.07	0.68
micro-PTFE	1090	420	0.31	0.46	1.10	0.59
meso-PTFE	1140	930	0.67	0.47	1.16	0.29

micro-PTFE: Porous carbons derived from the defluorination of PTFE film with $LiC_{10}H_8$ (Heat-treatment of 800 °C for 1 h in N_2 after removal of LiF).
meso-PTFE: Porous carbons derived from the defluorination of PTFE film with $LiC_{10}H_8$ (Heat-treatment of 700 °C for 1 h in N_2 before removal of LiF).
ACF: Activated carbon fiber derived from phenolic resin with steam activation (800 °C).
S_{BET}: BET specific surface area.
S_{meso}: Specific surface area of mesopores calculated by DH method.
V_{meso}: Pore volume of mesopore calculated by DH method.
V_{micro}, w_{micro}: Pore volume and average width of micropore.
d: Electrode bulk density.

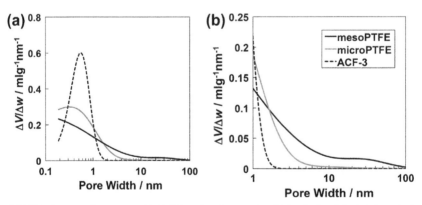

FIGURE 17.11 PSD curves of PTFE-based carbons and ACF-3 calculated by DFT method, (a) 0.1~100 nm and (b) 1~100 nm.

treatment has a PSD peak in the micropore region, while the mesoporous type (meso-PTFE), in which the mesopore structure was developed by the preheat treatment, has a wide PSD from the micropore to mesopore region. Figure 17.12 shows the potential-time curves measured in the organic electrolyte

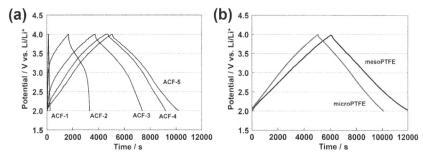

FIGURE 17.12 Chronopotentiograms (potential-time curves) of ACFs and PTFE-based carbons in 1.0 M LiClO$_4$/propylene carbonate (PC) by galvanostatic method (40 mAg^{-1}) using three-electrode cell.

for the PTFE-based porous carbons identical to those in Figure 17.11 and Table 17.1. The curves in the higher region around 3V versus Li/Li$^+$ are mainly governed by the anion adsorption/desorption, while those in the lower region are related to the cation adsorption/desorption. The curves of the ACF samples are distorted by the ion sieving effect of the narrow micropores, but the PTFE-based porous carbon shows a typical linear relationship for the capacitor electrode.

Figure 17.13 is the correlation between the specific surface area and the gravimetric capacitance calculated from Figure 17.12. The capacitance of the PTFE-based porous carbons is higher than that of the ACF samples. Especially, the mesoporous type (meso-PTFE) has a very high capacitance, in which the cation capacitance is more than twice as high as that of the ACF-3 with a comparable surface area. These enhanced capacitance of the PTFE-based porous carbon is due to the presence of abundant mesopores and wide micropores that can cancel the ion sieving effect.

The PTFE-based porous carbon is also noted for the rate performance of EDLC applications [17]. For the meso-PTFE and the ACF sample with a comparable surface area, the capacitance dependence on the current density when using various electrolyte salts is shown in Figure 17.14. The higher capacitance is obtained using the smaller cation size ((C_2H_5)$_4$N$^+$: 0.68 nm < (C_4H_9)$_4$N$^+$: 0.86 nm < (C_6H_{13})$_4$N$^+$: 0.96 nm) in the case of both the meso-PTFE and the ACF. However, the capacitance of the ACF sample decreases by increasing the current density, while the capacitance of the meso-PTFE is maintained even at a high current density and the bulky cation. These results indicated that the presence of a developed mesopore structure enhances the kinetics of the ion adsorption/desorption, which is important for the rate performance of the EDLC.

The defluorination technique of fluorocarbons by alkali metals can also be applied to the preparation of heteroatom-doped porous carbons. Nitrogen-enriched porous carbons with higher than 1000 m^{2}g^{-1} surface area cannot be generally easily prepared by a conventional pyrolyzing technique of the

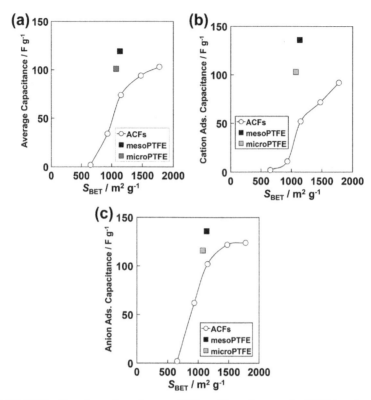

FIGURE 17.13 Capacitance dependence on BET-specific surface area for PTFE-based carbons and ACFs. (a) Average capacitance of cation desorption and anion adsorption, (b) cation adsorption capacitance, and (c) anion adsorption capacitance. Capacitance was measured in 1.0 M LiClO$_4$/PC by galvanostatic method (40 mAg^{-1}) using three-electrode cell.

polymer precursor, but they were successfully prepared by the defluorination of nitrogen-containing fluorocarbon molecules as shown in Figure 17.15. The pore structure and the capacitance are summarized in Table 17.2 [18]. The specific surface area of the N-doped porous carbon derived from defluorination is not very high compared to the conventional activated carbons, but the specific capacitance per surface area is higher than that of the nitrogen-free porous carbons. The nitrogen doping effect on the capacitance is well known for the porous carbon electrode, which is still discussed from the viewpoint of its wettability, pseudo-capacitance, and electronic effect [7,19–21].

A unique carbon material can be prepared from graphite fluoride (GF) by the defluorination technique. The GF is a covalent type graphite intercalation compound with fluorine atoms, which are commercialized as a lubricant or cathode material for primary batteries. The lamella-like carbon material derived from the defluorination of the GF with a Li$^+$–naphthalene anion radical complex has only a surface area of about 200 m^{2}g^{-1}, but the capacitance is comparable to

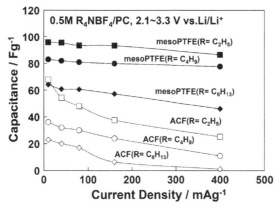

FIGURE 17.14 Capacitance dependence on current density for PTFE-based carbon and ACF. Capacitance was measured in 0.5 M R_4NBF_4/PC (R = C_2H_5, C_4H_9, or C_6H_{13}) by galvanostatic method (1.7~3.7 V vs Li/Li+) using three-electrode cell. Influence of IR-drop on capacitance was eliminated by using chronopotentiograms between 2.1~3.3 V versus Li/Li+.

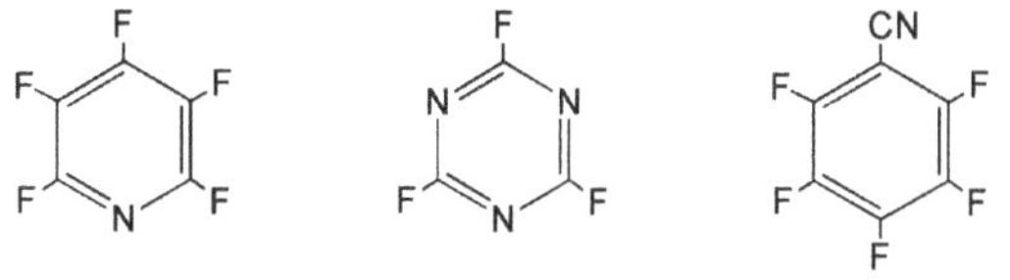

Pentafluoropyridine (PFP) Cyanuric fluoride (CYF) Pentafluorobenzonitrile (PFBN)

FIGURE 17.15 Various perfluoro-organics containing nitrogen atom as precursor for nitrogenous porous carbons.

that of the conventional activated carbon with a surface area of about 2000 m^2g^{-1} as shown in Figure 17.16 [22]. The origin of the large capacitance of the GF-based carbon can be attributed to the electrochemical activation mechanism, which is often observed for KOH-activated graphitizing carbons [23,24], due to the low specific surface area, relatively wide d-space (~0.37 nm), and voltage-dependence of the capacitance. In this regard, it has also been recently revealed that the cathode discharge product (carbon and LiF nano-composite) of the GF-Li primary battery operates as capacitor electrode [25].

17.4 CONCLUSION

This chapter reviews the features of porous carbons obtained from PTFE defluorinated by alkali metals or alkali cation–naphthalene anion radical complexes and their capacitance performance. The PTFE-based carbons possess high specific surface areas with abundant mesopores as well as micropores. The pore structure of these carbons can be controlled by the defluorination conditions such as defluorinating agent, fluorocarbons, and preheat treatment before the removal of the

TABLE 17.2 BET Specific Surface Area (S_{BET}), Average Micropore Width (w_{micro}), Atomic Ratio of Nitrogen and Carbon (N/C, %), and Specific Capacitance Per Surface Area (C_s) for Nitrogenous Porous Carbons (NPC) and Nitrogen-Free Porous Carbons

Sample	S_{BET} [m^{2}g^{-1}]	w_{micro} [nm]	C_{sa} [µFcm^{-3}]	C_{sc} [µFcm^{-3}]	N/C
ACF-a	1190	0.73	8.2	2.3	–
ACF-b	1700	1.05	6.7	5.8	–
ACF-c	2020	1.13	5.1	5.4	–
ACF-d	2200	1.21	5.4	4.7	–
KOH-MCMB-a	880	0.73	7.3	2.8	–
KOH-MCMB-b	1960	0.98	8.3	8.2	–
KOH-MCMB-c	2480	1.18	7.2	7.5	–
KOH-MCMB-d	2470	1.18	7.2	7.8	–
KB1	700	1.14	7.3	6.6	–
KB2	1120	1.33	7.3	7.0	–
NPC-PCP	1330	1.06	9.8	8.9	0.02
NPC-CYN-a	1280	1.08	8.3	7.7	0.15
NPC-CYN-b	1080	1.09	8.8	7.7	0.06
NPC-PFBN-a	840	0.86	10.8	5.6	0.07
NPC-PFBN-b	940	1.00	10.4	5.4	0.04

w_{micro}: calculated by DR method.
C_{sa}: Specific capacitance per surface area for anion adsorption capacitance.
C_{sc}: Specific capacitance per surface area for cation adsorption capacitance.
ACF: Phenolic resin based activated carbon fiber (steam activation, 800 °C).
KOH-MCMB: Mesocarbon microbeads activated by KOH (800 °C).
KB: Ketjen Black.
N/C was estimated by N1s and C1s XPS spectra.
Capacitance was measured in 0.5 M $(C_2H_5)_4NBF_4$/PC by galvanostatic method (40 mAg^{-1}) by three-electrode cell.

alkali fluoride. In the EDLC application, the PTFE-based porous carbons show a higher capacitance and excellent rate performance compared to the conventional activated carbons having comparable surface areas. The results are discussed on the basis of the cancellation of the strong ion sieving effect and the ion transfer promotion due to the presence of mesopores. The nitrogen-doped porous carbons and GF-based carbon prepared by the defluorination technique are also outlined.

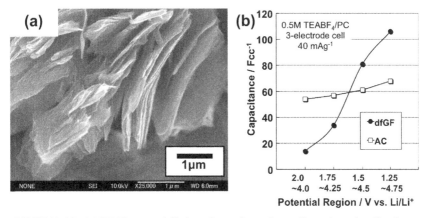

FIGURE 17.16 (a) SEM image and (b) dependence of capacitance (three-electrode cell, galvanostatic) on potential range for carbon prepared by defluorination of GF.

ACKNOWLEDGMENT

A part of this research is financially supported by NEDO Grant Program (No.00B58003c), the "Element Innovation" Project by MEXT, and JSPS KAKENHI Grant No. 26410250 in Japan.

REFERENCES

[1] J. Jansta, F.P. Dousek, F.P. Patzelova, Carbon 13 (1975) 377–380.
[2] H. Nishihara, T. Kyotani, Adv. Mater., 24 (2012) 4473–4498.
[3] L. Kavan, Chem. Rev. 97 (1997) 3061–3082.
[4] B.E. Conway, Electrochemical Supercapacitors: Scientific Fundamentals and Technological Applications, Kluwer Academic, New York, 1999.
[5] T. Pandolfo, V. Ruiz, S. Sivakkumar, J. Nerkar, General properties of electrochemical capacitors, in: F. Beguin, E. Frackowiak (Eds.), Supercapacitors: Materials, Systems and Applications, Wiley-VCH, Weinheim, 2013, pp. 69–109.
[6] S. Shiraishi, Electric double layer capacitors, in: E. Yasuda, M. Inagaki, K. Kaneko, M. Endo, A. Oya, Y. Tanabe (Eds.), Carbon Alloys, Elsevier, Oxford, 2003, pp. 447–457.
[7] Y. Yamada, O. Tanaike, S. Shiraishi, Tanso 215 (2004) 285–294.
[8] O. Tanaike, N. Yoshizawa, H. Hatori, Y. Yamada, S. Shiraishi, A. Oya, Carbon 40 (2002) 445–446.
[9] O. Tanaike, H. Hatori, Y. Yamada, S. Shiraishi, A. Oya, Carbon 41 (2003) 1759–1764.
[10] S. Shiraishi, H. Kurihara, H. Tsubota, A. Oya, S. Soneda, Y. Yamada, Electrochem. Solid-State Lett. 4 (2001) A5–A8.
[11] S. Shiraishi, D. Hiruma, Y. Onuma, A. Oya, T.-T. Liang, Y. Yamada, Tanso 195 (2000) 395–399.
[12] T. Liang, Y. Yamada, N. Yoshizawa, S. Shiraishi, A. Oya, Chem. Mat. 13 (2001) 2933–2939.
[13] Y. Yamada, H. Ohno, O. Tanaike, H. Hatori, Chem. Lett. 34 (2005) 1546–1547.
[14] M. Inagaki, H. Konno, O. Tanaike, J. Power Sources 195 (2010) 7880–7903.
[15] S. Shiraishi, Key Eng. Mater. 497 (2012) 80–86.

[16] Y. Yamada, O. Tanaike, T.-T. Liang, H. Hatori, S. Shiraishi, A. Oya, Electrochem. Solid-State Lett. 5 (2002) A283–A285.
[17] S. Shiraishi, Y. Aoyama, H. Kurihara, A. Oya, Y. Yamada, Mol. Cryst. Liq. Cryst. 388 (2002) 129–135.
[18] S. Shiraishi, M. Kibe, A. Oya (Eds.), Proceedings of Carbon, 2004, p. E022.
[19] E. Frackowiak, G. Lota, J. Machnikowski, C. Vix-Guterl, F. Béguin, Electrochim. Acta 51 (2006) 2209–2214.
[20] D. Hulicova-Jurcakova, M. Kodama, S. Shiraishi, H. Hatori, Z.H. Zhu, G.Q. Lu, Adv. Funct. Mater. 19 (2009) 1800–1809.
[21] T. Kwon, H. Nishihara, H. Itoi, Q.-H. Yang, T. Kyotani, Langmuir 25 (2009) 11961–11968.
[22] S. Shiraishi, D. Ajima (Eds.), Proceedings of Carbon, 2010, p. 349.
[23] M. Takeuchi, T. Maruyama, K. Koike, A. Mogami, T. Oyama, H. Kobayashi, Electrochem. 69 (2001) 487–492.
[24] S. Mitani, S.I. Lee, K. Saito, S.H. Yoon, Y. Korai, I. Mochida, Carbon 43 (2005) 2960–2967.
[25] S. Shiraishi, H. Fujimoto (Eds.), Proceedings of Carbon, 2012, p. 847.

Index

Note: Page numbers followed by "f" and "t" indicate figures and tables respectively.

A

Accelerated stress test (AST) protocols, 342–343
Activated carbon fibers (ACFs), 379–380
Activated carbon nanofibers (ACNFs), 380–382, 381t, 382f
Aluminum corrosion inhibitor, 401–403, 402f
Arkema, 234–235

B

Binary metal fluorides
  metal difluorides (MF_2)
    as-synthesized nanocomposites, 61–62, 62f
    cobalt (II) fluoride, 63
    copper (II) fluoride, 62–63
    lithiation and delithiation mechanism, 61–62
    nickel (II) fluoride, 63
  metal trifluorides (MF_3). *See* Metal trifluorides (MF_3)

C

Carbon anodes, surface modification
  electrochemical properties, 204
  electrochemical reduction, 203–204
  ethylene carbonate-based solvents, 203–204
  fluorinating gases, 204
  light fluorination, 204
  natural graphite samples
    surface chlorination, large surface areas, 210–214, 213t, 214f, 215t
    surface fluorination. *See* Surface fluorination
  plasma fluorination, 204
Carbon metal fluoride nanocomposites (CMFNCs), 56
CF_x powders, polypyrrole electrodeposition, 242f
  "core–shell" model, 239–240, 240f
  discharge rate, 240
  discharge reaction, 239
  electrochemical lithium insertion reaction, 242
  electron transport pathway, 240–241
  electropolymerization, 241–242
  galvanostatic discharge curves, 242
  graphite fluorides, 237–238, 238f
  limitations, 239
  lithium–graphite intercalation compound, 239
  PPy-CF_x sample preparation, acetonitrile
    CF–PPy-modified electrode, 245
    first cyclic voltammogram, 243–245, 244f
    impedance spectra, 245–246, 246f
    neighboring π-bonds, 243
    Nyquist diagrams, 245–246, 245f
    pyrrole monomer, 243, 243f
    Warburg behavior, 245–246
  PPy-coated samples, physical–chemical characterizations, 246–251, 247f–250f
  primary lithium battery
    charge transfer resistances, 253
    current density, 257
    discharge curves, 251, 251f, 257–258, 258f
    discharge rates, 254, 256f
    energy density, 257
    galvanostatic intermittent titration method, 252
    half-discharge capacities, 251
    impedance spectra, 252–253, 255t
    Nyquist diagrams, 252–253, 253f–254f
    OCV and CCV curves, 252, 252f
    Ragone plots, 254–257, 257f

CF$_x$ powders, polypyrrole electrodeposition (*Continued*)
  SCE, 242
  subfluorinated carbon nanofibers/re-fluorinated graphite fluorides, 240–241
  three-dimensional sp^3 hybridization, 239
  XRD measurements, 259

D

Density function theory (DFT), 7
Differential scanning calorimetry
  metallic lithium/lithium-intercalated graphite reactions, 154–155, 154f
  cyclic and linear carbonate solvents, 151–154, 152f, 153t, 154f
  delithiated graphite with SEI film, 157–161, 157f, 159f–160f
  fluorocarbonate-containing electrolyte solution and lithiated graphite, 155–157, 156f
  lithiated graphite with SEI film, 155, 155f, 157–161, 157f, 159f–160f
  molecular structures, 161
Direct methanol fuel cells (DMFCs)
  battery charging applications, 360, 361f
  500-W DMFC generator system, 361, 362f
  electrical efficiency, 360, 361f
  electrode reactions, 357, 358f
  Faraday's law, 359–360
  gasoline-/diesel-fueled generators, 360
  high-technology electronic applications, 357
  hydrogen–air systems, 361–362
  MEA durability test, 362, 363f
  methanol cross-over, 358, 359f
  methanol oxidation reaction, 357–358
  OorjaPac Models, 363–364
  "permanent" performance losses, 362–363
  power densities, 356
  "recoverable" performance losses, 362–363
  transport mechanisms, 358

E

Electric double layer capacitance (EDLC), 416
  BET-specific surface area, 425, 426f
  charge/discharge process, 422, 422f
  chronopotentiograms, 423–425, 425f
  current density, 425, 427f
  energy density, 422
  GF defluorination, 426–427, 429f
  heteroatom-doped porous carbons, 425–426
  lamella-like carbon material, 426–427
  nitrogen doping effect, 425–426
  nitrogen-free porous carbons, 425–426, 428t
  NPC, 425–426, 427f, 428t
  pore structure parameters, 423–425, 424t
  PSD curves, 423–425, 424f
  specific surface area and gravimetric capacitance, 423, 423f
Electrochemical capacitor
  carbyne structures, 415–416
  EDLC, 416
  mesoporous carbons, 416
  porous carbon
    fluorocarbon. *See* Fluorocarbon mechanism, 415–416, 416f
    pyrolyzing organic materials, 415–416
    template method, 415–416

F

Faradaic process, 375
Faradic yield
  discharge mechanism, 281
  electrochemical defluorination, 277
  ^{19}F NMR MAS spectra, 277–279, 278f
  galvanostatic discharge curves, 277, 278f, 279–281, 282f–283f
  graphene and porous carbons/postfluorination process, 283–284
  LiF particles, 281–282
  partial exfoliation, 279–281
  pore size distribution, 283–284
  raw HTGF, 277–279, 280f
  shrinking core model, 277
  soft-templated mesoporous carbons, 282–283
  XRD diagrams, 277–279, 278f
Fluorinated activated carbons, 376–379, 377t, 378f–380f, 380t
Fluorinated carbon nanotubes, 382–383, 384f–385f
Fluorinated and nonfluorinated additives
  GC-MS analysis, 200
  Li-ion principle, 174–175, 175f
  SEI. *See* Solid electrolyte interphase (SEI)
  TiSnSb anodes
    conversion reaction, 181–182
    cyclic voltammograms, 183–184, 184f
    EIS results, 192–193, 193f
    electrochemical impedance spectroscopy, 200
    electrochemical tests, 198–199
    at first cycle, 194–197, 196f
    galvanostatic curve, 184–185, 185f

Index 433

galvanostatic cycling, 185–186, 185f
lithium insertion/extraction, 182–183
long-chain flexible polycarbonates, 183
scanning electron microscopic analysis, 193–194, 194f, 199
at 20th cycle, 197–198, 199t
XPS, 186–192, 187t–188t, 189f–190f, 191t, 199
Fluorinated solvents
  electrolyte systems
    fluorinated 1,2,4-thiadiazianan-3,5-dion-1,1-dioxides, 138, 138f–139f
    hexafluorbutandiol derivatives, 135–138, 136f–137f
    propylene-carbonate-based electrolytes, alkyl tetrafluoro-2-(alkoxy)propionate, 132–135, 132f–134f
  PF_5-carbene adducts
    constant current cycling, 141–142, 142f
    cyclic voltammetry, 141, 141f
    nitrogen-containing carbene complexes, 140, 140f
    overcharge additives, 139
    shutdown additives, 139
    stability window, 140, 141f
  polyfluorinated lithium sulfonates, lithium conductive salts, 125–131, 126f–131f
  SEIs, 125
Fluorocarbon
  alkali metals–naphthalene anion radicals, 418–419, 419f
  diluted hydrogen fluoric/chloric acid, 417
  disadvantages, 418
  EDLC
    BET-specific surface area, 425, 426f
    charge/discharge process, 422, 422f
    chronopotentiograms, 423–425, 425f
    current density, 425, 427f
    energy density, 422
    GF defluorination, 426–427, 429f
    heteroatom-doped porous carbons, 425–426
    lamella-like carbon material, 426–427
    nitrogen doping effect, 425–426
    NPC, 425–426, 427f, 428t
    pore structure parameters, 423–425, 424t
    PSD curves, 423–425, 424f
    specific surface area and the gravimetric capacitance, 423, 423f
  ether-type solvent, 418–419
  nitrogen adsorption/desorption isotherms, 419–421, 420f
  nitrogen physisorption isotherms and pore size distribution, 417–418, 417f
  PTFE defluorination, potassium metal, 417, 417f
  raw fluorinated materials, 421–422
  scanning and transmission electron micrograph, 421–422, 421f
  sodium metal, 417–418, 418f
  X-ray diffraction patterns, 419–421, 420f
Fluorohydrogenate ions
  anions
    KF–HF system, phase diagram, 103–104, 104f
    Lewis acid–base reactions, 103
    organic fluorohydrogenate salts, 104
    properties, 104–106, 105f
  liquid crystals, 116–118, 117f–118f
  liquids
    aliphatic cation, 111–113
    cation structure, 106, 107f
    crystal packings, 111, 112f
    cyclic voltammograms, glass-like carbon, 111–113, 112f
    dielectric spacer, 111
    electrochemical capacitors, 114–115
    1-fluoro-2-iodo-1-phenylpropane, 116
    imidazolium-based salts, 111
    ion hopping mechanism, 111
    liquid–polymer composite electrolyte membrane, 114
    nonvolatile electrolyte, 113–114, 114f
    physical properties, 107–111, 108t–110t
    polarized optical microscopy, 113
    pulsed gradient spin-echo nuclear magnetic resonance, 111
    thermal stability, 113–114
    voltage dependence of capacitance, 114–115, 115f
    Walden plot, 107–111, 110f
    X-ray diffraction analysis, 113
  plastic crystals, 118–120, 119f–120f
Fluorophosphates
  fluorine-doped $LiFePO_4$, 81–83
  LiBr, 91–92
  Li_2FePO_4F, 88–89
  $LiMPO_4F$
    crystallographic parameters, 81, 82t
    high-voltage cathode materials, 87
    solid-state/ionothermal techniques, 87–88
    triclinic tavorite-type structure, 87–88
  Li_2MPO_4F, 89–90
  $LiVPO_4F$

Fluorophosphates (*Continued*)
  charge–discharge profile and derivative capacity, 84–85, 85f
  delithiated phase, 85–86
  discharge coulombic efficiency, 87
  lithiated phase, 85–86
  lithium-bearing pegmatite family, 83–84
  phase diagram, 85–86, 86f
  reversible specific capacity, 84–85
  self-heating rate, 86–87
  tavorite structure, 83–84, 84f
  two-phase reaction mechanism, 84–85
  two-step CTR method, 86
  $Na_3V_2(PO_4)_2F_3$ hybrid ion cathode, 90–91, 90f–91f
Fuel cells (FCs), 326, 333
  advantages, 346
  automotive applications, 349–351, 351f–352f
  automotive systems, 345
  catalyst materials, 348–349
  cost reduction, 348
  life cycle analysis, 345–346
  PEMFC stack, 346–348, 347f–348f
  portable application, 356
  requirements, 340–341
  stationary application, 351–356, 354f–355f
Fumapem® F-940rf membrane, 333, 334f

H

Hard X-ray photoemission spectroscopy (HAXPES) spectra, 400–401, 400f
High-energy lithium nonaqueous batteries, 52
Highest occupied molecular orbital (HOMO) levels, 148, 150t–151t
High performance lithium-ion batteries
  chemical–electrochemical reactions, 1–2
  Fermi energy, 1–2
  fluorinated redox shuttles
    battery pack charge, 13–14, 14f
    cell-level overcharge protection mechanism, 13–14
    chemical and electrochemical reactions, 13–14
    discharge capacity, 16–18, 17f–18f
    fluorinated lithium borate cluster salt, 16
    in situ high-energy X-ray diffraction, 18–19
    overcharge protection, 16, 17f
    PFPTFBB, 15–16, 15f–16f
    redox shuttle molecule, 14–15
    voltage profile, 18, 19f
  graphitic carbons, 1–2
  high voltage electrolytes
    Coulombic efficiency, 25–26
    cycling capacity retention, 22, 24, 25f–26f
    cycling performance, 25–26, 27f
    differential capacity analysis, 27
    electrolyte stability, 21
    fluorinated carbonates and ethers, chemical structures, 19, 20f
    fluorinated cyclic and linear carbonate compounds, 19
    Fourier transform infrared spectra, 28, 29f
    Gen 2 and E1 to E6 electrolytes, conductivity measurement, 21, 22f
    Gen 2 and fluorinated electrolytes E1 to E6, electrochemical stability, 21, 23f
    harvested LNMO cathodes, 27, 28f
    potentiostatic profiles, 22, 24f
    structure, oxidation potential, HOMO/LUMO energies, 19, 20t, 21
    sulfone-based electrolyte, 23–24
  lithiated anode stabilization
    differential scanning calorimetry, 4–5, 4f
    in situ high-energy X-ray diffraction patterns, 4–5, 4f
    SEI, 2–4, 3f. *See also* Solid electrolyte interphase (SEI)
    thermal instability, 2–4
  smart grids, 1

K

Kynar®, 228, 229f

L

Lewis acid–base reactions, 103
Lithium ion batteries (LIBs). *See also* High performance lithium-ion batteries
  advantages, 78
  carbon anodes, surface modification. *See* Carbon anodes, surface modification
  fluorophosphates. *See* Fluorophosphates
  fluorosulfates
    crystal chemistry, 92, 94f
    $LiFeSO_4F$, 92–95, 95f
    $LiMSO_4F$, 95–96
    structural properties, 92, 93t
  green transportation and grid storage systems, 77–78
  inductive effect, 80–81, 80f, 81t
  Li insertion cathodes, average voltage, 79
  oxides, 77–78

Index **435**

polyanion-based framework, 78
PVDF binders. *See* Polyvinylidene fluoride (PVDF) binders
silicon, 407–409, 410f
surface-fluorinated cathode materials. *See* Surface-fluorinated cathode materials
Lowest occupied molecular orbital (LUMO) energies, 148, 150t–151t

M

Membrane electrode assembly (MEA), 293, 328, 329f–330f
Metal trifluorides (MF_3)
  BiF_3, 59
  CrF_3, 61
  FeF_3
    CMFNCs, 56
    cost and toxicity, 53
    current density, 57–58
    cycling behavior, 56–57, 57f
    galvanostatic charge/discharge voltage *vs.* specific capacity profile, 58–59, 60f
    graphene nanocomposites, 57–58, 58f
    ionic liquid assisted synthesis, 58
    mixed conducting matrix, 56
    polystyrene colloidal crystals, 57
    specific capacity *vs.* cycle number, 58–59, 60f
    Swagelok cell reactor, 58–59
  MnF_3, 61
  TiF_3, 61
  VF_3, 61

N

Nafion® membrane, 290, 292f, 326–327, 332
  concentration, 334
  equivalent weight (EW), 334
  extrusion-cast membrane manufacturing process, 335
  properties, 335, 335t
  structure, 333–334, 334f
  thin-film electrode layers, 335
Negative electrode
  aluminum corrosion inhibitor, 401–403, 402f
  electrochemical formation, 387–388
  fluorinated electrolyte and additive
    energy/power density and cycle life, 392
    film-forming additive, 394–395
    galvanostatic reduction/oxidation curves, 394–395, 395f
    hard-carbon electrodes, capacity retention, 393, 394f
    Lewis acidity, 392
    molecular structure, electrolytes additives, 394, 395f
    organic electrolyte solution, 393
    reduction/oxidation curves, 394–395, 396f, 397, 398f
    reversible capacity variation, 394–395, 396f, 397, 398f
    XPS measurement, 396–397, 397f
  hard-carbon, 390–392, 391f
  lithium-ion battery, silicon, 407–409, 410f
  Na alloys and compounds
    Li-graphite system, 403
    Li-Sn systems, 403
    Na insertion materials, 403
    Na-Pb binary alloy, 403–404
    red and black phosphorus, 405–407, 406f, 408f
    tin and antimony, 404–405, 405f
  Na-ion battery, 388–390, 388t, 389f
  poly vinylidene fluoride and CMC-based binder
    capacity retention and Coulomb efficiency, 399
    electrode performance, 398–399, 399f
    HAXPES spectra, 400–401, 400f
    volatile organic solvents, 397–398
  sodium-insertion negative electrodes, 409
Nitrogen-free porous carbons, 425–426, 428t
Nitrogenous porous carbons (NPC), 425–426, 427f, 428t
Nonrechargeable lithium batteries
  applications, 261–262
  catalytic fluorination process, 262–263
  C–F bond energies, 261–262
  energy density, nanomaterials
    carbon lattice, curvature effect, 266–267, 267f
    ^{19}F chemical shifts, 266–267, 268f
    fluorinated fullerenes, 266
    fluorinated MWCNTs, 269–270
    fluorine adsorption, 263–264
    fluorine–graphite intercalation compounds, 263–264
    ^{19}F NMR chemical shift, 269, 270f
    Fourier transform infrared spectroscopy experiments, 266
    galvanostatic curves, 267, 268f
    IR wave numbers and chemical shifts, 266–267

Nonrechargeable lithium batteries (*Continued*)
    magic angle spinning, 267
    nanocarbon synthesis methods and fluorination conditions, 264–266, 265t
        Raman excitation wavelengths, 264–266
        σ–π rehybridization act, 264–266
    Faradic yield
        discharge mechanism, 281
        electrochemical defluorination, 277
        ^{19}F NMR MAS spectra, 277–279, 278f
        galvanostatic discharge curves, 277, 278f, 279–281, 282f–283f
        graphene and porous carbons/postfluorination process, 283–284
        LiF particles, 281–282
        partial exfoliation, 279–281
        pore size distribution, 283–284
        raw HTGF, 277–279, 280f
        shrinking core model, 277
        soft-templated mesoporous carbons, 282–283
        XRD diagrams, 277–279, 278f
    high-temperature fluorination process, 263
    Li/CF$_x$ battery system, 262
    nanostructured electrode materials, 263
    primary lithium battery
        next-generation carbon fluorides, 284
        power density. *See* Primary lithium battery
    sp^2 carbon hybridization, 262–263
    sp^3 hybridization, 261–262
NPC. *See* Nitrogenous porous carbons (NPC)

O

Organofluorine compounds, 149f–150f
    charge/discharge behavior, 167–169, 169f
        Coulombic efficiencies, 167–169, 168t, 170f
        electrochemical reduction, 165–166
        ethylene carbonate, 153t, 165–166
        first charge/discharge curves, 165–167, 166f–168f, 169–170, 170f
        natural graphite electrodes, 165–166
    cyclic carbonates, 147–148
    differential scanning calorimetry. *See* Differential scanning calorimetry
    electrochemical oxidation stability, 161–164, 162f–165f
    high thermal and oxidation stability, 147–148
    HOMO levels, 148, 150t–151t
    LUMO energies, 148, 150t–151t
    phosphorus compounds, 147–148

P

2-(Pentafluorophenyl)-tetrafluoro-1,3,2-benzodioxaborole (PFPTFBB), 15–16, 15f–16f
Perfluorinated sulfonic acid (PFSA) membrane
    advantage, 333
    automotive systems, 327
    degradation and durability aspects
        AST protocols, 342–343
        crack formation and propagation, 341–342
        Fenton-active transition metal ion impurities, 342
        fluoride ion release, 343
        hydrogen cross-over, 343–344, 344f–345f
        start–stop and humidity cycles, 341
        steady-state open cell voltage, 343, 344f
        unzipping mechanism, 342
        viscoelastic creep, 342
    degradation mechanism, 327
    DMFCs, 326. *See also* Direct methanol fuel cells (DMFCs)
    FCs, 326, 333
        advantages, 346
        automotive applications, 349–351, 351f–352f
        automotive systems, 345
        catalyst materials, 348–349
        cost reduction, 348
        life cycle analysis, 345–346
        PEMFC stack, 346–348, 347f–348f
        portable application, 356
        requirements, 340–341
        stationary application, 351–356, 354f–355f
    fluorinated membrane, 331–332
    Fumapem® F-940rf membrane, 333, 334f
    morphology and proton conductivity, 336–339, 337f, 339f
    Nafion® membrane, 326–327, 332
        concentration, 334
        equivalent weight, 334
        extrusion-cast membrane manufacturing process, 335
        properties, 335, 335t
        structure, 333–334, 334f
        thin-film electrode layers, 335
    polymer electrolyte fuel cells
        catalyst surface, 329
        ionic conductivity, 331

Index 437

MEA, 328, 329f–330f
PEMFC electrode reactions, 328, 329f
power density, 330, 331f
single-cell PEMFC, 330, 330f
SPFC, 326
structure, 332, 333f
synthesis and preparation, 332
water and methanol transport, 339–340, 340f
PF_5–carbene adducts
constant current cycling, 141–142, 142f
cyclic voltammetry, 141, 141f
nitrogen-containing carbene complexes, 140, 140f
overcharge additives, 139
shutdown additives, 139
stability window, 140, 141f
Polymer electrolyte fuel cells
catalyst surface, 329
ionic conductivity, 331
MEA, 328, 329f–330f
PEMFC electrode reactions, 328, 329f
power density, 330, 331f
single-cell PEMFC, 330, 330f
Polyvinylidene fluoride (PVDF) binders
advantages, 225
disadvantages, 226
electrochemical stability, 230–231, 230f
electrode performance test, 233, 234f
electrode preparation method, 231–232, 231f, 232t
electrolyte solvent, binder swelling, 230
fluorinated waterborne binders, 234–235, 235f
fluorine-containing binder
adhesive properties, 226–227
alpha, 227, 227f
beta, 227, 228f
delta, 228
dipole moment, 228
gamma, 228, 228f
greenhouse gases, 225
high-performance binders, 226
peel strength measurement, 232, 233f
properties, 228–230, 229f, 229t
types, 226
Primary lithium battery
charge transfer resistances, 253
current density, 257
discharge curves, 251, 251f, 257–258, 258f
discharge rates, 254, 256f
energy density, 257
galvanostatic intermittent titration method, 252

half-discharge capacities, 251
impedance spectra, 252–253, 255t
next-generation carbon fluorides, 284
Nyquist diagrams, 252–253, 253f–254f
OCV and CCV curves, 252, 252f
power density
^{13}C MAS NMR spectra, 271–272, 272f
Contour Energy System society, 273–274
controlled fluorination, 274
curvature effect, 273–274
discharge mechanism, 276
D-T_F and C-T_F series, 275–276, 276t
electrochemical process, 272–273
^{19}F MAS NMR spectra, 275
galvanostatic discharge curves, 273–274, 273f
intrinsic electrical conductivity, 271
solid-state NMR, 276
subfluorination, 271
TbF_4 decomposition, 274–275
unfluorinated nanofibers CNFs, 271, 272f
Ragone plots, 254–257, 257f
Proton exchange membrane fuel cells (PEMFCs), 326

Q

Quasianhydrous fuel cell membranes
advantages, 290
aromatic and heteroaromatic hydrocarbon polymer membranes, 290–293
high temperature PEMFC, 293
hydrogen purification process, 293
low temperature PEMFCs, 293
MEA, 293
membrane crosslinking, nitrogenous heterocycles
A/B-*ret*-DiA and A/B-*ret*-TEPA membranes *vs.* uncured membranes, 316–317, 317f, 318t
amine function/cyclocarbonate groups, 307–308, 308f
chlorosulfonic acid, 305
dynamical mechanical analyses, 311, 312t–313t
elastic modulus, 314
gylcerine carbonate vinyl ether, 306, 307f
hydroxyurethane bond, 308
mechanic–dynamic analyses, 310–311, 310f
membrane preparation process, 309, 309f
molar ratios, 308
protonic conductivities, 315–316, 316f

Quasianhydrous fuel cell membranes (*Continued*)
  radical terpolymerization, 306, 306f
  strain–stress curves, 310–311, 310f, 314, 314f
  swelling rates, 314–315, 315f
  TEPA, 307, 307f
  terpolymers, 306
  thermal stabilities, 309
  membrane/electrode interfaces, 320
  Nafion® membrane, 290, 292f
  nitrogen heterocycles, fluorinated copolymers
    degradation temperature, 296, 298t
    imidazole, benzimidazole and 1H-1,2,4-triazole, structures and chemical properties, 296, 297t
    membrane type *vs.* temperature range, 295–296, 295f
    oxidoreduction reaction, 295
    PBIs, 296
    PFSA copolymers, 295–296
    telechelic oligomers, 296
    triazole groups and sulfonated poly(ether ether ketone), 297–302, 299f–301f
  oxidoreduction reaction, 290
  PEMFCs, 290, 317–319
  PFSA polymers, 292t
  proton-acceptor/proton-donor interaction, 319
  proton mobility, nitrogenous heterocycles and s-PEEK
    2D ^{1}H MAS NMR EXchange SpectroscopY (EXSY) spectra, 302
    2D H MAS solid-state NMR (SSNMR) spectroscopy, 302
    1D Single pulse ^{1}H MAS SSNMR spectra, 302, 303f
    ^{1}H magic angle spinning (MAS), 302
    ^{1}H MAS SSNMR EXSY spectra, 303–304, 304f
    proton transport, 305
    secondary amine function, 304–305
    triazole and triazolium ring protons, 304
  transportation fuel cell system cost, 293–294, 294f
  types, 290, 291t

S
Saturated calomel electrode (SCE), 242
Secondary batteries
  advantages, 73
  electrochemical energy storage, 52–53, 54t–55t
  energy density systems, 73
  fluoride ion batteries
    architecture, 65, 66f
    charge process, 65
    chemical reaction, 65
    electropositive elements, 65–66
    liquid electrolyte, 72–73, 73f
    solid fluoride ion conductors. *See* Solid fluoride ion conductors
    solid state fluoride ion batteries, 69–72, 70t, 71f
  high-energy lithium nonaqueous batteries, 52
  lithium batteries
    binary metal fluorides. *See* Binary metal fluorides
    electron diffusion path lengths, 53
    ternary metal fluorides, 63–65, 64f
Sodium-insertion negative electrodes, 409
Solid electrolyte interphase (SEI), 125
  artificial layer
    capacity retention, 7–8
    charge capacity, 12–13, 12f
    differential capacity profiles, 8–9, 9f
    differential scanning calorimetry profiles, 10–13, 11f, 13f
    lithium bis[oxalato]borate and lithium difluoro[oxalato]borate additives, 9–10, 10f
    nonaqueous electrolyte, 7–8
    normalized discharge capacity, 10–12, 11f
    oxalato-based electrolyte additives, 8, 8f
  components, 180–181, 180t, 181f–182f
  electrode surface, 178–180
  formation, 175–178, 178t
    five-membered oxygen-containing heteroring additives, 176
    galvanostatic cycling measurements, 176–177, 176f
    Li/graphite half-cell, galvanostatic cycling, 177, 177f
    non-SEI-forming solvents, 176
  modification
    boron-based anion receptors and LiF, 6, 6f
    carbonaceous materials, 5–6
    DFT, 7
    electrochemical polymerization, 7
    graphitic anodes, 5–6, 5f
    physical–chemical properties, 5–6
    3,3,3-trifluoropropyltrimethoxysilane, 5–6, 6f
  structure and role, 178, 179f

Solid fluoride ion conductors
  fluorite-type fluoride ion conductors, 67–68
  structure types, 66
  tysonite-type fluoride ion conductors, 68–69
Solid polymer fuel cell (SPFC), 326
Supercapacitor
  ACFs, 379–380
  ACNFs, 380–382, 381t, 382f
  carbon materials, 375–376
  charge storage mechanisms, 375
  Faradaic process, 375
  fluorinated activated carbons, 376–379, 377t, 378f–380f, 380t
  fluorinated carbon nanotubes, 382–383, 384f–385f
  non-faradaic process, 375
  surface modification methods, 376
Surface-fluorinated cathode materials
  $LiFePO_4$
    advantages, 33–34
    charge/discharge capacity and cycle ability, 34
    charge–discharge test, 35
    classification, 33–34
    discharge curves, 40, 40f
    DSC curves, 42, 43f
    electrode kinetic parameters, 41, 42t
    electrode polarization, 41–42
    exchange current density, 41
    ion chromatography, 34
    lithium *ortho*-phosphates, 38–39
    Nyquist plots, 40–41, 41f
    organic compounds fluorination, 39, 39f
    powder X-ray diffraction, 34
    reaction conditions, 35, 36t
    scanning electron microscope, 34
    SEM images, 35–37, 38f
    thermal behavior, 35, 42
    TOM Cell, 34–35, 35f
    XPS spectra, 34–35, 37f
    XRD profiles, 35, 38f
  $LiNi_{0.5}MN_{1.5}O_4$
    charge/discharge cycles, 43–44
    electrochemical properties, 46–49, 48f
    peak intensity and FWHM ratio, 44–46, 46t
    two-electrode test cell, 44
    untreated and fluorinated samples, 44–46, 45f
    XPS spectra, 44–46, 47f
    X-ray diffraction patterns, 44–46, 45f
Surface fluorination
  large surface areas, 208–210, 208t–209t, 210f, 211t
  petroleum cokes
    charge/discharge potential curves, 214–219, 219f
    first Coulombic efficiencies, 214–219, 215t, 220t
    Raman shifts, peak intensity ratios, 214–219, 218t
    surface concentrations, 214–219, 217t
    surface-fluorinated PC2800, 214–219, 216f
    transmission electron microscopic analysis, 214–219
  small surface areas
    BET Method, 204–207, 206t
    elemental analysis, 204–207, 205t
    NG40μm fluorinated sample, 204–207, 207f
    NG25μm fluorinated sample, charge capacities, 204–207, 207f
    pore volume distributions, 204–207, 206f
    XPS, 204–207, 205t
Surface modification methods, 376

T

Two-electrode test cell (TOM cell), 34–35, 35f

CPSIA information can be obtained at www.ICGtesting.com
Printed in the USA
BVOW10*1354110815

412255BV00005B/1/P